VASCULARIZATION

Regenerative Medicine and Tissue Engineering

T0239952

VASCULARIZATION

Regenerative Medicine and Tissue Engineering

Edited by
ERIC M. BREY

CRC Press
Taylor & Francis Group
Boca Raton London New York

CRC Press is an imprint of the
Taylor & Francis Group, an **informa** business

CRC Press
Taylor & Francis Group
6000 Broken Sound Parkway NW, Suite 300
Boca Raton, FL 33487-2742

First issued in paperback 2017

© 2015 by Taylor & Francis Group, LLC
CRC Press is an imprint of Taylor & Francis Group, an Informa business

No claim to original U.S. Government works

ISBN-13: 978-1-4665-8045-9 (hbk)
ISBN-13: 978-1-138-07603-7 (pbk)

Library of Congress Cataloging-in-Publication Data

Vascularization : regenerative medicine and tissue engineering / editor, Eric M. Brey.
    p. ; cm.
    Includes bibliographical references and index.
    ISBN 978-1-4665-8045-9 (hardback : alk. paper)
    I. Brey, Eric M., editor.
    [DNLM: 1. Vascular Surgical Procedures--methods. 2. Blood Vessels--anatomy & histology. 3. Regenerative Medicine--methods. 4. Tissue Engineering--methods. 5. Tissue Transplantation. WG 170]

    RC691
    616.1'3--dc23   2014008076

Visit the Taylor & Francis Web site at
http://www.taylorandfrancis.com

and the CRC Press Web site at
http://www.crcpress.com

To Niko, Zak, and Sylvia

Contents

Section I   Cells

Section II   Biomaterials

Section III   Models

Section IV   Imaging

Section V   Vascularized Tissues

Preface

The first federally funded meeting on tissue engineering was in 1988. At this and subsequent meetings, the fields of tissue engineering and regenerative medicine (TERM) have been described as having the potential to revolutionize clinical approaches to the replacement, reconstruction, or regeneration of organs and tissues. While TERM research has resulted in new patient treatments, the broad impact on clinical practice envisioned has not been achieved. The development of new TERM therapies that have significant clinical impact requires the ability to control vascularization, the process of new vessel assembly.

The circulatory system plays a number of vital roles in regenerating and functioning tissues. It supplies oxygen and nutrients, removes wastes, and is a source of multiple cell types required to respond to changing physiological conditions. For nearly every TERM application, the ability to enhance, regenerate, or engineer new tissues requires spatial and temporal control over the process of vascularization. While vascularization is being studied in a number of physiological and pathological processes, TERM applications present distinct challenges. For example, unique microenvironmental conditions result from biomaterial and cell combinations used in TERM applications that are not encountered in any other system. In addition, clinical applications require vascularization of large tissue volumes within time frames that are much shorter than those found during vascularization in development and typical physiological processes. These requirements place significant constraints on the design of TERM therapies. The study of vascularization in TERM applications is a complex and growing field.

The goal of this book is to provide a broad overview of vascularization in TERM applications. Chapter contributions from leading engineers, basic scientists, and clinicians provide their insight on how to address important issues in the field. These chapters summarize the state-of-the-art research in a number of areas. Complex challenges related to multicellular processes in vascularization and cell-source issues are discussed. Advanced biomaterial design strategies for control of vascular network formation are presented as are *in silico* models designed to provide insight not possible in experimental systems. An overview of imaging methods critical to understanding vascularization in engineered tissues is also provided. Finally, vascularization issues are discussed within the context of specific tissue applications. This book is designed for researchers in TERM fields to guide and inform their research programs. In addition, it can be used for senior undergraduate and graduate level courses focused on TERM.

Editor

Eric Michael Brey, PhD is a professor of biomedical engineering at the Illinois Institute of Technology and a research health scientist at the Hines Veterans Administration Hospital. He received BS and MEng degrees in chemical engineering from the University of Louisville, and a PhD in chemical engineering from Rice University in Houston, Texas. Following his doctoral studies, he was awarded a National Institutes of Health (NIH) Fellowship for postdoctoral studies in the Departments of Surgery and Cell Biology at Loyola University Medical Center in Maywood, Illinois.

Professor Brey's research is focused on the areas of tissue engineering, vascularization, and biomaterials. Specifically, his group has developed and investigated new biomaterial and surgical approaches to control vessel assembly. These strategies are investigated using *in vitro*, *in silico*, and *in vivo* models and have progressed to clinical application for engineering vascularized bone. His laboratory has also investigated novel imaging methods for analysis and monitoring of engineered tissues. Dr. Brey's research has resulted in over 85 peer-reviewed publications, 9 book chapters, and over 200 presentations and invited talks. Professor Brey has been awarded an International Society of Applied Cardiovascular Biology Young Investigator Award, Coleman Faculty Scholar Award, Sigma Xi Award for Excellence in Research, and a visiting professorship at Chang Gung Memorial Hospital. His research has received support from a variety of sources, including the National Science Foundation (NSF), NIH, Pritzker Institute, Department of Defense, the Veterans Administration, and industrial partners.

In addition to research, Professor Brey has made significant contributions to engineering education, specifically in the area of research mentoring and its influence on education and career trajectories. He has led an NSF-funded undergraduate research experiences program for 9 years, developed new curricula for ethics training in research, and proposed guidelines for graduate mentor training. His education research has resulted in multiple presentations, 6 papers, and over $1.4 million in external support.

Contributors

Nasim Annabi
Department of Medicine
Brigham and Women's Hospital
Harvard Medical School
and
Harvard-MIT Division of Health Sciences
and Technology
Massachusetts Institute of Technology
Cambridge, Massachusetts
and
Wyss Institute for Biologically Inspired
Engineering
Harvard University
Boston, Massachusetts

Shruti Balaji
Wake Forest Institute for Regenerative
Medicine
Wake Forest University School of Medicine
Winston-Salem, North Carolina

Sebastian F. Barreto-Ortiz
Department of Chemical and
Biomolecular Engineering
Johns Hopkins Physical Sciences
Oncology Center
and
Institute for NanoBioTechnology
Johns Hopkins University
Baltimore, Maryland

Eric Michael Brey
Pritzker Institute of Biomedical
Science and Engineering
and
Department of Biomedical Engineering
Illinois Institute of Technology
Chicago, Illinois
and
Hines V.A. Hospital
Hines, Illinois

Gulden Camci-Unal
Department of Medicine
Brigham and Women's Hospital
Harvard Medical School
and
Harvard-MIT Division of Health Sciences
and Technology
Massachusetts Institute of Technology
Cambridge, Massachusetts

Ali Cinar
Department of Chemical and Biological
Engineering
Illinois Institute of Technology
Chicago, Illinois

Tracy Criswell
Wake Forest Institute for Regenerative
Medicine
Wake Forest University School of
Medicine
Winston-Salem, North Carolina

Ronald N. Cohen
Department of Medicine
University of Chicago
Chicago, Illinois

Mehmet R. Dokmeci
Department of Medicine
Brigham and Women's Hospital
Harvard Medical School
and
Harvard-MIT Division of Health Sciences
and Technology
Massachusetts Institute of Technology
Cambridge, Massachusetts
and
Wyss Institute for Biologically Inspired
Engineering
Harvard University
Boston, Massachusetts

John P. Fisher
Fischell Department of Bioengineering
University of Maryland
College Park, Maryland

Steven C. George
Department of Biomedical Engineering
and
Department of Chemical Engineering and
    Materials Science
and
Department of Medicine
and
The Edwards Lifesciences Center
    for Advanced Cardiovascular
    Technology
University of California
Irvine, California

Sharon Gerecht
Department of Chemical and
    Biomolecular Engineering
Johns Hopkins Physical Sciences
    Oncology Center
and
Institute for NanoBioTechnology
and
Department of Materials Science and
    Engineering
Johns Hopkins University
Baltimore, Maryland

Anjelica L. Gonzalez
Department of Biomedical Engineering
Yale University
New Haven, Connecticut

James B. Hoying
Cardiovascular Innovation Institute
Louisville, Kentucky

Christopher C.W. Hughes
Department of Biomedical Engineering
and
Department of Molecular Biology and
    Biochemistry
and

The Edwards Lifesciences Center for
    Advanced Cardiovascular Technology
University of California
Irvine, California

Ali Khademhosseini
Department of Medicine
Brigham and Women's Hospital
Harvard Medical School
and
Harvard-MIT Division of Health Sciences
    and Technology
Massachusetts Institute of Technology
Cambridge, Massachusetts
and
Wyss Institute for Biologically Inspired
    Engineering
Harvard University
Boston, Massachusetts

Aimal H. Khankhel
Department of Biomedical Engineering
Boston University
Boston, Massachusetts

Omaditya Khanna
Pritzker Institute of Biomedical
    Science and Engineering
Illinois Institute of Technology
Chicago, Illinois

Gisela A. Kuhn
Institute for Biomechanics
ETH Zurich
Zurich, Switzerland

Holly M. Lauridsen
Department of Biomedical Engineering
Yale University
New Haven, Connecticut

Byron Long
Department of Bioengineering
Rice University
Houston, Texas

John Patrick McQuilling
Wake Forest Institute for Regenerative
    Medicine
Wake Forest University School of Medicine
Winston-Salem, North Carolina

Hamidreza Mehdizadeh
Department of Chemical and Biological
    Engineering
Illinois Institute of Technology
Chicago, Illinois

Geraldine M. Mitchell
O'Brien Institute
University of Melbourne
and
Department of Surgery
St Vincent's Hospital
and
Australian Catholic University
Melbourne, Victoria, Australia

Wayne A. Morrison
O'Brien Institute
University of Melbourne
and
Department of Surgery
St Vincent's Hospital
and
Australian Catholic University
Melbourne, Victoria, Australia

Monica L. Moya
Department of Biomedical Engineering
University of California
Irvine, California

Ralph Müller
Institute for Biomechanics
ETH Zurich
Zurich, Switzerland

Laura Nebuloni
Institute for Biomechanics
ETH Zurich
Zurich, Switzerland

Bao-Ngoc B. Nguyen
Fischell Department of Bioengineering
University of Maryland
College Park, Maryland

David Noren
Department of Bioengineering
Rice University
Houston, Texas

Emmanuel C. Opara
Wake Forest Institute for Regenerative
    Medicine
Wake Forest University School of Medicine
Winston-Salem, North Carolina

Georgia Papavasiliou
Department of Biomedical
    Engineering
Illinois Institute of Technology
Chicago, Illinois

Andrew J. Putnam
Department of Biomedical
    Engineering
University of Michigan
Ann Arbor, Michigan

Amina Ann Qutub
Department of Bioengineering
Rice University
Houston, Texas

Rahul Rekhi
Institute of Biomedical Engineering
Oxford University
Oxford, United Kingdom

Matthew R. Richardson
Wells Center for Pediatric Research
and
Department of Pediatrics
Indiana University School of
    Medicine
Indianapolis, Indiana

Šeila Selimović
Department of Medicine
Brigham and Women's Hospital
Harvard Medical School
and
Harvard-MIT Division of Health Sciences
    and Technology
Massachusetts Institute of Technology
Cambridge, Massachusetts

Rahul K. Singh
Department of Biomedical Engineering
University of Michigan
Ann Arbor, Michigan

Quinton Smith
Department of Chemical and
    Biomolecular Engineering
Johns Hopkins Physical Sciences
    Oncology Center
and
Institute for NanoBioTechnology
Johns Hopkins University
Baltimore, Maryland

Shay Soker
Wake Forest Institute for Regenerative
    Medicine
Wake Forest University School of
    Medicine
Winston-Salem, North Carolina

Joe Tien
Department of Biomedical Engineering
Boston University
Boston, Massachusetts

James G. Truslow
Department of Biomedical Engineering
Boston University
Boston, Massachusetts

Michael V. Turturro
Department of Biomedical Engineering
Illinois Institute of Technology
Chicago, Illinois

Marcella K. Vaicik
Department of Biomedical Engineering
Illinois Institute of Technology
Chicago, Illinois

Marina Vigen
Department of Biomedical Engineering
University of Michigan
Ann Arbor, Michigan

Zhan Wang
Wake Forest Institute for Regenerative
    Medicine
Wake Forest University School of
    Medicine
Winston-Salem, North Carolina

Stuart K. Williams
Cardiovascular Innovation Institute
Louisville, Kentucky

Keith H. K. Wong
Department of Biomedical Engineering
Boston University
and
Center for Engineering in Medicine
Massachusetts General Hospital
Harvard Medical School
Boston, Massachusetts

Lara Yildirimer
Department of Medicine
Brigham and Women's Hospital
Harvard Medical School
and
Harvard-MIT Division of Health Sciences
    and Technology
Massachusetts Institute
    of Technology
Cambridge, Massachusetts

Mervin C. Yoder
Wells Center for Pediatric Research
and
Department of Pediatrics
and
Department of Biochemistry and
    Molecular Biology
Indiana University School
    of Medicine
Indianapolis, Indiana

Xin Zhao
Department of Medicine
Brigham and Women's Hospital
Harvard Medical School
and
Harvard-MIT Division of Health Sciences
    and Technology
Massachusetts Institute of Technology
Cambridge, Massachusetts

Yu Zhou
Wake Forest Institute for Regenerative
    Medicine
Wake Forest University School of
    Medicine
Winston-Salem, North Carolina

Section I

Cells

1

Endothelial Progenitor Cells: Current Status

Matthew R. Richardson and Mervin C. Yoder

CONTENTS

1.1 Introduction

In 1997, Asahara and colleagues [1] reported that some circulating blood cells possessed the potential to differentiate *in vitro* into adherent cells displaying cell surface antigens typical for endothelial cells and simultaneous loss of many typical hematopoietic markers. The injection of these adherent endothelial-like cells promoted vascular repair and restoration of blood flow in animals with induced hindlimb ischemia. The cells displaying these properties were called circulating endothelial progenitor cells (EPCs). EPC was proposed to be derived from circulating bone marrow-derived cells expressing the cell surface proteins CD34 and/or the kinase insert domain receptor (KDR; the human vascular endothelial growth factor 2 receptor). The authors further proposed that circulating EPC could integrate into the areas of ischemic tissue in the experimentally injured animals and directly assist in new blood vessel formation, thus, displaying postnatal vasculogenic activity. Thousands of papers have been published under the term EPC since that first publication in 1997. EPC has been evaluated as a biomarker for a variety of cardiovascular diseases [2–4], stroke [5,6], autoimmune disease [7], cancer [8–11], diabetes [12–14], pulmonary diseases [15,16], and infections [17,18]. EPC plays roles in vascular repair and regeneration in a host of preclinical animal models [19–21] and in numerous clinical trials of cardiovascular disorders in human subjects (see www.clinicaltrials.gov). One would expect that this vast number of papers would provide a clear indication of the unique EPC-identifying markers; however, no unique identifier for this cell has been reported for human subjects. This has led to great confusion, controversy, and debate regarding the best methods for defining human EPC [22–24]. Nonetheless, investigators have pushed forward to test the safety and benefit of these potentially useful cells in human subjects in an attempt to treat serious systemic illnesses that lack highly effective current therapies [23,25]. Given modest improvements in patient outcomes, a trend in recent clinical trials is to isolate more specific subsets of bone marrow cells to enrich the autologous cell product in hopes of providing more benefit to the patients [25,26].

1.2 Methods to Define Human EPCs

Three general approaches have been employed to define EPC. First, circulating low-density mononuclear cells (MNCs) isolated from peripheral blood or cord blood may be placed into fibronectin-coated tissue-culture plates in commercial medium (endothelial growth medium 2 [EGM2], Lonza) containing fetal calf serum and growth factors that may include vascular endothelial growth factor (VEGF), fibroblast growth factor 2 (FGF2), insulin-like growth factor 1 (IGF1), and epidermal growth factor (EGF) in addition to other additives [27,28]. The nonadherent MNC may be removed from the cultures and those adherent cells that remain over the next 5–7 days that display the capacity to ingest acetylated low-density lipoprotein (ac-LDL) and bind the plant lectin *Ulex Europeaus agglutinin 1* (UEA1) have been defined as EPC. These putative EPC proliferate little during the culture period but undergo differentiation with an increase in endothelial-like appearance and properties and diminished hematopoietic characteristics. The cultured adherent MNC isolated by this method demonstrates a variety of angiogenic-promoting properties *in vitro* and *in vivo* [23,24]. However, this method has been shown to be fraught with potential limitations. Circulating platelets commonly contaminate the MNC fraction and frequently disintegrate during the MNC culture period with release of platelet membrane particles that incorporate into the cell membrane of any cell adherent to the fibronectin-coated culture wells [29]. Since the platelet membranes contain many of the same cell surface molecules as found on vascular endothelial cells, it is not uncommon for adherent cells to display the platelet–endothelial protein markers on their cell surface even though they are not actively transcribing any of these genes [25]. Thus, platelet contamination in this assay provides a robust source for false-positive identification of adherent putative EPC. Furthermore, adhesion of peripheral blood monocytes to fibronectin-coated dishes is a well-recognized method for monocyte isolation and purification [30]. Culture of blood monocytes with VEGF stimulates the expression of a variety of endothelial-like cell surface proteins (VEGF receptors, CD144, CD31, and von Willebrand factor [vWF]) on the monocytes as a result of modulating gene expression. Even molecules such as endothelial nitric oxide synthase (eNOS) thought to be endothelial lineage specific can be generated by the cultured monocytes [31,32]. The cultured adherent MNC isolated by this method demonstrates a variety of proangiogenic properties *in vitro* and *in vivo* [11–22] but the retained hematopoietic nature of these cells has led to their redefinition as circulating angiogenic cells (CACs) [33]. Very recent data suggest that CAC may revert to an inflammatory macrophage phenotype and eventually accumulates lipid droplets (cholesterol laden) to become foam cells when challenged *in vitro* [34]. This observation suggests that the proangiogenic activity and functional state of the CAC may not be fixed, but could be modulated by the microenvironment within which the CAC lodges and interacts.

A second method used to define EPC has relied upon the detection of a variety of cell surface proteins on circulating blood cells using monoclonal antibodies and fluorescence-activated cell sorting (FACS). As originally defined by Asahara and colleagues [1] and later modified by Peichev and colleagues [35], human peripheral blood cells expressing CD34,

AC133, and KDR may represent a phenotype to define circulating EPC. The percentage of putative EPC expressing KDR may be dependent on the presence of activated platelets *in vitro* and *in vivo*. Activated platelets attract and facilitate immobilization of CD34+ cells and cause a mobilization of KDR from the endosomal compartment to the cell surface [36]. However, neither CD34, AC133, nor KDR individually or any combination of these proteins is specifically restricted in expression to the EPC. In fact, hematopoietic stem and progenitor cells are highly enriched in the CD34 or AC133 subset [37,38]. While numerous other antigens (CD105, CD144, CD106, CD117, and CD45) and some enzymatic activities (aldehyde dehydrogenase) have been proposed as markers for EPC, none of them have been able to discriminate the EPC from the hematopoietic lineage cells in the circulating bloodstream [24]. Thus, the inability to define putative EPC as distinct from hematopoietic cells suggests that hematopoietic cells may comprise the majority of circulating cell types previously referred to as EPC [33]. A recent report that both murine and human hematopoietic common myeloid progenitor cell subsets demonstrate the greatest proangiogenic activity among all hematopoietic cell subsets both *in vitro* and *in vivo* lends support to this paradigm [39,40]. The common myeloid progenitor cells purified by FACS were potent in enhancing the revascularization of the ischemic hindlimb injury in immunodeficient mice, but the human hematopoietic progenitor cells did not directly morph into the blood vessel endothelium. Thus, the use of standard protocols for the isolation of hematopoietic subsets permits identification of cells that mirror the function of prior circulating MNC subsets thought to represent EPC. Even more recent improvements in the approaches to using FACS to identify EPC have not changed the apparent evidence that circulating cells expressing AC133, CD34, and KDR also coexpress markers reflective of the hematopoietic origin of this subset [41–44]. For example, the use of polychromatic flow cytometry has permitted a more detailed description of the proangiogenic AC133 expressing cells that coexpress CD34, CD31, and CD45 and have provided some novel approaches to discriminate proangiogenic from nonangiogenic hematopoietic subsets [45]. The ratio of proangiogenic to nonangiogenic hematopoietic cells was elevated in newly diagnosed patients with malignancies, but returned to control levels following treatment suggesting that this ratio may serve as a systemic biomarker for the malignant angiogenic state [46].

Finally, human EPC has also been defined through the use of two colony-forming assays. Human peripheral blood MNCs may be added to fibronectin-coated culture dishes and 1–2 days later, the nonadherent cells are removed and then replated on fibronectin-coated dishes. Clusters of phase contrast bright cells emerging in 4–9 days that demonstrate the growth of spindle-shaped cells emerging from the base of the clusters are referred to as colony-forming unit-Hill (CFU-Hill) EPC [27]. The frequency of these clusters in human peripheral blood highly predicts adverse cardiovascular risk in human subjects with cardiovascular disease [27]. Detailed analysis of the cellular composition of these clusters has revealed that both myeloid progenitor and lymphoid cells participate in the formation of these characteristic putative EPC colonies [34,47,48]. It has been suggested that the myeloid progenitor cells initially comprise the clustered cells that differentiate in the presence of T lymphoid-derived chemokines and cytokines (in addition to the added growth factors) to mature into the spindle-shaped macrophages that emerge from the base of the clustered cells. Following 9 days in culture, all the clusters disappear as the cells have matured into the adherent macrophage populations. As with the CAC, CFU-Hill-derived cells fail to proliferate extensively and do not display the replating potential. Thus, CFU-Hill colonies reflect the presence of a heterogeneous hematopoietic population (which differentiates to become alternatively activated M2 macrophages and T lymphoid cells) that play a role in angiogenesis, but do not represent EPCs that are capable of postnatal vasculogenic activity [34,47,48].

The second colony-forming assay identifies the circulating viable endothelial cells that possess proliferative potential. While the vast majority of circulating endothelial cells (CECs) represent sloughed senescent or apoptotic cells, reports over four decades have indicated that some CECs exist that display the capacity to attach to synthetic materials (vascular grafts or patches) sewn into human or animal vessels and proliferate to form a complete endothelial monolayer covering the intravascular-exposed surfaces [49]. To isolate these rare viable endothelial cells, adult peripheral blood MNCs may be placed in culture plates coated with type 1 collagen and culture medium with added growth factors (EGM2) similar to the CAC assay [50]. Culture medium is replaced every few days to replenish the added growth factors. After 2–3 weeks in culture, colonies of adherent endothelial cells become visible. These colonies may be isolated and enzymatically released from the collagen-coated plates. The isolated cells display cell surface antigens analogous to those of vascular endothelial cells but not hematopoietic cells. Analysis of the individual cells comprising the colonies that emerge from the blood MNCs indicates that some of the cells display clonal proliferative potential [50]. The proliferative potential is heterogeneous, with some of the single cultured cells forming small clusters of progeny (2–50 cells) and some of the cells displaying high proliferative potential (>2000 progenies per colony) in a 2-week culture period. The use of clonal analysis to compare the colonies of endothelial cells derived from human adult peripheral blood or cord blood has permitted delineation of an age-related decline in the proliferative potential of the circulating endothelial colony-forming cells (ECFCs) in human subjects [50], which have been recently replicated in nonhuman primates across the entire monkey life span (newborn to aged) [51]. The ECFC with high proliferative potential displays robust replating potential and high telomerase activity and appears to generate more human vessels when implanted into immunodeficient mice than cells with lower proliferative potential [52]. Of interest, ECFC derived from the peripheral blood of aged rhesus monkeys displays a defect in vasculogenesis upon implantation into immunodeficient mice that may result from a diminished capacity to undergo cytoplasmic vacuolation, a known key step in vasculogenesis [51]. ECFC derived from adult human peripheral blood has been reported to synergize with adult peripheral blood CAC to rescue blood flow to an ischemic limb in experimentally instrumented nude mice [53]. Autologous ECFC was also reported to rescue myocardial capillary morphogenesis in the infarct zone and diminish the decline in cardiac function following an experimentally induced acute myocardial infarction in pigs [54]. ECFCs derived from adult human peripheral blood have also been noted to engraft in the superficial retinal vasculature to enhance normal vascular repair, reduce avascular areas, and prevent damaging neovascular growth following a hyperoxic exposure to young mice, an established model of proliferative retinopathy [55]. Thus, ECFC appears to display many features of circulating cells that are consistent with an EPC; circulating cells give rise *in vitro* to colonies of endothelial cells with inherent heterogeneous proliferative potential, high replating activity, high telomerase activity, endothelial cell surface phenotype, and *in vivo* vessel-forming ability, and demonstrated the ability to play a role in vascular repair or regeneration in preclinical animal models of human vascular disorders. However, given the requirement that ECFC is typically isolated from cultured human adult peripheral blood or cord blood MNC, and that plating of the MNC may also contain the CAC or CFU-Hill, more detailed analytical methods have been recently employed to examine the relationship between these putative EPC subsets.

1.3 mRNA Expression Pattern Analysis of EPC Subsets

Gene expression profiling of adult peripheral blood MNC in the CFU-Hill assay has been performed to assess whether analysis of whole-genome mRNA (messenger ribonucleic acid) transcripts might delineate better the types of cells comprising the colonies thought to be EPC. Desai and colleagues [56] reported that the CFU-Hill colonies displayed a gene expression profile more highly enriched for T lymphocyte transcripts ($r = 0.87$) than the transcripts isolated from human umbilical cord blood endothelial cells (HUVEC) ($r = 0.67$). Analysis of the cells in the CFU-Hill colonies by flow cytometric detection of labeled mono-clonal antibodies confirmed that cell surface expression was more enriched for proteins typically expressed on T lymphocytes rather than endothelial cells (HUVEC). Placing the group of volunteer subjects on an exercise program for 3 months followed by repeated gene expression profiling of the CFU-Hill colonies revealed a significant increase in colony number; however, the gene expression pattern in the CFU-Hill colonies remained skewed toward that of a T lymphoid population rather than endothelial cells. These data support prior work [34,47,48] that had identified an enrichment of T lymphocytes and monocytes, but not endothelial cells, in the CFU-Hill assay.

Medina and colleagues [57] performed an mRNA expression comparison of cells iso-lated by adhesion to fibronectin (CAC) with the outgrowth of endothelial colonies (OEC, also called ECFC) grown on collagen-coated tissue culture dishes. They also examined the pattern of mRNA expression in peripheral blood monocytes and dermal microvascular endothelial cells (DMECs) as control populations. Numerous differences in specific gene transcript levels between the CAC and OEC were identified and the patterns of gene expres-sion poorly correlated ($r = 0.77$) between these groups. In fact, principle component analysis revealed that the CAC and OEC were segregated into two distinct groups when analyzing a single component. Of interest, the CAC more closely displayed a gene expression pattern to peripheral blood monocytes while the OEC displayed a gene expression pattern similar to the DMEC. Thus, the transcriptomic comparison of two putative EPC assays identified CAC as cells displaying a molecular phenotype like that of a hematopoietic cell subset, while OEC was highly correlated with mature endothelial gene expression patterns.

Furuhata and colleagues [58] compared the mRNA expression of cord blood CD34+ cells cultured on fibronectin to the transcriptomes of mature human endothelial cells-isolated cord blood (HUVEC), adult lung microvasculature (LMEC), and adult dermal tissue (DMEC). The cultured CD34 cells displayed a significant increase in the level of expression of >150 genes and significantly lower expression of >150 genes compared to the HUVEC, LMEC, and DMEC. Quantitative RNA analysis validated the differences in expression between the putative EPC and the mature endothelial cell populations.

Many other groups have analyzed the transcriptome of cultured putative EPC and iden-tified changes in mRNA expression levels over time *in vitro* [59–61] or identified differ-ences in EPC gene expression in patients with diabetes compared to control subjects [62]. Of interest, none of the above studies have led to identification of a unique set of molecular markers that have been validated as biomarkers for isolating EPC or that have been use-ful to distinguish this proangiogenic subset from other circulating blood cell elements. However, the data from these transcriptomic approaches have clearly confirmed that the assays used to isolate circulating EPC enrich for two major types of cells, hematopoietic and endothelial cells. The data also confirm that the various circulating EPC assays pri-marily enrich for different subsets of hematopoietic cells or ECFC.

1.4 MicroRNA Expression Pattern Analysis of EPC Subsets

MicroRNAs (miR) are small noncoding RNA that is postulated to fine tune the level of gene expression by binding to mRNA to inhibit translation into protein or enhance mRNA degradation [63,64]. These RNA molecules can be transcribed as single transcripts or in clusters. Several proteolytic cleavage steps process the nascent transcripts into pre-miR that are transported out of the nucleus and into the cytoplasm, where the endonuclease dicer further cleaves the pre-miR into functional 22-base pair RNA duplexes. A group of proteins called the RNA-induced silencing complex take up the miR and one of the miR strands is retained to bind to a specific sequence within the target mRNA to inhibit that mRNA from translation. Thousands of human miRs have been identified and each miR may have numerous target mRNAs. Thus, cellular gene expression patterns are not only regulated at the transcriptional level for each gene within a cell, but the mRNA transcripts are also subject to modification by numerous miR that may be cell, tissue, and state specific. Over the past decade, many miRs have been identified that play important roles in development [65], cancer [66–71], cardiovascular disease [72–75], and other disorders [76,77].

Differentiation of human embryonic stem (hES) cells into endothelial cells via progenitor intermediates is known to be associated with changes in the levels of expression of numerous miR [78–80]. However, none of the miRs that have been shown to be dramatically increased or decreased with differentiation have been found to be required for endothelial cell emergence from hES [81]. Thus, while miR plays important roles in directly reprogramming somatic cells into pluripotent stem cells [82], no specific miRs have been isolated that direct the differentiation of hES specifically into EPC or more mature endothelium.

miR profiling of OEC (also called ECFC) has revealed the expression of numerous miRs that have been shown to play important angiogenic roles in mature human endothelial cells, such as miR-126 [83,84]. MNC isolated from patients with cardiovascular disease was identified to differentiate into CAC (called adherent EPC) that was dysfunctional in number and function and displayed different miR expression patterns compared to control CAC [85–87]. Of interest, patients who suffered from ischemic cardiomyopathy and displayed circulating proangiogenic cells (identified as CD34+KDR+ EPC) with elevated levels of miR-126 were found to have a significantly increased survival over a 24-month study period. In contrast, patients suffering from nonischemic cardiomyopathy who possessed circulating proangiogenic cells with higher levels of miR-508-5p had a lower 24-month survival than patients with lower levels of miR-508-5p in their circulating proangiogenic cells.

Goretti and colleagues [88] recently directly examined differences in the miRNome of proangiogenic hematopoietic cells (identified as early outgrowth EPC) when compared to ECFC (called late outgrowth EPC). The proangiogenic hematopoietic cells selectively expressed several miR including five members of the miR-16 family, which were not expressed by the ECFC, while the ECFC expressed certain miRs that were not expressed by the proangiogenic cells. The miR-16 family members in the proangiogenic cells were shown to directly regulate the expression of target genes that were also differentially expressed in the proangiogenic cells versus the ECFC. Antagonism of the miR-16 family members in the proangiogenic cells permitted some cells to reenter the cell cycle, a feature not found in control proangiogenic cells. Antagonism of the miR-16 family members in the proangiogenic cells also stimulated upregulation of several endothelial cell surface proteins and enhanced secretion of the proangiogenic molecule interleukin-8. In addition, miR-16 antagonism in the proangiogenic cells caused them to display greater paracrine angiogenic effects when placed in coculture with HUVEC; the HUVEC displayed greater capillary tube formation. Thus, Goretti and colleagues [88] have identified some mutually exclusive expression patterns of miRs that may be regulating proangiogenic properties of the proangiogenic hematopoietic cells and the ECFC. These exciting data suggest that further study may unravel the biological differences in the mechanisms through which these EPCs function to promote vascular repair.

1.5 Proteomic Approaches to Comparison of EPC Subsets

The promise of proteomics has always been to achieve accurate identification and quantitation of every expressed protein from any tissue, cell type, or biological fluid. Although not currently routinely possible, improvements in liquid chromatographic mass spectrometric (LC–MS) and computational analytical tools for protein identification and quantitation have made this goal feasible [89,90]. Regulation of lineage specification and cellular differentiation from progenitor cell states is multifactorial; so, monitoring changes in miR and mRNA levels alone may be insufficient to gain a complete understanding of EPC biology as protein expression levels are known to be regulated separately and are not necessarily correlated with mRNA levels [91,92]. In addition to elucidating mechanisms of physiology and pathophysiology, proteomic analyses should provide biomarkers or protein level expression fingerprints to discriminate various cell types and cell states as has been discovered in similar systems such as the hematopoietic system [93,94]. Perhaps, the best utility of proteomic approaches is to gain a comprehensive systems-level picture of the overall similarity or dissimilarity of one cell type to other cell types and cell states. Indeed, Medina et al. [57] used a combination of transcriptional profiling and two-dimensional electrophoresis (2-DE), a staple of proteomics, to clearly differentiate two populations of putative EPCs. They found that the molecular fingerprint of the adherent MNC cultured on fibronectin (CAC) more resembled that of monocytes than endothelial cells, whereas the fingerprint of OECs (ECFCs) resembled that of mature ECs. This study provided firm evidence that these two EPC subsets were distinct populations at the molecular level that further explained their overall dissimilar phenotypes and functions. In 2009, a one-dimensional electrophoresis (1-DE) and LC–MS/MS proteomic analysis of microparticles derived from CAC cultures by Prokopi et al. [29] revealed that platelet contamination in low-density MNC preparations leads to transfer of cell surface proteins typically expressed by endothelial cells, such as CD31, vWF, and UEA-1 lectin binding onto any adherent cells present in the culture well. These preliminary data led the authors to probe deeper in the proteomic characterization of CAC. In this subsequent analysis, Urbich et al. characterized the proteome and secretome of human CAC using difference-in-gel electrophoresis (DIGE) and shotgun proteomics (LC–MS/MS). They found that the CAC molecular fingerprint again was more consistent with cells of a hematopoietic origin. Myeloid markers such as the alternative macrophage marker CCL18, the hemoglobin scavenger receptor CD163, platelet factor 4 (CXCL4), and platelet basic protein (CXCL7) were enriched in CAC. The top ontological category of the CAC proteomic analysis was "platelet alpha granule" [95]. Mourino-Alvarez et al. [96] used flow cytometry to compare EPCs and CECs directly isolated from fresh peripheral blood MNCs based on differences in cell surface marker expression. EPCs were defined as CD45[-/dim] CD31[+] CD34[bright] CD133[+] cells and CECs were defined as CD45[-/dim] CD31[bright] CD34[+] CD133[-] cells. Although the functional behavior of these two populations was not delineated, 282 proteins unique to CECs and 243 unique to

EPCs were identified. Such discriminating markers between these populations supported the previous work that had used polychromatic flow cytometry to distinguish CEC from proangiogenic hematopoietic cells [97]. A number of other proteomic studies focused on analysis of endothelial progenitors have been published [98–101], and although they do not directly compare various EPC subtypes, it may be beneficial to compare the resulting datasets to determine if any consistent trends emerge that could be used to understand better the hematopoietic versus endothelial nature and state of the putative EPC. Increases in the sensitivity, resolution, and speed of LC–MS analyses as well as improved methods for sample complexity reduction, such as multidimensional protein identification technology (MudPIT) and more recent developments such as hydrophilic interaction liquid chromatography (HILIC), will continue to facilitate the depth of proteomic analyses [89]. These advances will increase the utility of proteomics toward a systems-level understanding of cell types and cell states when combined with other omics-level investigations. As biomarker discovery becomes more routine, it is likely that proteomics will be of great utility in providing clear definitions of these various EPC subtypes in health and disease and facilitate making an appropriate EPC choice for use as a cell therapy.

1.6 Summary

Methods to identify EPCs have been hampered by the lack of identification of any unique or specific cell surface marker to discriminate these cells from other lineage progeny. Most of the proangiogenic cells called EPCs display hematopoietic features and functions. While these hematopoietic cells display blood cell properties, they also appear to play roles in the paracrine support of blood vessel repair or regeneration. ECFC is unique among all the other putative EPC subsets in displaying clonal proliferative potential with derivation of endothelial progeny that displays a cell surface phenotype similar to vascular endothelial cells, forms capillary-like structures with lumens *in vitro*, and spontaneously generates a human vascular plexus upon implantation *in vivo* in immunodeficient mice. Whether examining the various assays of EPC for mRNA, miR, or proteomic expression patterns, accumulating evidence supports the resolution of the EPC into proangiogenic hematopoietic cells and ECFC. It is time to focus on understanding the specific molecular and cellular functions of these proangiogenic cells in health and disease to elucidate the most helpful subsets for use as cell therapy in various disease states. It will be interesting to focus future studies on how the hematopoietic system produces cells with proangiogenic activity. The discovery of the underlying mechanisms for proangiogenic cell generation may permit more detailed protocols for isolating and expanding this subset for use as a cell therapy. Going forward, the most clarity for the field of vascular repair and regeneration may be achieved by referring to the specific ECFC or proangiogenic subsets under investigation, such as cord blood ECFC or CD34+ proangiogenic cells, rather than the term EPC.

References

1. Asahara T, Murohara T, Sullivan A, Silver M, van der Zee R, Li T et al. Isolation of putative progenitor endothelial cells for angiogenesis. *Science*. 1997 Feb 14;275(5302):964–7.

2. Eizawa T, Ikeda U, Murakami Y, Matsui K, Yoshioka T, Takahashi M et al. Decrease in circulating endothelial progenitor cells in patients with stable coronary artery disease. *Heart*. 2004 Jun;90(6):685–6.

3. Fadini GP, Sartore S, Albiero M, Baesso I, Murphy E, Menegolo M et al. Number and function of endothelial progenitor cells as a marker of severity for diabetic vasculopathy. *Arterioscler Thromb Vasc Biol*. 2006 Sep;26(9):2140–6.

4. Hughes AD, Coady E, Raynor S, Mayet J, Wright AR, Shore AC et al. Reduced endothelial progenitor cells in European and South Asian men with atherosclerosis. *Eur J Clin Invest*. 2007 Jan;37(1):35–41.

5. Moubarik C, Guillet B, Youssef B, Codaccioni JL, Piercecchi MD, Sabatier F et al. Transplanted late outgrowth endothelial progenitor cells as cell therapy product for stroke. *Stem Cell Rev*. 2010 Jun; 5:1558–6804.

6. Navarro-Sobrino M, Rosell A, Hernandez-Guillamon M, Penalba A, Ribo M, Alvarez-Sabin J et al. Mobilization, endothelial differentiation and functional capacity of endothelial progenitor cells after ischemic stroke. *Microvasc Res*. 2010 Dec;80(3):317–23.

7. Szekanecz Z, Besenyei T, Szentpetery A, Koch AE. Angiogenesis and vasculogenesis in rheumatoid arthritis. *Curr Opin Rheumatol*. 2010 May;22(3):299–306.

8. Bertolini F, Shaked Y, Mancuso P, Kerbel RS. The multifaceted circulating endothelial cell in cancer: Towards marker and target identification. *Nat Rev Cancer*. 2006 Nov;6(11):835–45.

9. Lyden D, Hattori K, Dias S, Costa C, Blaikie P, Butros L et al. Impaired recruitment of bone-marrow-derived endothelial and hematopoietic precursor cells blocks tumor angiogenesis and growth. *Nat Med*. 2001 Nov;7(11):1194–201.

10. Mancuso P, Colleoni M, Calleri A, Orlando L, Maisonneuve P, Pruneri G et al. Circulating endothelial-cell kinetics and viability predict survival in breast cancer patients receiving metronomic chemotherapy. *Blood*. 2006 Jul 15;108(2):452–9.

11. Nolan DJ, Ciarrocchi A, Mellick AS, Jaggi JS, Bambino K, Gupta S et al. Bone marrow-derived endothelial progenitor cells are a major determinant of nascent tumor neovascularization. *Genes Dev*. 2007 Jun 15;21(12):1546–58.

12. Desouza CV, Hamel FG, Bidasee K, O'Connell K. Role of inflammation and insulin resistance in endothelial progenitor cell dysfunction. *Diabetes*. 2011 Apr;60(4):1286–94.

13. Fadini GP, Avogaro A. Potential manipulation of endothelial progenitor cells in diabetes and its complications. *Diabetes Obes Metab*. 2010 Jul;12(7):570–83.

14. Jarajapu YP, Grant MB. The promise of cell-based therapies for diabetic complications: Challenges and solutions. *Circ Res*. 2010 Mar 19;106(5):854–69.

15. Diller GP, Thum T, Wilkins MR, Wharton J. Endothelial progenitor cells in pulmonary arterial hypertension. *Trends Cardiovasc Med*. 2010 Jan;20(1):22–9.

16. Fadini GP, Avogaro A, Ferraccioli G, Agostini C. Endothelial progenitors in pulmonary hypertension: New pathophysiology and therapeutic implications. *Eur Respir J*. 2010 Feb;35(2):418–25.

17. Becchi C, Pillozzi S, Fabbri LP, Al Malyan M, Cacciapuoti C, Della Bella C et al. The increase of endothelial progenitor cells in the peripheral blood: A new parameter for detecting onset and severity of sepsis. *Int J Immunopathol Pharmacol*. 2008 Jul–Sep;21(3):697–705.

18. Yamada M, Kubo H, Ishizawa K, Kobayashi S, Shinkawa M, Sasaki H. Increased circulating endothelial progenitor cells in patients with bacterial pneumonia: Evidence that bone marrow derived cells contribute to lung repair. *Thorax*. 2005 May;60(5):410–3.

19. Critser PJ, Yoder MC. Endothelial colony-forming cell role in neoangiogenesis and tissue repair. *Curr Opin Organ Transplant*. 2010 Feb;15(1):68–72.

20. Jujo K, Ii M, Losordo DW. Endothelial progenitor cells in neovascularization of infarcted myocardium. *J Mol Cell Cardiol*. 2008 Oct;45(4):530–44.

21. Tongers J, Roncalli JG, Losordo DW. Role of endothelial progenitor cells during ischemia-induced vasculogenesis and collateral formation. *Microvasc Res*. 2010 May;79(3):200–6.

22. Asahara T, Kawamoto A, Masuda H. Concise review: Circulating endothelial progenitor cells for vascular medicine. *Stem Cells*. 2011 Nov;29(11):1650–5.

23. Fadini GP, Avogaro A. Diabetes impairs mobilization of stem cells for the treatment of cardio-vascular disease: A meta-regression analysis. *Int J Cardiol.* 2012 Nov; 21:1522–7.
24. Hirschi KK, Ingram DA, Yoder MC. Assessing identity, phenotype, and fate of endothelial progenitor cells. *Arterioscler Thromb Vasc Biol.* 2008 Sep;28(9):1584–95.
25. Losordo DW, Kibbe MR, Mendelsohn F, Marston W, Driver VR, Sharafuddin M et al. A randomized, controlled pilot study of autologous CD34+ cell therapy for critical limb ischemia. *Circ Cardiovasc Interv.* 2012 Dec;5(6):821–30.
26. Mackie AR, Losordo DW. CD34-positive stem cells: In the treatment of heart and vascular disease in human beings. *Tex Heart Inst J.* 2011;38(5):474–85.
27. Hill JM, Zalos G, Halcox JPJ, Schenke WH, Waclawiw MA, Quyyumi AA et al. Circulating endothelial progenitor cells, vascular function, and cardiovascular risk. *N Engl J Med.* 2003 Feb 13;348(7):593–600.
28. Ito H, Rovira II, Bloom ML, Takeda K, Ferrans VJ, Quyyumi AA et al. Endothelial progenitor cells as putative targets for angiostatin. *Cancer Res.* 1999 Dec;59(23):5875–7.
29. Prokopi M, Pula G, Mayr U, Devue C, Gallagher J, Xiao Q et al. Proteomic analysis reveals presence of platelet microparticles in endothelial progenitor cell cultures. *Blood.* 2009 Jul 16;114(3):723–32.
30. Hassan NF, Campbell DE, Douglas SD. Purification of human monocytes on gelatin-coated surfaces. *J Immunol Methods.* 1986 Dec 24;95(2):273–6.
31. Schmeisser A, Garlichs CD, Zhang H, Eskafi S, Graffy C, Ludwig J et al. Monocytes coexpress endothelial and macrophagocytic lineage markers and form cord-like structures in Matrigel(R) under angiogenic conditions. *Cardiovasc Res.* 2001 Feb 16;49(3):671–80.
32. Schmeisser A, Graffy C, Daniel WG, Strasser RH. Phenotypic overlap between monocytes and vascular endothelial cells. *Adv Exp Med Biol.* 2003;522:59–74.
33. Rehman J, Li J, Orschell CM, March KL. Peripheral blood endothelial progenitor cells are derived from monocyte/macrophages and secrete angiogenic growth factors. *Circulation.* 2003 Mar 4;107(8):1164–9.
34. Rohde E, Schallmoser K, Reinisch A, Hofmann NA, Pfeifer T, Frohlich E et al. Pro-angiogenic induction of myeloid cells for therapeutic angiogenesis can induce mitogen-activated protein kinase p38-dependent foam cell formation. *Cytotherapy.* 2011 Apr;13(4):503–12.
35. Peichev M, Naiyer AJ, Pereira D, Zhu Z, Lane WJ, Williams M et al. Expression of VEGFR-2 and AC133 by circulating human CD34+ cells identifies a population of functional endothelial precursors. *Blood.* 2000 Feb 1;95(3):952–8.
36. de Boer HC, Hovens MM, van Oeveren-Rietdijk AM, Snoep JD, de Koning EJ, Tamsma JT, Huisman MV, Rabelink AJ, van Zonneveld AJ. Human CD34+/KDR+ cells are generated from circulating CD34+ cells after immobilization on activated platelets. *Arterioscler Thromb Vasc Biol.* 2011 Feb;31(2):408–15.
37. Alaiti MA, Ishikawa M, Costa MA. Bone marrow and circulating stem/progenitor cells for regenerative cardiovascular therapy. *Transl Res.* 2010 Sep;156(3):112–29.
38. Case J, Mead LE, Bessler WK, Prater D, White HA, Saadatzadeh MR et al. Human CD34 + AC133 + VEGFR-2+ cells are not endothelial progenitor cells but distinct, primitive hematopoietic progenitors. *Exp Hematol.* 2007 Jul;35(7):1109–18.
39. Wara AK, Croce K, Foo S, Sun X, Icli B, Tesmenitsky Y et al. Bone marrow-derived CMPs and GMPs represent highly functional proangiogenic cells: Implications for ischemic cardiovascular disease. *Blood.* 2011 Dec 8;118(24):6461–4.
40. Wara AK, Foo S, Croce K, Sun X, Icli B, Tesmenitsky Y et al. TGF-beta1 signaling and Kruppel-like factor 10 regulate bone marrow-derived proangiogenic cell differentiation, function, and neovascularization. *Blood.* 2011 Dec 8;118(24):6450–60.
41. Bertolini F, Mancuso P, Benayoun L, Gingis-Velitski S, Shaked Y. Evaluation of circulating endothelial precursor cells in cancer patients. *Methods Mol Biol.* 2012;904:165–72.
42. Fadini GP, Losordo D, Dimmeler S. Critical reevaluation of endothelial progenitor cell phenotypes for therapeutic and diagnostic use. *Circ Res.* 2012 Feb 17;110(4):624–37.

43. Hristov M, Schmitz S, Nauwelaers F, Weber C. A flow cytometric protocol for enumeration of endothelial progenitor cells and monocyte subsets in human blood. *J Immunol Methods*. 2012 Jul 31;381(1–2):9–13.

44. Kuwana M, Okazaki Y. Quantification of circulating endothelial progenitor cells in systemic sclerosis: A direct comparison of protocols. *Ann Rheum Dis*. 2012 Apr;71(4):617–20.

45. Estes ML, Mund JA, Mead LE, Prater DN, Cai S, Wang H et al. Application of polychromatic flow cytometry to identify novel subsets of circulating cells with angiogenic potential. *Cytometry A*. 2010 Sep;77(9):831–9.

46. Pradhan KR, Mund JA, Johnson C, Vik TA, Ingram DA, Case J. Polychromatic flow cytometry identifies novel subsets of circulating cells with angiogenic potential in pediatric solid tumors. *Cytometry B Clin Cytom*. 2011 Sep;80(5):335–8.

47. Hur J, Yang HM, Yoon CH, Lee CS, Park KW, Kim JH et al. Identification of a novel role of T cells in postnatal vasculogenesis: Characterization of endothelial progenitor cell colonies. *Circulation*. 2007 Oct 9;116(15):1671–82.

48. Rohde E, Malischnik C, Thaler D, Maierhofer T, Linkesch W, Lanzer G et al. Blood monocytes mimic endothelial progenitor cells. *Stem Cells*. 2006 Feb;24(2):357–67.

49. Yoder MC. Is endothelium the origin of endothelial progenitor cells? *Arterioscler Thromb Vasc Biol*. 2010 Jun;30(6):1094–103.

50. Ingram DA, Mead LE, Tanaka H, Meade V, Fenoglio A, Mortell K et al. Identification of a novel hierarchy of endothelial progenitor cells using human peripheral and umbilical cord blood. *Blood*. 2004 Nov 1;104(9):2752–60.

51. Shelley WC, Huang L, Critser PJ, Mead LE, Zeng P, Prater D, Ingram DA, Tarantal AF, Yoder MC. Changes in the frequency and *in vivo* vessel forming ability of rhesus monkey circulating endothelial colony forming cells (ECFC) across the lifespan (birth to aged). *Peds Res*. In Press. 2011.

52. Yoder MC, Mead LE, Prater D, Krier TR, Mroueh KN, Li F et al. Redefining endothelial progenitor cells via clonal analysis and hematopoietic stem/progenitor cell principles. *Blood*. 2007 Mar 1;109(5):1801–9.

53. Yoon YS, Uchida S, Masuo O, Cejna M, Park JS, Gwon HC et al. Progressive attenuation of myocardial vascular endothelial growth factor expression is a seminal event in diabetic cardiomyopathy: Restoration of microvascular homeostasis and recovery of cardiac function in diabetic cardiomyopathy after replenishment of local vascular endothelial growth factor. *Circulation*. 2005 Apr 26;111(16):2073–85.

54. Dubois C, Liu X, Claus P, Marsboom G, Pokreisz P, Vandenwijngaert S et al. Differential effects of progenitor cell populations on left ventricular remodeling and myocardial neovascularization after myocardial infarction. *J Am Coll Cardiol*. 2010 May 18;55(20):2232–43.

55. Medina RJ, O'Neill CL, Sweeney M, Guduric-Fuchs J, Gardiner TA, Simpson DA et al. Molecular analysis of endothelial progenitor cell (EPC) subtypes reveals two distinct cell populations with different identities. *BMC Med Genomics*. 2010 May 13;3(1):18.

56. Desai A, Glaser A, Liu D, Raghavachari N, Blum A, Zalos G et al. Microarray-based characterization of a colony assay used to investigate endothelial progenitor cells and relevance to endothelial function in humans. *Arterioscler Thromb Vasc Biol*. 2009 Jan;29(1):121–7.

57. Medina RJ, O'Neill CL, Sweeney M, Guduric-Fuchs J, Gardiner TA, Simpson DA et al. Molecular analysis of endothelial progenitor cell (EPC) subtypes reveals two distinct cell populations with different identities. *BMC Med Genomics*. 2010;3:18.

58. Furuhata S, Ando K, Oki M, Aoki K, Ohnishi S, Aoyagi K et al. Gene expression profiles of endothelial progenitor cells by oligonucleotide microarray analysis. *Mol Cell Biochem*. 2007 Apr;298(1–2):125–38.

59. Ahrens I, Domeij H, Topcic D, Haviv I, Merivirta RM, Agrotis A et al. Successful *in vitro* expansion and differentiation of cord blood derived CD34+ cells into early endothelial progenitor cells reveals highly differential gene expression. *PLoS One*. 2011;6(8):e23210.

60. Gremmels H, Fledderus JO, van Balkom BW, Verhaar MC. Transcriptome analysis in endothelial progenitor cell biology. *Antioxid Redox Signal*. 2011 Aug 15;15(4):1029–42.

61. Igreja C, Fragoso R, Caiado F, Clode N, Henriques A, Camargo L et al. Detailed molecular characterization of cord blood-derived endothelial progenitors. *Exp Hematol*. 2008 Feb;36(2):193–203.

62. van Oostrom O, de Kleijn DP, Fledderus JO, Pescatori M, Stubbs A, Tuinenburg A et al. Folic acid supplementation normalizes the endothelial progenitor cell transcriptome of patients with type 1 diabetes: A case-control pilot study. *Cardiovasc Diabetol*. 2009;8:47.

63. Bartel DP. MicroRNAs: Genomics, biogenesis, mechanism, and function. *Cell*. 2004 Jan 23;116(2):281–97.

64. Slezak-Prochazka I, Durmus S, Kroesen BJ, van den Berg A. MicroRNAs, macrocontrol: Regulation of miRNA processing. *RNA*. 2010 Jun;16(6):1087–95.

65. Ambros V. MicroRNAs and developmental timing. *Curr Opin Genet Dev*. 2011 Aug;21(4):511–7.

66. Baer C, Claus R, Plass C. Genome-wide epigenetic regulation of miRNAs in cancer. *Cancer Res*. 2013 Jan 15;73(2):473–7.

67. Ceppi P, Peter ME. MicroRNAs regulate both epithelial-to-mesenchymal transition and cancer stem cells. *Oncogene*. 2014 Jan;33(3):269–78.

68. Chen B, Li H, Zeng X, Yang P, Liu X, Zhao X et al. Roles of microRNA on cancer cell metabolism. *J Transl Med*. 2012;10:228.

69. Kong YW, Ferland-McCollough D, Jackson TJ, Bushell M. microRNAs in cancer management. *Lancet Oncol*. 2012 Jun;13(6):e249–58.

70. Nana-Sinkam SP, Croce CM. Clinical applications for microRNAs in cancer. *Clin Pharmacol Ther*. 2013 Jan;93(1):98–104.

71. Wang Y, Taniguchi T. MicroRNAs and DNA damage response: Implications for cancer therapy. *Cell Cycle*. 2013 Jan 1;12(1):32–42.

72. Anand S. A brief primer on microRNAs and their roles in angiogenesis. *Vasc Cell*. 2013;5(1):2.

73. Han M, Toli J, Abdellatif M. MicroRNAs in the cardiovascular system. *Curr Opin Cardiol*. 2011 May;26(3):181–9.

74. Small EM, Frost RJ, Olson EN. MicroRNAs add a new dimension to cardiovascular disease. *Circulation*. 2010 Mar 2;121(8):1022–32.

75. Topkara VK, Mann DL. Role of microRNAs in cardiac remodeling and heart failure. *Cardiovasc Drugs Ther*. 2011 Apr;25(2):171–82.

76. Hulsmans M, De Keyzer D, Holvoet P. MicroRNAs regulating oxidative stress and inflammation in relation to obesity and atherosclerosis. *FASEB J*. 2011 Aug;25(8):2515–27.

77. McGregor RA, Choi MS. microRNAs in the regulation of adipogenesis and obesity. *Curr Mol Med*. 2011 Jun;11(4):304–16.

78. Kane NM, Howard L, Descamps B, Meloni M, McClure J, Lu R et al. Role of microRNAs 99b, 181a, and 181b in the differentiation of human embryonic stem cells to vascular endothelial cells. *Stem Cells*. 2012 Apr;30(4):643–54.

79. Kane NM, Meloni M, Spencer HL, Craig MA, Strehl R, Milligan G et al. Derivation of endothelial cells from human embryonic stem cells by directed differentiation: Analysis of microRNA and angiogenesis *in vitro* and *in vivo*. *Arterioscler Thromb Vasc Biol*. 2010 Jul;30(7):1389–97.

80. Yoo JK, Kim J, Choi SJ, Noh HM, Kwon YD, Yoo H et al. Discovery and characterization of novel microRNAs during endothelial differentiation of human embryonic stem cells. *Stem Cells Dev*. 2012 Jul 20;21(11):2049–57.

81. Heinrich EM, Dimmeler S. MicroRNAs and stem cells: Control of pluripotency, reprogramming, and lineage commitment. *Circ Res*. 2012 Mar 30;110(7):1014–22.

82. Bao X, Zhu X, Liao B, Benda C, Zhuang Q, Pei D et al. MicroRNAs in somatic cell reprogramming. *Curr Opin Cell Biol*. 2013 Apr;25(2):208–14.

83. Berezikov E, van Tetering G, Verheul M, van de Belt J, van Laake L, Vos J et al. Many novel mammalian microRNA candidates identified by extensive cloning and RAKE analysis. *Genome Res*. 2006 Oct;16(10):1289–98.

84. van Solingen C, Seghers L, Bijkerk R, Duijs JM, Roeten MK, van Oeveren-Rietdijk AM et al. Antagomir-mediated silencing of endothelial cell specific microRNA-126 impairs ischemia-induced angiogenesis. *J Cell Mol Med*. 2009 Aug;13(8A):1577–85.

85. Minami Y, Satoh M, Maesawa C, Takahashi Y, Tabuchi T, Itoh T et al. Effect of atorvastatin on microRNA 221/222 expression in endothelial progenitor cells obtained from patients with coronary artery disease. *Eur J Clin Invest.* 2009 May;39(5):359–67.

86. Zhang Q, Kandic I, Kutryk MJ. Dysregulation of angiogenesis-related microRNAs in endothelial progenitor cells from patients with coronary artery disease. *Biochem Biophys Res Commun.* 2011 Feb 4;405(1):42–6.

87. Zhang X, Mao H, Chen JY, Wen S, Li D, Ye M et al. Increased expression of microRNA-221 inhibits PAK1 in endothelial progenitor cells and impairs its function via c-Raf/MEK/ERK pathway. *Biochem Biophys Res Commun.* 2013 Feb 15;431(3):404–8.

88. Goretti E, Rolland-Turner M, Leonard F, Zhang L, Wagner DR, Devaux Y. MicroRNA-16 affects key functions of human endothelial progenitor cells. *J Leukoc Biol.* 2013 May;93(5):645–55.

89. Altelaar AF, Munoz J, Heck AJ. Next-generation proteomics: Towards an integrative view of proteome dynamics. *Nat Rev Genet.* 2013 Jan;14(1):35–48.

90. Nagaraj N, Wisniewski JR, Geiger T, Cox J, Kircher M, Kelso J et al. Deep proteome and transcriptome mapping of a human cancer cell line. *Mol Syst Biol.* 2011;7:548.

91. Maier T, Guell M, Serrano L. Correlation of mRNA and protein in complex biological samples. *FEBS Lett.* 2009 Dec 17;583(24):3966–73.

92. Schwanhausser B, Busse D, Li N, Dittmar G, Schuchhardt J, Wolf J et al. Global quantification of mammalian gene expression control. *Nature.* 2011 May 19;473(7347):337–42.

93. Cadeco S, Williamson AJ, Whetton AD. The use of proteomics for systematic analysis of normal and transformed hematopoietic stem cells. *Curr Pharm Des.* 2012;18(13):1730–50.

94. Whetton AD, Williamson AJ, Krijgsveld J, Lee BH, Lemischka I, Oh S et al. The time is right: Proteome biology of stem cells. *Cell Stem Cell.* 2008 Mar 6;2(3):215–7.

95. Urbich C, De Souza AI, Rossig L, Yin X, Xing Q, Prokopi M et al. Proteomic characterization of human early pro-angiogenic cells. *J Mol Cell Cardiol.* 2011 Feb;50(2):333–6.

96. Mourino-Alvarez L, Calvo E, Moreu J, Padial LR, Lopez JA, Barderas MG et al. Proteomic characterization of EPCs and CECs *in vivo* from acute coronary syndrome patients and control subjects. *Biochimica et Biophysica Acta.* 2013 Apr;1830(4):3030–53.

97. Mund JA, Estes ML, Yoder MC, Ingram DA Jr., Case J. Flow cytometric identification and functional characterization of immature and mature circulating endothelial cells. *Arterioscler Thromb Vasc Biol.* 2012 Apr;32(4):1045–53.

98. Appleby SL, Cockshell MP, Pippal JB, Thompson EJ, Barrett JM, Tooley K et al. Characterization of a distinct population of circulating human non-adherent endothelial forming cells and their recruitment via intercellular adhesion molecule-3. *PLoS One.* 2012;7(11):e46996.

99. Brea D, Rodriguez-Gonzalez R, Sobrino T, Rodriguez-Yanez M, Blanco M, Castillo J. Proteomic analysis shows differential protein expression in endothelial progenitor cells between healthy subjects and ischemic stroke patients. *Neurol Res.* 2011 Dec;33(10):1057–63.

100. Kim J, Jeon YJ, Kim HE, Shin JM, Chung HM, Chae JI. Comparative proteomic analysis of endothelial cells progenitor cells derived from cord blood- and peripheral blood for cell therapy. *Biomaterials.* 2013 Feb;34(6):1669–85.

101. Wei J, Liu Y, Chang M, Sun CL, Li DW, Liu ZQ et al. Proteomic analysis of oxidative modification in endothelial colony-forming cells treated by hydrogen peroxide. *Int J Mol Med.* 2012 Jun;29(6):1099–105.

2

Role of Pericytes in Tissue Engineering

Holly M. Lauridsen and Anjelica L. Gonzalez

CONTENTS

2.1 Introduction

2.1.1 Overview of the Human Vasculature

The vasculature is an essential component of every living tissue. Blood circulating through the vasculature delivers oxygen and nutrients while removing metabolic waste to sustain cell growth and function. The circulation additionally facilitates distal intercellular communication; soluble factors and proteins released from cells are delivered beyond the diffusion barrier to enable endocrine communication. Pumping blood from the heart and the lungs to peripheral tissue for exchange requires vascular structures to vary spatially. The high-pressure large arteries proximal to the heart (aorta, carotid, etc.) are tasked with transporting oxygenated blood to distal regions of the body, and therefore require a thick, mechanically stable structure. These vessels are comprised of three distinct layers: the *tunica intima*, the luminal layer, the *tunica media*, the middle region, and the *tunica adventitia*, the outermost layer (Figure 2.1). An endothelial cell (EC) monolayer covers the luminal surface of the tunica intima, which is critical to prevent the formation of aberrant thrombi. The underlying layers, the tunica media and the tunica adventitia, are composed of extracellular matrix proteins including collagen and elastin, proteoglycans (PGs) and glycosaminoglycans (GAGs), with vascular smooth muscle cells (in the media) and fibroblasts (in the adventitia). Although there are differences in the organization and size of each of these layers when comparing the various classes of large vessels (elastic arteries, muscular arteries, veins, etc.), all large vessels maintain this characteristic three-layer structure (Humphrey 2002).

The term "microvasculature" describes the smaller vessels of the circulatory system including precapillaries, capillaries, and postcapillary venules. While the large veins and arteries are responsible for distributing blood throughout the body, the microvasculature is required for nutrient, gas, and waste diffusion between the blood and the cellular and matrix components of vascularized tissues. Thus, unlike the mechanically stronger large vessels, the microvasculature is characterized by thin walls that facilitate molecular diffusion (Mitchell and Schoen 2009). As in large vessels, the luminal surface of the microvessel is a single EC monolayer; the abluminal tissue, however, differs significantly between the large and microvessels (Movat and Fernando 1964). Basement membrane (BM), composed of a heterogeneous protein matrix composed of collagen (type IV), laminin-8, nidogen, perlecan, and other proteins surrounds

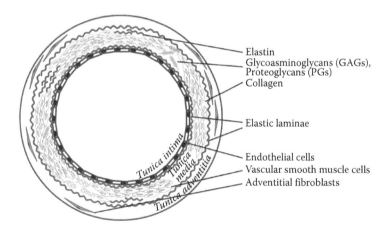

FIGURE 2.1
Common components of large blood vessels. Large vessels are composed of three layers: tunica intima, tunica media, and tunica adventitia. The structure of large vessels varies with location, but in general the cellular components are ECs (red), vascular smooth muscle cells (light blue) and fibroblasts (dark blue). Mural composition and ultrastructure vary spatially as well, but in most cases are composed of elastin (purple squiggle), collagen (green triple helix), and GAGs/PGs (orange feathers), among other things. Muscular arteries (with the exception of cerebral arteries) have two elastic laminae dividing the layers. (Adapted from Humphrey, J. D., 2002. Cardiovascular solid mechanics: Cells, tissues, and organs. 1st ed. New York: Springer.)

the EC (Wang et al. 2006, San Antonio and Iozzo 2006). Both EC (during angiogenesis) and pericytes (PCs) contribute to the BM formation (Stratman et al. 2009).

2.1.2 PCs: Origin, Characterization, and Location

PCs are believed to be of mesenchymal origin and related by lineage to vascular smooth muscle cells (vSMCs) (Sims 1986, Armulik et al. 2005). PC populations are distinguished from other cell types through expression of chondroitin sulfate proteoglycan 4 (NG2), CD90 (Thy-1), melanoma cell adhesion molecule (MCAM, CD146), α-smooth muscle actin (α-SMA), platelet-derived growth factor beta (PDGFβ), and calponin while lacking vSMC-, EC-, and leukocyte-specific markers including transgelin (SM22-α), smooth muscle myosin heavy chain, platelet EC adhesion molecule-1 (PECAM-1, CD31), CD34, leukocyte common antigen (protein tyrosine phosphatase receptor type C, CD45), and CD14 (Maier et al. 2010).

Historically, PCs were defined exclusively based on morphology; their defining characteristic being long primary processes extending along the direction of the vessel with secondary perpendicular processes circumferentially wrapping around the vessel (Allt and Lawrenson 2001). Secondary PC processes interdigitate to form functional intercellular bonds for communication and barrier functions (Allt and Lawrenson 2001). Since this initial definition, exploration of the microvasculature has revealed that PC morphology differs with vessel type and location in the body. As shown in Figure 2.2, PC in the

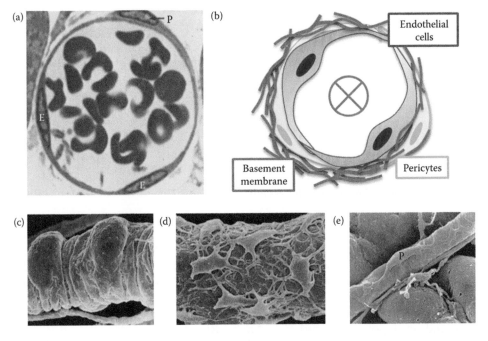

FIGURE 2.2
EC–PC orientations in the vasculature. (a) Cross section of a microvessel showing red blood cells in the center, two endothelial cells (E) and one pericyte (P) wrapping around the circumference (Sims 1986). (b) Schematic representation of a microvessel cross section similar to the image in panel A. Two ECs (pink) are surrounded by a mix of BM (purple; exaggerated for effect) and PC (blue). (c–e) Scanning electron micrographs of the varying structures of mural cells in different parts of the body, such as in a muscle arteriole (c; Fujiwara 1984), in a postcapillary venule (d; Fujiwara 1984) and in the rat submandibular gland (e; Shimada 1981).

precapillary arteriole are circumferentially oriented and densely packed around the vessel, whereas in the postcapillary venule and the venule, the PC possess a more stellate-like morphology (Armulik et al. 2011). Interestingly, PC expressing stellate-like processes are not restricted to a single vessel, but they may reach to other nearby capillaries or venules as well, creating a continuous network of communicating cells between vascular structures (Hirschi and D'Amore 1996). PC morphology is also variable between organ systems; in the central nervous system (CNS) for example, PC are spread with stellate-like features, whereas renal PC display a rounder and less spread morphology (Sims 1986, Armulik et al. 2005).

As PC morphology varies with location in the body, so does PC density. In the skin, the lungs, the eyes, and the brain, for example, there is a high concentration of PC, resulting in an EC to PC ratio of approximately 1:1 (Armulik et al. 2005, Takata et al. 2011). This dense network of PC is responsible for maintaining a tight vascular barrier, which is of particular importance in specialized tissues such as the brain and the eye (Armulik et al. 2005). In other tissues, like highly metabolic skeletal muscle in which rapid diffusion of oxygen and nutrients is paramount, the number of PC is drastically lower. In these tissues the EC to PC ratio is between 1:10 and 1:100 (Armulik et al. 2005). The variable presence of PCs in different tissues is believed to be associated with the heterogeneous functions of PC throughout the body; these diverse functions will be described in the next section.

2.2 Functional Roles of PC

Unlike the extensive imaging studies to classify PC morphology, studies to determine PC function were limited until recently. It has since been shown that EC and PC are highly interdependent and that PC serve as critical mediators in numerous vascular functions ranging from vessel formation and stabilization to regulating vascular permeability in inflammation (Allt and Lawrenson 2001, Armulik et al. 2005).

2.2.1 PC in Angiogenesis

2.2.1.1 PC-Recruitment to Sprouting Vessels

Starting in angiogenesis, the reliance of EC and PC on one another is critical for the appropriate formation of new blood vessels (Figure 2.3). During the initiation of angiogenesis, proangiogenic factors enable the ECs to degrade surrounding extracellular matrix and subsequently invade new void spaces; these voids can be referred to as vascular guidance tunnels (Stratman et al. 2009a,b). EC and PC communicate via paracrine signaling with soluble molecules including platelet derived growth factor beta (PDGFβ), transforming growth factor beta (TGFβ), vascular endothelial growth factor (VEGF), and angiopoietins (Ang1 and Ang2) (Ribatti et al. 2011). PDGFβ is required in order to recruit PC expressing both NG2 and PDGFR-β (the receptor for PDGFβ) to the site of sprouting (Lu and Sood 2008, Ribatti et al. 2011). Only select EC at the tip of the sprout will secrete PDGFβ, and the amount of PDGFβ secreted must be tightly regulated; underexpression results in a lack of PC recruitment and unstable EC in the new vessels, whereas overexpression can lead to embryonic lethality (Gerhardt and Betsholtz 2003).

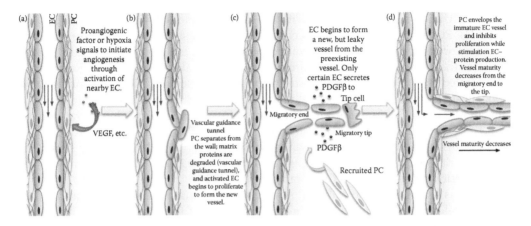

FIGURE 2.3
EC–PC interactions in angiogenesis. (a) Angiogenesis is initiated when proangiogenic stimuli (orange) activate nearby EC (pink cells) to begin forming a new vessel. (b) PC (blue cells) detach and activated EC degrade surrounding basement membrane (purple) in order to form a vascular guidance tunnel. After proliferating, new EC begin to populate the void in the extracellular matrix. (c) EC form an immature and leaky vessel capped with a specialized tip cell. Certain EC secrete platelet derived growth factor-β (PDGFβ) to recruit nearby PC. (d) PC arrive at the vessel and envelope the stalk of the newly formed vessel. The PC signal to the EC to inhibit EC proliferation and BM production. The PC can locally stabilize the vessel while in other regions it remains unstable.

VEGF is also able to attract PC to sites of hypoxia (low oxygenation) via nitric oxide signaling, contributing to the development of stabilized microvasculature; in areas where PC are already present, VEGF supports angiogenesis by promoting PC proliferation and migration (Ribatti et al. 2011). Once recruited, PCs envelop the stalk of the new vessel known as the migratory end. The confined presence of PC in the migratory end locally inhibit EC proliferation, EC in the sprouting tip continually proliferate and extend the vessel (Lu and Sood 2008). Thus, vessel maturation and growth are regulated spatially by the PC such that both processes occur simultaneously. It should be noted that although in most tissues, the EC are the first to invade and to initiate the paracrine signaling cascade that takes place during vessel formation, there are several instances in which PC initiate this process. As seen in the corpus luteum and several other tissues, PCs can invade tissues alone, and secrete VEGF while forming a new tube, which then enables the subsequent penetration of EC into the newly formed tube (Bergers and Song 2005, Ribatti et al. 2011).

2.2.1.2 PC Roles in Vessel Stabilization and Maturation

In most cases of angiogenesis, when EC initiate vessel formation, the ability of the PC to stabilize the vessel is derived from their secretion of Ang-1, which activates the EC angiopoietin receptor Tie-2 and allows maturation processes to begin (Ribatti et al. 2011). Ang-2, on the other hand, is secreted by the EC at the leading edge of the sprout and maintains a destabilized environment such that the growth of the vessel can continue (Ribatti et al. 2011). During the maturation phase of angiogenesis, physical contact becomes a critical mediator as well, as EC and PC form gap junctions comprised of connexins for intercellular transfer of soluble signaling molecules (Hall 2006). EC–PC contacts and contact-mediated signaling are not only limited to angiogenesis, but also serve many roles in mature vessels that will be discussed later.

The stabilization process continues with the formation of the BM around the EC (Ribatti et al. 2011). PC are critical in this process not only because they secrete matrix proteins that are incorporated into the vessel wall (Herman 2006), but also because their presence stimulates the luminal EC to secrete proteins as well (Stratman et al. 2009). Research has demonstrated that when EC and PC are in contact in vascular guidance tunnels during angiogenesis EC produce the extracellular matrix proteins fibronectin, nidogen-1, perlecan, and laminin; EC do not produce these proteins when PC are absent (Stratman et al. 2009). The extracellular matrix can also signal to EC and PC to mediate cell function. For example, during angiogenesis the presence of the aforementioned PC-induced fibronectin, nidogen, perlecan, and laminin corresponds to an increase in $\alpha_5\beta_1$, $\alpha_3\beta_1$, $\alpha_6\beta_1$, and $\alpha_1\beta_1$ integrin expression on both EC and PC (Stratman et al. 2009).

The BM is additionally influential in regulating paracrine signaling and macroscopic vessel structure. The heterogeneous structure of the matrix results in variable diffusion properties along the length of the vessel. Intramural paracrine signals can therefore become trapped in some areas leading to a more sustained effect (Davis and Senger 2005). EC-derived components of the BM are capable of altering the contractile phenotype of PC (discussed further later), which may be linked to certain pathologies (Herman 2006). The contractile phenotype of PC results, in part, from the presence of α-SMA. When EC-protein production is deregulated, the balance of BM proteins is disrupted (e.g., elevated collagen levels or increased heparin and fibroblast growth factor 2) resulting in the loss of PC-αSMA expression and therefore loss of PC contractility (Herman 2006).

2.2.1.3 Functional Blocking of PCs Results in Abnormal Vessel Formation

Genetically altered mice either lacking PC entirely or lacking a specific functional aspect of PC have been used to clarify the necessity of PC in appropriate vessel formation. Disruption of any of the aforementioned vascular mural cell signaling pathways, such as TGFβ, can lead to embryonic lethality and defects in vascular formation such as abnormal remodeling of the yolk sac vasculature and malformed mural cells (Armulik et al. 2011). When PCs are improperly recruited to the vessel via interference with the PDGFβ signaling pathway, numerous vessel abnormalities result; these abnormalities include endothelial hyperplasia, hypervariable vessel diameter, luminal membrane folds, and leaky vessels, among others (Lu and Sood 2008). Furthermore, the vascular phenotypes in PDGFβ-deficient mice replicate the pathophysiology that is associated with the loss of PC in mature human vessels. This has been useful in identifying signaling pathways that are potentially deregulated in pathologies. One example of this parallel is seen with the role of PDGFβ in diabetic retinopathy, which will be described in more detail later (Lindahl et al. 1997, Hammes et al. 2002).

2.2.2 EC/PC Interactions in Mature Vessels

2.2.2.1 Physical Interactions between EC and PC

Even when the vessel is stable and mature, the EC and PC continue to influence one another. Within the vascular wall, EC and PC are often within 20 nm of one another (Sims 1986). Physical EC–PC contacts form in voids in the BM structure; these intercellular contacts are of both mechanical and chemical significance (Courtoy and Boyles 1983, Allt and Lawrenson 2001, Armulik et al. 2005). EC–PC contacts are known as peg–socket contacts (Figure 2.4),

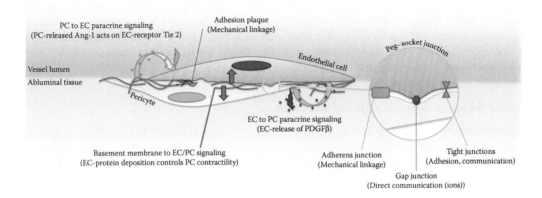

FIGURE 2.4

EC–PC communication in the vascular wall. EC–PC communication occurs in the wall via paracrine signaling and direct and indirect contact. Paracrine signaling is bidirectional; for example, EC (pink) release PDGFβ to attract PC (blue), whereas PC release Ang-1 which binds to EC Tie 2 receptors. Contact-mediated communication occurs indirectly through fibronectin-rich adhesion plaques, or directly in peg–socket junctions which may contain adherens-, gap-, or tight junctions.

which are characterized by an invagination of the EC or PC membrane that is filled by an extension from the other cell type. N-cadherin and β-catenin are found in the resulting tight-, gap-, or adherens-junctions formed in the peg–socket contact (Sims 1986, Allt and Lawrenson 2001, Armulik et al. 2005, Lu and Sood 2008). It has been reported that up to 1000 of these contact points may exist on a single EC (Armulik et al. 2011).

EC–PC contact is not limited to voids in the BM; the two cells can be linked mechanically through large deposits of the BM protein fibronectin (Armulik et al. 2005). These contact points are known as adhesion plaques and transmit PC contractions through the vessel to regulate vascular tone (Courtoy and Boyles 1983). Although the contractile role of PC was initially debated, it has since been confirmed that PC are in fact capable of contraction, even if not to the same extent as vascular smooth muscle cells (Kelley et al. 1987). The contractile phenotype of PC is derived not only from the previously mentioned α-SMA marker, but also from the presence of tropomyosin and myosin. Further, PC express cholinergic and adrenergic (α-2 and β-2) receptors that lead to contraction and relaxation, respectively, among other vasoactive signaling molecules (Bergers and Song 2005). As one may expect, adhesion plaques are therefore very common in the eye and brain, two tissues where regulation of vascular tone by PCs is required for proper tissue functions.

2.2.2.2 Paracrine Signaling within the Vascular Wall

Many paracrine signals existing within mature vessel have been identified and can either act synergistically or independently of contact-based communication (Bergers and Song 2005, Armulik et al. 2011). For example, vascular tone is not regulated exclusively by PC; rather the overall vessel diameter is the result of PC integrating a number of chemical signals from EC and from the surrounding tissues. In the presence of nitric oxide, EC produce endothelin-1 which subsequently binds to the PC and results in the reorganization of actin fibers into larger bundles oriented along the long axis of the cell (Chakravarthy et al. 1992, Bergers and Song 2005). This reorganization of actin fibers is thought to be related to the changes seen in vascular tone.

Oxygen content in the surrounding tissues is the primary environmental control over PC contraction (Bergers and Song 2005). Hyperoxia (an excessively oxygenated condition) signals to contract PC, whereas high levels of CO_2 relax PC contraction, indicating that maintaining a homeostatic level of gas exchange is one of the many functions of PC in the microvasculature (Bergers and Song 2005).

2.2.2.3 *Vascular Regulation of Inflammation*

Appropriate inflammatory responses are critical for maintaining normal health and for overcoming current challenges in immunological disorders and in mitigating transplant/ biomaterial implant rejection. The inflammatory reaction is initiated with the migration of white blood cells, or leukocytes, along a chemotactic gradient from the blood stream through the vascular wall to the site of injury or inflammation; this process is known as the leukocyte adhesion cascade (Engelhardt and Vestweber 2006, Ley et al. 2007). In brief, the leukocyte adhesion cascade is defined by a series of interactions between the leukocyte and the activated endothelium and BM: (1) leukocyte capture, (2) leukocyte rolling, (3) slow rolling/activation, (4) arrest, (5) adhesion strengthening, (6) intraluminal crawling, (7) paracellular and transcellular migration, and (8) migration through the BM (Figure 2.5,

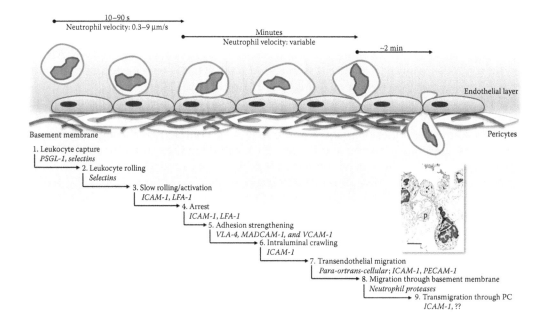

FIGURE 2.5

The leukocyte adhesion cascade. The interactions between the vasculature and the immune system have implications in tissue engineering and wound healing, as well as immunological pathologies. Leukocytes follow a series of well-defined steps to move from the circulation to peripheral tissue. Starting with leukocyte capture (far left), the steps of the leukocyte adhesion cascade (and the molecules that mediate each step) are described in sequence, ending with migration through the vascular wall components (far right). The time scale of these events is detailed on top. Although currently being research, the role of the PC in mediating this process are still not fully known. (Adapted from Ley, K. et al. 2007. *Nature Immunology*, 7, 678–689 and Engelhardt, B. and Vestweber, D., 2006. *Microvascular Research: Biology and Pathology*. Burlington, MA: Elsevier Academic Press, pp. 303–307. SEM imaging (inset) reveals the morphological changes in neutrophils during the migration processes (used with permission from Feng et al. 1998).)

Ley et al. 2007). Notably, transmigration across PC has been left out of the paradigm, and only recently considered as an additional step in the cascade.

The first two steps, catchment and leukocyte rolling, are largely governed by selectins: L-Selectin is present on leukocytes and E- and P-selectin are found on activated endothelium near the site of inflammation (Ley et al. 2007). P-selectin glycoprotein ligand-1 (PSGL-1), which is also a ligand for L-selectin, is expressed on EC and other leukocytes (Ley et al. 2007). PSGL-1–L-selectin-mediated interactions between EC and leukocytes facilitate initial leukocyte tethering to the vascular lumen, whereas PSGL-1–L-selectin interactions between leukocytes enables secondary leukocyte catchment (Engelhardt and Vestweber 2006, Ley et al. 2007). Following rolling, slow rolling, and leukocyte chemokine activation, integrin-mediated attachment becomes the primary mechanism facilitating leukocyte arrest on the EC surface (Ley et al. 2007). The β_2-integrins mediate the rolling process; neutrophil binding of EC-expressed E-selectin induces a conformational change in lymphocyte function-associated antigen-1 (LFA-1; aLb2 integrin) (Ley et al. 2007). The induced conformational change in LFA-1 enables LFA-1 to interact with ICAM-1 on the EC (Ley et al. 2007). In addition to LFA-1–ICAM-1 interactions, firm adhesion of the leukocyte to the luminal EC is mediated by leukocyte $\alpha_4\beta_7$ and $\alpha_4\beta_1$ integrin (very late antigen 4;VLA-4) interactions with EC mucosal vascular addressin cell-adhesion molecule 1 (MADCAM-1) and luminal vascular adhesion molecule 1 (VCAM-1), respectively (Ley et al. 2007). Once firmly adhered to the luminal surface, leukocytes crawl across the luminal surface to a preferential site of migration, which is commonly, but not exclusively a tricellular junction in the EC monolayer (Burns et al. 1997). At the site of transmigration, ICAM-1-rich microvilli encapsulate and guide the migrating leukocyte through the endothelial layer (Carman and Springer 2004).

The final outlined step in the current leukocyte adhesion cascade is leukocyte migration through the BM, which consists of a heterogeneous mix of numerous proteins. Tracking leukocyte migration in the mouse cremaster muscle has revealed that areas of low collagen and laminin expression within the vascular wall serve as paths for neutrophil migration (Wang et al. 2006, 2012). In addition to the physical guidance provided by the proteins in the BM, the composition of the BM will also dictate the ease with which neutrophils or other leukocytes can move through the wall, based on the leukocyte-specific expression of integrins and the availability of integrin-specific binding sites on BM proteins (Woodfin et al. 2010). Furthermore, different leukocytes are able to secrete different proteases that can improve their ability to move within the BM. Neutrophils, for example, are able to secrete elastase to degrade elastin, a key protein responsible for the elastic nature of the vasculature (Woodfin et al. 2010).

The role of PC embedded in the BM, has been entirely neglected from the currently accepted leukocyte adhesion cascade. Recent research, however, has demonstrated that PC do in fact interact with migrating leukocytes to influence inflammation. Within the past several years, the ability to study the role of PCs in inflammation has increased dramatically with improved culturing techniques for *in vitro* studies and advanced imaging techniques for *in vivo* studies. Neutrophil-tracking studies have revealed that PC also guide migrating neutrophils in addition to the guidance they receive from the ICAM-1-microvilli on EC. Rather than forming new microvilli, PC morphologically change and utilize certain preexisting processes to lead neutrophils out of the wall (Proebstl et al. 2012). During inflammation (cytokine stimulation) the gap between some PC processes opens to enable to the migrating leukocytes to pass through the densely packed cells into the perivascular tissue (Proebstl et al. 2012). Of these gaps, some appear to either be more instructive or to be "preferred" by neutrophils for transmigration. In a recent study 28.9% of PC gaps were used by only a single neutrophil exiting the vasculature, whereas 59.7% of the PC gaps used were breached by 3–9 neutrophils, suggesting that preferential sites for leukocyte transpericyte migration exist (Proebstl et al. 2012; Ayres-Sander et al. 2013).

As was the case during luminal crawling, movement of the leukocytes on the PCs was determined to be mediated by ICAM-1–integrin interactions; when ICAM-1 was functionally blocked with antibodies the migration velocity and total distance were decreased (Proebstl et al. 2012).

The heterogeneity of PC once again becomes critical in understanding the complex interactions between the immune system and the vasculature. PCs found in the postcapillary venules are NG2-negative (NG2–) whereas those found in the capillaries and the arterioles are NG2 positive (NG2+). PCs expressing NG2 are uniquely capable of detecting damage associated molecular patterns (DAMPs), compared to NG2– PC (Stark et al. 2013). Following DAMP detection, NG2+ PC increase surface molecule expression of ICAM-1 and macrophage migration-inhibitory factor (MIF), a chemoattractant. The increased adhesion molecule presentation in the vasculature and the production of chemokines by the EC results in leukocyte activation and recruitment. Therefore, although neutrophils transmigrate through the NG2– PC of the postcapillary venule during inflammation, they continue to migrate through the interstitial tissue along a chemoattractant gradient, leading them to the NG2+ PC. NG2+ PC then "instruct" activated neutrophils and efficiently guide them to the center of the inflammatory reaction. As a result, the NG2+ PC have been likened to a leukocyte "highway" (Stark et al. 2013). Additionally, NG2+ PC–leukocyte interactions ameliorate the ability of the innate immune system to detect other surround tissue damage and to act against any peripheral tissues during their journey to the origin of the inflammatory response (Stark et al. 2013). Clearly, PCs directly influence the ability of the leukocytes to navigate to the site of inflammation.

Beyond the instructive and barrier roles of PC in regulating inflammation and wound healing, studies have indicated that PC are capable of phenotypic changes to more directly contribute to wound healing. It has been shown that PC detach from the vascular wall, and exhibit more fibroblast-like behavior, including secreting large amounts of collagen-1. This phenotypic change results in the formation of scar tissue or in inflammation-induced tissue fibrosis (Gerhardt and Betsholtz 2003).

2.2.3 Tissue-Specific Roles of PCs

In addition to being involved in angiogenesis and inflammation throughout the body, PCs are involved in numerous tissue-specific roles. These are especially important in tissues such as the skin, the lungs, the eyes, and the brain that maintain an equal ratio of EC and PC in the vasculature. Identifying these specific roles also elucidates how the loss or dysfunction of PC is often linked to tissue-specific damage or disease such as increased endothelial transcytosis in vessels of the blood–brain barrier (BBB) lacking PC (Armulik et al. 2011). The following sections will therefore focus on the roles of PC in both healthy and diseased tissues in the skin, the lungs, the eye, and the brain (Figure 2.6).

2.2.3.1 PCs in Skin

The microvasculature in the skin is primarily confined to the dermis; postcapillary venules are the most common vessels in the papillary dermis layer (Agapitov and Haynes 2006). Although the PC layer in the skin is discontinuous, PC still play an essential role in blood flow regulation, wound healing, and maintenance of the dermal layer (Kuntsfeld

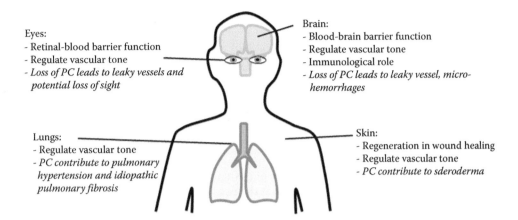

FIGURE 2.6
Tissue-specific roles of PC. The eyes, brain, lungs, and skin all have a high ratio of PC to EC. The roles of PC in each of these tissues are listed above with associated pathologies listed below in italics.

and Detmar 2006, Paquet-Fifield et al. 2009). As demonstrated in organotypic cultures of human neonatal foreskin, dermal PC indirectly enhance the ability of human epidermal cells to regenerate tissue, independent of their role in angiogenesis. This ability stems from the PC expression of the *LAMA5* gene that encodes for the α5 subunit of Laminin-511/521, an extracellular matrix component. Both *in vitro* and *in vivo*, LAMA5 has been known to promote skin regeneration, and its production is critical in injury repair. Co-culture of dermal PC and human epidermal cells revealed that the presence of the PC and the α5 subunit expression lead to an increased amount of laminin-511/521 being deposited in the dermal–epidermal junction of human foreskin cultures (Paquet-Fifield et al. 2009). A similar trend was seen in co-cultures of primary human keratinocytes, which were also unable to regenerate epithelial tissue in the absence of PCs in a similar culture condition as previously noted (Paquet-Fifield et al. 2009).

2.2.3.2 PCs in Skin Pathologies

Autoimmune destruction or deregulation of appropriate PC functions can result in dermal pathologies, such as scleroderma. Morphological studies of scleroderma patient skin have demonstrated that an increase in the basal lamina of the microvasculature, a destruction of the ECs, and an influx of immune cells and fibroblasts are primarily responsible for the increased collagen deposition. Although the exact mechanisms leading to scleroderma are not entirely understood, PC are thought to be a critical factor in the pathogenesis of this disease due to their ability to differentiate into matrix-depositing fibroblasts and immune cells, as well as their ability to produce TBGβ and other cytokines implicated in initiation and exacerbation of scleroderma (Helmbold et al. 2004).

Biopsies of tissues from patients with scleroderma have also revealed that PC located near affected regions were more numerous, indicating hyperplasia. As a result, the PC:EC ratio in tissues proximal to scleroderma-damaged tissues was twice that of healthy patients; these PC-dense regions mark an "active front" of disease progression (Helmbold et al. 2004). This increase in PC may also be linked to an elevated expression of TGFβ-2, seen in scleroderma-affected microvasculature (Helmbold et al. 2004). These findings corroborate previous studies revealing that activated microvascular PC in patients with early systemic scleroderma express PDGFβ and may be a critical component in autoimmune disorders leading to chronic fibrosis and alterations in the microvascular structure (Rajkumar et al. 1999).

PC plasticity in the skin is also thought to be of potential importance in initiating ectopic calcification (Armulik et al. 2005). The ability of microvascular PC to differentiate into osteoblasts, chondrocytes, and adipocytes, which are important for wound healing, can also lead to deleterious tissue effects that may play a role in larger vessel disorders such as atherosclerosis (Doherty et al. 1998, Farrington-Rock et al. 2004, Armulik et al. 2005).

2.2.3.3 PCs in the Lung

Mammalian pulmonary PCs were initially elusive; it was not until the extensive imaging work published in 1974 by Ewald Weibel that the presence of PCs in the pulmonary microvasculature was confirmed (Weibel 1974). The fine sections of the rat lung that were provided in his research revealed that PC in the mammalian pulmonary capillary maintained many of the characteristic features that have since been demonstrated in more detail in other areas of the body: PC are capable of spanning multiple vessels; EC and PC form physical contacts for cellular communication in gaps in the BM; and PC are enveloped in the BM and extend to processes that circumferentially surround the endothelial tube with larger primary processes and smaller, secondary processes (Weibel 1974). The lungs are an extremely fragile organ; their complex structure and thin walls enable efficient gas exchange but leave pulmonary tissue susceptible to damage. PCs are hypothesized to act as a barrier in the lungs to reduce the permeability of the tissue. In conjunction with their contractile roles, PC may help prevent fluid accumulation in the lungs, also known as pulmonary edema (Speyer et al. 2000).

2.2.3.4 PCs in Pulmonary Pathologies

In hypoxia-induced pulmonary hypertension, structural analysis has demonstrated that there is a significant change in vessel structure and an associated increase in PC proliferation. The expansion of the pulmonary PC population is hypothesized to be a result of hypoxia-induced EC death. As previously described, in angiogenesis PC have an antiproliferative effect on EC; similarly, in the lungs, EC tend to have an antiproliferative effect on PC, as PC grown on pulmonary EC matrix *in vitro* did not rapidly expand as expected. This reduced PC proliferation supports the hypothesis that EC loss in pulmonary microvasculature initiates PC proliferation and deleterious remodeling of the hypertensive blood vessels (Reid 1986).

PC have also been suspected to be involved in pulmonary fibrosis for a number of reasons, including their role in other tissues and their ability to escape confinement in the capillary bed in pulmonary fibrosis (determined through imaging studies) (Rock et al. 2011). Recent work to identify PC in pulmonary fibrotic lesions has been complicated by the fact there is no absolute definition for or means of characterizing PC. Fate mapping studies have suggested that a subpopulation of PC that is NG2–, Forkhead box J1–, and α-SMA positive (α-SMA+) may be major players in idiopathic pulmonary fibrosis (IPF), but more work is needed on this topic (Rock et al. 2011).

2.2.3.5 PCs in the Eye and the Brain

The eye and the brain share the title of possessing the densest concentration of PCs in the body (Bergers and Song 2005). As the functions of PC in these two tissues overlap significantly, PC functions will be covered for these two organ systems simultaneously, although pathological information for these two tissues is addressed independently later. In both of these tissues, the need to maintain a tight barrier is critical as both organs possess unique tissue environments with highly regulated chemical properties. In the eye this occurs in the vascular network known as the retinal–blood barrier, and in the brain this occurs in the BBB (Bergers and Song 2005, Antonetti 2009). Additionally, due to the dense network of PC in both of these systems, the PC are highly involved in the regulation of vascular tone.

In the brain specifically, the PC have an additional immunological role. PC are able to act in a macrophage-like manner to take up small molecules via pinocytosis and clean the extracellular space in the BBB (Goodwin and D'Amore 2008).

2.2.3.6 PCs in Pathologies of the Eye

Genetic manipulation of the PDGFβ pathway in animal models resulted both in PC loss and microvascular phenotypes mimicking those seen in pathologies such as diabetic retinopathy. Diabetic retinopathy is the leading cause of blindness among working-age adults and is very prevalent in those with diabetes (Hammes et al. 2002, Antonetti 2009). This disease has since been linked to an increase in glucose in the eye; hyperglycemia induces activation of nuclear factor kappa-light-chain-enhancer of activated B cells (NF-κB) and PDGFβ resistance pathways, which simultaneously leads to prolonged activation of protein kinase C-delta (PKC-δ) to induce apoptosis (Geraldes et al. 2009). The subsequent loss of PC in the eye can lead to the formation of microaneurysms. Together, the combined effect of microaneurysms and unsupported EC result in a weakened vascular structure and an increased susceptibility to rupture.

2.2.3.7 PCs in Cerebral Pathologies

As PC are critical for maintenance of the BBB and for regulation of vascular tone, pathologies affecting these cells present in patients as changes in vessel integrity and blood flow regulation. Changes in vessel integrity exist within a spectrum based on the degree of PC loss. In mild cases, PC retraction can result in reduced EC–PC contact points, subsequently resulting in a decrease in the barrier ability of the BBB (Herman 2006). In more extreme cases, where there is loss of PC, the vessel integrity as a whole is compromised, often leading to diseases such as hereditary cerebral hemorrhage with amyloidosis. In this disease, for example, vessel degradation and abnormal amyloid protein deposition in the vascular walls lead to severe damage including stroke, dementia, and other neurologic trauma (Bergers and Song 2005, Office of Rare Diseases Research, National Institutes of Health 2009).

Conversely, an increase in PCs has been associated with cerebral hypertension, although whether this association or causative nature or not has not been established. In spontaneously hypertensive rat models, the PC:EC ratio increases with blood pressure, ultimately plateauing when blood pressure remains constant (Herman et al. 1987).

2.2.3.8 Overview

Thus far, the roles of the PC in angiogenesis, inflammation, and tissue-specific pathologies (skin, lungs, eyes, and brain) have been discussed. The regulatory roles in angiogenesis and inflammation highlight the need to examine and incorporate mural cells in future tissue-engineered constructs, as vascularization and immune rejection continue to negate advances in polymer science and stem cell research, among other fields (these challenges will be discussed further in the chapter). The aforementioned tissues and the pathologies described within them highlight the numerous currently unmet medical needs that effective tissue engineering has the potential to eradicate. Therefore, the remainder of this chapter will focus on the current state of tissue engineering, initial attempts to incorporate PCs into tissue engineering, and the potential future directions of the field.

2.3 PC in Tissue Engineering

2.3.1 Current Limitations in Tissue Engineering

While the field of tissue engineering has made significant advances in producing therapies for patients in need of large vessels, dermis, and cartilage, it has struggled to successfully create or commercialize any complex tissues. The main reason for this failure lies within the ability to recreate the microvasculature (Rouwkema et al. 2008). Unlike large vessels that have been successfully engineered (Hibino et al. 2010), the microvasculature poses a number of additional challenges including the complex arrangement of small vessels as well as the fragility of the vessels themselves (San Antonio and Iozzo 2006). Although it would be ideal to avoid replicating the microvasculature in tissue-engineered constructs, the microvasculature is critical for the success of any thick implant.

Essential nutrient delivery is required for appropriate cell ingrowth and differentiation; without vascularization, nutrient delivery is limited to diffusion, which cannot exceed 100–200 μm of tissue, or even less in highly metabolic organs. Therefore, vascular integration of tissue-engineered constructs is essential to ensure cell survival (Rouwkema et al. 2008, Miller et al. 2012). In the last decade, numerous attempts have been made to vascularize engineered tissues, with varying degrees of success. The methods that have been attempted to engineer scaffolds that support microvascular structure include bioengineering new scaffolds, using growth factors to induce angiogenesis, and tissue prevascularization (both *in vivo* and *in vitro*) (Rouwkema et al. 2008). A brief overview of the advantages and disadvantages of these methods can be found here.

2.3.2 Previous Methodologies for Vascular Engineering

Scaffold engineering methods have advanced significantly in the past several years to incorporate complex mechanical and biological cues. Substrate surface topography (surface shape that the cell "sees") can influence cell morphology and therefore overall cell function. For example, with the desire of creating a vascularized network, it would be advantageous to use a substrate patterned with 10 μm protein stripes as opposed to 30-μm protein stripes because vascular-like cords with lumens are known to form when the stripes are 10 μm, but not 30 μm (San Antonio and Iozzo 2006). With regard to biological cues, it has been suggested that designing a so-called "super-polymer" scaffold that selectively mimics desirable characteristics of collagen Type 1 or fibrin in the BM while avoiding associated negative characteristics, could be used for tissue engineering (San Antonio and Iozzo 2006). Such a polymer would ideally have appropriate EC integrin binding sites, such as $\alpha_2\beta_1$ distributed every 10–50 nm for proper polymer–cell interactions. This "super-polymer" could also include binding sites for angiogenic growth factors such as

VEGF but be designed to exclude unwanted sites such as ligands for proteases and toxins. Additionally, this scaffold would ideally be designed to exclude epitopes that are immunogenic to avoid any rejection from the host (San Antonio and Iozzo 2006).

Proangiogenic techniques have had mixed success in the past. Following the discovery in the 1970s that EC alone, in the presence of proangiogenic factors, would form complex vascular networks, many have tried simple injection methods to initiate angiogenesis by exogenously increasing levels of VEGF and other proangiogenic factors (Niklason 2011). For the most part, these attempts have failed because the integrity of the resulting vessels. Without perivascular support from either vascular smooth muscle cells or PCs, the microvessels that are formed are leaky, fragile, and lack any contractility (Niklason 2011, Frontini et al. 2011). More recent advances in staggered injections of angiogenic factors to more closely replicate endogenous signaling in angiogenesis have showed some success. More specifically, proangiogenic substances such as VEGF are initially injected into tissues, followed by the injection of fibroblast growth factor 9 (FGF9). The inclusion of FGF9 ultimately leads to an increase in mural support cell (vascular smooth muscle cell or PC) recruitment and subsequently vessel stabilization and vasoreactivity (Frontini et al. 2011).

The attempts to induce vascularization in a tissue-engineered construct prior to true implantation have taken two forms: *in vivo* and *in vitro*. In the former, the tissue construct is placed in the patient near a major artery such that endogenous angiogenic processes can take place. This poses the problem of time. Although the tissue is not serving its intended function at this primary implantation site, the seeded cells still require nutrients and face the challenge of limited nutrient delivery until vessel ingrowth (days to weeks). Ultimately, this may lead to loss of cells in the construct or inappropriate/undesired cell differentiation. The benefit of this method is that with the advance of microsurgery, when the vascularized implant is moved to the final site, the surrounding vessels can be anastomosed immediately to supply the cells that survived the original vascularization process (Rouwkema et al. 2008). Additionally, since this method utilizes endogenous angiogenesis mechanisms, it is possible that PC or vSMC are incorporated into the vessels.

In vitro perfusion and prevascularization of a tissue-engineered construct is an alternative approach. Although this method does not allow for microsurgery to immediately connect the engineered vasculature to the host vasculature it does have several advantages. First, it avoids the need for an additional surgery to implant the construct for prevascularization. Second, the perfusion of the construct *in vitro* ensures a higher rate of cell survival since the endogenous vascularization process *in vivo* (days to weeks) is bypassed with artificial perfusion. Thus, the cells that are ultimately transplanted at the final site have a higher chance of maintaining their intended structure. Lastly, the vascular density from *in vitro* perfusion is high, such that vascular ingrowth from host tissue only needs to innervate superficial layers before connecting to the tissue-engineered blood vessels (Rouwkema et al. 2008).

2.3.3 PC: A New Direction for Vascularizing Tissue-Engineered Constructs

It is clear from the previous discussion of PCs and their role in angiogenesis, vessel stabilization and the regulation of vessel tone, integrity, and inflammation, that PCs are a critical component for appropriate tissue functions. The inclusion of PCs in tissue-engineered vasculature has the potential to improve the physiology of tissue-engineered constructs in order to better replicate native tissue, specifically behavior. Especially in engineering for the skin, lungs, eyes, and brain, the presence of PCs is critical to ensure the formation of unique vascular structures such as the BBB.

As demonstrated by the advancement of proangiogenic treatment (addition of FGF9 following VEGF administration), previously tested methods of tissue vascularization can be altered to include human PCs. Although these new methods are still in their infancy, given new methods for human PC isolations (see Maier et al. 2010) and the wealth of information on EC–PC communications in angiogenesis, it is likely that their use will increase. Tissue engineering successes have already demonstrated that the formation of EC vessels is possible, despite their fragile nature. Therefore, the future of engineering most likely entails manipulating the known mechanisms of angiogenesis to stabilize and to mature these EC vessels to form a mechanically stable and biologically relevant microvasculature.

The possibilities for using PC in tissue engineering and regenerative medicine are endless, but many of the same concepts that underlie the four vascularization methods described earlier may be used. As polymer chemistry has advanced, polymers can be tuned in time and in space to expose numerous signaling mechanisms through degradation or other external controls (Griffith and Naughton 2002, DeForest et al. 2009). Therefore, in the example of vascularization through novel scaffold engineering, not only could the mechanical and biological cues for endothelial lumen formation be used, but they could be paired with similar cues for PC migration *in vitro*, such as expression of PDGFβ expression.

2.4 Conclusion

Tissue engineering has the promise to advance patient treatment and to ultimately restore health and function to countless people. In order to fully reach that potential, however, the issue of proper vascularization must be addressed. As detailed here, PCs are a critical, but often overlooked component of the vascular wall. These cells are intimately connected with EC starting in angiogenesis, when they are recruited to vascular guidance tunnels through paracrine signaling. EC-PC interdependence continues throughout the life of the vessel via communication in peg–socket contacts and paracrine signaling. Without PC, the microvasculature would be incapable of maintaining a barrier function, such as the retinal–blood barrier, and of regulating vascular tone. Additionally, the ability of the vessel to mediate inflammatory reactions and to act appropriately in wound healing scenarios would seriously be hindered without PC. By drawing on the current understanding of PC physiology and intramural signaling, and by building upon the existing methods of EC vascularization, the production of tissue-engineered constructs using both endothelial cells and PCs stands to significantly improve the state of vascularized tissue-engineered constructs.

References

Agapitov, A. V. and Haynes, W. G., 2006. Impaired skin microcirculatory function in human obesity. In: D. Shepro, ed. *Microvascular Research: Biology and Pathology.* Burlington, MA: Elsevier Academic Press, pp. 921–924.

Allt, G. and Lawrenson, J., 2001. Pericytes: Cell biology and pathology. *Cells Tissues Organs,* 169, 1–11.

Antonetti, D., 2009. Eye vessels saved by rescuing their pericyte partners. *Nature Medicine,* 15(11), 1248–1249.

Armulik, A., Abramsson, A., and Betsholtz, C., 2005. Endothelial/pericyte interactions. *Circulation Research*, 97, 512–523.

Armulik, A., Genove, G., and Betsholtz, C., 2011. Pericytes: Developmental, physiological, and pathological perspectives, problems, and promises. *Developmental Cell*, 21, 193–215.

Ayres-Sander, C. E., Lauridsen, H., Maier, C. L., Sava, P., Pober, J. S. and Gonzalez, A. L., 2013. Transendothelial Migration Enables Subsequent Transmission of Neutrophils through Underlying Pericytes. *PLoS ONE*, 8(3), e60025.

Bergers, G. and Song, S., 2005. The role of pericytes in blood-vessel formation and maintenance. *Neuro-Oncology*, 7, 452–464.

Burns, A. R. et al., 1997. Neutrophil transendothelial migration is independent of tight junctions and occurs preferentially at tricellular corners. *Journal of Immunology*, 159(6), 2893–2903.

Carman, C. V. and Springer, T. A., 2004. A transmigratory cup in leukocyte diapedesis both through individual vascular endothelial cells and between them. *Journal of Cell Biology*, 167(2), 377–388.

Chakravarthy, U. et al., 1992. The effect of endothelin 1 on retinal microvascular pericyte. *Microvascular Research*, 43, 241–254.

Courtoy, P. J. and Boyles, J., 1983. Fibronectin in the microvasculature: Localization in the pericyte-endothelial interstitium. *Journal of Ultrastructure Research*, 83, 258–273.

Davis, G. E. and Senger, D. R., 2005. Endothelial extracellular matrix: Biosynthesis, remodeling and functions during vascular morphogenesis and neovascular stabilization. *Circulation Research*, 97, 1093–1107.

DeForest, C. A., Polizzotti, B. D., and Anseth, K. S., 2009. Sequential click reactions for synthesizing and patterning three-dimensional cell microenvironments. *Nature Materials*, 8, 659–664.

Doherty, M. J. et al., 1998. Vascular pericytes express osteogenic potential *in vitro* and *in vivo*. *Journal of Bone and Mineral Research*, 13(5), 828–838.

Engelhardt, B. and Vestweber, D., 2006. The multistep cascade of leukocyte extravasation. In: D. Shepro, ed. *Microvascular Research: Biology and Pathology*. Burlington, MA: Elsevier Academic Press, pp. 303–307.

Farrington-Rock, C. et al., 2004. Chondrogenic and adipogenic potential of microvascular pericytes. *Circulation*, 110, 2226–2232.

Frontini, M. J. et al., 2011. Fibroblast growth factor 9 delivery during angiogenesis produces durable, vasoresponsive microvessels wrapped by smooth muscle cells. *Nature Biotechnology*, 29(5), 421–427.

Geraldes, P. et al., 2009. Activation of PKC-delta and SHP-1 by hyperglycemia causes vascular cell apoptosis and diabetic retinopathy. *Nature Medicine*, 15(11), 1298–1306.

Gerhardt, H. and Betsholtz, C., 2003. Endothelial-pericyte interactions in angiogenesis. *Cellular and Tissue Research*, 314, 15–23.

Goodwin, A. M. and D'Amore, P. A., 2008. Vessel maturation and perivascular cells. In: D. Marmé and N. Fusenig, eds. *Tumor Angiogenesis: Basic Mechanisms and Cancer Therapy*. New York: Springer, pp. 273–288.

Griffith, L. G. and Naughton, G., 2002. Tissue engineering—Current challenges and expanding opportunities. *Science*, 2955557, 1009–1014.

Hall, A. P., 2006. Review of the pericyte during angiogenesis and its role in cancer and diabetic retinopathy. *Toxicologic Pathology*, 34(6), 763–775.

Hammes, H.-P. et al., 2002. Pericytes and the pathogenesis of diabetic retinopathy. *Diabetes*, 51, 3107–3112.

Helmbold, P., Fiedler, E., Fischer, M., and Marsch, W. C., 2004. Hyperplasia of dermal microvascular pericytes in scleroderma. *Journal of Cutaneous Pathology*, 31, 431–440.

Herman, I. M., 2006. Pericytes and microvascular morphogenesis. In: D. Shepro, ed. *Microvascular Research: Biology and Pathology*. Burlington, MA: Elsevier Academic Press, pp. 111–116.

Herman, I. M., Newcomb, P. M., Coughlin, J. E., and Jacobson, S., 1987. Characterization of microvascular cell cultures from normotensive and hypertensive rat brains: Pericyte-endothelial cell interactions in vitro. *Tissue and Cell*, 19(2), 197–206.

Hibino, N. et al., 2010. Late-term results of tissue engineered vascular grafts in humans. *The Journal of Thoracic and Cardiovascular Surgery*, 139(2), 431–436.

Hirschi, K. K. and D'Amore, P. A., 1996. Pericytes in the microvasculature. *Cardiovascular Research*, 32, 687–698.

Humphrey, J. D., 2002. Cardiovascular solid mechanics: Cells, tissues, and organs. 1st ed. New York: Springer.

Kelley, C., D'Amore, P., Hectman, H. B., and Shepro, D., 1987. Microvascular pericyte contractility *in vitro*: Comparison with other cells of the vascular wall. *Journal of Cell Biology*, 104, 480–490.

Kuntsfeld, R. and Detmar, M., 2006. Angiogenic diseases of the skin. In: D. Shepro, ed. *Microvascular Research: Biology and Pathology*. Burlington, MA: Elsevier Academic Press, pp. 929–932.

Ley, K., Laudanna, C., Cybulsky, M., and Nourshargh, S., 2007. Getting to the site of inflammation: The leukocyte adhesion cascade updated. *Nature Immunology*, 7, 678–689.

Lindahl, P., Johnansson, B. R., Levee, P., and Betsholtz, C., 1997. Pericyte loss and microaneurysm formation in PDGF-B-deficient mice. *Science*, 227, 242–245.

Lu, C. and Sood, A. K., 2008. Role of pericytes in angiogenesis. In: B. A. Teicher, ed. *Cancer Drug Discovery and Development: Antiangiogenic Agents in Cancer Therapy*. Totowa, NJ: Humana Press, pp. 117–132.

Maier, C., Shepherd, B., Yi, T., and Pober, J. S., 2010. Explant, outgrowth, propagation and characterization of human pericytes. *Microcirculation*, 17(5), 367–380.

Miller, J. S. et al., 2012. Rapid casting of patterned vascular networks for perfusable engineered three-dimensional tissues. *Nature Materials*, 11, 768–774.

Mitchell, R. N. and Schoen, F. J., 2009. Blood vessels. In V. Kumar, A. K. Abbas, and N. Fausto, eds.: *Robbins and Cotran: Pathologic Basis of Disease*. St. Louis Missouri: Saunders, p. Online Edition.

Movat, H. Z. and Fernando, N. V., 1964. The fine structure of the terminal vascular bed: IV. The venules and the perivascular cells (pericytes, adventitial cells). *Experimental and Molecular Pathology*, 3, 98–114.

Niklason, L. E., 2011. Building stronger microvessels. *Nature Biotechnology*, 29(5), 405–406.

Office of Rare Diseases Research, National Institutes of Health, 2009. *Hereditary Cerebral Hemorrhage with Amyloidosis*. [Online] Available at: http://rarediseases.info.nih.gov/GARD/Condition/10266/Hereditary_cerebral_hemorrhage_with_amyloidosis.aspx [Accessed 1 3 2013].

Paquet-Fifield, S. et al., 2009. A role for pericytes as microenvironmental regulators of human skin tissue regeneration. *The Journal of Clinical Investigation*, 119(9), 2795–2806.

Proebstl, D. et al., 2012. Pericytes support neutrophil subendothelial cell crawling and breaching of venular walls *in vivo*. *Journal of Experimental Medicine*, 209(6), 1219–1234.

Rajkumar, V. S. et al., 1999. Activation of microvascular pericytes in autoimmune Raynaud's phenomenon and systemic sclerosis. *Arthritis* and *Rheumatism*, 42(5), 930–941.

Reid, L. M., 1986. Structure and function in pulmonary hypertension: New perceptions. *Chest*, 89(2), 279–288.

Ribatti, D., Nico, B., and Crivellato, E., 2011. The role of pericytes in angiogenesis. *The International Journal of Developmental Biology*, 55, 261–268.

Rock, J. R. et al., 2011. Multiple stomal populations contribute to pulmonary fibrosis without evidence for epithelial to mesenchymal transition. *Proceedings of the National Academy of Sciences, USA*, 108(52), E1475–E1483.

Rouwkema, J., Rivron, N. C., and van Bitterswijk, C. A., 2008. Vascularization in tissue engineering. *Trends in Biotechnology*, 26(8), 434–441.

San Antonio, J. D. and Iozzo, R., 2006. The two-phase model for angiogenesis regulation by the extracellular matrix. In: D. Shepro, ed. *Microvascular Research: Biology and Pathology*. Burlington, MA: Elsevier Academic Press, pp. 127–136.

San Antonio, J. D. and Iozzo, R. V., 2006. The two-phase model for angiogenesis regulation by the extracellular matrix. In: D. Shepro, ed. *Microvascular Research: Biology and Pathology*. Burlington, MA: Elsevier Academic Press, pp. 127–136.

Sims, D. E., 1986. The pericyte-A review. *Tissue* and *Celll*, 18(2), 153–174.

Speyer, C. L., Steffes, C. P., Tyburski, J. G., and Ram, J. L., 2000. Lipopolysaccharide induces relazation in lung pericytes by an iNOS-independent mechanism. *American Journal of Molecular Physiology-Lung Cellular and Molecular Physiology*, 278(5), L880–L887.

Stark, K. et al., 2013. Capillary and arteriolar pericytes attract innate leukocytes exiting through venules and 'instruct' them with pattern-recognition and motility programs. *Nature Immunology*, 14(1), 41–51.

Stratman, A. N., Malotte, K. M., Mahan, R. D., Davis, M. J., and Davis, G. E. 2009. Pericyte recruitment during vasculogenic tube assembly stimulates endothelial basement membrane formation. *Blood*, 119(4), 5091–5101.

Stratman, A. N. et al., 2009a. Endothelial cell lumen and vascular guidance tunnel formation requires MT1-MMP-dependent proteolysis in 3-D collagen matrices. *Blood*, 114(2), 237–247.

Stratman, A. N. et al., 2009b. Pericyte recruitment during vasculogenic tube assembly stimulates endothelial basement memebrane matrix formation. *Blood*, 114(24), 5091–5101.

Takata, F. et al., 2011. Brain pericytes among cells constituting the brain pericytes among cells constituting the blood–brain barrier are highly sensitive to tumor necrosis factor-a, releasing matrix metalloproteinase-9 and migrating in vitro. *Journal of Neuroinflammation*, 106(8), 1–12.

Wang, S. et al., 2012. Pericytes regulate vascular basement membrane remodeling and govern neutrophil extravasation during inflammation. *PLoS One*, 7(9), 2–21.

Wang, S. et al., 2006. Venular basement membranes contain specific matrix protein low expression regions that act as exit points for emigrating neutrophils. *Journal of Experimental Medicine*, 203(6), 1519–1532.

Weibel, E. R., 1974. On Pericytes, Particularly their existence on lung capillaries. *Microvascular Research*, 8, 218–235.

Woodfin, A., Voisin, M.-B., and Nourshargh, S., 2010. Recent developments and complexities in neutrophil transmigration. *Current Opinion in Hematology*, 17(1), 9–17.

3

Mesenchymal Support Cells in the Assembly of Functional Vessel Networks

Rahul K. Singh, Marina Vigen, and Andrew J. Putnam*

CONTENTS

3.1 Introduction

The field of regenerative medicine has witnessed impressive advances over the past 25–30 years, moving us ever closer to the goal of translating engineered tissue constructs into human patients. However, despite an exponentially expanding literature documenting advances in biomaterials and stem cell biology, the two biggest factors limiting the clinical applicability of engineered tissues 20 years ago continue to be the biggest hurdles today: the ability to generate tissues that function equivalently to the native tissues they are intended to replace, and the ability to vascularize these tissues to sustain their metabolic demands. With respect to the latter of these two hurdles, an improved fundamental understanding of the process of blood vessel assembly in development and disease may lead to new strategies to vascularize tissues.

---

* Rahul K. Singh and Marina Vigen both contributed equally on this book chapter.

Blood vessels are responsible for the convective delivery of oxygen, nutrients, and other large macromolecules, as well as immune cells, to all tissues in the human body. Vessels form primarily via two morphogenetic programs: vasculogenesis and angiogenesis (Figure 3.1). Vasculogenesis involves the *de novo* assembly of vessels from progenitor cells, and primarily occurs in development. Angiogenesis involves the sprouting and growth of new capillaries from existing vessels into a previously avascular tissue. Both are complex and dynamic processes that depend on the interplay of soluble factors and insoluble cues from the extracellular matrix (ECM) (Ingber and Folkman, 1989). Fundamental studies of these processes have led to the identification of numerous proangiogenic factors, the delivery of which has been explored to promote the development of new blood vessels to restore blood supply to ischemic tissue. However, clinical trials relying on bolus injection of individual factors have been disappointing (Simons and Ware, 2003), perhaps due to the limited half-life of most protein growth factors, the lack of temporal and spatial control over growth factor release, and the inability of single factors to properly regulate neovascularization (Chen et al. 2007, Sun et al. 2010). Newer strategies involving sustained

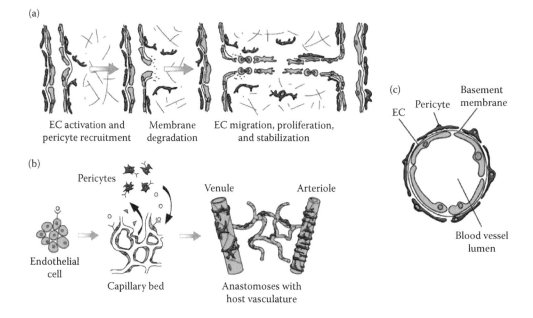

FIGURE 3.1
Formation of new blood vessels and the role of pericytes. (a) Angiogenesis is the process by which new blood vessels sprout from preexisting blood vessels. Pericytes (red) are key players from the beginning. Dissociation of the pericytes from the vessel wall coincides with EC activation, followed by degradation of the vascular basement membrane. The ECs then migrate into the surrounding interstitial ECM, proliferate, and organize into immature vessels. Nascent vessels are stabilized via the deposition of basement membrane and association of pericytes with these tubules. (b) In development, ECs originate from angioblasts or hemangioblasts in the embryo, while pericytes are derived from mesenchymal stem cells or neural crest cells. These two populations of cells then cooperate to form a primitive vascular plexus via a process known as vasculogenesis. Many tissue engineering approaches to vascularization have attempted to recreate this process either by prevascularizing a scaffold or via the direct injection of the cells within a carrier matrix. Networks formed via this method have been demonstrated to inosculate with host vasculature following implantation *in vivo*. (c) Mature vasculature is characterized by the presence of a basement membrane and a pericyte coat surrounding the endothelial cell tubule. (Adapted from Bergers, G. and Song, S. 2005. The role of pericytes in blood-vessel formation and maintenance. *Neuro Oncol*, 7, 452–64, with permission from Oxford University Press.)

delivery of proangiogenic factors or genes from biodegradable scaffolds to overcome protein stability issues (Lee et al. 2000, Murphy and Mooney, 1999, Sun et al. 2005, Zisch et al. 2003), as well as delivery of multiple proangiogenic factors in a time-dependent fashion to mimic the process of natural vessel development (Richardson et al. 2001, Sun et al. 2010), have been shown to induce the formation of vascular networks. However, even combinations of multiple factors may not fully recapitulate the complex mileiu of proangiogenic signals presented to cells *in vivo*.

Cell-based therapies have also been explored to more completely mimic the cascade of signals needed to promote stable vasculogenesis. These approaches involve delivering (an) appropriate cell type(s) that can directly differentiate into capillary structures or provide a physiological mixture of proangiogenic cues to accelerate the recruitment of host vessels. A variety of cell types have been shown to form new capillary networks and/or induce collateral blood vessel development *in vivo* (Iba et al. 2002, Kinnaird et al. 2004, Pesce et al. 2003, Rehman et al. 2004). In addition, cells have been implanted using scaffold materials and ECM proteins to improve cell retention and engraftment (Nor et al. 1999, 2001). However, delivery of endothelial cells (ECs) alone within a scaffold has led to mixed results, with some reports of leaky unstable vessels. There is now general consensus that codelivery of an appropriate supporting cell type can stabilize nascent capillaries (Koike et al. 2004), and that these cells are essential for the formation of mature and functional vessel networks (Au et al. 2008b). Despite the consensus regarding this paradigm, the choice of which cells to codeliver with ECs remains an open question. This chapter will focus primarily on the use of mesenchymal support cells, and discuss key studies that document their potential roles in vascular assembly.

3.2 Initial Vascular Engineering Approaches Focused on Endothelial Cell Delivery

Early cell-based approaches to vascularize tissues *in vivo* concentrated primarily on the implantation of ECs without a supportive cell type, and yielded variable success. One of the first studies to report the formation of functional vasculature *in vivo* used genetically modified human umbilical vein endothelial cells (HUVECs) and assessed their ability to assemble into vascular structures in a subcutaneous mouse model (Schechner et al. 2000). HUVECs overexpressing Bcl-2 (to enhance their survival) supported the formation of a dense vascular network in collagen–fibronectin gels, while wild-type HUVECs did not. Host erythrocytes were found within the HUVEC-derived vessels, indicating their functional perfusion. Furthermore, mouse-derived mural cells positive for smooth muscle α-actin (αSMA) associated with the nascent vasculature within the implant region only in constructs containing Bcl-2 transduced HUVECs. While the use of viral gene delivery approaches to overexpress Bcl-2 and enhance EC survival upon implantation presents translational limitations, there is no doubt that this study spawned greater interest in cell transplantation approaches to promote vascularization.

Another study from about the same time showed that functional new vasculature can also be created by seeding human dermal microvascular endothelial cells (HDMECs) in scaffolds of poly (L-lactic acid) (PLLA) and Matrigel and subsequently implanting these tissue constructs in subcutaneous pockets in SCID mice (Nor et al. 2001). Host erythrocytes were found to perfuse the implant vasculature within 7–10 days, indicating functional inosculation with host vessels. However, the association of mouse-derived cells positive for αSMA with the HDMEC tubules did not occur until 21 days postimplantation; after 4 weeks, some HDMECs apoptosed, presumably due to the lack of perivascular support from mural cells. This apoptosis could be partially overcome by overexpressing Bcl-2 in the HDMECs, which also led to increased vessel density in the constructs. Collectively, these results suggest that the delivery of ECs alone can lead to the formation of functional blood vessels, but the stability of these vessel networks depends on the survival of the ECs.

3.3 Pericytes Support the Formation and Maturation of Capillary Blood Vessels

In the body, supporting cells of mesenchymal origin known as pericytes (also known as Rouget cells or mural cells) closely encircle ECs in capillaries and microvessels and are believed to be responsible for stabilizing capillary blood vessels (Figure 3.1c). Pericytes are generally described as perivascular cells embedded within the basement membrane of the microvasculature where they closely associate with ECs (Armulik et al. 2005). While no single marker identifies all pericytes, common markers include αSMA, desmin, neuronglial antigen 2 (NG-2), platelet-derived growth factor receptor (PDGF)-β, aminopeptidase A & N, and RGS5 (Armulik et al. 2005). Not all pericytes display these markers, and their expression patterns change with the tissue and stage of development. This ambiguity has confounded a unifying and consistent definition of a pericyte.

Despite this ambiguity in their identity, the general consensus from two decades of research is that pericytic stabilization of EC-lined vessel structures is critical both in development and in physiological and pathological processes in the adult organism (Bergers and Song, 2005). Pericytes associate with EC tubules *in vivo* and this association modulates pruning of the vasculature, vessel permeability, and basement membrane deposition (Thurston, 2002), suggesting that their predominant function is to mediate vascular stabilization and maturation. A number of recent studies utilizing *in vitro* cocultures of bovine retinal pericytes and ECs in 3D collagen gels have revealed important new mechanistic insights regarding the reciprocal interactions between these two cell types. In one such study, ECs were shown to remodel existing ECM and create guidance tunnels that served as conduits for the recruitment and motility of pericytes (Stratman et al. 2009a). These recruited pericytes then induced ECs to produce basement membrane components involved in ECM crosslinking, including fibronectin, nidogen-1/2, perlecan, and laminin. Pericytes also induced ECs to upregulate integrins capable of binding to the basement membrane, including $\alpha_5\beta_1$, $\alpha_3\beta_1$, $\alpha_6\beta_1$, and $\alpha_1\beta_1$. Inhibiting those integrins, disrupting fibronectin assembly, or suppressing pericyte TIMP-3 expression all decreased basement membrane deposition and led to pathological, increased lumenal diameter (Stratman et al. 2009a). Furthermore, the recruitment of the pericytes to the nascent vasculature has been shown to result from EC-derived PDGF–BB and heparin-binding EGF-like growth factor (HB-EGF); disrupting these signals in quail embryogenesis results in vascular pathologies that may have relevance for human congenital abnormalities as well (Stratman et al. 2010). However, because of the still ambiguous definition of a pericyte, many investigators have utilized a variety of different supporting mesenchymal cell types capable of supporting capillary morphogenesis. A great deal of mechanistic information about their ability to regulate angiogenesis has now been documented in the literature. Although the remainder of this chapter will primarily focus on the use of mesenchymal stem cells (MSCs) as perivascular support cells, we will also briefly discuss insights from studies that have utilized fibroblasts or vascular smooth muscle cells (SMCs) to compare and contrast the mechanistic similarities and differences.

3.4 Mesenchymal Stem/Stromal/Support Cells

Mesenchyme refers to a type of undifferentiated loose connective tissue derived from the mesoderm during development. The term stroma is used almost interchangeably with mesenchyme, though strictly speaking stroma refers to the supportive framework of adult tissues in which functional (parenchymal) cells reside, and mesenchyme is used more often in a developmental context. Regardless, MSCs are a population of adult tissue-derived adherent cells that were first discovered in the bone marrow by Friedenstein et al. (1968), who described them as "osteogenic stem cells." These cells were initially identified based on their ability to form clonal adherent colonies of fibroblastic cells ("colony-forming unit-fibroblast"), and later shown to possess the capacity to differentiate into bone, cartilage, and fat (Pittenger et al. 1999). This latter capability is why they were initially dubbed "stem" cells by Caplan (2010). MSCs from bone marrow and a variety of other adult tissues are already the focus of numerous human clinical trials (Giordano et al. 2007, Wagner et al. 2009), and have shown enormous promise in preclinical studies to facilitate bone regeneration (Simmons et al. 2004), promote tissue neovascularization (Kinnaird et al. 2004, Nagaya et al. 2005, Silva et al. 2005), and reduce inflammation (Caplan, 2007). Much of their therapeutic benefit seems to be related to their trophic effects, that is, through the secretion of numerous growth factors (Caplan, 2007), and thus many in the literature now refer to them as mesenchymal stromal cells, marrow stromal cells, or (more recently) "medicinal signaling cells" (Caplan, 2010).

MSCs reside within a perivascular niche *in vivo* and can support vascular stability and development. *In vivo*, MSCs have been identified in multiple human organs and tissues, including bone marrow, skeletal muscle, pancreas, placenta, white adipose tissue, and others (Figure 3.2). In each of these tissues, MSCs reside next to capillaries larger than 10 μm in diameter and arterioles ranging from 10 to 100 μm in diameter (Crisan et al. 2008). Further studies suggest that the perivascular niche for MSCs may aid in their ability to home to sites of stroke (Kokovay et al. 2005) and cancer (Beckermann et al. 2008) and to produce paracrine effectors (Doorn et al. 2011). In response to injury, MSCs localize to the injured tissue and produce factors that destabilize existing vessels, promote angiogenesis, and contribute to the maturation and stabilization of nascent capillaries (da Silva Meirelles et al. 2009). Despite the increasing evidence that all MSCs may in fact be pericytes, their exact location(s) and function(s) *in vivo* remain open questions. Nevertheless, both *in vitro* and *in vivo* studies of MSCs and ECs have revealed promising clues as to their roles in vascular support.

3.4.1 Angiogenic Factors Secreted by MSCs

To induce angiogenesis, MSCs secrete numerous soluble factors that modulate the behavior of normally quiescent ECs. Of note are the factors VEGF, IGF-1, PlGF, MCP-1, bFGF, and IL-6 (daSilva Meirelles et al. 2009). VEGF (vascular endothelial growth factor) is a crucial

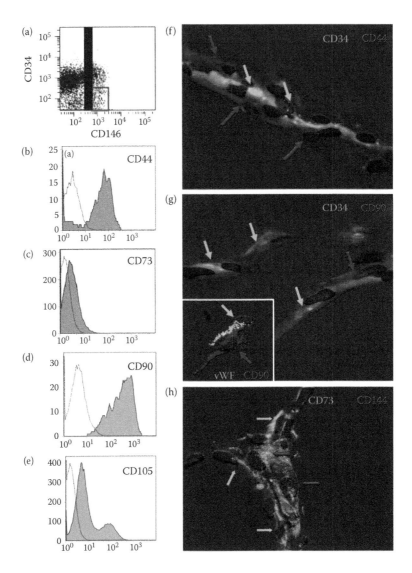

FIGURE 3.2

Perivascular cells natively express MSC markers. The stromal vascular fraction isolated from human white adipose tissue was stained simultaneously with antibodies to perivascular cells and MSCs and analyzed by flow cytometry. Cells negative for CD45, CD56, and CD34 and high for CD146 expression were gated (a) and analyzed for coexpression of CD44 (b), CD73 (c), CD90 (d), and CD105 (e). Clear histograms in (b)–(e) represent control cells incubated with unrelated isotype-matched antibodies. (f)–(h) Frozen sections of human white adipose tissue were costained with antibodies to CD34 (green) or von Willebrand factor (vWF, green) to reveal endothelial cells (green arrows) and either CD44 (red, [f]) or CD90 (red, [g]). Pericytes lining small blood vessels express both CD44 (f) and CD90 ([g], main) (red arrows). Pericytes surrounding capillaries also strongly express CD90 ([g], inset ×1000) (red arrow). (H) Frozen sections of adult human muscle were costained with antibodies to CD144 (red) to reveal endothelial cells (red arrows) and CD73 (green). Pericytes lining the small blood vessel express CD73 ([h], ×600, green arrows). (Adapted from *Cell Stem Cell*, 3, Crisan, M. et al., A perivascular origin for mesenchymal stem cells in multiple human organs, 301–13. Copyright 2008, with permission from Elsevier.)

promoter of angiogenesis, and the most widely studied of the proangiogenic growth factors. Through binding to its receptor tyrosine kinase, VEGFR-2, VEGF is able to trigger signaling cascades that induce endothelial tubulogenesis. Further, VEGF induces tip cell migration and proliferation of ECs, while through NOTCH activation, it leads to the downregulation of VEGFR-2 on stalk cells, thereby providing spatial regulation of angiogenesis. Levels of VEGF also modulate vascular permeability, with higher levels leading to fenestrations and leaky vessels (Carmeliet and Jain 2011, Olsson et al. 2006). Another factor that activates VEGF receptors is PlGF (placental growth factor). It binds to and activates VEGF-R1; further, it enhances the effects of VEGF in promoting angiogenesis. However, it is not normally found in adult tissues, but rather is found in ischemic tissues and is associated with pathological recruitment of vasculature (Carmeliet et al. 2001).

Other factors modulate the proliferation of ECs. IGF-1 (insulin-like growth factor)-1 binds to a corresponding receptor, IGF1R, which is highly expressed on microvascular ECs (Chisalita and Arnqvist, 2004). It plays a key role in the formation of new vasculature in adult tissues (Aghdam et al. 2012) and is found on proliferating ECs. While an overabundance of IGF-1 is often implicated in tumorogenesis, a lack of it has been associated with diabetic retinopathy and poor dermal wound healing (Shaw et al. 2006). IL-6 (interleukin 6) binds to its corresponding receptor IL-6R on ECs to activate ERK1/2 and STAT pathways and induce proliferation, migration, and tube formation. Although this factor acts in a dose-dependent manner, ECs have a high threshold for IL-6 and require high levels of the factor, suggesting it largely has a localized role (Fan et al. 2007).

Factors that promote vascular stability are also important to angiogenesis. bFGF (basic fibroblast growth factor) was the first proangiogenic factor identified. It binds with high affinity to heparin sulfate proteoglycans where it is sequestered until binding to its associated cell-surface receptor FGFR. bFGF is crucial to vascular maintenance and may serve as a trophic factor for ECs (Cross and Claesson-Welsh, 2001). MCP-1/CCL2 (monocyte chemotactic protein-1, also known as CCL2) binds to CCR2 on ECs and can induce migration and angiogenic invasion *in vitro*. The addition of MCP-1 increases angiogenic invasion *in vivo*, but knockouts of MCP-1 or CCR2 form normal vasculature, suggesting only a modulatory role (Salcedo et al. 2000). Collectively, these key angiogenic factors (VEGF, IGF-1, PlGF, MCP-1, bFGF, and IL-6) represent a major component of MSC's ability to modulate vascular assembly. While this collection of MSC-derived or proangiogenic factors is not exhaustive, these six factors provide a means by which MSCs can drive the proliferation, migration, and organization of ECs toward the production of vasculature.

3.4.2 Hypoxia-Induced Activation of MSCs

Soluble factors play central roles in the inducement of vasculature in a hypoxic environment. These factors provide a means for MSCs to communicate over a distance to recruit ECs for revascularization. The loss of vasculature leads to a hypoxic environment within a tissue, and revascularization approaches have investigated hypoxia as a cue for the recruitment of vasculature. Microarray analysis of 2232 genes revealed changes in MSC gene expression during hypoxic culture, with notable changes in the levels of VEGF, IGF, hepatocyte growth factor (HGF), and other factors implicated in angiogenesis. In comparison to mononuclear cells, which upregulate inflammatory and chemotactic genes, MSCs responded to hypoxia through genes involved in development, morphogenesis, cell adhesion, and proliferation (Ohnishi et al. 2007).

A separate study evaluated the secretion of VEGF, HGF, and transforming growth factor-β (TGF-β) from MSCs derived from adipose tissue. It was found that only the secretion of

VEGF increased significantly under hypoxia. Further, conditioned media from these hypoxic MSCs increased EC proliferation and decreased apoptosis (Rehman et al. 2004). A later study assessed the molecular pathways activated in EC by conditioned medium from MSC cultured under hypoxic conditions. The medium decreases hypoxic EC apoptosis, increases survival, and increases tube formation and has higher levels of proangiogenic factors IL-6 and MCP-1 in addition to VEGF. These were shown to activate the PI3K-Akt pathway in EC, which regulates apoptosis. Blocking this pathway with inhibitors of PI3KT or expression of dominant negative genes for PI3K attenuated the proangiogenic effect. While IL-6 promoted angiogenesis in a dose-dependent manner and activated the ERK1/2 pathway, inhibition of this pathway did not attenuate angiogenesis, suggesting that these factors promote angiogenesis via the PI3K pathway (Hung et al. 2007).

In response to hypoxic stress, MSCs have been shown to upregulate the expression of hypoxia-inducible factor-1α (HIF-1α), which drives increased expression of VEGFR1 (Okuyama et al. 2006). This increased the migration response of MSC to VEGF, and provides a feedback mechanism wherein MSC not only secrete VEGF but are also more sensitive to VEGF gradients, thereby enabling MSCs to efficiently home to hypoxic regions. In addition to immediate responses to hypoxia, MSCs can also be preconditioned to a hypoxic environment. This was found to drive changes in the Wnt signaling pathway, which is responsible for self-renewal and morphogenesis in stem cells. Preconditioned MSCs were more potent at restoring vasculature to an ischemic limb *in vivo* (Leroux et al. 2010). These studies reveal that MSCs under hypoxic conditions produce factors that both reduce cell death and promote revascularization.

3.4.3 Interactions with the Extracellular Matrix

Although soluble cues from MSCs can signal ECs to initiate angiogenesis, insoluble cues from the ECM can also influence MSCs and their proangiogenic capacity. These ECM cues can be chemical and/or mechanical in nature, as has now been shown in a number of studies (Discher et al. 2009). With respect to the former, the interactions between MSCs and the endothelial basement membrane are likely to be important, consistent with published evidence for other adult stem cell populations that can interact with the vasculature (Shen et al. 2008). It has been previously reported that the recruitment of pericytes stimulates EC basement membrane assembly (Stratman et al. 2009a), and that MSCs utilize the $\alpha_6\beta_1$ integrin to bind to it (Carrion et al. 2010). Whether this adhesive mechanism influences the secretion of proangiogenic cues from the MSCs or not remains to be seen, but that seems to be a plausible hypothesis given the evidence in the literature.

With respect to the latter, the intrinsic mechanical properties of the ECM are known to affect cell phenotypes in general (Discher et al. 2005), and MSC behavior specifically (Engler et al. 2006). A 2009 study by Mammoto et al. (2009) showed that ECs modulate the expression of their VEGF receptor (VEGF-R2) in response to the elastic properties of their substrate, and that this mechanism is critical in ECM-based control of angiogenesis. Another study investigated the secretome of MSCs cultured on soft and hard substrates. They measured the levels of >90 angiogenic proteins and found that MSC secrete higher levels of IL-8, uPA, and VEGF on stiff substrates than on soft substrates (Seib et al. 2009).

Externally applied mechanical forces provide an additional means to influence the proangiogenic potential of MSCs. In one study, MSCs cultured in a fibrin gel compressed in a bioreactor responded to mechanical loading by changing the profile of secreted growth factors (Kasper et al. 2007). Application of conditioned medium from compressed MSCs to ECs enhanced 2D tubulogenesis and 3D spheroid sprouting assays via a mechanism that depends on FGF and VEGF receptor activation in ECs. Analysis of the MSC-conditioned media revealed increased soluble MMP-2, TGF-β1, and bFGF, but not VEGF, as a result of the mechanical stimulation (Kasper et al. 2007). In a later study, the same group demonstrated that cyclic strain disrupted endothelial organization on 2D Matrigel assays, with elevated levels of VEGF and unchanged levels of MMP-2 and MMP-9 in response to stretching (Wilson et al. 2009). By repeating the assay with the addition of conditioned media from MSCs cultivated in similarly dynamic mechanical conditions, paracrine stimuli were shown to increase network lengths, but not to alter the negative effect of cyclic stretching (Wilson et al. 2009). Collectively, these findings suggest that changes in MSCs in response to altered mechanical stimuli may influence their abilities to induce angiogenesis.

3.4.4 Stromal Cell Control of ECM Remodeling

ECs must break through their basement membrane and invade the surrounding interstitial matrix to sprout a new blood vessel. Essential to this process is proteolytic remodeling of the ECM. It is now well established that capillary invasion in type I collagen matrices depends critically on membrane-bound MMP14 (also known as MT1-MMP) (Chun et al. 2004, Stratman et al. 2009b), while invasion of the provisional clot (composed primarily of the plasma protein fibrinogen in its cleaved form, fibrin) present in wounds, sites of inflammation, and tumors may depend less on MMPs and more on the serine proteases that comprise the plasminogen activator (PA)/plasmin axis (Collen et al. 2003, Kroon et al. 1999). However, in more complex tissue explant cultures, capillary invasion in fibrin matrices proceeds independent of the PA/plasmin axis and relies instead on MT-MMP activity (Hiraoka et al. 1998). These results imply that in addition to the ECM, stromal cell types have the potential to bias the proteases utilized by ECs to undergo capillary morphogenesis.

Research from our own laboratory has utilized a 3D angiogenic model in which ECs invade a fibrin clot in coculture with stromal cells (Ghajar et al. 2006) to investigate the influence of stromal cell identity on ECM remodeling. When MSCs from the bone marrow were used as the stromal cell type, we found that ECM proteolysis during angiogenesis requires MMPs, and that new sprout formation was almost completely inhibited in the presence of small-molecule inhibitors of MMPs (Ghajar et al. 2010). In contrast, when fibroblasts were used as the supporting cell type, angiogenic sprouting could only be inhibited when both plasmin-mediated and MMP-mediated fibrinolysis were blocked, suggesting a proteolytic plasticity dictated by the identity of the supporting cells (Ghajar et al. 2010). When adipose-derived MSCs (ASCs) were used instead, angiogenic sprouting could be achieved when either MMPs or plasmin were inhibited individually, but not when both proteolytic axes were inhibited simultaneously (Kachgal and Putnam, 2011). In this fashion, angiogenesis induced by ASCs mirrored that induced by fibroblasts, despite their similar multipotency to MSCs from bone marrow. Furthermore, both fibroblasts and ASCs produced higher levels of HGF and tumor necrosis factor-α (TNFα) while inducing EC to upregulate urokinase plasminogen activator (uPA) when compared to cocultures containing ECs and MSCs from bone marrow (Kachgal and Putnam, 2011). A more definitive study used RNA interference to demonstrate that ECs require MT1-MMP, but not MMP-2 or MMP-9, to form new sprouts in the presence of bone marrow-derived MSCs in a fibrin matrix (Kachgal et al. 2012). These data suggest that the mechanism(s) by which ECs form new vessel networks depend(s) not only on the ECs and the matrix but also on the identity of supporting stromal cells present in the interstitial matrix.

Several studies now also suggest that pericytes regulate the expression of MT1-MMP in endothelial tips cells at the leading ends of sprouting capillaries. Among these, Lafleur et al. (2001) used cocultures of ECs with SMCs or pericytes to first show that the perivascular support cells regulate the activity of MT1-MMP and subsequent MMP-2 activation in ECs via the secretion of tissue inhibitor of metalloproteinase-2 (TIMP-2) (Lafleur et al. 2001). Saunders et al. (2006) then showed that EC–pericyte interactions in 3D collagen gels induced TIMP-3 expression by the pericytes and TIMP-2 expression by the ECs, while Yana et al. (2007) showed that the mural cells limit the expression of MT1-MMP to endothelial tip cells by restricting its expression in stalk cells. By controlling the specification of ECs into tip cells and stalk cells via local control of proteolysis, these mechanisms can potentially explain one possible way in which pericytes stabilize nascent capillaries.

3.4.5 Differentiation of MSCs to SMCs Supports Vascular Stability

After recruitment of ECs, MSCs begin to display vascular SMC markers and take on a new role in the maintenance of EC structures. *In vitro*, MSCs can be induced to differentiate directly into SMCs by exposure to TGF-β. The MSCs then express SMC markers such as αSMA, calponin 1, and myocardin. TGF-β acts by the induction of Jagged 1 (JAG1), a ligand for Notch. Interestingly, the direct activation of the Notch signaling pathway induced differentiation of MSCs to SMCs (Kurpinski et al. 2010). Further, ECs express the Notch ligand delta-like-4 (DLL4) and are themselves modulated by Notch signaling, providing a means for reciprocal interaction between ECs and SMCs (Taylor et al. 2002).

Another means of driving MSC differentiation toward an SMC is through interactions with ECs. Using 10T½ cells (a mouse multipotent cell line widely used as a model mural cell type), Hirschi et al. (1998) controlled their interactions with ECs *in vitro* using a coculture system in which the two cell types could interact directly through cell–cell junctions or indirectly through soluble, paracrine factors. When plated in adjacent agarose wells to permit only soluble communication, the 10T½ cells migrated toward the ECs mediated by PDGF-BB, while the ECs produced TGF-β that induced the 10T½s to express several SMC markers (i.e., SM-myosin, SM22α, and calponin) (Hirschi et al. 1998). In the absence of heterotypic cell–cell contacts, the ECs were found to secrete PDGF-B that increased 10T½ cell proliferation. However, coculture with direct cell–cell contact decreased the proliferation of both cell types independent of TGF-β and PDGF (Hirschi et al. 1999).

In another study, direct contact coculture between ECs and 10T½s was shown to upregulate SM22α to a greater extent than direct TGF-β treatment alone (Ding et al. 2004). Likewise, in a transwell coculture system, MSCs were shown to migrate in response to PDGF signaling from ECs (Au et al. 2008b). When placed in direct contact with the ECs, the MSCs upregulated the SMC marker myocardin, which could not be induced in separated cocultures or by TGF-β alone. Together, these studies suggest that ECs utilize soluble cues to induce MSC proliferation and migration toward EC; once the MSCs come into contact with the ECs, they stop proliferating and begin to differentiate down an SMC lineage.

3.5 Stimulation of Capillary Morphogenesis by Other Supporting Cell Types

Up to this point, this chapter has focused primarily on the use of MSCs with multilineage potential as a source of stromal cells capable of supporting the formation of vascular networks. However, tremendous mechanistic insights have been achieved through the use of other terminally differentiated mesenchymal cell types, most notably fibroblasts and SMCs. The choice of fibroblasts is justified based on their known roles in wound healing, and the argument that nearly every tissue in the body contains resident fibroblasts that are likely to influence vessel morphogenesis. Like MSCs, fibroblasts can occupy a perivascular location proximal to blood vessels, both *in vitro* and *in vivo*, and attain some characteristics of pericytes (Figure 3.3). The justification for using SMCs is also pretty clear, given their roles in maintaining vessel tone and contractility in arterioles and arteries, and the fact that pericytes are widely considered to be primitive SMCs. Early papers focused on the development of model systems to study the paracrine induction of angiogenic sprouting by fibroblasts (Montesano et al. 1993, Nehls and Drenckhahn 1995). Using one of these model systems, Hughes and colleagues have published extensively on the control of angiogenic sprouting by fibroblasts (Nakatsu et al. 2003a,b, Sainson et al. 2005, 2008). This group has recently and exhaustively investigated the "secretome" of fibroblasts for factors that enhance the angiogenic activity of ECs, and zeroed in on the importance of high levels of fibronectin and HGF (Newman et al. 2013).

To further understand the influence of stromal fibroblasts on capillary formation, Newman et al. (2011) performed a series of experiments to determine the relative importance of fibroblast-derived soluble factors and insoluble factors. They identified a cocktail of soluble factors that could induce sprouting but not lumen development in the absence

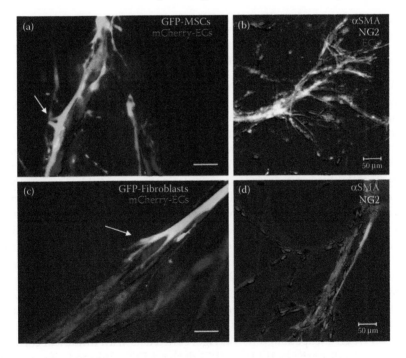

FIGURE 3.3
MSCs and fibroblasts both stimulate ECs to form mature capillary networks in model 3D cultures. MSCs (a,b) or fibroblasts (c,d) interspersed throughout fibrin ECMs in the presence of mCherry-transduced ECs induce capillary formation and occupy perivascular locations (white arrows in a and c). Cultures containing mCherry-transduced ECs and either MSCs (b) or fibroblasts (d) were fixed and IF stained for pericyte markers α SMA (aqua) and NG2 (white). Both MSCs and fibroblasts expressed α SMA, but only the MSCs expressed NG2. Nuclei (DAPI) are visible in the blue channel. In panels a and c, scale = 25 µm; in panels b and d, scale = 50 µm. (Adapted from *Experimental Cell Research*, 316, Ghajar, C. M. et al., Mesenchymal cells stimulate capillary morphogenesis via distinct proteolytic mechanisms, 813–25. Copyright 2010, with permission from Elsevier.)

of stromal support cells: angiopoietin-1, angiogenin, HGF, transforming growth factor-α (TGF-α), and TNF. Further studies revealed that the addition of fibroblast-conditioned medium restored the lumenogenesis. The authors showed that five genes expressed in fibroblasts—collagen I, procollagen C endopeptidase enhancer 1, secreted protein acidic and rich in cysteine (SPARC), TGF-β-induced protein IG-h3, and insulin growth factor-binding protein 7 (IGFBP7)—are necessary for lumen formation. When purified collagen I, SPARC, and IGFBP7 were added back to cultures containing fibroblasts deficient in the production of these proteins, lumen formation was recovered and the matrix became stiffer (Newman et al. 2011). This very elegant loss-of-function/gain-of-function approach provides some of the most compelling mechanistic evidence to date to explain how supporting fibroblasts regulate the formation of vascular network, and is consistent with a prior study implicating ECM deposition (Berthod et al. 2006). It remains to be seen if multipotent MSCs or fully committed SMCs facilitate capillary growth and development via the same mechanisms as fibroblasts, but it seems likely that at least some of the mechanisms will be conserved.

3.6 Delivery of Stromal Cells with ECs Enhances Vessel Formation and Stability *In Vivo*

3.6.1 Terminally Differentiated Adult Cells

In the past 10 years, a number of studies have focused on the potential of terminally differentiated adult cell types, particularly the aforementioned fibroblasts and SMCs, to promote vessel formation and stabilization upon implantation *in vivo*. Building on the numerous contributions in the literature demonstrating the ability of fibroblasts to support angiogenic sprouting *in vitro* (discussed above), a 2009 study by Chen et al. demonstrated that preformed vessel networks assembled in fibrin gels from HUVECs and normal human lung fibroblasts (NHLFs) *in vitro* prior to implantation *in vivo* could successfully inosculate with host vessels and become perfused with red blood cells. Increased EC proliferation and perfusion was dependent on the presence of NHLFs. A subsequent study comparing endothelial progenitor cells (EPCs) with HUVECs demonstrated that prevascularized constructs containing an order of magnitude higher density of NHLFs (2 million/mL) yielded more mature vessels that inosculated with host vessels 2–3 days more quickly than those with a lower density of support cells (0.2 million/mL) (Chen et al. 2010).

SMCs, commonly found *in vivo* in the medial layer of larger-diameter vessels, are another adult cell type that has been used in several studies to support implanted endothelial networks. In one study, human saphenous vein SMCs (HSVSMCs) were shown to promote vascularization when encapsulated in Matrigel constructs with umbilical cord blood (CB)-derived EPCs or adult peripheral blood (PB)-derived EPCs (termed cb- and adult-EPCs, respectively) and then implanted subcutaneously in nude mice (Melero-Martin et al. 2007). Vessel density depended on EPC source, as seen from sections stained for hCD31 and αSMA. Constructs containing EPCs alone did not vascularize, a finding consistent with prior work from others (Au et al. 2008a). Robust αSMA staining was seen both around vessel lumens as well as throughout the Matrigel in implants with SMCs, suggesting that a percentage of these cells colocalize with the EPC-derived vasculature (Melero-Martin et al. 2007).

Human aortic SMCs (HASMCs) have also been shown to accelerate vessel formation *in vivo* when coimplanted with Bcl-2-transduced HUVECs within a 3D PGA–collagen–fibronectin scaffold in a SCID mouse model (Shepherd et al. 2009). Grafts incorporating both HUVECs and HASMCs had more αSMA-positive cells associated with the vasculature compared to grafts containing either cell type alone at all time points. By 60 days, grafts incorporating both cell types had vessels of significantly larger caliber than grafts containing either cell type alone. Similar observations have been reported with Matrigel-enriched PLLA scaffolds containing human microvascular endothelial cells (HMVECs) and human pulmonary artery SMCs (hPASMCs) implanted subcutaneously in NOD/SCID mice (Hegen et al. 2011). Colocalization of structures staining positive for hCD31 and αSMA increased from 7 to 21 days postimplantation in constructs containing both cell types, suggesting SMCs assume a perivascular location over time. Fluorescence angiography using systemically injected fluorescein isothiocyanate (FITC)-labeled lectin demonstrated perfusion of implant vasculature and successful inosculation with host vessels. SMC association with the vascular network was confirmed by multiphoton fluorescence microscopy with labeled SMCs and UEA-1 stained human microvessels.

Collectively, these studies demonstrate that delivering fibroblasts or SMCs with ECs is more effective than ECs alone at creating stable vasculature capable of inosculating with host-derived vessels over time, supporting the general consensus that codelivery of a supporting cell type can yield better results. However, the feasibility of acquiring fibroblasts from the lung or SMCs from the walls of large vessels may limit the clinical translatability of these cells types, which thus motivates the investigation of alternate stromal cell sources. Furthermore, new evidence (discussed below) suggests that fibroblasts may not regulate vessel quality and quantity to the same degree as other populations of supporting cells, perhaps because they are not truly bona fide pericytes.

3.6.2 Multipotent Cells Enhance Vessel Network Functionality *In Vivo*

ECs have the capacity to support the formation of functional, mature vasculature *in vivo* when codelivered with human embryonic stem cells (hESCs) and MSCs derived from CB, bone marrow, or adipose tissue. This area of research has prompted significant interest, as stem cells from various sources act as pericytes *in vivo* (Traktuev et al. 2008). Additionally, they may be a more clinically viable option for therapeutic vascularization as many of these cell types can be derived autologously (Melero-Martin et al. 2008) and easily manipulated *in vitro* (Au et al. 2008b).

One of the earliest and most influential studies to use a supporting mesenchymal cell type to drive the formation of functional vasculature *in vivo* involved the coimplantation of HUVECs with 10T½s in a fibronectin–collagen gel in a SCID mouse cranial window (Koike et al. 2004). Vasculature in constructs containing HUVECs alone regressed within 60 days, likely due to delays in the recruitment of mouse-derived mural cells. In contrast, vasculature formed via the codelivery approach was functional for up to 1 year, a result that was hypothesized to result from the pericytic support. Several metrics were used to assess functionality. Fluorescent dextran was delivered systemically to mice, allowing visual confirmation via intravital microscopy of the inosculation of implant vasculature with host-derived vessels. Engineered vessels were more permeable than quiescent vasculature, but less permeable than tumor vasculature. Further, like native vasculature, the implant's vessels responded to the vasoconstrictor endothelin-1. Overall, this study demonstrated that the addition of stromal cells improved vessel functionality, and highlighted the importance of pericytic support for engineering vasculature *in vivo*.

In a later paper, the same investigators compared the ability of umbilical CB-derived and PB-derived EPCs to form functional, long-lasting vasculature when combined with 10T½s *in vivo* (Au et al. 2008a). In addition to the metrics of functionality used in their previous work, red blood cell velocity and leukocyte rolling in response to cytokine stimulation were monitored. Vasculature derived from PB-EPCs underwent regression within 3 weeks, regardless of the presence of 10T½s, while the stability of the vasculature derived from the CB-EPCs was dependent on the presence of 10T½s, a result that highlights the importance of pairing appropriate endothelial and stromal cell sources to develop engineered vascular networks.

Numerous studies have also explored combinations of bone marrow-derived MSCs with ECs in vascularization strategies. In one such study, HUVECs were coencapsulated with MSCs in a collagen–fibronectin gel and implanted into SCID mice in a cranial window (Au et al. 2008b). The resultant engineered vasculature was stable and perfused for >130 days *in vivo*. In this work, both 10T½s and MSCs were used as support cells to assess the ability of each to support vascular development. While both cell types supported vascular networks of comparable length, it was not clear if both supporting cell types yielded vasculature with similar functional properties *in vivo*. Nevertheless, in constructs containing MSCs, systemically administered dextran localized to the lumens of the engineered vessel networks, demonstrating functional inosculation with host vessels. MSCs localized to a periendothelial location and expressed αSMA, SM22α, and desmin. Furthermore, the nascent vessels circumscribed by the MSCs were capable of vasoconstriction in response to endothelin-1. Taken together, these data suggest MSCs differentiate into perivascular cells. Contrary to reports from other research groups (Oswald et al. 2004), the MSCs contributed to vascularization *in vivo* but did not directly differentiate into ECs.

Recent work from another leading team investigated the pericytic capacity of mesenchymal progenitor cells (MPCs) derived from bone marrow or umbilical CB. It is unclear if these cells are the same MSCs used by other investigators or not, since there are no consensus markers to prospectively isolate progenitors from the bone marrow. Regardless, like MSCs, these MPCs appeared to act like pericytes when combined with CB-derived EPCs in Matrigel and subsequently implanted subcutaneously in immunocompromised mice (Melero-Martin et al. 2008). Vascularization was apparent using both MPC sources 1 week after implantation, and was maintained for more than four weeks *in vivo*, as evidenced by the perfusion of vasculature with host erythrocytes. Additionally, αSMA and hCD31 staining confirmed that EPCs remained on the lumenal side of the tubule while the MPCs assumed a periendothelial location consistent with pericytes. Perfusion through the implant was assessed using a bioluminescent assay, in which luciferin delivered via intraperitoneal injection was detected in samples containing luciferase-expressing EPCs and unlabeled MPCs, but not in those samples containing EPCs alone. These data suggest that inosculation with host vasculature depends on the presence of stromal cells, which corroborates previous data suggesting that functional perfusion is compromised in the absence of some type of mesenchymal support cell (Chen et al. 2009, Koike et al. 2004, Melero-Martin et al. 2007, Shepherd et al. 2009).

In an approach that moves away from the use of adult stem cells, vascular progenitor cells were isolated from human embryoid bodies and differentiated into endothelial-like (EL) or smooth muscle-like (SML) cells (Ferreira et al. 2007). Subsequent transplantation of either EL cells alone or a combination of EL and SML cells in a Matrigel scaffold into nude mice resulted in microvessel formation. Microvessel density did not differ substantially between the EL only and EL and SML-containing scaffolds, which deviates somewhat from previous studies that emphasize codelivery for robust vascularization. However, these data are confounded by heterogeneous cell populations and unclear assessment of the functionality of the hollow lumens, which were quantified to determine vessel density.

ASCs are seen as a particularly attractive clinical source for vascular therapies since they can be easily harvested in large numbers from liposuction procedures. ASCs associate with capillaries *in vivo* and have been demonstrated to assume a pericytic role both *in vitro* and *in vivo* (Traktuev et al. 2008). In one important study, collagen constructs containing EPCs and ASCs, alone and in combination, were implanted into NOD-SCID mice (Traktuev et al. 2009). The density of perfused vessels increased in constructs containing both EPCs and ASCs. Multilayered structures were found more frequently in implants containing both cell types. To investigate other clinical applications, the constructs were implanted with either pancreatic islets or adipocytes and the organization of the parenchymal cells into neoorgans was demonstrated.

3.6.3 Comparing the Vasculature Formed with Distinct Mesenchymal Support Cells

The studies reviewed so far provide strong support for the argument that codelivery of a supporting mesenchymal cell type with ECs can yield stable vascular networks that inosculate with host vessels. However, the choice of stromal cell type has varied widely over these studies, and has included MSCs from bone marrow, adipose tissue, CB, as well as fibroblasts (from human lung) and both venous and arterial SMCs. It is clear that despite the consensus that the delivery of two cell types is better than one in terms of building vasculature, there is still no consensus as to which population of mesenchymal support cells will be the "winner" in terms of a viable clinical strategy.

Recent work from our own laboratory aimed to partially address this issue. Building on some of our prior *in vitro* observations, and the pioneering work of many other investigative teams highlighted earlier in this chapter, we set out to systematically compare the functionality of vasculature formed *in vivo* using various mesenchymal support cells (Grainger et al. 2013). In our study, HUVECs suspended in fibrin were injected subcutaneously with one of three stromal cell types: NHLFs, bone marrow-derived MSCs, or ASCs. The formation of vasculature was monitored over 14 days via laser Doppler perfusion imaging and standard histological assessments. Furthermore, vessel "leakiness" was assessed using a modified dextran tracer assay. Previous studies assayed for tracer in the lumens of engineered vessels, but we also quantified the fraction of dye outside the capillary lumen as a metric of permeability. As was expected, vessels formed via delivery of ECs alone were of a fairly poor quality and showed signs of leakiness via the presence of fluorescent dextran in the interstitial matrix (Figure 3.4). Interestingly, despite the fact that the EC-NHLF implants yielded the highest vessel density, the vessels created from this combination of cells were also of a rather poor quality. In contrast, vessels formed via codelivery of ECs with MSCs from either the bone marrow or adipose tissue exhibited decreased vessel leakiness (Figure 3.4) and increased the expression of mature smooth muscle markers. These data suggest that the multipotent stromal cells may in fact yield vasculature of superior quality because they indeed can differentiate into pericytes and eventually into SMCs.

In translational therapies, it will be critical to determine an appropriate combination of cells and matrix to use to revascularize ischemic tissues. The studies presented are pivotal in reaching this end; however, the gaps in the current data highlight the work that still needs to be performed. In particular, vascular elements do not exist in isolation within the body, but rather are incorporated within tissues. As a result, there is significant interest in studying vascularization in the presence of other tissues.

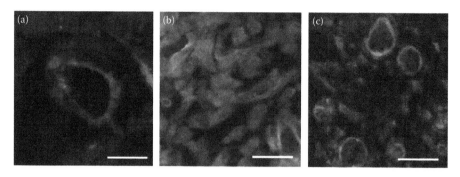

FIGURE 3.4
Coinjection of ECs with MSCs yields neovasculature with superior functional properties. SCID mice bearing implants containing only (a) ECs, (b) ECs + fibroblasts, or (c) ECs + MSCs were subjected to tail vein injections of a 70-kDa Texas Red–dextran (red) tracer to visualize inosculation and characterize vessel leakiness. Tissues were counterstained with antihuman CD31 antibodies (green) to verify the human origins of the vessels. The red tracer was found in the interstitial matrix outside of the new vessels when ECs were delivered alone (a) or with fibroblasts (b), but was contained within the vessels formed from ECs + MSCs (c). Scale bars = 20 μm. (Adapted from Grainger, S. J. et al. 2013. Stromal cell identity influences the *in vivo* functionality of engineered capillary networks formed by co-delivery of endothelial cells and stromal cells. *Tissue Engineering. Part A,* 19, 1209–22, with permission from Mary Ann Liebert, Inc.)

3.7 Vascularization in the Context of Complex Engineered Tissues

Vascularization is of substantial clinical value for cardiovascular therapies and is a key step in the formation of large, complex engineered tissues. Toward that goal, researchers seeking to create tissue constructs with complex functions have begun to combine parenchymal cells with vascular and stromal cells to facilitate construct vascularization. One such demonstration of this approach created vascularized skeletal muscle *in vitro* by combining myoblasts, mouse embryonic fibroblasts (MEFs), and HUVECs in an oriented poly (L-lactic acid)/poly(lactide-*co*-glycolide) (PLLA/PLGA) scaffold, which was then implanted in SCID mice *in vivo* (Levenberg et al. 2005). Human CD31 and vWF staining of histological sections confirmed the human origin of implant vasculature. Additionally, fluorescent lectin was administered systemically and localized to the lumen of construct vasculature, suggesting the successful inosculation of host and implant tissue. Both vascularization and differentiation of muscle tissue were shown to be optimal in constructs combining all three cell types.

A similar combination strategy has also been utilized to create vascularized cardiac tissue constructs, formed from hESC-derived cardiomyocytes, human umbilical vein and hESC-derived ECs, and MEFs (Stevens et al. 2009). Vessel number, contractile force, collagen content, and β-myosin heavy-chain (β-MHC) positive cells increased in constructs containing all three cell types as compared with constructs composed of cardiomyocytes alone. Constructs lacking the vascular element did not display physiologically appropriate active contraction, electrical pacing, and mechanical properties, suggesting that the provascular milieu facilitates both vascular and cardiac development. Although many challenges remain, these two examples demonstrate that successful vascularization of engineered tissues requires not only the ECs but also a supporting stromal cell type that serves as a stabilizing pericyte.

3.8 Conclusions and Future Directions

Consensus within the research community suggests that the vascularization of engineered constructs is significantly enhanced when ECs are not delivered alone, but rather in combination with mesenchymal support cells. This finding has been confirmed across several murine models as well as with various mesenchymal support cells. Nevertheless, the most appropriate choice of mesenchymal support cell for eventual therapeutic applications remains unclear due to both concerns about cell sourcing and the current gaps in the literature addressing how vessel quality and function vary with the identity of the support cells. Though the data remain incomplete, multiple studies suggest that multipotent MSCs from various sources provide vascular support in a manner similar to native pericytes. Further research is needed to justify the choice of cells for any eventual cell therapy to treat ischemic diseases or for the creation of robust engineered tissues, and to accelerate the translation of these research cell-based approaches to the clinic.

References

Aghdam, S. Y., Eming, S. A., Willenborg, S. et al. 2012. Vascular endothelial insulin/IGF-1 signaling controls skin wound vascularization. *Biochem Biophys Res Commun*, 421, 197–202.

Armulik, A., Abramsson, A., and Betsholtz, C. 2005. Endothelial/pericyte interactions. *Circ Res*, 97, 512–23.

Au, P., Daheron, L. M., Duda, D. G. et al. 2008a. Differential *in vivo* potential of endothelial progenitor cells from human umbilical cord blood and adult peripheral blood to form functional long-lasting vessels. *Blood*, 111, 1302–5.

Au, P., Tam, J., Fukumura, D., and Jain, R. K. 2008b. Bone marrow-derived mesenchymal stem cells facilitate engineering of long-lasting functional vasculature. *Blood*, 111, 4551–8.

Beckermann, B. M., Kallifatidis, G., Groth, A. et al. 2008. VEGF expression by mesenchymal stem cells contributes to angiogenesis in pancreatic carcinoma. *Br J Cancer*, 99, 622–31.

Bergers, G., and Song, S. 2005. The role of pericytes in blood-vessel formation and maintenance. *Neuro Oncol*, 7, 452–64.

Berthod, F., Germain, L., Tremblay, N., and Auger, F. A. 2006. Extracellular matrix deposition by fibroblasts is necessary to promote capillary-like tube formation *in vitro*. *J Cell Physiol*, 207, 491–8.

Caplan, A. I. 2007. Adult mesenchymal stem cells for tissue engineering versus regenerative medicine. *J Cell Physiol*, 213, 341–7.

Caplan, A. I. 2010. What's in a name? *Tissue Eng Part A*, 16, 2415–7.

Carmeliet, P., and Jain, R. K. 2011. Molecular mechanisms and clinical applications of angiogenesis. *Nature*, 473, 298–307.

Carmeliet, P., Moons, L., Luttun, A. et al. 2001. Synergism between vascular endothelial growth factor and placental growth factor contributes to angiogenesis and plasma extravasation in pathological conditions. *Nat Med*, 7, 575–83.

Carrion, B., Huang, C. P., Ghajar, C. M. et al. 2010. Recreating the perivascular niche *ex vivo* using a microfluidic approach. *Biotechnol Bioeng*, 107, 1020–8.

Chen, R. R., Silva, E. A., Yuen, W. W. et al. 2007. Integrated approach to designing growth factor delivery systems. *FASEB J*, 21, 3896–903.

Chen, X., Aledia, A. S., Ghajar, C. M. et al. 2009. Prevascularization of a fibrin-based tissue construct accelerates the formation of functional anastomosis with host vasculature. *Tissue Eng Part A*, 15, 1363–71.

Chen, X., Aledia, A. S., Popson, S. A. et al. 2010. Rapid anastomosis of endothelial progenitor cell-derived vessels with host vasculature is promoted by a high density of cotransplanted fibroblasts. *Tissue Eng Part A*, 16, 585–94.

Chisalita, S. I., and Arnqvist, H. J. 2004. Insulin-like growth factor I receptors are more abundant than insulin receptors in human micro- and macrovascular endothelial cells. *Am J Physiol Endocrinol Metab*, 286, E896–901.

Chun, T. H., Sabeh, F., Ota, I. et al. 2004. MT1-MMP-dependent neovessel formation within the confines of the three-dimensional extracellular matrix. *J Cell Biol*, 167, 757–67.

Collen, A., Hanemaaijer, R., Lupu, F. et al. 2003. Membrane-type matrix metalloproteinase-mediated angiogenesis in a fibrin-collagen matrix. *Blood*, 101, 1810–7.

Crisan, M., Yap, S., Casteilla, L. et al. 2008. A perivascular origin for mesenchymal stem cells in multiple human organs. *Cell Stem Cell*, 3, 301–13.

Cross, M. J., and Claesson-Welsh, L. 2001. FGF and VEGF function in angiogenesis: signalling pathways, biological responses and therapeutic inhibition. *Trends Pharmacol Sci*, 22, 201–7.

Da Silva Meirelles, L., Fontes, A. M., Covas, D. T., and Caplan, A. I. 2009. Mechanisms involved in the therapeutic properties of mesenchymal stem cells. *Cytokine Growth Factor Rev*, 20, 419–27.

Ding, R., Darland, D. C., Parmacek, M. S., and D'amore, P. A. 2004. Endothelial-mesenchymal interactions *in vitro* reveal molecular mechanisms of smooth muscle/pericyte differentiation. *Stem Cells Dev*, 13, 509–20.

Discher, D. E., Janmey, P., and Wang, Y. L. 2005. Tissue cells feel and respond to the stiffness of their substrate. *Science*, 310, 1139–43.

Discher, D. E., Mooney, D. J., and Zandstra, P. W. 2009. Growth factors, matrices, and forces combine and control stem cells. *Science*, 324, 1673–7.

Doorn, J., Moll, G., Le Blanc, K., Van Blitterswijk, C., and De Boer, J. 2011. Therapeutic applications of mesenchymal stromal cells: Paracrine effects and potential improvements. *Tissue Eng Part B Rev*, 18, 101–15.

Engler, A. J., Sen, S., Sweeney, H. L., and Discher, D. E. 2006. Matrix elasticity directs stem cell lineage specification. *Cell*, 126, 677–89.

Fan, Y., Ye, J., Shen, F. et al. 2007. Interleukin-6 stimulates circulating blood-derived endothelial progenitor cell angiogenesis in vitro. *J Cereb Blood Flow Metab*, 28, 90–8.

Ferreira, L. S., Gerecht, S., Shieh, H. F. et al. 2007. Vascular progenitor cells isolated from human embryonic stem cells give rise to endothelial and smooth muscle like cells and form vascular networks *in vivo*. *Circ Res*, 101, 286–94.

Friedenstein, A. J., Petrakova, K. V., Kurolesova, A. I., and Frolova, G. P. 1968. Heterotopic of bone marrow. Analysis of precursor cells for osteogenic and hematopoietic tissues. *Transplantation*, 6, 230–47.

Ghajar, C. M., Blevins, K. S., Hughes, C. C. W., George, S. C., and Putnam, A. J. 2006. Mesenchymal stem cells enhance angiogenesis in mechanically viable prevascularized tissues via early matrix metalloproteinase upregulation. *Tissue Eng*, 12, 2875–88.

Ghajar, C. M., Kachgal, S., Kniazeva, E. et al. 2010. Mesenchymal cells stimulate capillary morphogenesis via distinct proteolytic mechanisms. *Exp Cell Res*, 316, 813–25.

Giordano, A., Galderisi, U., and Marino, I. R. 2007. From the laboratory bench to the patient's bedside: An update on clinical trials with mesenchymal stem cells. *J Cell Physiol*, 211, 27–35.

Grainger, S. J., Carrion, B., Ceccarelli, J., and Putnam, A. J. 2013. Stromal cell identity influences the *in vivo* functionality of engineered capillary networks formed by co-delivery of endothelial cells and stromal cells. *Tissue Eng Part A*, 19, 1209–22.

Hegen, A., Blois, A., Tiron, C. E. et al. 2011. Efficient *in vivo* vascularization of tissue-engineering scaffolds. *J Tissue Eng Regen Med*, 5, e52–62.

Hiraoka, N., Allen, E., Apel, I. J., Gyetko, M. R., and Weiss, S. J. 1998. Matrix metalloproteinases regulate neovascularization by acting as pericellular fibrinolysins. *Cell*, 95, 365–77.

Hirschi, K. K., Rohovsky, S. A., Beck, L. H., Smith, S. R., and D'amore, P. A. 1999. Endothelial cells modulate the proliferation of mural cell precursors via platelet-derived growth factor-BB and heterotypic cell contact. *Circ Res*, 84, 298–305.

Hirschi, K. K., Rohovsky, S. A., and D'amore, P. A. 1998. PDGF, TGF-β, and heterotypic cell–cell interactions mediate endothelial cell–induced recruitment of 10T1/2 cells and their differentiation to a smooth muscle fate. *J Cell Biol*, 141, 805–14.

Hung, S.-C., Pochampally, R. R., Chen, S.-C., Hsu, S.-C., and Prockop, D. J. 2007. Angiogenic effects of human multipotent stromal cell conditioned medium activate the PI3K-Akt pathway in hypoxic endothelial cells to inhibit apoptosis, increase survival, and stimulate angiogenesis. *Stem Cells*, 25, 2363–70.

Iba, O., Matsubara, H., Nozawa, Y. et al. 2002. Angiogenesis by implantation of peripheral blood mononuclear cells and platelets into ischemic limbs. *Circulation*, 106, 2019–25.

Ingber, D. E., and Folkman, J. 1989. How does extracellular matrix control capillary morphogenesis? *Cell*, 58, 803–5.

Kachgal, S., Carrion, B., Janson, I. A., and Putnam, A. J. 2012. Bone marrow stromal cells stimulate an angiogenic program that requires endothelial MT1-MMP. *J Cell Physiol*, 227, 3546–55.

Kachgal, S., and Putnam, A. 2011. Mesenchymal stem cells from adipose and bone marrow promote angiogenesis via distinct cytokine and protease expression mechanisms. *Angiogenesis*, 14, 47–59.

Kasper, G., Dankert, N., Tuischer, J. et al. 2007. Mesenchymal stem cells regulate angiogenesis according to their mechanical environment. *Stem Cells*, 25, 903–10.

Kinnaird, T., Stabile, E., Burnett, M. S. et al. 2004. Local delivery of marrow-derived stromal cells augments collateral perfusion through paracrine mechanisms. *Circulation*, 109, 1543–9.

Koike, N., Fukumura, D., Gralla, O. et al. 2004. Tissue engineering: Creation of long-lasting blood vessels. *Nature*, 428, 138–9.

Kokovay, E., Li, L., and Cunningham, L. A. 2005. Angiogenic recruitment of pericytes from bone marrow after stroke. *J Cereb Blood Flow Metab*, 26, 545–55.

Kroon, M. E., Koolwijk, P., Van Goor, H. et al. 1999. Role and localization of urokinase receptor in the formation of new microvascular structures in fibrin matrices. *Am J Pathol*, 154, 1731–42.

Kurpinski, K., Lam, H., Chu, J. et al. 2010. Transforming growth factor-β and notch signaling mediate stem cell differentiation into smooth muscle cells. *Stem Cells*, 28, 734–42.

Lafleur, M. A., Forsyth, P. A., Atkinson, S. J., Murphy, G., and Edwards, D. R. 2001. Perivascular cells regulate endothelial membrane type-1 matrix metalloproteinase activity. *Biochem Biophys Res Commun*, 282, 463–73.

Lee, K. Y., Peters, M. C., Anderson, K. W., and Mooney, D. J. 2000. Controlled growth factor release from synthetic extracellular matrices. *Nature*, 408, 998–1000.

Leroux, L., Descamps, B., Tojais, N. F. et al. 2010. Hypoxia preconditioned mesenchymal stem cells improve vascular and skeletal muscle fiber regeneration after ischemia through a Wnt4-dependent pathway. *Mol Ther*, 18, 1545–52.

Levenberg, S., Rouwkema, J., Macdonald, M. et al. 2005. Engineering vascularized skeletal muscle tissue. *Nat Biotechnol*, 23, 879–84.

Mammoto, A., Connor, K. M., Mammoto, T. et al. 2009. A mechanosensitive transcriptional mechanism that controls angiogenesis. *Nature*, 457, 1103–8.

Melero-Martin, J. M., De Obaldia, M. E., Kang, S. Y. et al. 2008. Engineering robust and functional vascular networks *in vivo* with human adult and cord blood-derived progenitor cells. *Circ Res*, 103, 194–202.

Melero-Martin, J. M., Khan, Z. A., Picard, A. et al. 2007. *In vivo* vasculogenic potential of human blood-derived endothelial progenitor cells. *Blood*, 109, 4761–8.

Montesano, R., Pepper, M. S., and Orci, L. 1993. Paracrine induction of angiogenesis *in vitro* by Swiss 3T3 fibroblasts. *J Cell Sci*, 105 (Pt 4), 1013–24.

Murphy, W. L., and Mooney, D. J. 1999. Controlled delivery of inductive proteins, plasmid DNA and cells from tissue engineering matrices. *J Periodontal Res*, 34, 413–9.

Nagaya, N., Kangawa, K., Itoh, T. et al. 2005. Transplantation of mesenchymal stem cells improves cardiac function in a rat model of dilated cardiomyopathy. *Circulation*, 112, 1128–35.

Nakatsu, M. N., Sainson, R. C., Aoto, J. N. et al. 2003a. Angiogenic sprouting and capillary lumen formation modeled by human umbilical vein endothelial cells (HUVEC) in fibrin gels: The role of fibroblasts and angiopoietin-1. *Microvasc Res*, 66, 102–12.

Nakatsu, M. N., Sainson, R. C., Perez-Del-Pulgar, S. et al. 2003b. VEGF(121) and VEGF(165) regulate blood vessel diameter through vascular endothelial growth factor receptor 2 in an *in vitro* angiogenesis model. *Lab Invest J Tech Methods Pathol*, 83, 1873–85.

Nehls, V., and Drenckhahn, D. 1995. A microcarrier-based cocultivation system for the investigation of factors and cells involved in angiogenesis in three-dimensional fibrin matrices in vitro. *Histochem Cell Biol*, 104, 459–66.

Newman, A. C., Chou, W., Welch-Reardon, K. M. et al. 2013. Analysis of stromal cell secretomes reveals a critical role for stromal cell–derived hepatocyte growth factor and fibronectin in angiogenesis. *Arterioscler Thromb Vascular Biol, 33,* 513–22.

Newman, A. C., Nakatsu, M. N., Chou, W., Gershon, P. D., and Hughes, C. C. W. 2011. The requirement for fibroblasts in angiogenesis: fibroblast-derived matrix proteins are essential for endothelial cell lumen formation. *Mol Biol Cell*, 22, 3791–800.

Nor, J. E., Christensen, J., Mooney, D. J., and Polverini, P. J. 1999. Vascular endothelial growth factor (VEGF)-mediated angiogenesis is associated with enhanced endothelial cell survival and induction of Bcl-2 expression. *Am J Pathol*, 154, 375–84.

Nor, J. E., Peters, M. C., Christensen, J. B. et al. 2001. Engineering and characterization of functional human microvessels in immunodeficient mice. *Lab Invest*, 81, 453–63.

Ohnishi, S., Yasuda, T., Kitamura, S., and Nagaya, N. 2007. Effect of hypoxia on gene expression of bone marrow-derived mesenchymal stem cells and mononuclear cells. *Stem Cells*, 25, 1166–77.

Okuyama, H., Krishnamachary, B., Zhou, Y. F. et al. 2006. Expression of vascular endothelial growth factor receptor 1 in bone marrow-derived mesenchymal cells is dependent on hypoxia-inducible factor 1. *J Biol Chem*, 281, 15554–63.

Olsson, A.-K., Dimberg, A., Kreuger, J., and Claesson-Welsh, L. 2006. VEGF receptor signalling ? in control of vascular function. *Nat Rev Mol Cell Biol*, 7, 359–71.

Oswald, J., Boxberger, S., Jorgensen, B. et al. 2004. Mesenchymal stem cells can be differentiated into endothelial cells in vitro. *Stem Cells*, 22, 377–84.

Pesce, M., Orlandi, A., Iachininoto, M. G. et al. 2003. Myoendothelial differentiation of human umbilical cord blood-derived stem cells in ischemic limb tissues. *Circ Res*, 93, e51–62.

Pittenger, M. F., Mackay, A. M., Beck, S. C. et al. 1999. Multilineage potential of adult human mesenchymal stem cells. *Science*, 284, 143–7.

Rehman, J., Traktuev, D., Li, J. et al. 2004. Secretion of angiogenic and antiapoptotic factors by human adipose stromal cells. *Circulation*, 109, 1292–8.

Richardson, T. P., Peters, M. C., Ennett, A. B., and Mooney, D. J. 2001. Polymeric system for dual growth factor delivery. *Nat Biotechnol*, 19, 1029–34.

Sainson, R. C., Aoto, J., Nakatsu, M. N. et al. 2005. Cell-autonomous notch signaling regulates endothelial cell branching and proliferation during vascular tubulogenesis. *FASEB J Off Publ Fed Am Soc Exp Biol*, 19, 1027–9.

Sainson, R. C., Johnston, D. A., Chu, H. C. et al. 2008. TNF primes endothelial cells for angiogenic sprouting by inducing a tip cell phenotype. *Blood*, 111, 4997–5007.

Salcedo, R., Ponce, M. L., Young, H. A. et al. 2000. Human endothelial cells express CCR2 and respond to MCP-1: Direct role of MCP-1 in angiogenesis and tumor progression. *Blood*, 96, 34–40.

Saunders, W. B., Bohnsack, B. L., Faske, J. B. et al. 2006. Coregulation of vascular tube stabilization by endothelial cell TIMP-2 and pericyte TIMP-3. *J Cell Biol*, 175, 179–91.

Schechner, J. S., Nath, A. K., Zheng, L. et al. 2000. *In vivo* formation of complex microvessels lined by human endothelial cells in an immunodeficient mouse. *Proc Natl Acad Sci U S A*, 97, 9191–6.

Seib, F. P., Prewitz, M., Werner, C., and Bornhäuser, M. 2009. Matrix elasticity regulates the secretory profile of human bone marrow-derived multipotent mesenchymal stromal cells (MSCs). *Biochem Biophys Res Commun*, 389, 663–7.

Shaw, L. C., Pan, H., Afzal, A. et al. 2006. Proliferating endothelial cell-specific expression of IGF-I receptor ribozyme inhibits retinal neovascularization. *Gene Ther*, 13, 752–60.

Shen, Q., Wang, Y., Kokovay, E. et al. 2008. Adult SVZ stem cells lie in a vascular niche: A quantitative analysis of niche cell-cell interactions. *Cell Stem Cell*, 3, 289–300.

Shepherd, B. R., Jay, S. M., Saltzman, W. M., Tellides, G., and Pober, J. S. 2009. Human aortic smooth muscle cells promote arteriole formation by coengrafted endothelial cells. *Tissue Eng Part A*, 15, 165–73.

Silva, G. V., Litovsky, S., Assad, J. A. et al. 2005. Mesenchymal stem cells differentiate into an endothelial phenotype, enhance vascular density, and improve heart function in a canine chronic ischemia model. *Circulation*, 111, 150–6.

Simmons, C. A., Alsberg, E., Hsiong, S., Kim, W. J., and Mooney, D. J. 2004. Dual growth factor delivery and controlled scaffold degradation enhance *in vivo* bone formation by transplanted bone marrow stromal cells. *Bone*, 35, 562–9.

Simons, M., and Ware, J. A. 2003. Therapeutic angiogenesis in cardiovascular disease. *Nat Rev Drug Discov*, 2, 863–71.

Stevens, K. R., Kreutziger, K. L., Dupras, S. K. et al. 2009. Physiological function and transplantation of scaffold-free and vascularized human cardiac muscle tissue. *Proc Natl Acad Sci U S A*, 106, 16568–73.

Stratman, A. N., Malotte, K. M., Mahan, R. D., Davis, M. J., and Davis, G. E. 2009a. Pericyte recruitment during vasculogenic tube assembly stimulates endothelial basement membrane matrix formation. *Blood*, 114, 5091–101.

Stratman, A. N., Saunders, W. B., Sacharidou, A. et al. 2009b. Endothelial cell lumen and vascular guidance tunnel formation requires MT1-MMP-dependent proteolysis in 3-dimensional collagen matrices. *Blood*, 114, 237–47.

Stratman, A. N., Schwindt, A. E., Malotte, K. M., and Davis, G. E. 2010. Endothelial-derived PDGF-BB and HB-EGF coordinately regulate pericyte recruitment during vasculogenic tube assembly and stabilization. *Blood*, 116, 4720–30.

Sun, Q., Chen, R. R., Shen, Y. et al. 2005. Sustained vascular endothelial growth factor delivery enhances angiogenesis and perfusion in ischemic hind limb. *Pharm Res*, 22, 1110–6.

Sun, Q., Silva, E. A., Wang, A. et al. 2010. Sustained release of multiple growth factors from injectable polymeric system as a novel therapeutic approach towards angiogenesis. *Pharm Res*, 27, 264–71.

Taylor, K. L., Henderson, A. M., and Hughes, C. C. W. 2002. Notch activation during endothelial cell network formation *in vitro* targets the basic HLH transcription factor HESR-1 and downregulates VEGFR-2/KDR expression. *Microvasc Res*, 64, 372–83.

Thurston, G. 2002. Complementary actions of VEGF and angiopoietin-1 on blood vessel growth and leakage. *J Anat*, 200, 575–80.

Traktuev, D. O., Merfeld-Clauss, S., Li, J. et al. 2008. A population of multipotent CD34-positive adipose stromal cells share pericyte and mesenchymal surface markers, reside in a periendothelial location, and stabilize endothelial networks. *Circ Res*, 102, 77–85.

Traktuev, D. O., Prater, D. N., Merfeld-Clauss, S. et al. 2009. Robust functional vascular network formation *in vivo* by cooperation of adipose progenitor and endothelial cells. *Circ Res*, 104, 1410–20.

Wagner, J., Kean, T., Young, R., Dennis, J. E., and Caplan, A. I. 2009. Optimizing mesenchymal stem cell-based therapeutics. *Curr Opin Biotechnol*, 20, 531–6.

Wilson, C. J., Kasper, G., Schütz, M. A., and Duda, G. N. 2009. Cyclic strain disrupts endothelial network formation on Matrigel. *Microvasc Res*, 78, 358–63.

Yana, I., Sagara, H., Takaki, S. et al. 2007. Crosstalk between neovessels and mural cells directs the site-specific expression of MT1-MMP to endothelial tip cells. *J Cell Sci*, 120, 1607–14.

Zisch, A. H., Lutolf, M. P., Ehrbar, M. et al. 2003. Cell-demanded release of VEGF from synthetic, biointeractive cell ingrowth matrices for vascularized tissue growth. *FASEB J*, 17, 2260–2.

4

Adipose Stromal Vascular Fraction Cells for Vascularization of Engineered Tissues

Stuart K. Williams and James B. Hoying

CONTENTS

4.1 Microcirculation and Tissue-Engineered Constructs

4.1.1 Function of the Microcirculation

Microcirculation plays a central role in maintaining tissue homeostasis. The major functions include removal of waste products and delivery of nutrients to the perfused parenchymal tissue. For this reason, the formation of a functional microcirculation within tissue-engineered constructs is essential for composite tissue function. The complexity of specialized microvascular beds was identified by developmental biologists and morphologists who established the unique structure/function relationships of the microcirculation in various tissues. Complete restoration of microvascular function in engineered tissues following their implantation in a patient will require careful consideration of the morphology of the microcirculation constructed in the engineered tissue.

For example, the microcirculation present in the brain, muscle, kidney, and liver represent quite disparate structural complexities. At the level of the capillary endothelium, brain capillaries exhibit tight junctional associations to maintain the blood–brain barrier. Dichotomously, the liver microcirculation exhibits extensive discontinuities between the endothelium to support unrestricted exchange of solutes between the blood and parenchyma. The endothelium present in the kidney glomerulus exhibit fenestra that provide selective permeability to serum proteins of molecular weight >60 kD while completely excluding cellular entrance into the urinary space. These structural/functional relationships must be established in tissue-engineered constructs based on the type of engineered tissue that is constructed.

Another consideration when constructing a new tissue is the ability of the new microcirculation to respond to changing metabolic needs of the tissue. Many "angiogenic" therapies have successfully stimulated the formation of a new vasculature; however, these blood vessels are often limited to immature capillaries with limited formation of a complete, mature microcirculation (Dvorak et al. 1995). Operationally, a mature microcirculation is considered to be feeding arterioles, capillaries, and draining venules. The formation of functional arterioles is of particular importance as these microvascular components regulate blood flow through constriction and relaxation of smooth muscle cells. Also of importance are perivascular cells that include mural cells, often described as pericytes, and stromal cells present in the tissue surrounding the vasculature (Majno and Palade 1961; Bruns and Palade 1968; Wagner and Matthews 1975). These cells provide both direct signals affecting microvascular function and indirect signals that include paracrine and autocrine factors (e.g., NO) and produce extracellular matrix proteins that regulate microvascular function. Arguably, the heart microcirculation is one of the most complex tissues to construct as the microvascular function varies widely based on cardiac work. The creation of simply a plexus of capillaries will not provide the microvascular functional fidelity necessary for cardiac tissue function.

4.2 Adipose Tissue and the Stromal Vascular Fraction

4.2.1 Source of Vascular Cells to Build Vascularized Tissue

It is well recognized that the most critical cell necessary for the creation of a new microcirculation in a tissue-engineered construct is the endothelium. The sources of endothelium used in the construction of new vasculatures has paralleled technologies that have established techniques for first the isolation of endothelium from various tissues and then culture techniques, including media and growth factors necessary to maintain and expand endothelial cell numbers in culture. In 1921, Warren Lewis first reported the successful primary culture of human vascular endothelial cells in a blood-clot culture (Lewis and Webster 1921). For the following 40+ years, the study of endothelial cells as a participant in normal and pathologic conditions was not actively studied. Several discoveries establishing the endothelium as a metabolically active and protein-/cytokine-producing cell resulted in an explosion in studies (Wagner et al. 1972; Keirns et al. 1975; Moncada et al. 1976; Furchgott and Zawadzki 1980; Palmer et al. 1987). The need for *in vitro* models to study endothelial cell function led to methods for the isolation of large vessel endothelium as well as microvascular endothelium (Wagner and Matthews 1975; Sherer et al. 1980; Williams et al. 1980; Rupnick et al. 1988; Stamer et al. 1995). The culture of human adult endothelial cells was accomplished by many investigators (Maruyama 1963; Gimbrone et al. 1973; Jaffe et al. 1973; Lewis et al. 1973; Gospodarowicz and Zetter 1976; Haudenschild et al. 1976; Zetter and Antoniades 1979); however, the proliferative capacity of these human endothelia was quite limited with rarely more than a few population doublings (Haudenschild et al. 1976). The historical progression of methods to culture endothelial cells from human tissue sources is provided in Table 4.1.

TABLE 4.1

Progress in Culturing Human Endothelial Cells *In Vitro*

Tissue of Origin	First Reported	Culture Conditions	Relative Culture Life Span	Total Cell Yield
Lymph node	1921	Blood clot	Very short	$10–10^2$
Umbilical vein	1974	Serum	Very short	$10–10^2$
Umbilical vein	1981	Serum + ECGF	Short	10^{10}
Adult vessel (post mortem)	1982	Serum + ECGF	Very short	10^6
Adult vessel	1983	Serum + ECGF + heparin	Very long (single-cell cloning)	10^{18}
Adipose	1984	Serum + ECGF + heparin	Long	10^{12}

Initial difficulties toward culture expansion of human endothelial cells was solved with the discovery that heparin- binding growth factors (e.g., fibroblast growth factor and vascular endothelial cell growth factor) and heparin provided the stimulus for human endothelial cell proliferation in culture (Thornton et al. 1983; Jarrell et al. 1984). The ability to study human endothelium in isolation from other cells established this cell as the central player in pathologic and normal physiologic blood vessel function. The source of large blood vessel-derived human endothelial cells used in *in vitro* studies has included human umbilical cord vein and artery and adult artery and vein segments from essentially every peripheral vascular component. Examples include the human iliac artery and vein and human saphenous vein endothelium isolated from vessels obtained during surgery to procure organs for transplant or unused segments of saphenous vein used for coronary artery bypass.

The isolation and subsequent culture of microvascular endothelial cells was based on the studies of Wagner and Matthews who established methods to isolate and culture microvascular endothelium from fat (Wagner and Matthews 1975). The methods developed by Wagner were based on the original enzyme digestion protocols established by Rodbell for the isolation of adipocytes (Rodbell 1966). The general method for the isolation of regenerative cells from fat is illustrated in Figure 4.1. The pellet of cells that results from the centrifugation of the completely digested adipose tissue has been referred to as the stromal vascular fraction (SVF), adipose stromal cells, endothelial cell product, and adipose-derived stromal cells (Figure 4.2). All these terms describe the same heterogeneous cell population. The term SVF has been most commonly used to describe this population of cells. The most critical element of this cell-isolation process is the enzyme(s) used for the digestion of the intact fat tissue. The preparation and purification of enzymes for adipose digestion has been studied extensively (Peterkofsky 1982; Williams et al. 1989, 1995; McCarthy et al. 2008, 2011; Breite et al. 2010). The population of cells in the SVF remains an area of intense study with the agreement that the population includes endothelium, smooth muscle cells, pericytes, fibroblasts, mesenchymal stem cells (MSCs), monocytes, macrophages, lymphocytes, and stem cells with, to date, a lack of consensus regarding the relative contribution of each of these cell types to the makeup of the entire SVF cell population.

The study of endothelium from different anatomic sources collectively established that although the endothelium from all these sources shared some common traits, significant differences exist in the function and phenotype of the endothelium between artery and vein, large vessel to microvessel, and microvessel endothelium from different organs (Rupnick et al. 1988; Augustin et al. 1994; Aird et al. 1997; Rajotte et al. 1998; Thurston et al. 2000;

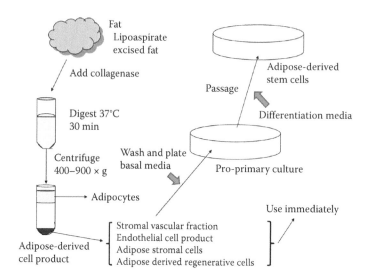

FIGURE 4.1
Method for the isolation of regenerative and stem cells from adipose tissue. As shown, the initial pellet of cells formed after enzymatic digestion of fat has been identified by different authors using different descriptive terms. This population of cells is the same and is generally called the stromal vascular fraction or SVF.

Gale et al. 2001; Shin et al. 2001; Chi et al. 2003; Hendrickx et al. 2004; Balestrieri and Napoli 2007; Frontczak-Baniewicz et al. 2007; Richardson et al. 2010; Nunes et al. 2011). The phrase "endothelium is not endothelium is not endothelium" is commonly used to describe these phenotypic and functional differences.

This body of knowledge regarding the isolation and culture of endothelium serves as a foundation for technologies to build a macro- or microcirculation. As a practical example, a tissue- engineered blood vessel for use as a bypass graft was first conceptualized as a

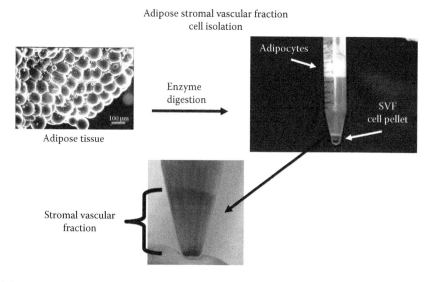

FIGURE 4.2
The digestion of fat tissue using proteolytic enzymes followed by centrifugation results in a slurry of buoyant adipocytes and a pellet of SVF cells.

scaffold treated with endothelium on the luminal surface (Debakey et al. 1964). Since the use of this construct is for large vessel bypass, the source of endothelium was theorized, by necessity, to originate from a large blood vessel. The first preclinical and then clinical use of this technique utilized saphenous vein endothelium to coat the inner surface of synthetic grafts (Herring et al. 1978; Graham et al. 1979; Stanley et al. 1982; Herring et al. 1984; Williams et al. 1985). An example of a Dacron® graft coated with a low density of human endothelial cells is shown in Figure 4.3. The fibers of the Dacron wee were coated with collagen to stimulate endothelial cell interaction. Culture expansion of autologous vein endothelium, made possible by the discovery of heparin and heparin-binding growth factor mitogenic activity, provided the cell density or dose necessary to cover the entire surface of a synthetic graft with endothelium prior to implantation. This leads to two terms to describe endothelial cell treatment of the luminal surface of vascular grafts; endothelial cell "seeding" where a low density of endothelial cells is mixed with blood or plasma and placed on the luminal surface and endothelial cell "sodding" where a density of cells equal to or greater than the cell density on a normal human artery is directly placed on the luminal surface of a graft (Herring et al. 1978; Rupnick et al. 1989). Human trials of this technique have provided proof of concept that endothelial cell "seeding" of synthetic vascular grafts using large vessel-derived and large vessel-cultured endothelium results in improved graft patency when the devices are implanted as a bypass conduit in the legs of patients with arterial occlusive disease (Zilla et al. 1994; Deutsch et al. 1999, 2009).

Concerns regarding the effect of tissue culture expansion on endothelial cell function (Nichols et al. 1987) and the cost and time necessary to culture expand cells resulted in a search for an endothelial cell source that could be used immediately after isolation to tissue engineer a vascular conduit. Although microvascular endothelial cell isolation from fat was established, the use of these cells to build a macrocirculation was at first considered impractical. The microcirculation in fat, and specifically fat-derived endothelium, was considered to be a terminally differentiated cell. Transplantation of these cells onto the luminal surface of a large-diameter conduit was possible, but it was hypothesized that these cells would not form a monolayer of endothelium with large blood vessel characteristics. Extensive studies proved this hypothesis incorrect as fat-derived

FIGURE 4.3
Dacron vascular graft sodded with human endothelial cells. The adherence and spreading of cells is evident on the Dacron fibers.

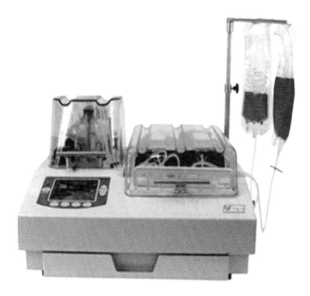

FIGURE 4.4
Automated adipose tissue digestion and SVF cell isolation system.

microvascular endothelium, transplanted onto the luminal surface of synthetic grafts with internal diameters of 1 mm or greater resulted in the formation of a complete and functional neointima (Jarrell et al. 1986; Williams et al. 1989). The endothelium that lined the surface of the graft exhibited the phenotype and function of large vessel endothelium (Williams et al. 1991; Williams and Jarrell 1996). Moreover, the use of fat-derived endothelium for graft "sodding" could be accomplished in the operating room in a process defined as point of care. These studies have now progressed to human clinical trials (Park et al. 1990). Translation and commercialization of this technology has progressed with the emergence of instrumentation that automates the process for the isolation of the SVF from adipose in the operating room, point of care (Williams et al. 2013). An example of one of these automated adipose digestion and SVF cell isolation systems is shown in Figure 4.4.

4.3 Three-Dimensional Cultures

The development of technologies to support the formation of a microcirculation within tissue-engineered constructs has benefitted from studies evaluating cells placed in three-dimensional (3-D) culture. *In vitro* studies of endothelial function were initially dominated by two-dimensional (2-D) culture systems. These studies began to evaluate the effect of external agonists on endothelial cell function, most notably, the effect of extracellular matrix molecules on endothelial cells. Madri and Williams reported the differential effect of collagen type I as compared to collagen type IV on both endothelial cell proliferation *in vitro* and endothelial cell phenotype (Madri and Williams 1983). As shown in Figure 4.5, adipose- derived microvascular endothelial cells exhibit a more

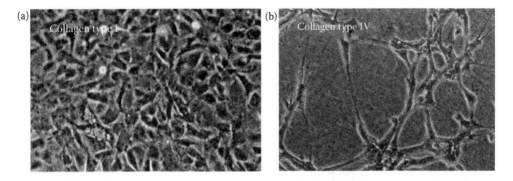

FIGURE 4.5
Differing morphology of microvascular endothelial cells grown on either collagen type I (a) or collagen type IV (b).

cobblestone-like morphology on collagen type I-coated tissue culture polystyrene plates while this same cell population exhibits the formation of tube-like structures when placed on collagen IV-treated plastic. This formation of this tube-like structure has often been described as *"in vitro* angiogenesis" but is more correctly described as neo-vasculogenesis or de novo formation of endothelial cell tubes in 2-D culture. First described using adipose-derived microvascular endothelial cells, it is now established that this tube-like morphology can be established in endothelial cells from large vessel sources (Lawley and Kubota 1989; Deroanne et al. 2001) as well as nonendothelial cell types (Gual et al. 2000; Murai et al. 2004). These studies have also established that endothelial cells in culture do not need to exhibit a cobblestone morphology as a confirmation that they are endothelial cells.

Since all tissues exist in a 3-D structure, the use of 3-D cell cultures has been seen as a necessity to study cells in a more physiologic condition. Numerous cell types have been studied in 3-D culture systems (Sato et al. 1977; Allen and Rankin 1990; Obradovic et al. 1999; Levenberg et al. 2003; Caspi et al. 2007; Duarte Campos et al. 2013). These studies have established that in many instances, the function of the cells in 3-D culture are different than the cells directly cultured on plastic (Pineda et al. 2013). These differences are more profound in mixed cultures of cells where two or more cell systems are cocultured. With regard to 3-D endothelial cell cultures, the majority of initial studies evaluated either neo-vasculogenesis of endothelial cells placed in gels or the outgrowth of capillary-like tubular structures from tissues or large vessel segments placed in 3-D gels. In most cases, phase contract microscopy was used to evaluate tube-like structure formation. In some instances, the tubular nature of the structures formed was confirmed by electron microscopy (Madri and Williams 1983; Nangia-Makker et al. 2000). In essentially all morphological studies, the endothelial cells present in the cultures exhibited a luminal and ablumenal plasma membrane with evidence of a junction between cells. However, the luminal space was consistently found to be filled with proteinaceous content. Nevertheless, these studies established that endothelial cells placed in 3-D structures exhibited the ability to undergo a similar neo-vasculogenesis observed when these cells are placed in 2-D culture. This provided the initial feasibility that a tissue-engineered construct might be constructed from cellular components, including endothelium with the goal of producing a functional microcirculation within the construct. Missing from these studies was the evidence that the endothelium that formed the tube-like structures could undergo subsequent expansion of the meshwork of tubular structures. The cells would undergo a migration and assembly but progression of the meshwork beyond the initial density of structures was not reported. The goal of prevascularization of tissue-engineered structures with a microcirculation that could both expand in complexity (e.g., sprouting) and integrate with other microvascular beds (e.g., inosculation) remained a major goal.

In 1996, Hoying and colleagues reported that microvascular fragments from adipose tissue placed in collagen gels would subsequently undergo sprouting and expansion of tube-like structure density (Hoying et al. 1996). The isolation of microvascular fragments, defined as intact, albeit short, segments of arterioles, capillaries, and venules, was achieved using a more limited collagenase digestion (~8 min) compared to conditions used to isolate single-cell populations from adipose tissue (>30 min collagenase digestion). This procedure is diagrammed in Figure 4.6.

The fragments contained, in addition to the luminal endothelium, vascular smooth muscle cells and perivascular cells that include pericytes. This technique removes a majority of stromal cells including tissue-resident macrophages, white cells, and red blood cells. The morphologic heterogeneity of these fragments is shown in Figure 4.7. The fragments remain essentially intact and do not dissociate with cellular migration and dispersion within the collagen gel during subsequent 3-D culture. The initiation of the sprouting activity commences after approximately 8 days in culture and the tube-like structures continue to form through extensive sprouting and cellular proliferation. The 3-D *in vitro* angiogenesis of the microvascular fragments is shown in Figure 4.8. Of interest, the luminal dimensions of the tube-like structures become somewhat uniform with an average diameter of approximately 23 µm (Chang and Hoying 2006). There is no perfusion through these structures and the mechanism(s) underlying the formation of tube-like structures with dimensions greater than expected for capillary segments (e.g., 4–10 µm i.d.) remains unknown. These studies provide the initial proof of feasibility that a microvascularized tissue can be engineered using microvascular fragments. It is presumed but awaits further study that the heterogeneous cell population present in the fragments is necessary for the sprouting of these cells and the formation of a complex plexus of tube-like structures.

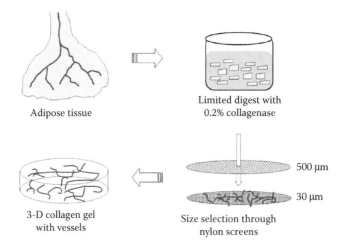

Adipose tissue

Limited digest with
0.2% collagenase

500 µm

30 µm

3-D collagen gel
with vessels

Size selection through
nylon screens

FIGURE 4.6
Method for the isolation of microvascular fragments from adipose tissue.

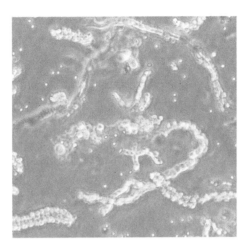

FIGURE 4.7
Primary isolate of microvascular fragments on tissue culture plastic from limited digestion of fat tissue.

A working hypothesis remains that a pure population of endothelial cells will not be sufficient to tissue engineer a microvascularized tissue.

On the basis of our current understanding of the structure and function of specialized microcirculations and the availability of cells and techniques to form 3-D tissue constructs numerous research groups that are exploring tissue engineering of the microcirculation (Skalak and Price 1996; Neumann et al. 2003; Watzka et al. 2004; Arosarena 2005; Chrobak et al. 2006; Wang et al. 2007; Moon et al. 2010; Athanassopoulos et al. 2012). These studies represent a multifaceted approach that includes the use of different matrices to provide the super structure of the construct. These matrices include synthetic biomaterials synthesized to provide architecture to the proposed construct (Wesolowski et al. 1961; Sterpetti et al. 1992; Salzmann et al. 1997; Sharkawy et al. 1998; Whang et al. 1999; Williams et al. 2005; James et al. 2011; Stabenfeldt et al. 2012). Studies have also explored the use of different combinations of cells with endothelium to support new vessel formation.

The most commonly used solution/gel formulation used for the construction of 3-D microvascular constructs has been collagen. This ubiquitous extracellular matrix protein

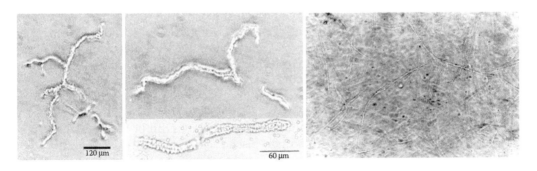

FIGURE 4.8
The initial morphology of adipose microvascular fragments in 3-D collagen gels is illustrated on the left. The right panel illustrates the branching *in vitro* angiogenesis that occurs when these gels are cultured.

can be easily isolated from numerous mammalian sources and has a long history of use as a biomedical implant (Battista 1949; Copenhagen 1965; Bazin et al. 1976; Quatela and Chow 2008; Ferreira et al. 2012). Collagen gels can be easily cast in various forms based on the ability of solutions of collagen types I and III to undergo polymerization from solutions by increasing the calcium concentration, pH, and temperature of the solution. The ability of cells to migrate within collagen gels and thus the ability of cells to form tube- like structures is influenced by the density of collagen gels and the orientation of the collagen fibrils (Hoying et al. 1996; Hoying and Williams 1996; Grant and Kleinman 1997; Telemeco et al. 2005). Fibrin gels have also been used as a matrix for endothelial cell migration and *in vitro* angiogenesis (Nicosia et al. 1993; Aplin et al. 2008). Fibrin gels were originally used by Lewis in the early 1920s as a matrix to support the outgrowth of vascular endothelium from tissue samples. The use of both collagen and fibrin as a scaffold for tissue engineering of constructs is based, in part, on the known recognition sites between cellular integrin proteins and specific binding sites within collagen and fibrin. These matrix-forming molecules not only act as a scaffold for the development of vascular structures but also provide signals to encapsulated cells that trigger cellular migration as well as cellular maturation.

Numerous synthetic polymers have been evaluated for use as a scaffold for tissue-engineering vascular structures. The porosity of biomaterials has been known for years as an important factor in the appropriate healing of synthetic devices (Wesolowski et al. 1961). Porosity in synthetic materials can be established in a variety of ways including variation in knitting and weaving techniques, extrusion conditions, the use of leachable materials to leave pores in nondegradable materials, and more recently, electrospinning of materials to vary porosity. Porous biomaterials have been used as a scaffold with adipose SVF cells to tissue engineer blood vessels with internal diameters >1 mm (Ahlswede and Williams 1994; Williams et al. 1994; Williams et al. 1994; Williams and Jarrell 1996; Williams et al. 2013). These biomaterials include Dacron, expanded as polytetrafluoroethylene and polyurethane.

Cells used in vascularized tissue-engineered constructs are diverse. Boyd and colleagues have explored the role of human embryonic stem cells, and specifically MSCs in the stabilization of tube-like structures (Boyd et al. 2011, 2013). Both umbilical vein endothelium and microvascular endothelium were used in these studies. Their work establishes the critical role that perivascular cells play in the formation of tube-like structures *in vitro* as well as the stability of these structures. Mixed cultures of endothelial cells and smooth muscle cells have been extensively studied toward the development of both arteriole and venule elements of the microcirculation (Hayward et al. 2013). The role of MSCs in tissue regeneration has been well established and the role of these cells in the maturation of the microcirculation within *in vitro*- formed gels is under investigation (Wu et al. 2007; Sorrell et al. 2009). These studies have a goal of determining whether MSCs alone or in conjunction with other cells including endothelium can be used to build tissue-engineered microcirculations. Of interest, the presence of a cell population defined as MSCs was first identified in bone marrow aspirates (Caplan 1991). Subsequent studies have identified these MSCs in numerous other tissues (Corradetti et al. 2013; Hayward et al. 2013; Utsunomiya et al. 2013; Zhou et al. 2013; Zimmerlin et al. 2013). And, most recently, studies have established that adipose tissue is a rich source of MSCs (Zhou et al. 2013; Zimmerlin et al. 2013). Some reports suggest that adipose tissue contains a significantly higher density of MSCs as bone marrow (Zhou et al. 2013). If MSCs play a critical role in tissue-engineered microvascular constructs, adipose tissue may represent an optimum source of cells.

4.4 *In Vivo* Studies of Tissue-Engineered Microvascularized Constructs

4.4.1 Reestablishment of a Functional Microcirculation in Ischemic Tissue through a Tissue-Engineering Approach

Bringing a new blood supply to the ischemic tissue remains a major goal of many tissue-engineering approaches. Ischemic disease provides some of the best evidence that lack of a functional microcirculation leads to dysfunctional tissue and for this reason, any tissue-engineered tissue construct must be designed with a plan for the incorporation of an adequate blood supply. The most critical example of the relationship between poor microvascular blood supply and poor tissue function is the ischemic heart. With reduced microvascular supply, the heart muscle exhibits reduced contractility and is clinically seen as reduced cardiac output. A cardiac surgeon, Claude S. Beck, working in his clinic at Western Reserve hospital first hypothesized that engineering a new microcirculation on the heart of patients with ischemic myocardial tissue may bring new blood flow to the ischemic tissue and provide relief to the patient (Beck 1935). In November, 1934, Beck noted during an operation on a patient's heart that a blood supply had formed between the patient's myocardium and scar tissue that had formed on the surface of the heart in continuum with the pericardium (Beck 1935). He hypothesized that the ischemic myocardium might be revascularized from the epicardial surface through an approach that involved developing new microvascular-rich tissue on the surface of the heart. He was particularly interested in the pericardial fat as a source of new blood vessels. On February 13, 1935, Claude Beck performed an operation on a 46-year-old patient with severe angina. The heart of this patient was exposed and a dental bur was used to roughen the epicardial surface to make the myocardial vessel susceptible to interaction with a pedicle of pectoral muscle brought in contact with the roughened epicardium. The patient was subsequently recovered from surgery and did well postoperatively with Dr. Beck noting his improved demeanor, and the patient's claim that he could do the exercise without pain. The heroism of this patient is noteworthy. Moreover, this is the first reported clinical case of a vascularized tissue construct; in this case, the pectoral muscle is being applied to the surface of the heart overlying a region of severe ischemia. This simple operation provides a simple path toward vascularized tissue-engineered implants (Beck 1943, 1949). Provide the cells or signals to support the development of a new microcirculation in the tissue implanted into a tissue space.

The concept of a tissue-engineered heart patch that stimulates the formation of a microcirculation following placement on the surface of an ischemic heart remained essentially unexplored for the next 60 years. In 2001, Kellar and colleagues published a method for using a tissue-engineered patch, designed to stimulate angiogenesis in peripheral ischemic tissue, as an epicardial patch on the ischemic heart (Kellar et al. 2001). This work and subsequent studies (Kellar et al. 2002, 2005, 2009; Thai et al. 2009) established the feasibility and practicality of placing a living tissue patch on the heart to stimulate angiogenesis in the ischemic tissue. In these studies, the cells used to manufacture the tissue-engineered construct were fibroblasts derived from human neonatal foreskin samples. These cells produced a variety of angiogenic factors *in vitro*, including vascular endothelial cell growth factor and hepatocyte growth factor (Jiang and Harding 1998). The vascular system that is created is derived from the host vessels present in the ischemic tissue and the cells in the heart patch simply provide a source of proteins and cytokines to create a proangiogenic environment. Studies in both rodent

models and large animal models established that the fibroblast patches not only stimulate the formation of new microvessels (arterioles, capillaries, and venules) but also
protect the heart from further loss of function when placed on a heart subjected to
a chronic ischemic injury (Kellar et al. 2005, 2011). These studies did not establish a
tissue-engineering approach to use transplanted cells or tissue constructs to provide a
premicrovascularized tissue.

4.4.2 Adipose Stromal Vascular Cells and the Micro-Bypass Tissue-Engineered Graft

A common shortcoming of any tissue-engineered construct is the inability to create a
structure that is >1 mm in thickness that will remain viable once implanted into the host
tissue. This thickness limitation is based on the need for a vascular system to provide
nutrients and remove waste products. Even a 1-mm-thick tissue construct will have a
hypoxic core. This has prompted many investigators to consider the incorporation of a
microcirculation in tissue-engineered constructs to support the development of thicker
constructs. More complex microvascular systems can now be created in tissue constructs
but interconnecting this preformed vasculature with the host vasculature becomes a
new roadblock. The most talented plastic surgeons can perform anastomoses of blood
vessels in the range of 500 µm; however, a microvascular anastomosis involves vessels
with internal diameters of an order of magnitude smaller, in the range of 50 µm or less.
A biological solution is therefore necessary to interconnect very small microvascular
components.

In 2004, Shepherd and colleagues reported the ability of a tissue-engineered prevascularized construct, engineered using microvascular fragments from adipose tissue in a
collagen gel, to, following implantation into the subcutaneous tissue, undergo vessel-to-
vessel reconnection resulting in blood flow into and within the construct (Shepherd et al.
2004). The constructs used in these implants were constructed from the adipose-derived
microvascular fragments cultured for 14 days in a 3-D collagen gel under conditions that
supported the expansion of the microvascular network into an extensive system of interconnected tube-like structures. The average diameter of the vessels in the gels prior to
implantation was approximately 23 µm. Thus, this engineered construct did not represent
a mature microcirculation (e.g., arterioles, capillaries, and venules). Blood flow into the
construct was evaluated at the time of explants using India ink to identify patent vessels.
As early as 1 day after implantation of the microvascularized constructs, patent vessels
were observed within the implanted collagen gels (Figure 4.9). These results support the
conclusion that the microvessels present within the construct have the ability to anastomose or "inosculate" with the recipient microcirculation. This represents a biological interconnection of these vascular systems. It also indicates that the tube-like structures in the
tissue-engineered construct have the capacity to carry blood indicating that they are tubes.
The flow patterns in the constructs were observed to expand rapidly following observations at day 2 and day 3 after implantation. By 3 days, the vessels within the construct
exhibited morphologies that support the conclusion that arterioles, venules, and capillaries were being formed. Cell-tracking methods established that the newly formed vessels
were not the result of ingrowth of recipient vessels; rather, the cells present originated from
the adipose-derived microvascular fragment cell population (Figure 4.10). These studies
established the feasibility of preforming a microvascular system in tissue-engineered constructs, transplanting the construct into a target tissue site and reestablishing a vascular
supply to the construct.

(a)　　　　　　　　　　　　　　　　(b)

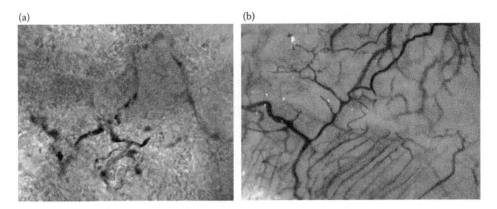

FIGURE 4.9
Microvascular fragment—3-D gel implants after 1 day (a) and 3 days (b) of implantation. India ink was injected into the circulation to identify patent vessels. As shown, there are patent vessels in the construct as early as 1 day after implantation.

4.4.3 Micro-Coronary Artery Bypass Grafts

The studies by Shepherd and colleagues established that prevascularized tissue-engineered constructs can inosculate with a competent microcirculation at implantation. One of the applications of tissue engineering is to provide a new vascular supply to the ischemic tissue, and for this reason, the adipose microvascular fragment tissue-engineered construct has been evaluated in a model of tissue ischemia. The specific model used was a mouse model of cardiac ischemia wherein a segment of the left anterior descending coronary artery is ligated creating a significant zone of ischemia. Shepherd and colleagues evaluated the prevascularized construct, composed of adipose microvascular fragments in a collagen gel, in a mouse model of cardiac ischemia placing the construct on the epicardial surface (Shepherd et al. 2007). The hypothesis tested was that the microcirculation in the tissue-engineered construct would inosculate with both well-perfused

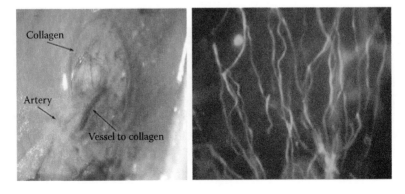

FIGURE 4.10
A microvascular fragment 3-D gel construct stimulates extensive revascularization following implantation as illustrated in the photomicrograph on the left. The cells in this construct were isolated from an animal expressing GFPs in all cells. The image on the right establishes that the new microvessels formed originate from the cells in the gel and not from ingrowth of vessels from the recipient.

microvascular beds at the edges of the construct and would also inosculate with vessels in the ischemic tissue. The term micro-coronary artery bypass graft or micro-CABG has been used to describe this application. Results of these studies permitted several conclusions to be drawn. First, similar to results seen in the subcutaneous tissue, the vessels in the tissue-engineered construct undergo inosculation with the recipient microcirculation. Inosculation of vessels is not limited to subcutaneous sites. Second, the transplanted microcirculation undergoes a similar maturation on the epicardial surface resulting in extensive development of arterioles, venules, and capillaries. Third, the use of this tissue-engineered micro-CABG slows the process of loss of cardiac function seen in untreated controls. This micro-CABG appears to restore blood flow to the ischemic myocardium and thus maintains a more normal ventricular contractility. And, although not a goal of these studies, the use of adipose-derived cells provides the opportunity to create these tissue-engineered constructs from cells taken from the same patient who will receive the construct providing an autologous tissue. There will be no need for immunosuppressive therapies using an autologous approach.

The ability to perform microcirculations *in vitro* using regenerative cells from adipose tissue has been actively studied to determine the cell populations in fat that participate in neovascularization. Nunes and colleagues evaluated whether specific components of the adipose microcirculation are predetermined to form the same component of the microcirculation following implantation (Nunes et al. 2010a,b). These studies have established, for example, that adipose-derived arterioles will undergo *in vitro* "sprouting" angiogenesis when placed in collagen gels. Nunes and colleagues also established that regenerative cells from adipose tissue will provide the building blocks for microvascular tissue engineering and these vascularized constructs will undergo maturation once implanted. This maturation includes the incorporation of perivascular cells in association with the maturing microvascular elements. The cell population isolated from adipose tissue is truly heterogeneous and these cells act together to form a new microcirculation both *in vitro* and *in vivo*. An example of the ability of the adipose SVF cells to form competent microvascular beds *in vivo* is illustrated in Figure 4.11. In this example, the SVF cells were isolated from a rat where all the cells expressed GFP (green fluorescent protein). As shown in Figure 4.11a, once implanted, the SVF forms vessels of varying internal diameter. Panel 4.11b establishes these vessels as being patent. Fluorescent dextran was injected into the

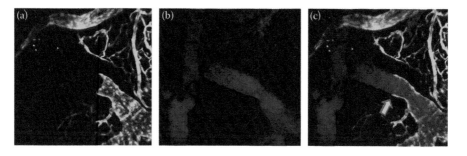

FIGURE 4.11
(a) Microcirculation that forms following the implantation of GFP-labeled SVF cells. (b) The patency of the new circulation formed is illustrated in this panel. A rhodamine dextran solution was injected intravascularly prior to visualization. (c) Merged image of panels (a) and (b). The arrow identifies the site of inosculation between implanted, GFP-labeled, cells and the recipient microcirculation.

animals prior to sacrifice and visualization. Panel 4.11c is the merged image of panels a and b. The most striking characteristic of this image is the point of transplanted vessel to recipient vessel interaction or inosculation. This inosculation point is highlighted with arrows. Moreover, this technology provides the microvascular solution to building more complex tissue-engineered constructs with biological functions beyond just providing blood flow. In 2008, Hiscox and colleagues reported the use of adipose-derived regenerative cells to tissue engineer a prevascularized islet construct (Hiscox et al. 2008). Upon implantation of this islet construct, the tissue- engineered vasculature inosculated with the recipient circulation and provided a microcirculation to the islets. Islet function was maintained *in vivo.*

These previous examples of tissue-engineered microvascular constructs using adipose stromal and vascular cells all involved the use of microvascular fragments isolated from adipose. LeBlanc and colleagues questioned whether adipose-derived SVF cells, isolated as essentially a single-cell preparation, could be used to construct a tissue patch for use in the treatment of myocardial infarction (Leblanc et al. 2012a,b). The adipose SVF cell population was grown on Vicryl® meshes similar to techniques used to develop 3-D fibroblast patches commercially sold under the trade name Dermagraft®. The results reported support the conclusion that the adipose-SVF patch not only maintains heart function in mice after an acute myocardial infarction (MI) and supports microvascular perfusion, but the microvessels formed exhibit the ability to regulate increases and decreases in blood flow based on the metabolic needs of the tissue perfused. Not only does the SVF tissue-engineered patch support vessel formation but the vessels formed exhibit a mature phenotype and function.

4.5 Bioprinting Adipose-Derived Cells to Form Tissue-Engineered Constructs

All tissues in the body have a specific 3-D orientation of cellular components. Cells organize into these defined 3-D tissue structures during development, providing tissue-/organ-specific functions to maintain the homeostasis of the body. As tissue-engineering technology previously moved from 2-D to 3-D configurations, the specific cell orientation of cells within a tissue-engineered construct will require a new generation of approaches to the complexity of tissues. 3-D printing of tissues has emerged as one approach to build complex patterns of cells in fabricated tissue constructs (Mironov et al. 2003, 2006; Smith et al. 2004; Jakab et al. 2006; Campbell and Weiss 2007; Marga et al. 2007; Chang et al. 2012; Tasoglu and Demirci 2013). The first attempt to use bioprinting to build vascular structures utilized a direct-write printing technology termed the bioassembly tool (BAT) (Smith et al. 2004). These first systems and approaches utilized endothelial cells in various gel-forming systems. The structures created were spatially oriented in two dimensions and the 3-D printing involved layer-by-layer assembly of vascular components. The major utility of these earliest studies of vascular bioprinting have been toward an understanding of vascular cell-to-cell interaction.

Recent studies using bioprinting have explored that the ability to organize vascular structures in more complex tissue constructs was followed by assessment of inosculation and maturation of the bioprinted constructs following implantation. Chang has utilized a

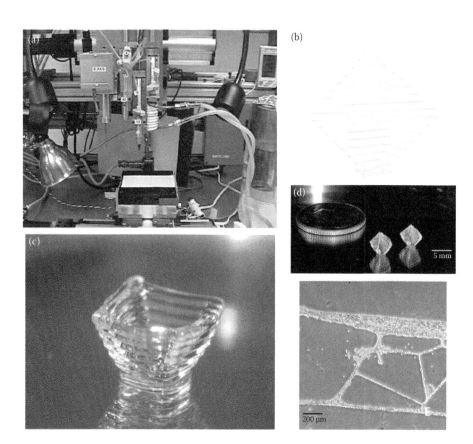

FIGURE 4.12
(a) BAT or "bioprinter" used to construct 3-D vascularized constructs. (b) Illustrates the script used to drive
the cell-/gel-delivery pens. This script produces a construct that resembles two hallow pyramids joined at their
bases. (c) Illustrates the printed constructs using this script. (d) Is a higher magnification image of a single hallow pyramid printed in an inverted form. (e) Printed endothelial cells placed in culture. Different interconnec-
tions can be obtained based on the script used.

direct-write BAT to print microvascular constructs using regenerative cells from adipose
tissue (Chang et al. 2011). Figure 4.12 illustrates the BAT utilized to perform 3-D printing and several examples of structures printed with the BAT. On the basis of the work
by Smith et al. (2004, 2007a,b), establishing the environmental conditions necessary to
maintain cell viability following bioprinting and patterns of microvascular components
were constructed in gels and implanted to establish whether these patterns are maintained
in vivo. Chang's work has established that vascular patterns in the microvascular tissue-
engineered constructs are maintained following implantation. Utilizing a Tie2-GFP transgenic mouse where the endothelial cells carry and express the Tie2-GFP label, the fate of
the transplanted cells can be followed. As shown in Figure 4.13, the printed fragments
maintain their orientation following implantation. Moreover, the vascular components in
the constructs inosculate with the recipient microcirculation providing blood perfusion to
the construct. This work provides the foundational proof of concept that complex tissue-
engineered constructs with significant tissue mass can be rapidly constructed with bioprinting technology and a functional microcirculation can be printed within the construct
to provide necessary vascularization.

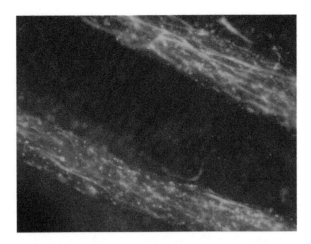

FIGURE 4.13
Microvascular constructs printed using GFP- labeled SVF cells and implanted in non-GFP animals. The new vessels that form are aligned in the direction of the printed constructs.

4.6 Clinical Applications

A major goal of tissue engineering is to create new therapies that impact human health. Tissue-engineered constructs with direct effect on the microcirculation have already reached widespread clinical use as exemplified by artificial skin-/wound-healing products (Jadlowiec et al. 2012). These current products have, as a mode of action, the stimulation of new blood vessel growth in treated tissues through the synthesis and secretion of growth factors and cytokines. The next generation of vascularized tissue-engineered products will have the endothelium as a component of the construct itself. Human clinical studies using adipose tissue-derived regenerative cells, and specifically the SVF have commenced with several past and ongoing clinical trials (Williams et al. 1989; Park et al. 1990; Williams 1995; Williams and Jarrell 1996). The initial use of the adipose SVF was for the formation of endothelial cell linings on tissue-engineered vascular grafts. More recently, human studies (identified on clinicaltrials.gov) using either the adipose SVF cells or a derivative cell population known as adipose stromal-/adipose-derived stem cells have begun for treatments that include cardiac myocardial infarction, congestive heart failure, critical limb ischemia, peripheral vascular graft sodding, wound repair, reconstructive plastic surgery, orthopedic applications (e.g., arthritis and spinal fusion), and erectile dysfunction. The use of adipose SVF cells to engineer complex tissue constructs is under active investigation in both *in vitro* and *in vivo* settings.

References

Ahlswede, K. M. and S. K. Williams. 1994. Microvascular endothelial cell sodding of 1-mm expanded polytetrafluoroethylene vascular grafts. *Arterioscler Thromb* **14**(1): 25–31.

Aird, W. C., J. M. Edelberg et al. 1997. Vascular bed-specific expression of an endothelial cell gene is programmed by the tissue microenvironment. *J Cell Biol* **138**(5): 1117–1124.

Allen, R. E. and L. L. Rankin. 1990. Regulation of satellite cells during skeletal muscle growth and development. *Proc Soc Exp Biol Med* **194**(2): 81–86.

Aplin, A. C., E. Fogel et al. 2008. The aortic ring model of angiogenesis. *Methods Enzymol* **443**: 119–136.

Arosarena, O. 2005. Tissue engineering. *Curr Opin Otolaryngol Head Neck Surg* **13**(4): 233–241.

Athanassopoulos, A., G. Tsaknakis et al. 2012. Microvessel networks in pre-formed in artificial clinical grade dermal substitutes *in vitro* using cells from haematopoietic tissues. *Burns* **38**(5): 691–701.

Augustin, H. G., D. H. Kozian et al. 1994. Differentiation of endothelial cells: Analysis of the constitutive and activated endothelial cell phenotypes. *Bioessays* **16**(12): 901–906.

Balestrieri, M. L. and C. Napoli. 2007. Novel challenges in exploring peptide ligands and corresponding tissue-specific endothelial receptors. *Eur J Cancer* **43**(8): 1242–1250.

Battista, A. F. 1949. The reaction of various tissues to implants of a collagen derivative. *Can J Res* **27**(2): 94–104.

Bazin, S., M. Le Lous et al. 1976. Collagen in granulation tissues. *Agents Actions* **6**(1–3): 272–276.

Beck, C. S. 1935. The development of a new blood supply to the heart by operation. *Ann Surg* **102**(5): 801–813.

Beck, C. S. 1943. Principles underlying the operative approach to the treatment of myocardial ischemia. *Ann Surg* **118**(5): 788–806.

Beck, C. S. 1949. Revascularization of the heart. *N Y State J Med* **49**(14): 1727–1729.

Boyd, N. L., S. S. Nunes et al. 2011. Microvascular mural cell functionality of human embryonic stem cell-derived mesenchymal cells. *Tissue Eng Part A* **17**(11–12): 1537–1548.

Boyd, N. L., S. S. Nunes et al. 2013. Dissecting the role of human embryonic stem cell-derived mesenchymal cells in human umbilical vein endothelial cell network stabilization in three-dimensional environments. *Tissue Eng Part A* **19**(1–2): 211–223.

Breite, A. G., F. E. Dwulet et al. 2010. Tissue dissociation enzyme neutral protease assessment. *Transplant Proc* **42**(6): 2052–2054.

Bruns, R. R. and G. E. Palade. 1968. Studies on blood capillaries. I. General organization of blood capillaries in muscle. *J Cell Biol* **37**(2): 244–276.

Campbell, P. G. and L. E. Weiss. 2007. Tissue engineering with the aid of inkjet printers. *Expert Opin Biol Ther* **7**(8): 1123–1127.

Caplan, A. I. 1991. Mesenchymal stem cells. *J Orthop Res: Official Publication Orthop Res Soc* **9**(5): 641–650.

Caspi, O., A. Lesman et al. 2007. Tissue engineering of vascularized cardiac muscle from human embryonic stem cells. *Circulation Res* **100**(2): 263–272.

Chang, C. C., E. D. Boland et al. 2011. Direct-write bioprinting three-dimensional biohybrid systems for future regenerative therapies. *J Biomed Mater Res B Appl Biomater* **98**(1): 160–170.

Chang, C. C. and J. B. Hoying. 2006. Directed three-dimensional growth of microvascular cells and isolated microvessel fragments. *Cell Transplant* **15**(6): 533–540.

Chang, C. C., L. Krishnan et al. 2012. Determinants of microvascular network topologies in implanted neovasculatures. *Arterioscler Thromb Vasc Biol* **32**(1): 5–14.

Chi, J. T., H. Y. Chang et al. 2003. Endothelial cell diversity revealed by global expression profiling. *Proc Natl Acad Sci USA* **100**(19): 10623–10628.

Chrobak, K. M., D. R. Potter et al. 2006. Formation of perfused, functional microvascular tubes *in vitro*. *Microvasc Res* **71**(3): 185–196.

Copenhagen, H. 1965. A new collagen tape in reconstructive surgery. *Brit J Surg* **52**: 697–699.

Corradetti, B., A. Meucci et al. 2013. Mesenchymal stem cells from amnion and amniotic fluid in bovine. *Reproduction* **145**(4): 391–400.

Debakey, M. E., J. P. Abbott et al. 1964. Endothelial lining of a human vascular prosthesis. *Cardiovasc Res Cent Bull* **24**: 7–12.

Deroanne, C. F., C. M. Lapiere et al. 2001. *In vitro* tubulogenesis of endothelial cells by relaxation of the coupling extracellular matrix–cytoskeleton. *Cardiovasc Res* **49**(3): 647–658.

Deutsch, M., J. Meinhart et al. 1999. Clinical autologous *in vitro* endothelialization of infrainguinal ePTFE grafts in 100 patients: A 9-year experience. *Surgery* **126**(5): 847–855.

Deutsch, M., J. Meinhart et al. 2009. Long-term experience in autologous *in vitro* endothelialization of infrainguinal ePTFE grafts. *J Vasc Surg* **49**(2): 352–362; discussion 362.

Duarte Campos, D. F., A. Blaeser et al. 2013. Three-dimensional printing of stem cell-laden hydrogels submerged in a hydrophobic high-density fluid. *Biofabrication* 5(1): 015003.

Dvorak, H. F., L. F. Brown et al. 1995. Vascular permeability factor/vascular endothelial growth factor, microvascular hyperpermeability, and angiogenesis. *Am J Pathol* 146(5): 1029–1039.

Ferreira, A. M., P. Gentile et al. 2012. Collagen for bone tissue regeneration. *Acta Biomater* 8(9): 3191–3200.

Frontczak-Baniewicz, M., M. Walski et al. 2007. Diversity of immunophenotypes of endothelial cells participating in new vessel formation following surgical rat brain injury. *J Physiol Pharmacol* 58 **Suppl 5**(Pt 1): 193–203.

Furchgott, R. F. and J. V. Zawadzki. 1980. The obligatory role of endothelial cells in the relaxation of arterial smooth muscle by acetylcholine. *Nature* 288(5789): 373–376.

Gale, N. W., P. Baluk et al. 2001. Ephrin-B2 selectively marks arterial vessels and neovascularization sites in the adult, with expression in both endothelial and smooth-muscle cells. *Dev Biol* 230(2): 151–160.

Gimbrone, M. A., Jr., R. S. Cotran et al. 1973. Endothelial regeneration: Studies with human endothelial cells in culture. *Ser Haematol* 6(4): 453–455.

Gospodarowicz, D. and B. R. Zetter. 1976. The use of fibroblast and epidermal growth factors to lower the serum requirement for growth of normal diploid cells in early passage: A new method for cloning. *Dev Biol Stand* 37: 109–130.

Graham, L. M., D. W. Vinter et al. 1979. Cultured autogenous endothelial cell seeding of prosthetic vascular grafts. *Surg Forum* 30: 204–206.

Grant, D. S. and H. K. Kleinman. 1997. Regulation of capillary formation by laminin and other components of the extracellular matrix. [Review] [64 refs]. *EXS* 79: 317–333.

Gual, P., S. Giordano et al. 2000. Sustained recruitment of phospholipase C-gamma to Gab1 is required for HGF-induced branching tubulogenesis. *Oncogene* 19(12): 1509–1518.

Haudenschild, C. C., D. Zahniser et al. 1976. Human vascular endothelial cells in culture. Lack of response to serum growth factors. *Exp Cell Res* 98(1): 175–183.

Hayward, C. J., J. Fradette et al. 2013. Harvesting the potential of the human umbilical cord: Isolation and characterisation of four cell types for tissue engineering applications. *Cells Tissues Organs* 197(1): 37–54.

Hendrickx, J., K. Doggen et al. 2004. Molecular diversity of cardiac endothelial cells *in vitro* and *in vivo*. *Physiol Genomics* 19(2): 198–206.

Herring, M., S. Baughman et al. 1984. Endothelial seeding of Dacron and polytetrafluoroethylene grafts: The cellular events of healing. *Surgery* 96(4): 745–755.

Herring, M., A. Gardner et al. 1978. A single-staged technique for seeding vascular grafts with autogenous endothelium. *Surgery* 84(4): 498–504.

Hiscox, A. M., A. L. Stone et al. 2008. An islet-stabilizing implant constructed using a preformed vasculature. *Tissue Eng Part A* 14(3): 433–440.

Hoying, J. B., C. A. Boswell et al. 1996. Angiogenic potential of microvessel fragments established in three-dimensional collagen gels. *In Vitro Cell Dev Biol Anim* 32(7): 409–419.

Hoying, J. B. and S. K. Williams. 1996. Effects of basic fibroblast growth factor on human microvessel endothelial cell migration on collagen I correlates inversely with adhesion and is cell density dependent. *J Cell Physiol* 168(2): 294–304.

Jadlowiec, C., R. A. Brenes et al. 2012. Stem cell therapy for critical limb ischemia: What can we learn from cell therapy for chronic wounds? *Vascular* 20(5): 284–289.

Jaffe, E. A., R. L. Nachman et al. 1973. Culture of human endothelial cells derived from umbilical veins. Identification by morphologic and immunologic criteria. *J Clin Invest* 52(11): 2745–2756.

Jakab, K., B. Damon et al. 2006. Three-dimensional tissue constructs built by bioprinting. *Biorheology* 43(3–4): 509–513.

James, R., S. G. Kumbar et al. 2011. Tendon tissue engineering: Adipose-derived stem cell and GDF-5 mediated regeneration using electrospun matrix systems. *Biomed Mater* 6(2): 025011.

Jarrell, B., E. Levine et al. 1984. Human adult endothelial cell growth in culture. *J Vascul Surg: Official Publication, the Soc Vascul Surg Int Soc Cardiovascul Surg*, North American Chapter 1(6): 757–764.

Jarrell, B. E., S. K. Williams et al. 1986. Use of an endothelial monolayer on a vascular graft prior to implantation. Temporal dynamics and compatibility with the operating room. *Ann Surg* **203**(6): 671–678.

Jarrell, B. E., S. K. Williams et al. 1986. Use of freshly isolated capillary endothelial cells for the immediate establishment of a monolayer on a vascular graft at surgery. *Surgery* **100**(2): 392–399.

Jiang, W. G. and K. G. Harding. 1998. Enhancement of wound tissue expansion and angiogenesis by matrix-embedded fibroblast (Dermagraft), a role of hepatocyte growth factor/scatter factor. *Int J Mol Med* **2**(2): 203–210.

Keirns, J. J., R. C. Wagner et al. 1975. Preparation of isolated epididymal capillary endothelial cells and their use in the study of cyclic AMP metabolism. *Methods Enzymol* **39**: 479–482.

Kellar, R. S., L. B. Kleinert et al. 2002. Characterization of angiogenesis and inflammation surrounding ePTFE implanted on the epicardium. *J Biomed Mater Res* **61**(2): 226–233.

Kellar, R. S., L. K. Landeen et al. 2001. Scaffold-based three-dimensional human fibroblast culture provides a structural matrix that supports angiogenesis in infarcted heart tissue. *Circulation* **104**(17): 2063–2068.

Kellar, R. S., B. R. Shepherd et al. 2005. Cardiac patch constructed from human fibroblasts attenuates reduction in cardiac function after acute infarct. *Tissue Eng* **11**(11–12): 1678–1687.

Kellar, R. S., S. K. Williams et al. 2011. Three-dimensional fibroblast cultures stimulate improved ventricular performance in chronically ischemic canine hearts. *Tissue Eng Part A* **17**(17–18): 2177–2186.

Lawley, T. J. and Y. Kubota. 1989. Induction of morphologic differentiation of endothelial cells in culture. *J Invest Dermatol* **93**(2 Suppl): 59S–61S.

Leblanc, A. J., L. Krishnan et al. 2012a. Microvascular repair: Post-angiogenesis vascular dynamics. *Microcirculation* **19**(8): 676–695.

Leblanc, A. J., J. S. Touroo et al. 2012b. Adipose stromal vascular fraction cell construct sustains coronary microvascular function after acute myocardial infarction. *Am J Physiol Heart Circ Physiol* **302**(4): H973–H982.

Levenberg, S., N. F. Huang et al. 2003. Differentiation of human embryonic stem cells on three-dimensional polymer scaffolds. *Proc Natl Acad Sci USA* **100**(22): 12741–12746.

Lewis, L. J., J. C. Hoak et al. 1973. Replication of human endothelial cells in culture. *Science* **181**(4098): 453–454.

Lewis, W. H. and L. T. Webster. 1921. Wandering cells, endothelial cells, and fibroblasts in cultures from human lymph nodes. *J Exp Med* **34**(4): 397–405.

Madri, J. A. and S. K. Williams. 1983. Capillary endothelial cell cultures: Phenotypic modulation by matrix components. *J Cell Biol* **97**(1): 153–165.

Majno, G. and G. E. Palade. 1961. Studies on inflammation. 1. The effect of histamine and serotonin on vascular permeability: An electron microscopic study. *J Biophys Biochem Cytol* **11**: 571–605.

Marga, F., A. Neagu et al. 2007. Developmental biology and tissue engineering. *Birth Defects Res C Embryo Today* **81**(4): 320–328.

Maruyama, Y. 1963. The human endothelial cell in tissue culture. *Z Zellforsch Mikrosk Anat* **60**: 69–79.

McCarthy, R. C., A. G. Breite et al. 2011. Tissue dissociation enzymes for isolating human islets for transplantation: Factors to consider in setting enzyme acceptance criteria. *Transplantation* **91**(2): 137–145.

McCarthy, R. C., B. Spurlin et al. 2008. Development and characterization of a collagen degradation assay to assess purified collagenase used in islet isolation. *Transplant Proc* **40**(2): 339–342.

Mironov, V., T. Boland et al. 2003. Organ printing: Computer-aided jet-based 3D tissue engineering. *Trends Biotechnol* **21**(4): 157–161.

Mironov, V., N. Reis et al. 2006. Review: Bioprinting: A beginning. *Tissue Eng* **12**(4): 631–634.

Moncada, S., R. Gryglewski et al. 1976. An enzyme isolated from arteries transforms prostaglandin endoperoxides to an unstable substance that inhibits platelet aggregation. *Nature* **263**(5579): 663–665.

Moon, S., S. K. Hasan et al. 2010. Layer by layer three-dimensional tissue epitaxy by cell-laden hydrogel droplets. *Tissue Eng Part C Methods* **16**(1): 157–166.

Murai, S., T. Umemiya et al. 2004. Expression and localization of membrane-type-1 matrix metal-loproteinase, CD 44, and laminin-5gamma2 chain during colorectal carcinoma tumor progression. *Virchows Arch* **445**(3): 271–278.

Nangia-Makker, P., Y. Honjo et al. 2000. Galectin-3 induces endothelial cell morphogenesis and angiogenesis. *Am J Pathol* **156**(3): 899–909.

Neumann, T., B. S. Nicholson et al. 2003. Tissue engineering of perfused microvessels. *Microvasc Res* **66**(1): 59–67.

Nichols, W. W., E. B. Buynak et al. 1987. Cytogenetic evaluation of human endothelial cell cultures. *J Cell Physiol* **132**(3): 453–462.

Nicosia, R. F., E. Bonanno et al. 1993. Fibronectin promotes the elongation of microvessels during angiogenesis *in vitro*. *J Cell Physiol* **154**(3): 654–661.

Nunes, S. S., K. A. Greer et al. 2010a. Implanted microvessels progress through distinct neovascularization phenotypes. *Microvasc Res* **79**(1): 10–20.

Nunes, S. S., L. Krishnan et al. 2010b. Angiogenic potential of microvessel fragments is independent of the tissue of origin and can be influenced by the cellular composition of the implants. *Microcirculation* **17**(7): 557–567.

Nunes, S. S., H. Rekapally et al. 2011. Vessel arterial–venous plasticity in adult neovascularization. *PloS One* **6**(11): e27332.

Obradovic, B., R. L. Carrier et al. 1999. Gas exchange is essential for bioreactor cultivation of tissue engineered cartilage. *Biotechnol Bioeng* **63**(2): 197–205.

Palmer, R. M., A. G. Ferrige et al. 1987. Nitric oxide release accounts for the biological activity of endothelium-derived relaxing factor. *Nature* **327**(6122): 524–526.

Park, P. K., B. E. Jarrell et al. 1990. Thrombus-free, human endothelial surface in the midregion of a Dacron vascular graft in the splanchnic venous circuit—Observations after nine months of implantation. *J Vasc Surg* **11**(3): 468–475.

Peterkofsky, B. 1982. Bacterial collagenase. *Methods Enzymol* **82**: 453–471.

Pineda, E. T., R. M. Nerem et al. 2013. Differentiation patterns of embryonic stem cells in two- versus three-dimensional culture. *Cells Tissues Organs* **197**(5): 399–410.

Quatela, V. C. and J. Chow. 2008. Synthetic facial implants. *Facial Plast Surg Clin North Am* **16**(1): 1–10.

Rajotte, D., W. Arap et al. 1998. Molecular heterogeneity of the vascular endothelium revealed by *in vivo* phage display. *J Clin Invest* **102**(2): 430–437.

Richardson, M. R., X. Lai et al. 2010. Venous and arterial endothelial proteomics: Mining for markers and mechanisms of endothelial diversity. *Expert Rev Proteomics* **7**(6): 823–831.

Rodbell, M. 1966. The metabolism of isolated fat cells. IV. Regulation of release of protein by lipolytic hormones and insulin. *J Biol Chem* **241**(17): 3909–3917.

Rupnick, M. A., A. Carey et al. 1988. Phenotypic diversity in cultured cerebral microvascular endothelial cells. *In Vitro Cell Dev Biol* **24**(5): 435–444.

Rupnick, M. A., F. A. Hubbard et al. 1989. Endothelialization of vascular prosthetic surfaces after seeding or sodding with human microvascular endothelial cells. *J Vasc Surg* **9**(6): 788–795.

Salzmann, D. L., L. B. Kleinert et al. 1997. The effects of porosity on endothelialization of ePTFE implanted in subcutaneous and adipose tissue. *J Biomed Mater Res* **34**(4): 463–476.

Sato, H., M. Goto et al. 1977. Growth behavior of ascites tumor cells in three-dimensional agar culture. *Tohoku J Exp Med* **122**(2): 155–160.

Sharkawy, A. A., B. Klitzman et al. 1998. Engineering the tissue which encapsulates subcutaneous implants. III. Effective tissue response times. *J Biomed Mater Res* **40**(4): 598–605.

Shepherd, B. R., H. Y. Chen et al. 2004. Rapid perfusion and network remodeling in a microvascular construct after implantation. *Arterioscler Thromb Vasc Biol* **24**(5): 898–904.

Shepherd, B. R., J. B. Hoying et al. 2007. Microvascular transplantation after acute myocardial infarction. *Tissue Eng* **13**(12): 2871–2879.

Sherer, G. K., T. P. Fitzharris et al. 1980. Cultivation of microvascular endothelial cells from human preputial skin. *In Vitro* **16**(8): 675–684.

Shin, D., G. Garcia-Cardena et al. 2001. Expression of EphrinB2 identifies a stable genetic difference between arterial and venous vascular smooth muscle as well as endothelial cells, and marks subsets of microvessels at sites of adult neovascularization. *Dev Biol* **230**(2): 139–150.

Skalak, T. C. and R. J. Price. 1996. The role of mechanical stresses in microvascular remodeling. *Microcirculation* **3**(2): 143–165.

Smith, C. M., J. J. Christian et al. 2007a. Characterizing environmental factors that impact the viability of tissue-engineered constructs fabricated by a direct-write bioassembly tool. *Tissue Eng* **13**(2): 373–383.

Smith, C. M., J. Cole Smith et al. 2007b. Automatic thresholding of three-dimensional microvascular structures from confocal microscopy images. *J Microsc* **225**(Pt 3): 244–257.

Smith, C. M., A. L. Stone et al. 2004. Three-dimensional bioassembly tool for generating viable tissue-engineered constructs. *Tissue Eng* **10**(9–10): 1566–1576.

Sorrell, J. M., M. A. Baber et al. 2009. Influence of adult mesenchymal stem cells on *in vitro* vascular formation. *Tissue Eng Part A* **15**(7): 1751–1761.

Stabenfeldt, S. E., M. Gourley et al. 2012. Engineering fibrin polymers through engagement of alternative polymerization mechanisms. *Biomaterials* **33**(2): 535–544.

Stamer, W. D., R. E. Seftor et al. 1995. Isolation and culture of human trabecular meshwork cells by extracellular matrix digestion. *Curr Eye Res* **14**(7): 611–617.

Stanley, J. C., W. E. Burkel et al. 1982. Enhanced patency of small-diameter, externally supported Dacron iliofemoral grafts seeded with endothelial cells. *Surgery* **92**(6): 994–1005.

Sterpetti, A. V., W. J. Hunter et al. 1992. Healing of high-porosity polytetrafluoroethylene arterial grafts is influenced by the nature of the surrounding tissue. *Surgery* **111**(6): 677–682.

Tasoglu, S. and U. Demirci. 2013. Bioprinting for stem cell research. *Trends Biotechnol* **31**(1): 10–19.

Telemeco, T. A., C. Ayres et al. 2005. Regulation of cellular infiltration into tissue engineering scaffolds composed of submicron diameter fibrils produced by electrospinning. *Acta Biomater* **1**(4): 377–385.

Thai, H. M., E. Juneman et al. 2009. Implantation of a three-dimensional fibroblast matrix improves left ventricular function and blood flow after acute myocardial infarction. *Cell Transplant* **18**(3): 283–295.

Thornton, S. C., S. N. Mueller et al. 1983. Human endothelial cells: Use of heparin in cloning and long-term serial cultivation. *Science* **222**(4624): 623–625.

Thurston, G., P. Baluk et al. 2000. Determinants of endothelial cell phenotype in venules. *Microcirculation* **7**(1): 67–80.

Utsunomiya, H., S. Uchida et al. 2013. Isolation and characterization of human mesenchymal stem cells derived from shoulder tissues involved in rotator cuff tears. *Am J Sports Med* **41**(3): 657–668.

Wagner, R. C., P. Kreiner et al. 1972. Biochemical characterization and cytochemical localization of a catecholamine-sensitive adenylate cyclase in isolated capillary endothelium. *Proc Nat Acad Sci USA* **69**(11): 3175–3179.

Wagner, R. C. and M. A. Matthews. 1975. The isolation and culture of capillary endothelium from epididymal fat. *Microvasc Res* **10**(3): 286–297.

Wang, G. J., K. H. Ho et al. 2007. Microvessel scaffold with circular microchannels by photoresist melting. *Biomed Microdev* **9**(5): 657–663.

Watzka, S. B., M. Steiner et al. 2004. Establishment of vessel-like structures in long-term three-dimensional tissue culture of myocardium: An electron microscopy study. *Tissue Eng* **10**(11–12): 1684–1694.

Wesolowski, S. A., C. C. Fries et al. 1961. Porosity: Primary determinant of ultimate fate of synthetic vascular grafts. *Surgery* **50**: 91–96.

Whang, K., K. E. Healy et al. 1999. Engineering bone regeneration with bioabsorbable scaffolds with novel microarchitecture. *Tissue Eng* **5**(1): 35–51.

Williams, S. K. 1995. Endothelial cell transplantation. *Cell Transplant* **4**(4): 401–410.

Williams, S. K., J. F. Gillis et al. 1980. Isolation and characterization of brain endothelial cells: Morphology and enzyme activity. *J Neurochem* **35**(2): 374–381.

Williams, S. K. and B. E. Jarrell. 1996. Tissue-engineered vascular grafts. *Nat Med* **2**(1): 32–34.

Williams, S. K. and B. E. Jarrell 1996. Tissue-engineered vascular grafts [comment]. *Nat Med* **2**(1): 32–34.

Williams, S. K., B. E. Jarrell et al. 1985. Adult human endothelial cell compatibility with prosthetic graft material. *J Surg Res* **38**(6): 618–629.

Williams, S. K., B. E. Jarrell et al. 1994. Endothelial cell transplantation onto polymeric arteriovenous grafts evaluated using a canine model. *J Invest Surg* **7**(6): 503–517.

Williams, S. K., B. E. Jarrell et al. 1989. Human microvessel endothelial cell isolation and vascular graft sodding in the operating room. *Ann Vascul Surg* **3**(2): 146–152.

Williams, S. K., P. Kosnik et al. 2013. Adipose stromal vascular fraction cells isolated using an automated point of care system improve the patency of ePTFE vascular grafts. *Tissue Eng Part A* **19**(11–12): 1295–1302.

Williams, S. K., S. McKenney et al. 1995. Collagenase lot selection and purification for adipose tissue digestion. *Cell Transplant* **4**(3): 281–289.

Williams, S. K., V. B. Patula et al. 2005. Dual porosity expanded polytetrafluoroethylene for soft-tissue augmentation. *Plast Reconstr Surg* **115**(7): 1995–2006.

Williams, S. K., D. G. Rose et al. 1994. Microvascular endothelial cell sodding of ePTFE vascular grafts: Improved patency and stability of the cellular lining. *J Biomed Mater Res* **1994/02/01**(2): 203–212.

Williams, S. K., T. Schneider et al. 1991. Formation of a functional endothelium on vascular grafts. *J Electron Microsc Tech* **19**(4): 439–451.

Wu, Y., L. Chen et al. 2007. Mesenchymal stem cells enhance wound healing through differentiation and angiogenesis. *Stem Cells* **25**(10): 2648–2659.

Zetter, B. R. and H. N. Antoniades. 1979. Stimulation of human vascular endothelial cell growth by a platelet-derived growth factor and thrombin. *J Supramol Struct* **11**(3): 361–370.

Zhou, Z., Y. Chen et al. 2013. Comparison of mesenchymal stem cells from human bone marrow and adipose tissue for the treatment of spinal cord injury. *Cytotherapy* **15**(4): 434–448.

Zilla, P., M. Deutsch et al. 1994. Clinical *in vitro* endothelialization of femoropopliteal bypass grafts: An actuarial follow-up over three years. *J Vascul Surg* **19**(3): 540–548.

Zimmerlin, L., V. S. Donnenberg et al. 2013. Mesenchymal markers on human adipose stem/progenitor cells. *Cytometry Part A: J Int Soc Anal Cytol* **83**(1): 134–140.

Section II

Biomaterials

5

Vascular Development and Morphogenesis in Biomaterials

Sebastian F. Barreto-Ortiz, Quinton Smith, and Sharon Gerecht

CONTENTS

5.1 Introduction

There are a vast number of materials that have been used to promote vascular development *in vitro*, in growth, and anastomosis *in vivo*, as well as to differentiate stem cells into vascular cells for blood vessel regeneration. Some of these materials have been useful to study several aspects of the vascular morphogenesis process, defining distinct steps of cell spreading, branching, vacuole formation, coalescence, and lumen formation. The developmental information gained from these studies has come hand in hand with the advancement of biomaterials for therapeutic purposes, with demonstrated progress toward biocompatible, biodegradable, bioadhesive, mechanically sound, nonimmunogenic scaffolds that support and enhance vascular network formation.

Generally, these scaffolds can be divided into natural and synthetic, with some overlap when two or more of these materials are used together. Among the natural polymers, the most extensively used for vascularization studies have been collagen, fibrin, fibronectin, Matrigel, chitosan, dextran, and hyaluronic acid (HA), all further discussed in this chapter. On the other hand, the reproducible design and well-defined properties of synthetic polymers, such as their dimensionality, degradation profiles, limited immunogenic reactivity, and other chemical characteristics, have made these an emerging alternative to natural biomaterials. Examples of synthetic biomaterials discussed in this chapter include commonly used polyesters and their copolymers, such as poly(lactic acid) (PLA), poly(glycolic acid) (PGA), and poly(lactic-*co*-glycolic acid) (PLGA), along with polyether compound polyethylene glycol (PEG) and poly(caprolactone) (PCL), a well-studied composite of polylactones.

Both types of materials have been shaped into scaffolds by different techniques, including electrospinning, hydrogel cross-linking, micropatterning, and particle leaching, among others. In this chapter, we will also describe the morphological features desired in engineered vascular networks, the morphological steps that have been identified toward the formation of these networks, the biomaterials that have made these findings possible and the techniques used to make them.

5.2 Human Vasculature: Development and Cell Functionality

5.2.1 Vasculogensis and Angiogenesis

During development, a scaffolding network of endothelial cells (ECs), arising from splanchnic mesodermal differentiation postgastrulation, marks the onset of a well-defined sequence of chemical, spatial, and temporal cues governing *de novo* vascular formation within the yolk sac and growing embryo. This process, known as vasculogenesis, is characterized by a lumen plexus construct of ECs structurally supported by recruited mural pericytes and smooth muscle cells (SMCs), and continually expands via sprouting into the somatic mesoderm during angiogenesis. The angiogenic molecular mediators that contribute to establishing this EC network include vascular endothelial growth factor (VEGF), fibroblast growth factor (FGF), placental growth factor (PGF), insulin-like growth factor (IGF), and transforming growth factor-$\beta2$ (TGF-$\beta2$). In addition, chemokines and proteins such as the stromal-cell-derived factor-1 (SDF-1) and angiopoietin-1 (Ang-1), respectively, contribute to angiogenesis stimulation. These molecular cues coax EC generation of proteolytic enzymes, such as matrix metalloproteinases (MMPs), that act to remodel the extracellular matrix (ECM) to support sprout propagation during angiogenesis (Klagsbrun and Moses 1999). Contrary to its name, the platelet-derived growth factor (PDGF) is also produced and secreted by ECs, acting as a potent mural cell chemoattractant (Dardik et al. 2005).

5.2.2 Phenotype and Functionality

ECs, SMCs, and pericytes have very distinctive morphologies and purposes in blood vessels. ECs are thin and elongated, and, particularly in arteries, align with the direction of flow. They are tightly packed and adhere firmly to each other to form the continuous layer that makes up the luminal surface of blood vessels, interacting directly with the blood stream as a semipermeable membrane that allows both controlled diffusion and active transport of different substances from the blood stream into the surrounding tissue and vice versa. ECs also take part in both the formation and dissolution of blood clots, in inflammatory responses and wound healing, in inducing vasoconstriction and vasodilatation through SMC stimulation, and in producing components of the basal lamina (Standring 2008).

Pericytes are only present at the surface of capillaries and small venules. They have an elongated morphology with long cytoplasmic extensions that wrap around the endothelium (Standring 2008). Pericytes form a discontinuous layer that is eventually replaced by a continuous SMC layer as the capillaries converge into bigger blood vessels. SMCs form most of the media of arteries, though they are also found in venules and veins in smaller quantities. They are also elongated, but they have a different orientation from that of ECs. Specifically, they wrap around arteries, perpendicular to the direction of flow (Standring, 2008). This allows SMCs to provide support against blood pressure as well as to produce the responses of vasoconstriction and vasodilatation, which are a direct result of vascular SMC contraction and relaxation (Standring, 2008). SMCs also secrete ECM components of the media such as elastin and collagen, balancing a complex array of mechanical properties that vary in different vessels, including their elasticity, rigidity, strength, and distensibility.

5.3 *In Vitro* Aims to Investigate Vascular Morphogenesis

5.3.1 Endothelial Cells

Recapitulating the complex vascular microenvironment for investigating this developmental process has emerged as an interdisciplinary challenge, requiring the precise control of chemical and mechanical stimuli to direct cellular phenotype. The stepwise morphogenesis process revealed through *in vitro* studies specifically demonstrates the morphologic sensitivity of mature ECs, endothelial progenitor cells (EPCs), and pluripotent stem cells to both chemical cues and topographical features. The first successful display of a flat polygonal-shaped EC morphology, arranged as a monolayer *in vitro*, arose from their isolation in human umbilical veins, and have similar morphology to endothelial colony-forming cells (ECFCs) (Nachman and Jaffe 2004, Egorova et al. 2011).

Known as HUVECs, these ECs demonstrate a variety of phenotypes when cultured on planar two-dimensional (2D) substrates in the presence of nonsoluble proteins, such as collagen, glycoproteins, proteoglycans, and laminin, which are present in the ECM. Although HUVECs retain an elongated nucleus, their cytoskeletal arrangements transition from a cobblestone appearance when plated on plastic or fibronectin, to rounded when plated on collagen. When grown on top of Matrigel, a matrix material richly composed of an assortment of basal membrane ECM-derived proteins, HUVECs align, cease proliferation, and differentiate into capillary lumen-like structures (Grant et al. 1989, Hughes et al. 2010). On the other hand, when HUVECs or ECFCs are grown in a two-dimensional (3D) environment such as collagen, fibrin, or modified hyaluronic acid (HA) hydrogels, they exhibit a stepwise morphogenesis process. Although a plethora of factors, including integrins, MMPs, and Rho GTPase signaling, directs angiogenesis (Sacharidou et al. 2012), it is well established that single (Davis and Camarillo 1996a, Davis et al. 2002) and EC aggregates (Davis and Camarillo 1996a, Meyer et al. 1997, Hanjaya-Putra et al. 2011, Sacharidou et al. 2012) are able to form vacuoles, which enlarge, coalesce, and fuse with the plasma membrane to form lumen structures *in vitro* (Figure 5.1). Furthermore, in multitubular assembly, these intracellular vacuoles are able to aggregate in concert with EC branching and sprouting events.

5.3.2 Mural Cells

During vascular remodeling and development, SMCs transition from a quiescent contractile state, into a heightened proliferative synthetic state, and demonstrate morphological plasticity that coincides with their multifunctional capabilities (Thakar et al. 2009). Contractile SMCs differ from their synthetic counterparts in that they contain an enlarged endoplasmic reticulum, a plethora of stress bundles and fibers, and contain a filamentous architecture of cytoskeletal proteins (Wanjare et al. 2013). When plated in 2D, SMCs isolated from their native 3D environment transform from a spindle-shaped, elongated conformation to a spread-out phenotype characterized by a branched cytoplasm (Owens 1995). Pericytes, on the other hand, exhibit cytoplasmic extrusions in a branched morphology, extending from a rounded cell body (Kelley et al. 1987).

5.3.3 Receptor–Ligand Guiding of Vascular Morphology

Transmembrane integrin receptors direct how a cell responds to its environment by relaying extracellular chemical composition and mechanical status to intracellular information (Hynes, 2002). Signal transduction is initiated by inside–out integrin activation, permitting the interaction with an array of ECM protein ligand motifs such as LVD, GFOGER, and RGD, which are named according to their sequence of amino acids (Rahmany and Van Dyke 2012). Dimerization of integrin subunits, allowing for outside–in signaling, is an essential property resulting in cell–cell cross talk, migration, tissue assembly, and vascular assembly.

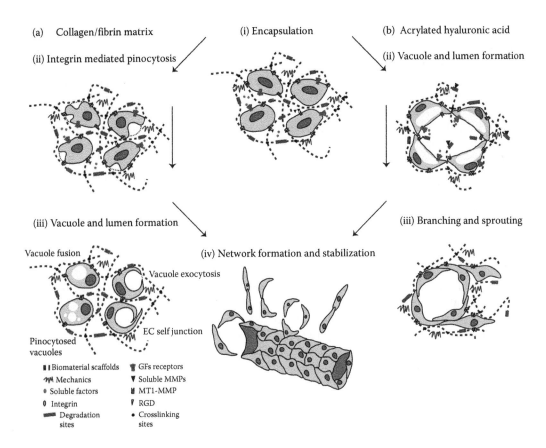

FIGURE 5.1
Vascular morphogenesis in endothelial cells. (a) Schematic of single EC vascular morphogenesis in 3D collagen and fibrin matrices leading to lumen formation and network assembly. (Adapted from Davis, G. E. and Bayless, K. J. 2003. *Microcirculation*, 10, 27–44.) Schematic of EC aggregate vascular morphogenesis in AHA hydrogels. (Adapted from Hanjaya-Putra, D. et al. 2011. *Blood*, 118, 804–815.)

5.4 Biomaterials in Vascular Tissue Engineering

5.4.1 Scaffold Properties

Besides the evident requirements of being nontoxic and biocompatible, there are several chemical and physical properties that can be tuned in biomaterials to enable vascular morphogenesis. This includes whether it is natural or synthetic, its chemistry, dimensionality, porosity, biodegradability, stiffness, topography, and the capability of supporting EC, mural cells, and other supporting cells at the same time for blood vessel maturation.

5.4.1.1 Composition

Nucleic acids, lignin, polysaccharides, proteins, polythioesters, polyisoprenoids, polyanhydrides, and polyoxoesters make up the primary classes of naturally derived polymers. Within these, polysaccharides and proteins comprise the most used biomaterials for angiogenesis studies. These polymers are advantageous in that they stem from renewable resources and recapitulate the *in vivo* niche that supports vascular morphogenesis. This is accomplished by unique properties such as reactive functional groups, pseudoplasticity, and capacity to from hydrogels. In regenerative applications, natural polymers favorably resorb by cell-secreted enzymes, allowing simultaneous cell infiltration and matrix degradation, and therefore supporting the substitution of implanted scaffolds with normal tissue (Hubbell 1995). Although recent advances in biopolymer manufacture in microorganisms has allowed the production of well-characterized functionally amenable materials, large-scale production and isolation of naturally occurring polymers from animal sources leads to inherent batch-to-batch variations with ill-defined properties (Basnett and Roy 2010). Furthermore, although the resemblance of biological substrates as materials is beneficial, they can cause an inflammatory response due to impurities and presence of endotoxins (Nakagawa et al. 2003, Blobel 2010).

5.4.1.2 Chemistry

Both natural and synthetic polymers are composed of monomeric residues, sequentially repeated small molecular units, which are covalently bound in a network of chains. The binding of these units transcribes the nanoscopic properties of the monomers into the bulk polymeric behavior. Polymers are typically long chains with high molecular weight, and can be formed biologically by enzymatically mediated processes or synthetically by several polymerization schemes. Polymers form three classes of conformations, namely linear, branched, and cross-linked (Jenkins 2004, Peacock and Calhoun 2006). Briefly, in addition to polymerization, unsaturated linear or cyclic monomeric residues such as ethylene and ethylene oxide, respectively, form radical reactive intermediates that are continuously restored and propagated to a growing chain until termination ensues. Unlike addition polymerization, natural polymers can be produced in step-growth polymerization in which the repeated condensation of functional groups serves as the reactive intermediate (Hubbell 1995, Jenkins 2004). In synthetic polymer design, material properties such as hydrophilicity, elasticity, stiffness, and cellular adherences can be finely tuned by physical copolymerization of monomeric units or through chemical cross-linkages with functional groups (Lutolf and Hubbell 2005, Unger et al. 2005).

5.4.1.3 Binding Motifs

As alluded to previously, vascular cell behavior is mediated by surface integrin receptors. Some of the naturally occurring peptide sequences recognized by integrins stem from laminin (e.g., IKLLI, LRE, LRGDN, PDGSR, IKVAV, LGTIPG, and YIGSR), collagen (e.g., DGEA and GFOGER), elastin (e.g., VAPG), and fibronectin (e.g., RGD, KQAGDV, REDV, and PHSRN) (Rahmany and Van Dyke 2012). In addition to plasma treatment, covalent tethering of functional groups, and photopolymerization schemes, treatment with interfacial biomaterial peptides (IFBM) has emerged as a novel method for immobilizing these binding domains without negatively affecting the mechanical properties of the polymer (Huang et al. 2010).

5.4.1.4 Degradation Rate

In general, materials should remain viable long enough to allow the cellular interactions desired in each study, but not endure past the point where it interferes with further vascular development and maturation. It is also desirable to control the degradation rate of the biomaterial independently of the stiffness (Naderi et al. 2011). Furthermore, scaffolds loaded with growth factors or drugs will have release rates proportional to their degradation rate, which highlights the importance of carefully tuning this property.

Cross-linking is an effective strategy to control the degradation rate of natural polymers (Naderi et al. 2011). Synthetic materials can be easily tuned to have specific degradation rates by using copolymers of fast and slow degradation rates at different relative concentrations. For example, PLGA is known to require a shorter degradation time at higher glycolic acid content. It is important to note, however, that even though the degradation by-products can be considered biocompatible, a fast degradation rate can lead to local toxicity. Indeed, the glycolic and lactic acid by-products of PLGA degradation have been shown to lower pH in a manner directly proportional to a decrease in cell viability (Sung et al. 2004).

For specific purposes, material-specific enzymes could be used *in vitro* to quickly degrade a natural scaffold if needed. For example, fibrin can be cleaved by plasmin (Kolev et al. 1997), though it is important to have in mind that a fast scaffold degradation can lead to developing vascular network collapse.

5.4.1.5 Physical Characteristics

The biomaterial's mechanical strength, often referred to as stiffness and described by its Young's modulus, is critical to EC morphology, growth, proliferation, and collective tubular morphogenesis (Korff and Augustin 1999, Sieminski et al. 2004, Hanjaya-Putra et al. 2010, Critser and Yoder 2011). It can also by modified by EC interactions over time (Hanjaya-Putra et al. 2011), and by the scaffold's degradation rate. Another critical physical characteristic of scaffolds is their porosity, which should mimic the ECM and simultaneously maintain mechanical integrity, support uniform cell distribution, while permitting vascularization for the effective transport of metabolites and nutrients. Scaffold porosity influences cell–cell interactions, guiding EC morphology and function (Unger et al. 2005, Annabi et al. 2010). Although natural biomaterials such as hydrogels consist of favorable chemical compositions, they have ill-defined porous structures with poor mechanical architectures. Even though freeze-drying, solvent casting/particle leaching, and electrospinning have evolved as tools to control the porosity of hydrogels (Chan and Leong 2008, Annabi et al. 2010), synthetic materials are more permissible to chemical modifications, resulting in highly characterized porosities (Risbud et al. 2001, Chan and Leong 2008).

The topography of the biomaterial also plays an important role in determining cell attachment, shape, orientation, and circularity (Deutsch et al. 2000). It also affects proliferation, migration, and capillary tube formation (Bettinger et al. 2008, Sun et al. 2010). In the human body, ECM provides physical cues that regulate these events. In biomaterials, some of these cues can be simulated by using scaffolds with well-defined topographies, such as aligned nanoridges (Bettinger et al. 2008) and nanofibers (Xu et al. 2004).

5.4.1.6 Dimensionality

In the context of mimicking *in vivo* physiological conditions through designing scaffolds arising from either natural or synthetic sources, it is important to realize the differences in cell behavior in 2D or 3D environments. Changes in cellular metabolism and heterogeneity in cell size results when cells are taken from a 3D environment and cultivated in 2D surfaces (Rubin 1997). This variability might arise because only the apical side of the cells is exposed to paracrine and autocrine signaling factors when cultured in 2D. In 2D,

ECs form chord-like structures in response to growth factor (GF) stimulation, but require additional adhesion junction, integrin, and MMP protein expression to form vasculature in 3D (Blobel 2010).

Proliferation, adhesion, spreading, and migration have been observed when SMCs and ECs are grown on high surface-area-to-volume ratio nanofibrous scaffolds, presenting yet another dimensional feature important in blood vessel engineering (Mo et al. 2004, Ma et al. 2005). Specifically, these nanofibers mirror the mechanical topographical and dimensional properties of native ECM at the nanoscale, providing a construct that promotes vascularization (Vasita and Katti 2006, Ravi and Chaikof 2010).

The ECM, consisting of a myriad of chemical compositions and mechanical topographies, acts to direct cell phenotype and homeostasis in a continually evolving extracellular microenvironment (Frantz et al. 2010). From a dimensional standpoint, the ECM primarily consisting of water can resist tensile and compressive forces through a hydrated network of glycosaminoglycan (GAG) chains interlaced with fibrous proteins. One goal in utilizing natural polymers or designing synthetic biomaterials is to simulate the soft, fibrous, constrained architectures found within the ECM. For example, in the context of vasculogenesis, the differentiation of vascular networks was substantially enhanced when pluripotent stem cells were grown in a 3D scaffold in comparison to unconfined dynamic and Petri-dish culturing environments (Gerecht-Nir et al. 2004).

5.4.2 Biomaterial Fabrication Techniques

5.4.2.1 Electrospinning

Electrospinning has proven to be a successful technique for the fabrication of nanofibers from a wide variety of materials. These ultrathin fibers, which have diameters in the range of 3 nm to 5 µm, can be used for several different applications. In the biomedical industry, it has been shown that nanofibers created with electrospinning can be tailored to resemble the ECM structure, which is composed mainly of collagen and elastin fibrils with nanoscale diameters (Pham et al. 2006). The spun nanofiber meshes can be used to grow cell cultures in a 3D matrix that provides the mechanical support and environmental cues similar to the ones found in the ECM (Li and Xia, 2004, Pham et al. 2006, Sefcik et al. 2008, Lim and Mao, 2009). Both ECs and SMCs have been cultured on electrospun scaffolds and shown to maintain viability, elongate, align, and ultimately form capillary networks with lumens (Zhang et al. 2008).

The typical electrospinning setup consists of a syringe pump with a syringe containing a polymer solution, a high-voltage source, and a collecting plate (Li and Xia 2004). The technique is based on applying an electric field to induce an electric charge on the polymer solution. As the solution is dispensed from the syringe, the electrostatic force opposes the surface tension of the polymer solution and eventually overcomes it to produce a liquid jet stream (Li and Xia 2004, Pham et al. 2006). As the jet travels, the electric forces cause the stream to spin around. At the same time, the solvent evaporates from the solution and the polymer fibers fall on the collecting plate forming an ultrathin fiber mesh. The technique can be easily modified to produce aligned or randomly deposited fibers.

Possible biocompatible polymers that have been successfully spun are PLA, PCL, PEO, PLLA, PLGA, collagen, silk, fibrinogen, chitosan, dextran, DNA, gelatin, HA, and oxidized cellulose (Li and Xia 2004, Pham et al. 2006). The technique has been modified to introduce two different materials at the same time (coelectrospinning), or with one material inside another with coaxial electrospinning, increasing its versatility and potential.

Another important advantage of these scaffolds compared to hydrogels is the ability to introduce different cell lines, growth factors, cytokines, and enzymes at different time points. This allows for the study of sequential introduction of different agents in the angiogenesis process. However, despite its many advantages, a common challenge in electrospun scaffolds is achieving a uniform cell distribution within the scaffold.

5.4.2.2 Hydrogels

Hydrogels comprise a vast family of 3D cross-linked polymeric networks with highly tunable biophysical and biochemical properties (Yee et al. 2011, Sun et al. 2011a). The high water content of these scaffolds supports cell viability while providing a stable 3D support for cellular interactions. Materials that have been used in hydrogel form for vascular studies include collagen (Seliktar et al. 2000), HA (Hanjaya-Putra et al. 2011), dextran (Sun et al. 2011b), and fibrin (Shaikh et al. 2008), among many others. The physical properties of the gels, such as stiffness, can be easily tuned by varying the level of cross-linking of their polymers. The scaffolds can also be biochemically modified by introducing different growth factors, enzymes, cytokines, or binding motifs (as detailed above). The broad range of properties that can be given to these scaffolds has made them a popular choice for angiogenesis studies.

Typically, hydrogels are made by mixing the main hydrogel material with a cell suspension, immediately followed by the addition of a gelling or cross-linking agent. The gelling process should be fast to avoid cells from sinking to the bottom of the scaffold.

Despite their great versatility, hydrogels possess some inherent limitations. Introducing different cell lines at different time points is challenging; once the gel forms, other cells would have to penetrate into the scaffold to interact with structures within the gel. In vascularization studies, this makes it difficult studying mural cell interactions with developing EC tubular structures, as the structures should form before the introduction of SMCs. Additionally, the introduction of growth factors after gel formation is limited by the diffusion rate of the growth factors into the gel, which limits the control of the biochemical cues presented to the cells. Furthermore, encapsulated growth factors diffuse out of the scaffolds unless they are bound to the polymer chains, further limiting the control of growth factor action within the gels. This is of critical importance since it has been shown that angiogenesis is dependent on a sequential presentation of different biochemical cues (Stratman et al. 2011).

5.4.2.3 Micropatterning

ECs have been shown to respond not only to chemical cues but also to mechanical stimulation and morphological confinement. ECM is an important source of these signals, and many studies have used collagen, fibronectin, and HA to fabricate micropatterned surfaces to elucidate the role of EC morphology in migration, ECM deposition, alignment, proliferation, lumen formation, and tube-like structure development (Li et al. 2001, Dickinson et al. 2010).

The process of micropatterning typically uses photolithography to make silicon masters by spin coating a silicon wafer with a photoresist and then exposing the wafer to UV through a carefully designed photomask. These masters are then used to make PDMS stamps that can be inked using solutions containing the ECM protein of interest. Finally, the stamps are pressed on the culture surface to create the micropatterns. This technique has been used with different materials for vascular tissue studies. For example,

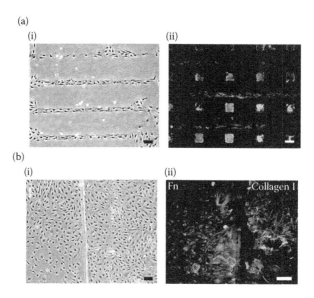

FIGURE 5.2
Preferential ECFC attachment on micropatterned ECM substrates. Cell culture on 2D micropatterned HA/Fn surfaces. (a) ECFCs on HA/Fn surfaces. (i) Light microscope and (ii) immunofluorescence (phalloidin—stripes/ HA—squares) images. (b) ECFCs cultured on collagen 1 and Fn-coated surfaces. (i) Light microscope and (ii) immunofluorescence (collagen—right, fibronectin—left) images. Scale bars, 100 μm. (From Dickinson et al. 2012a, *Lab on a Chip*, 12, 4244–4248.)

PCL micropatterns have been fabricated with micropatterning combined with particulate leaching of PLGA particles to make patterned porous scaffolds to study SMC alignment (Sarkar et al. 2006). SMC alignment and EC interactions have also been studied on HA micropatterned surfaces (Li et al. 2013), and gelatin has been used as a medium for immobilized, micropatterned, VEGF-directed growth of ECs (Ito et al. 2005).

A major limitation of this system is the introduction of a full 3D environment for the ECs and mural cells to interact. To overcome this, recent modifications have been developed to introduce a 3D milieu using micropatterned surfaces and channels that enable the study of matrix interactions of different cell types in 3D using conventional lithography techniques in an innovative way (Raghavan et al. 2010, Dickinson et al. 2012a,b) (Figure 5.2).

Collagen gels micropatterned using PDMS channels have been shown to guide EC tubulogenesis, allowing the creation of geometrically defined networks and permitting the study of the morphological steps leading to this process (Raghavan et al. 2010). Developing this study, the group proceeded to polymerize fibrin on top of the structures after collagen gel formation, after which the PDMS could be peeled away leaving organized cell-seeded cords on top of the fibrin substrate. A second fibrin gel polymerization on top of this structure resulted in fully embedded parallel cords seeded with vascular cells, which were later shown to become perfused with host blood postimplantation *in vivo*. Furthermore, these organized networks enhanced tissue integration and function *in vivo* compared to implants with randomly formed vascular networks, showcasing the importance of macroscopic multicellular organization in vascular tissue constructs (Baranski et al. 2013).

Other novel micropatterning approaches in 3D include using carbohydrate glass printing and thermal extrusion to create a cytocompatible sacrificial template forming a network that can then be embedded in a cellularized hydrogel. After embedding, the template can be dissolved, leaving behind an interconnected network of microchannels. This network can subsequently be perfused, supporting pulsatile flow, and allowing the seeding of ECs inside the channels to create the endothelium layer. Furthermore, the perfused networks were shown to support cell metabolic function in the core of thick densely populated tissue constructs, a development that has great prospects toward the creation of large tissues without the problem of necrotic cores (Miller et al. 2012).

5.4.2.4 Particle Leaching

This technique is very simple and commonly utilized for the fabrication of sponges of synthetic materials. It consists of adding particles (usually salts) to a polymer solution before it is placed in a mold to be cast into a scaffold. Once the solvent in the solution evaporates, the polymer left behind acquires the shape of the mold. A rinsing solvent (usually water) is then used to extract the particles embedded in the scaffold, creating a porous 3D scaffold that can be used for vascularization studies (Levenberg et al. 2002).

Materials commonly processed with this technique include PLA, PLGA, PGA, and PCL (Levenberg et al. 2002, Singh et al. 2011). PLA/PLGA porous scaffolds were created through salt leaching and used for *in vivo* implantation of ECs differentiated from human embryonic stem cells, demonstrating their ability to form microvasculature *in vivo* and suggesting anastomosis with the host vasculature (Levenberg et al. 2002). Sucrose is another compound often used for leaching. PCL scaffolds prepared through sucrose leaching have been utilized to promote vascularization in EPC-seeded implants activated with surface-immobilized heparin and VEGF. This study demonstrated the scaffold's ability to support EPC-mediated, accelerated blood vessel formation and anastomosis (Singh et al. 2011).

5.4.3 Natural Polymers

Common natural biomaterials can be roughly divided into proteins and polysaccharides (Sun et al. 2010). The most common proteins used in vascular development studies include collagen, fibrin, fibronectin, and decellularized ECM, which will be discussed in this section. Other proteins include silk fibroin, which has been electrospun to create scaffolds that supported EC capillary tube formation and SMC alignment, elongation, and ECM production (Zhang et al. 2008). Although there is a plethora of polysaccharides that serve as biomaterials for studying vascularization both *in vivo* and *in vitro*, the most common ones (chitosan, HA, and dextran) will be reviewed.

5.4.4 Proteins

5.4.4.1 Collagen

Collagen, the most abundant protein in mammals, has been used extensively to study angiogenesis due to its pronounced biocompatibility, bioreabsorbability, and stability arising from its self-aggregation and cross-linking, a property that can be used to tune its biodegradability (Lee et al. 2001). Collagen is naturally degraded by collagenase, among other enzymes. Even though there are at least 19 types of collagen (Lee et al. 2001), most studies use rat tail type I collagen to create hydrogels (Davis and Camarillo 1996b, Bayless and Davis 2003, Myers et al. 2011, Stratman et al. 2011). Other studies have used collagen to coat synthetic polymer electrospun scaffolds (He et al. 2006) or have used different collagen types to make electrospun meshes (Matthews et al. 2002). Though chemical modifications are not as common in collagen scaffolds, attempts have been made to covalently incorporate VEGF, achieving a modest increase in angiogenic potential (Koch et al. 2006).

For vascular morphogenesis studies, collagen hydrogels have been useful to elucidate the steps and mechanisms of vascular network formation. Researchers have used collagen hydrogels to identify and study key steps of the vascular morphogenesis process, showcasing the importance of vacuole formation and coalescence toward vascular lumen formation, a Cdc42- and Rac1-dependent event (Bayless and Davis 2002, Davis and Bayless 2003). More recently, the same system was used to distinguish the role of VEGF and FGF as primers for vascular tube morphogenesis regulated by hematopoietic stem cell cytokines IL-3, stem cell factor, and stromal derived factor-1α (Barreto-Ortiz and Gerecht 2011, Stratman et al. 2011).

5.4.4.2 Fibrin

Fibrin, one of the main components in hemostatic plugs or clots, has been shown to induce angiogenesis and is therefore an excellent material for vascular development studies (Bayless et al. 2000, van Hinsbergh et al. 2001, Bayless and Davis 2003, Rowe et al. 2007, Shaikh et al. 2008). It is biodegradable, biocompatible, bioadhesive, and can even be harvested from a patient's own blood to avoid immune or foreign body reactions (Aper et al. 2007). Typically, fibrin scaffolds are prepared by catalyzing the polymerization of fibrinogen into fibrin with thrombin (Grassl et al. 2002, Aper et al. 2007, Rowe et al. 2007, Lesman et al. 2011, Shaikh et al. 2008). Modifications of fibrin matrices have been performed to control the release of growth factors (Shireman and Greisler 2000, Schmoekel et al. 2005) and genes (Trentin et al. 2006) in these scaffolds. Like most other natural materials, however, it has a limited mechanical strength; it generally is too weak to resist dynamic physiological conditions (Kannan et al. 2005, Shaikh et al. 2008).

Still, fibrin has been used as an alternative to collagen I for the preparation of media equivalents with SMCs, showing increased collagen production by SMCs compared to collagen scaffolds (Grassl et al. 2002). Furthermore, this difference was accentuated after treatment with TGF-β and insulin. This resulted in increased mechanical properties (stiffness and ultimate tensile strength) compared to similarly treated collagen scaffolds, achieving ultimate tensile strengths similar to those of rat abdominal aorta (Grassl et al. 2003).

Fibrin hydrogels were also used to confirm the Cdc42 and Rac1 dependence of vacuole formation and coalescence toward lumen formation in 3D (Bayless and Davis 2002, Davis and Bayless 2003). Fibrin gels alone or in combination with PLA and PLGA sponges were used to create 3D capillary beds *in vitro* and enhance neovascularization in mice implants (Lesman et al. 2011). It was also found that fibrin concentration regulated the degree of vascularization, and that further blood vessel maturity was aided by the presence of PLA/PLGA sponges to provide mechanical support. The stiffness and ultimate tensile strength of fibrin gels was also shown to increase with decreasing thrombin concentration, and this change was proven to also affect smooth muscle cell morphology (Rowe et al. 2007).

5.4.4.3 Fibronectin

This glycoprotein has been shown to be necessary for blood vessel development (George et al. 1993) and to bind and increase the biological activity of VEGF, enhancing EC migration and suggesting a cooperative action of VEGF/fibronectin complexes (Wijelath et al. 2002). It has been used mostly in 2D studies as opposed to 3D hydrogels: fibronectin was used in early studies of EC morphogenesis as a coating for 2D surfaces (Ingber and Folkman 1989), and as previously mentioned, it has been extensively used to create 2D micropatterned surfaces. We previously used this technique to guide EC elongation in 2D, and to study the effects of different growth factors *in vitro* (Dickinson et al. 2010). Furthermore, we introduced a 3D environment by placing a fibrin gel on top of these patterned cells, which resulted in caveolae, lumen, and Weibel–Palade body formation, processes part of vascular morphogenesis and development (Dickinson et al. 2010).

5.4.4.4 Matrigel and Decellularized ECM

Matrigel, a heterogeneous ECM protein mixture produced by mouse sarcoma cells, remains one of the most widely used materials to assay the angiogenic potential of cells and growth factors (Levenberg et al. 2002, Vo et al. 2010). It resembles the native basement membrane, which surrounds the ECs in the lumen of blood vessels. ECs have been shown to create cobweb capillary structures when cultured on Matrigel (Levenberg et al. 2002), an effect that is not seen on regular cell culture surfaces (Arnaoutova et al. 2009). This assay encompasses several steps in the angiogenesis process: adhesion, spreading, migration, protease activity, alignment, elongation, and capillary tube formation (Arnaoutova et al. 2009). The assay protocol is widely used because it is very standardized, practical, and gives results in a short time. However, despite its advantages, Matrigel is not always well characterized and has a large degree of variability resulting from its production process, which limits its clinical application.

We previously utilized Matrigel to demonstrate the vascular potential of smooth muscle-like cells differentiated from human embryonic stem cells and their interactions with EPCs (Vo et al. 2010). Other studies have developed their own naturally derived ECM by using human fibroblasts, which were removed afterwards by detergent extraction (Soucy and Romer 2009), similarly to how different tissues and organs have been decellularized to remove xenogeneic and allogeneic antigens that typically induce undesired immunological responses (Gilbert et al. 2006). The ECM was found to be composed of different types of collagen, fibronectin, tenascin-C, versican, and decorin, and it was shown to support EC adhesion, elongation, and tubular network formation with distinct lumens (Soucy and Romer 2009).

5.4.5 Polysaccharides

Repeating carbohydrate monomeric units can covalently link together or with other chemical moieties through oxygen-tethering O-glycosidic bonds to form long polysaccharide molecules. Polysaccharides can be categorized within two domains according to their monosaccharide building blocks, whose naming scheme depends on the number of carbons they contain. Homopolysaccharides contain a single monosaccharide type, while heteropolysacchardes contain different subunits. In their natural conformations, polysaccharides have several functioning roles, including energy storage (e.g., starch and glycogen), adhesion (e.g., pullulan, xanthan gum, and dextran), and structural integrity (e.g., cellulose, arabinoxylans, pectins, and chitin). The linkages amidst polysaccharide ketoses or aldoses give rise to linear and branched soluble polymers with unique gelling potentials, interfacial properties, flow behaviors, and biological activities that are utilized in biomaterial development for tissue engineering applications (Rinaudo 2008).

5.4.5.1 Chitin/Chitosan

In its natural state, chitin, chemically described as nitrogen-containing polysaccharide *N*-acetylglucosamine, acts as an exoskeleton in insects and arthropods, or as a cell wall in fungi (Di Martino et al. 2005). Chitosan arises from the deacetylation of chitin, has a wide array of functionalities ranging from drug delivery systems to healing, and has been shown to be biocompatible, biodegradable, and bioadhesive (Ono et al. 2000). The linear monomeric glucosamine and *N*-acetylglucosamine units interconnected by β (1–4) glycosidic bonds of chitosan leave structural topographies similar to GAGs, which permit interactions with adhesion proteins, growth factors, and proteins (Deng et al. 2010). In addition to a primary and secondary hydroxyl functional groups, chitosan contains an amino moiety, making the polysaccharide tunable for many applications through chemical modification, and is suitable for tissue engineering. Chitosan polyvinyl alcohol (PVA) hydrogel scaffolds, fabricated through cross-linking and freeze–dry cycling techniques, were demonstrated to enhance proliferation and attachment of both bovine aortic SMCs and ECs (Vrana et al. 2008). The immune competence of ECs can be deleterious in that cytokine release can activate leukocytes, leading to islet graft rejection and malfunction. In addition, vasculature is needed for adequate oxygen and nutrient delivery in immunoisolation devices. Chitosan polyvinyl pryrrolidone (PVP) hydrogel blends suppress EC activation but sustain their functionality, permitting angiogenesis adjacent to the scaffold (Risbud et al. 2001, Chung et al. 2002). Adhesive peptides promoting EC attachment and proliferation can be photochemically grafted on chitosan scaffolds, further demonstrating the amenability of chitin and chitosan biopolymers for vascular engineering applications (Khor and Lim 2003).

5.4.5.2 Hyaluronic Acid

HAS1, HAS2, and HAS3 make up the three classes of integral membrane hyaluronan synthase proteins, which catalyze the formation of repeated *N*-acetyl-D-glucosamine and D-glucuronic acid disaccharide subunits constituting HA. As such, HA is in a special class of linear, nonsulfated GAGs that are not synthesized in the Golgi apparatus. In addition to promoting differentiation, wound healing, and cell motility through its recognition by cell surface protein CD44 and CD168, HA is a highly viscous soluble molecule that is an integral component in synovial movements, serving to structurally organize cartilage ECM compositions *in vivo* (Bulpitt and Aeschlimann 1999, Toole 2004). Although HA has several limitations in its natural state, namely its solubility in water and rapid degradation in the presence of hyaluronidase, several modification techniques have circumvented these hindrances (Campoccia et al. 1998).

We have extensively investigated the role of modified HA using dithiol and polyethylene glycol diacraylate (PEGDA) cross-linkers as a biomaterial for engineering vasculature structures both on 2D substrates and as a 3D milieu. We reinforced the importance of mechanical stimuli on endothelial tubulogenesis by growing ECFCs on HA cross-linked with varying concentrations of PEGDA to yield hard, firm, and soft substrates. The average capillary lumen tube area, length, and thickness, was significantly hindered when ECFCs were grown on stiffer HA gels in the presence of VEGF in comparison to softer gels, further elucidating the cointeraction between MMP expression and matrix stiffness (Hanjaya-Putra et al. 2010). The ability to from capillary-like structures was also governed by physical cues when ECFCs were encapsulated in HA–PEGDA scaffolds. In particular, HA gels exhibiting lower viscoelasticities and larger microporous structures resulted in a significant increase in tube formation in comparison to their stiff, small pore-size counterparts (Khetan et al. 2009, Yee et al. 2011).

Another chemical modification of HA results in multiple acrylated macromer groups within the HA construct. Angiogenic soluble factors, RGD-binding domains, and MMP-cleavable dithiol cross-linkers can be introduced to the acrylated HA (AHA) through the

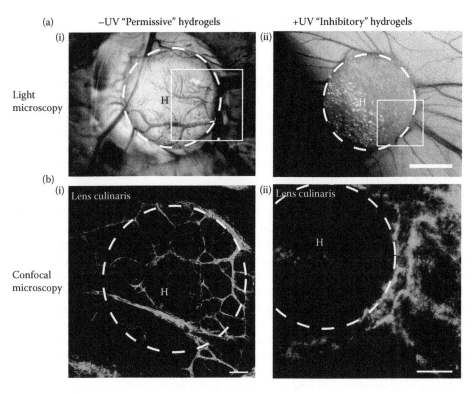

FIGURE 5.3
Acrylated hyaluronic acid hydrogels. (a) Light microscopy imaging and (b) confocal analysis of the boxed regions in (a) shows CAM vessels penetrating into the −UV but not into the +UV hydrogels. CAM vessels are stained with fluorescein-conjugated *Lens culinaris* lectin. Scale bars, A is 20 mm, B is 100 μm; H, hydrogels; dotted white lines indicate the boundaries of the hydrogels.

addition polymerization scheme in the presence of a photoinitiator, a process in which only a subset of acrylate groups are spent in the reaction process. Finally, regions that are insensitive to MMP degradation arise when the remaining acrylate groups are exposed to UV light, allowing for spatially confined regions that are either permissible or inhibitory to cellular hydrogel remodeling (Figure 5.3). Encapsulating ECFCs in AHA with a cocktail of growth factors known to promote blood vessel formation (VEGF, bFGF, Ang-1), soluble factors regulating MMP expression (SDF-1), tumor necrosis factor-1 (TNF-α), and varying concentrations of RGD with MMP-sensitive cross-linkers unveiled the dose-dependent role of RGD in vacuole lumen formation in the synthetic matrix (Khetan et al. 2009, Hanjaya-Putra et al. 2011). Finally, through micropatterning fabrication technology, we have demonstrated the behavior of ECFCs cocultured with breast cancer cells, in a spatially defined AHA and fibrin hydrogel microenvironment mirroring the angiogenic tumor niche (Dickinson et al. 2012a). These studies show the versatility in utilizing AHA that will allow further insights into the factors mediating vascular morphogenesis.

5.4.5.3 Dextran

In the presence of sucrose, *Leuconostoc mesenteroides* synthesize and secrete α-1,6-linked D-glucopyranose, consisting of a small composition of α-1,2-, α-1,3-, or α-1,4-side-chain residues (Ifkovits and Burdick 2007). This chemically amenable polysaccharide known as dextran can form biodegradable hydrogels of varying swelling properties due to the availability of three free hydroxyl groups in its monomeric components. One chemical modification, leading to bifunctionalization of dextran with methacrylate and aldehyde groups, was permissible to 2D culture of ECs and formed SMC cell-encapsulating interpenetrating polymer networks (Liu and Chan-Park 2009). Functionalized dextran acrylate hydrogels, integrated with Acr–PEG–RGD-binding motifs by copolymerization, loaded with microparticles containing VEGF, were shown to induce the differentiation of encapsulated human embryonic stem cells. In fact, the degree of differentiation as determined by VEGF receptor expression had a 20-fold increase in comparison to embryoid body spontaneous differentiation (Ferreira et al. 2007).

We have utilized PEGDA and allyl isocyanate-ethylamine (AE) to form Dex-AE/PEGDA hydrogels that exhibit favorable biocompatibilities allowing host tissue integration and promoting scaffold-mediated vascularization for biomaterial-assisted wound healing (Sun et al. 2009, 2011a,b). The cross-linking density of this dextran system, a property dependent on the number of carbon-containing double bonds within the construct (i.e., Dex-AE/Dex-PEGDA concentration), influences physical properties that are essential in the vascularization of scaffolds in regenerative medicine (e.g., mechanical strength and capacity to release GFs via diffusion). Both the number of cross-linking sites and cumulative release of GFs decreased with increasing AE concentration, resulting in hydrogels with enhanced water retention capabilities but limited GF diffusion over time. Five weeks subsequent to implanting Dex-AE/PEGDA (Dex 80/20) subcutaneously in rats, 40% and 70% of tissue ingrowth was observed in gels without and containing VEGF, respectively. Vascularization in Dex 80/20 hydrogel scaffolds was further enhanced when a combination of SDF-1, IGF, and Ang-1 were coencapsulated with VEGF (Sun et al. 2011a). These findings demonstrate the importance of reducing cross-linking for fabricating soft, porous hydrogels in dextran systems with VEGF to permit rapid vascularization. As the cells infiltrate soft hydrogels, they are able to easily degrade the porous matrix, and encounter growth factors over a prolonged period of time.

5.4.6 Synthetic Biomaterials

Poly(L-lactide), also referred to as PLLA or PLA, is the most common form of polylactic acid used to make scaffolds for vascular studies. Poly(lactide-*co*-glycolide) (PLGA) is the copolymer of this polyester with polyglycolic acid (PGA), and together, PLA and PLGA constitute the most extensively used synthetic polymers for tissue engineering (Naderi et al. 2011), microvascular cell cultures (Kannan et al. 2005), and drug delivery (Patel and Mikos 2004). Other common synthetic biopolymers include polycaprolactone (PCL), polyethylene glycol (PEG), and polyester amide (PEA), though various synthetic polymers have been used in creative ways. For example, a method of fabricating perfused microvessels by seeding SMCs on 80-μm-diameter nylon strands has been developed (Neumann et al. 2003). SMCs were shown to multiply forming concentric layers around the nylon strand, creating a layer thickness of 70–100 μm after 21 days. Nylon strands were then physically removed and the resulting tubular structure perfused for 7 days, with no leakage and showcasing significant changes in vessel diameter in response to pressure, which suggests effective vaso-responsiveness.

5.4.6.1 PEG

PEG hydrogels are hydrophilic, biocompatible, and have an intrinsic resistance to protein adsorption. Alone, PEG is incapable of supporting cell growth, but it has been chemically modified and functionalized to make it an attractive alternative for vascular tissue engineering (Mann et al. 2001, DeLong et al. 2005, Leslie-Barbick et al. 2009, 2010, Miller et al. 2010). RGDS-functionalized PEG hydrogels have been used to demonstrate TGF-β-induced ECM deposition by SMCs (Mann et al. 2001), in line with results on fibrin scaffolds stated above. Similarly, gradients of immobilized bFGF on these functionalized PEG scaffolds were shown to induce SMC alignment and migration (DeLong et al. 2005). Additionally, patterned regions of covalently bound RGDS and VEGF on the surface of PEG hydrogels resulted in EC tubule formation with observable lumens, further demonstrating the potential of this material for organized vascular tube formation (Leslie-Barbick et al. 2010).

5.4.6.2 *PLA, PGA, and PLGA*

PLA, PGA, and PLGA are biocompatible and biodegradable polyesters that have been used for tissue engineering for decades (Patel and Mikos 2004). As mentioned previously, they degrade by hydrolysis at different rates, which enables the tuning of a scaffold's degradation rate by changing the relative amounts of each copolymer. Their by-products are also biocompatible, though a fast degradation rate has been shown to lower the local pH to cytotoxic levels (Sung et al. 2004). They are either typically used as porous sponges fabricated by particulate leaching (Levenberg et al. 2002, Patel and Mikos 2004, Rowe et al. 2007) or electrospun into nonwoven meshes. These three polymers have been well studied for their potential in the delivery of several proangiogenic factors: VEGF (Murphy et al. 2000, Richardson et al. 2001, Linn et al. 2003, Elcin and Elcin 2006), bFGF (Lee et al. 2002) and PDGF-BB (Richardson et al. 2001).

PLA/PLGA sponges coated with either Matrigel or fibronectin aid in the differentiation of pluripotent stem cells to ECs, and were able to form 3D tubular structures *in vitro* (Levenberg et al. 2003). These scaffolds were also used to study the angiogenic potential of ESC-derived ECs in mouse implants, demonstrating their ability to form microvessels that anastomosed with the host's vasculature (Levenberg et al. 2002, 2003). Furthermore, these scaffolds aid in the construction of 3D vascularized and contractile human cardiac tissue consisting of cardiomyocytes, ECs, and embryonic fibroblasts, creating a model for cardiac tissue developmental studies and regenerative medicine (Caspi et al. 2007).

5.4.6.3 *Polycaprolactone*

Polycaprolactone (PCL) is a synthetic biodegradable polyester permitting the hydrolysis of its ester linkages, and is formed through a ring-opening ε-caprolactone polymerization scheme (Labet and Thielemans 2009). The crystallinity, tensile strength, stiffness, surface morphology, and dimensionality can all be finely tuned by a combination of techniques. By using wet spinning via minimal gravitation shear and cold drawing fabrication techniques, PCL fibers with varying physical dimensions and tensile properties were appraised based upon their ability to sustain HUVEC proliferation and inflammatory responses. Consequently, differences in HUVEC morphology and proliferation rates arose as a function of varying substrate properties, suggesting successful endothelialization of vascular grafts is dependent on fiber dimensionality (Williamson et al. 2006).

In comparison to cells grown on flat tissue culture plastic, HUVECs on a stiff cold drawn PCL fiber with 67-μm diameter had a dramatic increase in proliferation with notable spreading, orientation, and focal adhesion density replicating the contours of the fibers, even 9 days after initial seeding. On the contrary, fibers with similar dimensionality but made from thermoplastic polymer Dacron resulted in rounded HUVEC morphology, with poor proliferation. Furthermore, these PCL fibers were able to support normal immune response with similar expression levels of inflammatory intercellular adhesion molecule-1 (I-CAM1) when compared to cells grown on plastic tissue culture dishes (Rothlein et al. 1986). When compromised with endotoxins, the cells were able to express I-CAM1, suggesting that the architecture of the PCL fibers do not inhibit inflammatory response, and can be a feasible synthetic biomaterial for developing scaffolds in vascular tissue engineering.

5.5 Conclusion

Although there is a vast amount of literature on the molecular modulation governing vascular maturation, the specialization of vascular networks in various tissues and organs still remains elusive (Jain, 2003). As such, tissue engineering has emerged as a highly interdisciplinary field that seeks to address this issue by recapitulating the complex physiological niche *in vitro* to eventually generate functional vascular constructs. Through exploiting natural materials and developing synthetic polymers, research have begun to develop biomaterials permitting the investigation of the stepwise vascular morphogenesis process. Future directions in these efforts include utilizing recombinant protein technology toward the generation of a genetic template that can be used to produce a well-defined biomaterial monomeric composition, molecular weight, dimensionality, and stereochemistry; addressing the batch-to-batch variability in naturally derived biomaterials (Romano et al., 2011). Additionally, researchers are now integrating microfluidic technology with biomaterials to create dynamic cell-culturing microenvironments permitting the investigation of the milieu of factors influencing blood vessel development, paving the way for complex *in vitro* models (Domachuk et al. 2010). In summary, we demonstrate the use and fabrication of a variety of synthetic and natural materials for studying vasculogenesis, highlighting their effects on driving morphogenic processes.

References

Annabi, N., Nichol, J. W., Zhong, X., Ji, C., Koshy, S., Khademhosseini, A., and Dehghani, F. 2010. Controlling the porosity and microarchitecture of hydrogels for tissue engineering. *Tissue Engineering Part B: Reviews*, 16, 371–383.

Aper, T., Schmidt, A., Duchrow, M., and Bruch, H. P. 2007. Autologous blood vessels engineered from peripheral blood sample. *European Journal of Vascular and Endovascular Surgery*, 33, 33–39.

Arnaoutova, I., George, J., Kleinman, H., and Benton, G. 2009. The endothelial cell tube formation assay on basement membrane turns 20: State of the science and the art. *Angiogenesis*, 12, 267–274.

Baranski, J. D., Chaturvedi, R. R., Stevens, K. R., Eyckmans, J., Carvalho, B., Solorzano, R. D., Yang, M. T., Miller, J. S., Bhatia, S. N., and Chen, C. S. 2013. Geometric control of vascular networks to enhance engineered tissue integration and function. *Proceedings of the National Academy of Sciences*, 110, 7586–7591.

Barreto-Ortiz, S. F. and Gerecht, S. 2011. Dealing with immature blood vessels? Give them some FGF9. *Regenerative Medicine*, 6, 551–554.

Basnett, P. and Roy, I. 2010. Microbial production of biodegradable polymers and their role in cardiac stent development. *Current Research, Technology and Education Topics in Applied Microbiology and microbial Biotechnology*, 1405–1415.

Bayless, K. J. and Davis, G. E. 2002. The Cdc42 and Rac1 GTPases are required for capillary lumen formation in three-dimensional extracellular matrices. *Journal of Cell Science*, 115, 1123–1136.

Bayless, K. J. and Davis, G. E. 2003. Sphingosine-1-phosphate markedly induces matrix metalloproteinase and integrin-dependent human endothelial cell invasion and lumen formation in three-dimensional collagen and fibrin matrices. *Biochemical and Biophysical Research Communications*, 312, 903–913.

Bayless, K. J., Salazar, R., and Davis, G. E. 2000. RGD-dependent vacuolation and lumen formation observed during endothelial cell morphogenesis in three-dimensional fibrin matrices involves the $\alpha_v\beta_3$ and $\alpha_5\beta_1$ integrins. *The American Journal of Pathology*, 156, 1673–1683.

Bettinger, C. J., Zhang, Z., Gerecht, S., Borenstein, J. T., and Langer, R. 2008. Enhancement of *in vitro* capillary tube formation by substrate nanotopography. *Advanced Materials*, 20, 99–103.

Blobel, C. P. 2010. 3D trumps 2D when studying endothelial cells. *Blood*, 115, 5128–5130.

Bulpitt, P. and Aeschlimann, D. 1999. New strategy for chemical modification of hyaluronic acid: Preparation of functionalized derivatives and their use in the formation of novel biocompatible hydrogels. *Journal of Biomedical Materials Research*. John Wiley & Sons, Inc. 47, 152–169.

Campoccia, D., Doherty, P., Radice, M., Brun, P., Abatangelo, G., and Williams, D. F. 1998. Semisynthetic resorbable materials from hyaluronan esterification. *Biomaterials*, 19, 2101–2127.

Caspi, O., Lesman, A., Basevitch, Y., Gepstein, A., Arbel, G., Habib, I. H. M., Gepstein, L., and Levenberg, S. 2007. Tissue engineering of vascularized cardiac muscle from human embryonic stem cells. *Circulation Research*, 100, 263–272.

Chan, B. P. and Leong, K. W. 2008. Scaffolding in tissue engineering: General approaches and tissue-specific considerations. *European Spine Journal*, 17, 467–479.

Chung, T.-W., Lu, Y.-F., Wang, S.-S., Lin, Y.-S., and Chu, S.-H. 2002. Growth of human endothelial cells on photochemically grafted Gly-Arg-Gly-Asp (GRGD) chitosans. *Biomaterials*, 23, 4803–4809.

Critser, P. and Yoder, M. 2011. Biophysical properties of scaffolds modulate human blood vessel formation from circulating endothelial colony-forming cells. In: Gerecht, S. (ed.) *Biophysical Regulation of Vascular Differentiation and Assembly*. Springer, New York.

Dardik, A., Yamashita, A., Aziz, F., Asada, H., and Sumpio, B. E. 2005. Shear stress-stimulated endothelial cells induce smooth muscle cell chemotaxis via platelet-derived growth factor-BB and interleukin-1α. *Journal of Vascular Surgery*, 41, 321–331.

Davis, G. E. and Bayless, K. J. 2003. An integrin and Rho GTPase-dependent pinocytic vacuole mechanism controls capillary lumen formation in collagen and fibrin matrices. *Microcirculation*, 10, 27–44.

Davis, G. E., Bayless, K. J., and Mavila, A. 2002. Molecular basis of endothelial cell morphogenesis in three-dimensional extracellular matrices. *Anatomical Record*, 268, 252–275.

Davis, G. E. and Camarillo, C. W. 1996. An alpha 2 beta 1 integrin-dependent pinocytic mechanism involving intracellular vacuole formation and coalescence regulates capillary lumen and tube formation in three-dimensional collagen matrix. *Experimental Cell Research*, 224, 39–51.

Delong, S. A., Moon, J. J., and West, J. L. 2005. Covalently immobilized gradients of bFGF on hydrogel scaffolds for directed cell migration. *Biomaterials*, 26, 3227–3234.

Deng, C., Li, F., Griffith, M., Ruel, M., and Suuronen, E. J. 2010. Application of chitosan-based biomaterials for blood vessel regeneration. *Macromolecular Symposia*. Wiley-VCH Verlag, 297, 138–146.

Deutsch, J., Motlagh, D., Russell, B., and Desai, T. A. 2000. Fabrication of microtextured membranes for cardiac myocyte attachment and orientation. *Journal of Biomedical Materials Research*, 53, 267–275.

Di Martino, A., Sittinger, M., and Risbud, M. V. 2005. Chitosan: A versatile biopolymer for orthopaedic tissue-engineering. *Biomaterials*, 26, 5983–5990.

Dickinson, L. E., Lütgebaucks, C., Lewis, D. M., and Gerecht, S. 2012a. Patterning microscale extracellular matrices to study endothelial and cancer cell interactions *in vitro*. *Lab on a Chip*, 12, 4244–4248.

Dickinson, L. E., Moura, M. E., and Gerecht, S. 2010. Guiding endothelial progenitor cell tube formation using patterned fibronectin surfaces. *Soft Matter*, 6, 5109–5119.

Dickinson, L. E., Rand, D. R., Tsao, J., Eberle, W., and Gerecht, S. 2012b. Endothelial cell responses to micropillar substrates of varying dimensions and stiffness. *Journal of Biomedical Materials Research Part A*, 100A, 1457–1466.

Domachuk, P., Tsioris, K., Omenetto, F. G., and Kaplan, D. L. 2010. Bio-microfluidics: Biomaterials and biomimetic designs. *Advanced Materials*, 22, 249–260.

Egorova, A. D., Deruiter, M. C., Boer, H. C., Pas, S., Gittenberger-de Groot, A. C., Zonneveld, A. J., Poelmann, R. E., and Hierck, B. P. 2011. Endothelial colony-forming cells show a mature transcriptional response to shear stress. *In Vitro. Cellular & Developmental Biology—Animal*, 48, 21–29.

Elcin, A. E. and Elcin, Y. M. 2006. Localized angiogenesis induced by human vascular endothelial growth factor-activated PLGA sponge. *Tissue Engineering*, 12, 959–968.

Ferreira, L. S., Gerecht, S., Fuller, J., Shieh, H. F., Vunjak-Novakovic, G., and Langer, R. 2007. Bioactive hydrogel scaffolds for controllable vascular differentiation of human embryonic stem cells. *Biomaterials*, 28, 2706–2717.

Frantz, C., Stewart, K. M., and Weaver, V. M. 2010. The extracellular matrix at a glance. *Journal of Cell Science*, 123, 4195–4200.

George, E. L., Georges-Labouesse, E. N., Patel-King, R. S., Rayburn, H., and Hynes, R. O. 1993. Defects in mesoderm, neural tube and vascular development in mouse embryos lacking fibronectin. *Development*, 119, 1079–1091.

Gerecht-Nir, S., Cohen, S., Ziskind, A., and Itskovitz-Eldor, J. 2004. Three-dimensional porous alginate scaffolds provide a conducive environment for generation of well-vascularized embryoid bodies from human embryonic stem cells. *Biotechnology and Bioengineering*, 88, 313–320.

Gilbert, T. W., Sellaro, T. L., and Badylak, S. F. 2006. Decellularization of tissues and organs. *Biomaterials*, 27, 3675–3683.

Grant, D. S., Tashiro, K.-I., Segui-Real, B., Yamada, Y., Martin, G. R., and Kleinman, H. K. 1989. Two different laminin domains mediate the differentiation of human endothelial cells into capillary-like structures *in vitro*. *Cell*. Cell Press, 58, 933–943.

Grassl, E., Oegema, T., and Tranquillo, R. 2002. Fibrin as an alternative biopolymer to type-I collagen for the fabrication of a media equivalent. *Journal of Biomedical Materials Research*, 60, 607–612.

Grassl, E., Oegema, T., and Tranquillo, R. 2003. A fibrin-based arterial media equivalent. *Journal of Biomedical Materials Research Part A*, 66, 550–561.

Hanjaya-Putra, D., Bose, V., Shen, Y.-I., Yee, J., Khetan, S., Fox-Talbot, K., Steenbergen, C., Burdick, J. A., and Gerecht, S. 2011. Controlled activation of morphogenesis to generate a functional human microvasculature in a synthetic matrix. *Blood*, 118, 804–815.

Hanjaya-Putra, D., Yee, J., Ceci, D., Truitt, R., Yee, D., and Gerecht, S. 2010. Vascular endothelial growth factor and substrate mechanics regulate *in vitro* tubulogenesis of endothelial progenitor cells. *Journal of Cellular and Molecular Medicine*, 14, 2436–2447.

He, W., Yong, T., Ma, Z. W., Inai, R., Teo, W. E., and Ramakrishna, S. 2006. Biodegradable polymer nanofiber mesh to maintain functions of endothelial cells. *Tissue Engineering*, 12, 2457–2466.

Huang, X., Zauscher, S., Klitzman, B., Truskey, G., Reichert, W., Kenan, D., and Grinstaff, M. 2010. Peptide interfacial biomaterials improve endothelial cell adhesion and spreading on synthetic polyglycolic acid materials. *Annals of Biomedical Engineering*. Springer US, 38, 1965–1976.

Hubbell, J. A. 1995. Biomaterials in tissue engineering. *Nature Biotechnology*. Nature Publishing Group, 13, 565–576.

Hughes, C. S., Postovit, L. M., and Lajoie, G. A. 2010. Matrigel: A complex protein mixture required for optimal growth of cell culture. *Proteomics*, 10, 1886–1890.

Hynes, R. O. 2002. Integrins: Bidirectional, allosteric signaling machines. *Cell*. Elsevier, 110, 673–687.

Ifkovits, J. L. and Burdick, J. A. 2007. Review: Photopolymerizable and degradable biomaterials for tissue engineering applications. *Tissue Engineering*, 13, 2369–2385.

Ingber, D. E. and Folkman, J. 1989. Mechanochemical switching between growth and differentiation during fibroblast growth factor-stimulated angiogenesis *in vitro*: Role of extracellular matrix. *The Journal of Cell Biology*, 109, 317–330.

Ito, Y., Hasuda, H., Terai, H., and Kitajima, T. 2005. Culture of human umbilical vein endothelial cells on immobilized vascular endothelial growth factor. *Journal of Biomedical Materials Research Part A*, 74A, 659–665.

Jain, R. K. 2003. Molecular regulation of vessel maturation. *Nature Medicine*, 9, 685–693.

Jenkins, A. D. 2004. *Contemporary Polymer Chemistry* (3rd edition). Allcock, H. R., Lampe F. W., and Mark, J. E. Pearson Education, Inc (Pearson/Prentice Hall), Upper Saddle River, NJ, USA, 2003. ISBN 0-13-065056-0, pp, xviii + 814. *Polymer International*. John Wiley & Sons, Ltd.

Kannan, R. Y., Salacinski, H. J., Sales, K., Butler, P., and Seifalian, A. M. 2005. The roles of tissue engineering and vascularisation in the development of micro-vascular networks: A review. *Biomaterials*, 26, 1857–1875.

Kelley, C., D'amore, P., Hechtman, H. B., and Shepro, D. 1987. Microvascular pericyte contractility *in vitro*: Comparison with other cells of the vascular wall. *The Journal of Cell Biology*, 104, 483–490.

Khetan, S., Katz, J. S., and Burdick, J. A. 2009. Sequential crosslinking to control cellular spreading in 3-dimensional hydrogels. *Soft Matter*. The Royal Society of Chemistry, 5, 1601–1606.

Khor, E. and Lim, L. Y. 2003. Implantable applications of chitin and chitosan. *Biomaterials*, 24, 2339–2349.

Klagsbrun, M. and Moses, M. A. 1999. Molecular angiogenesis. *Chemistry and Biology*, 6, R217–R224.

Koch, S., Yao, C., Grieb, G., Prevel, P., Noah, E., and Steffens, G. C. 2006. Enhancing angiogenesis in collagen matrices by covalent incorporation of VEGF. *Journal of Materials Science: Materials in Medicine*, 17, 735–741.

Kolev, K., Tenekedjiev, K., Komorowicz, E., and Machovich, R. 1997. Functional evaluation of the structural features of proteases and their substrate in fibrin surface degradation. *Journal of Biological Chemistry*, 272, 13666–13675.

Korff, T. and Augustin, H. G. 1999. Tensional forces in fibrillar extracellular matrices control directional capillary sprouting. *Journal of Cell Science*, 112, 3249–3258.

Labet, M. and Thielemans, W. 2009. Synthesis of polycaprolactone: A review. *Chemical Society Reviews*. The Royal Society of Chemistry, 38, 3484–3504.

Lee, C. H., Singla, A., and Lee, Y. 2001. Biomedical applications of collagen. *International Journal of Pharmaceutics*, 221, 1–22.

Lee, H., Cusick, R. A., Browne, F., Kim, T. H., MA, P. X., Utsunomiya, H., Langer, R., and Vacanti, J. P. 2002. Local delivery of basic fibroblast growth factor increases both angiogenesis and engraftment of hepatocytes in tissue-engineered polymer devices1. *Transplantation*, 73, 1589–1593.

Leslie-Barbick, J. E., Moon, J. J., and West, J. L. 2009. Covalently-immobilized vascular endothelial growth factor promotes endothelial cell tubulogenesis in poly(ethylene glycol) diacrylate hydrogels. *Journal of Biomaterials Science, Polymer Edition*, 20, 1763–1779.

Leslie-Barbick, J. E., Shen, C., Chen, C., and West, J. L. 2010. Micron-scale spatially patterned, covalently immobilized vascular endothelial growth factor on hydrogels accelerates endothelial tubulogenesis and increases cellular angiogenic responses. *Tissue Engineering Part A*, 17, 221–229.

Lesman, A., Koffler, J., Atlas, R., Blinder, Y. J., Kam, Z., and Levenberg, S. 2011. Engineering vessel-like networks within multicellular fibrin-based constructs. *Biomaterials*, 32, 7856–7869.

Levenberg, S., Golub, J. S., Amit, M., Itskovitz-Eldor, J., and Langer, R. 2002. Endothelial cells derived from human embryonic stem cells. *Proceedings of the National Academy of Sciences*, 99, 4391–4396.

Levenberg, S., Huang, N. F., Lavik, E., Rogers, A. B., Itskovitz-Eldor, J., and Langer, R. 2003. Differentiation of human embryonic stem cells on three-dimensional polymer scaffolds. *Proceedings of the National Academy of Sciences*, 100, 12741–12746.

Li, D. and Xia, Y. 2004. Electrospinning of nanofibers: Reinventing the wheel? *Advanced Materials*, 16, 1151–1170.

Li, J., Li, G., Zhang, K., Liao, Y., Yang, P., Maitz, M. F., and Huang, N. 2013. Co-culture of vascular endothelial cells and smooth muscle cells by hyaluronic acid micro-pattern on titanium surface. *Applied Surface Science*, 273, 24–31.

Li, S., Bhatia, S., Hu, Y.-L., Shiu, Y.-T., Li, Y.-S., Usami, S., and Chien, S. 2001. Effects of morphological patterning on endothelial cell migration. *Biorheology*, 38, 101–108.

Lim, S. H. and Mao, H.-Q. 2009. Electrospun scaffolds for stem cell engineering. *Nanofibers in Regenerative Medicine and Drug Delivery*, 61, 1084–1096.

Linn, T., Erb, D., Schneider, D., Kidszun, A., Elçin, A. E., Bretzel, R. G., and Elçin, Y. M. 2003. Polymers for induction of revascularization in the rat fascial flap: Application of vascular endothelial growth factor and pancreatic islet cells. *Cell Transplantation*, 12, 769–778.

Liu, Y. and Chan-Park, M. B. 2009. Hydrogel based on interpenetrating polymer networks of dextran and gelatin for vascular tissue engineering. *Biomaterials*, 30, 196–207.

Lutolf, M. P. and Hubbell, J. A. 2005. Synthetic biomaterials as instructive extracellular microenvironments for morphogenesis in tissue engineering. *Nature Biotechnology*, 23, 47–55.

Ma, Z., Kotaki, M., Yong, T., He, W., and Ramakrishna, S. 2005. Surface engineering of electrospun polyethylene terephthalate (PET) nanofibers towards development of a new material for blood vessel engineering. *Biomaterials*, 26, 2527–2536.

Mann, B. K., Schmedlen, R. H., and West, J. L. 2001. Tethered-TGF-β increases extracellular matrix production of vascular smooth muscle cells. *Biomaterials*, 22, 439–444.

Matthews, J. A., Wnek, G. E., Simpson, D. G., and Bowlin, G. L. 2002. Electrospinning of collagen nanofibers. *Biomacromolecules*, 3, 232–238.

Meyer, G. T., Matthias, L. J., Noack, L., Vadas, M. A., and Gamble, J. R. 1997. Lumen formation during angiogenesis *in vitro* involves phagocytic activity, formation and secretion of vacuoles, cell death, and capillary tube remodelling by different populations of endothelial cells. *Anatomical Record*, 249, 327–340.

Miller, J. S., Shen, C. J., Legant, W. R., Baranski, J. D., Blakely, B. L., and Chen, C. S. 2010. Bioactive hydrogels made from step-growth derived Peg–peptide macromers. *Biomaterials*, 31, 3736–3743.

Miller, J. S., Stevens, K. R., Yang, M. T., Baker, B. M., Nguyen, D.-H. T., Cohen, D. M., Toro, E. et al. 2012. Rapid casting of patterned vascular networks for perfusable engineered three-dimensional tissues. *Nature Materials*, 11, 768–774.

Mo, X. M., Xu, C. Y., Kotaki, M., and Ramakrishna, S. 2004. Electrospun P(LLA-CL) nanofiber: A biomimetic extracellular matrix for smooth muscle cell and endothelial cell proliferation. *Biomaterials*, 25, 1883–1890.

Murphy, W. L., Peters, M. C., Kohn, D. H., and Mooney, D. J. 2000. Sustained release of vascular endothelial growth factor from mineralized poly (lactide-*co*-glycolide) scaffolds for tissue engineering. *Biomaterials*, 21, 2521–2527.

Myers, K. A., Applegate, K. T., Danuser, G., Fischer, R. S., and Waterman, C. M. 2011. Distinct ECM mechanosensing pathways regulate microtubule dynamics to control endothelial cell branching morphogenesis. *The Journal of Cell Biology*, 192, 321–334.

Nachman, R. L. and Jaffe, E. A. 2004. Endothelial cell culture: Beginnings of modern vascular biology. *Journal of Clinical Investigation*, 114, 1037–1040.

Naderi, H., Matin, M. M., and Bahrami, A. R. 2011. Review paper: Critical issues in tissue engineering: Biomaterials, cell sources, angiogenesis, and drug delivery systems. *Journal of Biomaterials Applications*, 26, 383–417.

Nakagawa, Y., Murai, T., Hasegawa, C., Hirata, M., Tsuchiya, T., Yagami, T., and Haishima, Y. 2003. Endotoxin contamination in wound dressings made of natural biomaterials. *Journal of Biomedical Materials Research*. Wiley Subscription Services, Inc., A Wiley Company, 66B, 347–355.

Neumann, T., Nicholson, B. S., and Sanders, J. E. 2003. Tissue engineering of perfused microvessels. *Microvascular Research*, 66, 59–67.

Ono, K., Saito, Y., Yura, H., Ishikawa, K., Kurita, A., Akaike, T., and Ishihara, M. 2000. Photocrosslinkable chitosan as a biological adhesive. *Journal of Biomedical Materials Research*. John Wiley & Sons, Inc, 49, 289–295.

Owens, G. K. 1995. Regulation of differentiation of vascular smooth muscle cells. *Physiological Reviews*. American Physiological Society, 75, 487–517.

Patel, Z. S. and Mikos, A. G. 2004. Angiogenesis with biomaterial-based drug-and cell-delivery systems. *Journal of Biomaterials Science, Polymer Edition*, 15, 701–726.

Peacock, A. J. and Calhoun, A. 2006. *Polymer Chemistry—Properties and Applications*. Munich: Hanser Publishers.

Pham, Q. P., Sharma, U., and Mikos, A. G. 2006. Electrospinning of polymeric nanofibers for tissue engineering applications: A review. *Tissue Engineering*, 12, 1197–1211.

Raghavan, S., Nelson, C. M., Baranski, J. D., Lim, E., and Chen, C. S. 2010. Geometrically controlled endothelial tubulogenesis in micropatterned gels. *Tissue Engineering Part A*, 16, 2255–2263.

Rahmany, M. B. and Van Dyke, M. 2012. Biomimetic approaches to modulate cellular adhesion in biomaterials: A review. *Acta Biomaterialia*, 9, 5431–5437.

Ravi, S. and Chaikof, E. L. 2010. Biomaterials for vascular tissue engineering. *Regenerative Medicine*, 5, 107–120.

Richardson, T. P., Peters, M. C., Ennett, A. B., and Mooney, D. J. 2001. Polymeric system for dual growth factor delivery. *Nature Biotechnology*, 19, 1029–1034.

Rinaudo, M. 2008. Main properties and current applications of some polysaccharides as biomaterials. *Polymer International*. John Wiley & Sons, Ltd, 57, 397–430.

Risbud, M. V., Bhonde, M. R., and Bhonde, R. R. 2001. Effect of chitosan-polyvinyl pyrrolidone hydrogel on proliferation and cytokine expression of endothelial cells: Implications in islet immunoisolation. *Journal of Biomedical Materials Research*, 57, 300–305.

Romano, N. H., Sengupta, D., Chung, C., and Heilshorn, S. C. 2011. Protein-engineered biomaterials: Nanoscale mimics of the extracellular matrix. *BBA—General Subjects*. Elsevier B.V, 1810, 339–349.

Rothlein, R., Dustin, M. L., Marlin, S. D., and Springer, T. A. 1986. A human intercellular adhesion molecule (ICAM-1) distinct from LFA-1. *The Journal of Immunology*. The American Association of Immunologists, 137, 1270–1274.

Rowe, S. L., Lee, S., and Stegemann, J. P. 2007. Influence of thrombin concentration on the mechanical and morphological properties of cell-seeded fibrin hydrogels. *Acta Biomaterialia*, 3, 59–67.

Rubin, H. 1997. Cell aging *in vivo* and *in vitro*. *Mechanisms of Ageing and Development*, 98, 1–35.

Sacharidou, A., Stratman, A. N., and Davis, G. E. 2012. Molecular mechanisms controlling vascular lumen formation in three-dimensional extracellular matrices. *Cells Tissues Organs*, 195, 122–143.

Sarkar, S., Lee, G. Y., Wong, J. Y., and Desai, T. A. 2006. Development and characterization of a porous micro-patterned scaffold for vascular tissue engineering applications. *Biomaterials*, 27, 4775–4782.

Schmoekel, H. G., Weber, F. E., Schense, J. C., Grätz, K. W., Schawalder, P., and Hubbell, J. A. 2005. Bone repair with a form of BMP-2 engineered for incorporation into fibrin cell ingrowth matrices. *Biotechnology and Bioengineering*, 89, 253–262.

Sefcik, L. S., Neal, R. A., Kaszuba, S. N., Parker, A. M., Katz, A. J., Ogle, R. C., and Botchwey, E. A. 2008. Collagen nanofibres are a biomimetic substrate for the serum-free osteogenic differentiation of human adipose stem cells. *Journal of Tissue Engineering and Regenerative Medicine*, 2, 210–220.

Seliktar, D., Black, R. A., Vito, R. P., and Nerem, R. M. 2000. Dynamic mechanical conditioning of collagen-gel blood vessel constructs induces remodeling *in vitro*. *Annals of Biomedical Engineering*, 28, 351–362.

Shaikh, F. M., Callanan, A., Kavanagh, E. G., Burke, P. E., Grace, P. A., and Mcgloughlin, T. M. 2008. Fibrin: A natural biodegradable scaffold in vascular tissue engineering. *Cells Tissues Organs*, 188, 333–346.

Shireman, P. K. and Greisler, H. P. 2000. Mitogenicity and release of vascular endothelial growth factor with and without heparin from fibrin glue. *Journal of Vascular Surgery*, 31, 936–943.

Sieminski, A. L., Hebbel, R. P., and Gooch, K. J. 2004. The relative magnitudes of endothelial force generation and matrix stiffness modulate capillary morphogenesis *in vitro*. *Experimental Cell Research*, 297, 574–84.

Singh, S., Wu, B. M., and Dunn, J. C. 2011. Accelerating vascularization in polycaprolactone scaffolds by endothelial progenitor cells. *Tissue Engineering Part A*, 17, 1819–1830.

Soucy, P. A. and Romer, L. H. 2009. Endothelial cell adhesion, signaling, and morphogenesis in fibroblast-derived matrix. *Matrix Biology*, 28, 273–283.

Standring, S. 2008. *Gray's Anatomy: The Anatomical Basis of Clinical Practice*, 40th edition, London: Churchill Livingstone.

Stratman, A. N., Davis, M. J., and Davis, G. E. 2011. VEGF and FGF prime vascular tube morphogenesis and sprouting directed by hematopoietic stem cell cytokines. *Blood,* 117, 3709–3719.

Sun, G., Kusuma, S., and Gerecht, S. 2010. The integrated role of biomaterials and stem cells in vascular regeneration. *Biomaterials as Stem Cell Niche,* 2, 195–223.

Sun, G., Shen, Y.-I., Ho, C. C., Kusuma, S., and Gerecht, S. 2009. Functional groups affect physical and biological properties of dextran-based hydrogels. *Journal of Biomedical Materials Research,* 93A, 1080–1090.

Sun, G., Shen, Y.-I., Kusuma, S., Fox-Talbot, K., Steenbergen, C. J., and Gerecht, S. 2011a. Functional neovascularization of biodegradable dextran hydrogels with multiple angiogenic growth factors. *Biomaterials.* Elsevier Ltd, 32, 95–106.

Sun, G., Zhang, X., Shen, Y.-I., Sebastian, R., Dickinson, L. E., Fox-Talbot, K., Reinblatt, M., Steenbergen, C., Harmon, J. W., and Gerecht, S. 2011b. Dextran hydrogel scaffolds enhance angiogenic responses and promote complete skin regeneration during burn wound healing. *Proceedings of the National Academy of Sciences of the United States of America,* 108, 20976–20981.

Sung, H.-J., Meredith, C., Johnson, C., and Galis, Z. S. 2004. The effect of scaffold degradation rate on three-dimensional cell growth and angiogenesis. *Biomaterials,* 25, 5735–5742.

Thakar, R. G., Cheng, Q., Patel, S., Chu, J., Nasir, M., Liepmann, D., Komvopoulos, K., and Li, S. 2009. Cell-shape regulation of smooth muscle cell proliferation. *Biophysical Journal.* Biophysical Society, 96, 3423–3432.

Toole, B. P. 2004. Hyaluronan: From extracellular glue to pericellular cue. *Nature Reviews Cancer,* 4, 528–539.

Trentin, D., Hall, H., Wechsler, S., and Hubbell, J. A. 2006. Peptide-matrix-mediated gene transfer of an oxygen-insensitive hypoxia-inducible factor-1α variant for local induction of angiogenesis. *Proceedings of the National Academy of Sciences of the United States of America,* 103, 2506–2511.

Unger, R. E., Peters, K., Huang, Q., Funk, A., Paul, D., and Kirkpatrick, C. J. 2005. Vascularization and gene regulation of human endothelial cells growing on porous polyethersulfone (PES) hollow fiber membranes. *Biomaterials,* 26, 3461–3469.

Van Hinsbergh, V. W. M., Collen, A., and Koolwijk, P. 2001. Role of fibrin matrix in angiogenesis. *Annals of the New York Academy of Sciences,* 936, 426–437.

Vasita, R. and Katti, D. S. 2006. Nanofibers and their applications in tissue engineering. *International Journal of Nanomedicine,* 1, 15–30.

Vo, E., Hanjaya-Putra, D., Zha, Y., Kusuma, S., and Gerecht, S. 2010. Smooth-muscle-like cells derived from human embryonic stem cells support and augment cord-like structures *in vitro. Stem Cell Reviews,* 6, 237–247.

Vrana, N. E., Liu, Y., Mcguinness, G. B., and Cahill, P. A. 2008. Characterization of poly(vinyl alcohol)/chitosan hydrogels as vascular tissue engineering scaffolds. *Macromolecular Symposia.* Wiley-VCH Verlag, 269, 106–110.

Wanjare, M., Kuo, F., and Gerecht, S. 2013. Derivation and maturation of synthetic and contractile vascular smooth muscle cells from human pluripotent stem cells. *Cardiovascular Research,* 97, 321–330.

Wijelath, E. S., Murray, J., Rahman, S., Patel, Y., Ishida, A., Strand, K., Aziz, S. et al. 2002. Novel vascular endothelial growth factor binding domains of fibronectin enhance vascular endothelial growth factor biological activity. *Circulation Research,* 91, 25–31.

Williamson, M. R., Woollard, K. J., Griffiths, H. R., and Coombes, A. G. A. 2006. Gravity spun polycaprolactone fibers for applications in vascular tissue engineering: Proliferation and function of human vascular endothelial cells. *Tissue Engineering,* 12, 45–51.

Xu, C., Inai, R., Kotaki, M., and Ramakrishna, S. 2004. Electrospun nanofiber fabrication as synthetic extracellular matrix and its potential for vascular tissue engineering. *Tissue Engineering,* 10, 1160–1168.

Yee, D., Hanjaya-Putra, D., Bose, V., Luong, E., and Gerecht, S. 2011. Hyaluronic acid hydrogels support cord-like structures from endothelial colony-forming cells. *Tissue Engineering Part A,* 17, 1351–1361.

Zhang, X., Baughman, C. B., and Kaplan, D. L. 2008. *In vitro* evaluation of electrospun silk fibroin scaffolds for vascular cell growth. *Biomaterials,* 29, 2217–2227.

6

Biophysical Mechanisms That Govern the Vascularization of Microfluidic Scaffolds

Keith H. K. Wong, James G. Truslow, Aimal H. Khankhel, and Joe Tien

CONTENTS

6.1 Introduction

Methods to engineer vascularized tissues have historically focused on chemical and/or biological strategies (Lovett et al. 2009). For instance, loading of scaffolds with vascular growth factors, functionalization of scaffolds with bioactive peptides, or incorporation of extracellular matrix (ECM) components into scaffolds can induce angiogenesis and vasculogenesis, the natural processes of vascular formation and growth that occur during development and wound healing [reviewed in (Bouhadir and Mooney 2001, Vailhé et al. 2001, Kaully et al. 2009)]. Likewise, cultures of differentiated vascular cells, mesenchymal stem cells, and/or blood-borne progenitor cells can reorganize to form long-lived vascular networks (Asahara et al. 1997, Koike et al. 2004, Au et al. 2008).

In comparison, *physical* approaches to vascularization (e.g., those that involve stress, stiffness, flow, and geometry) are much less well studied. The rationale for investigating such physical strategies lies in the possibility of obtaining complementary methods to tailor the formation and functionality of vascular networks. Recent work in vascular and nonvascular systems has indicated the importance of the physical microenvironment in biological function (Chen et al. 2004, Discher et al. 2005). While this work in mechanobiology has mostly focused on two-dimensional cultures, the underlying ideas also apply to three-dimensional systems (Alcaraz et al. 2004, Gjorevski and Nelson 2010). Studying how physical signals may affect or underlie vascularization of scaffolds will likely lead to a better understanding of how to manipulate the vascularization process for a desired outcome.

To this end, we and others have advocated forming microvessels in microfluidic scaffolds (those that contain open channels or networks), both as well-defined model systems for studying the role of physical signals on vascular function and as perfusable materials for potential implantation (Price and Tien 2009, Miller et al. 2012, Tien et al. 2013, Zheng et al. 2012). The high-resolution techniques that are used to create microfluidic scaffolds enable intricate control over the channel geometries and the perfusion rates (Wong et al. 2012); such control allows one to alter physical signals within a vascularized microfluidic scaffold precisely. Practically, a major advantage of microfluidic scaffolds is their capability to sustain immediate perfusion by virtue of the preexisting fluidic channels. In principle, this perfusion is required to maintain the viability of densely cellularized tissue constructs (Choi et al. 2007). This property of microfluidic scaffolds lies in sharp contrast to chemically- or biologically driven angiogenesis or vasculogenesis, both of which require at least several days to develop structures capable of sustaining perfusion (Shepherd et al. 2004, Tremblay et al. 2005, Chiu et al. 2012, Yeon et al. 2012).

This chapter summarizes our recent experimental studies of vascularization of microfluidic scaffolds, proposes a unified biomechanical theory of vascular stability in these scaffolds, and suggests some testable predictions for future investigation. It updates and elaborates upon the physical concepts introduced in a recent review (Tien et al. 2013).

6.2 Formation, Vascularization, and Perfusion of Microfluidic Scaffolds

Our experimental approach is to form microfluidic channels or networks inside hydrogels, and to culture endothelial cells within these channels under perfusion. The cultured cells will then grow to form confluent tubes with a circular cross section (Price and Tien 2009, 2011). We have chosen to use ECM hydrogels because they promote cell adhesion, in contrast to synthetic polymers that often require coupling with adhesive peptides before use.

The specific methods for forming the microfluidic structures vary with the channel geometry. For single channels, we typically rely on cylindrical needles as removable templates (Chrobak et al. 2006); for interconnected networks, we use micropatterned hydrogels as sacrificial materials that can be partially encapsulated and dissolved (Golden and Tien 2007) or as separate layers that can be bonded (Price et al. 2008). Once the channels have formed, their vascularization is straightforward albeit technically demanding. Below, we restrict the discussion to the vascularization of single channels (Chrobak et al. 2006) with and without accompanying drainage channels (Wong et al. 2013) because these simple geometries have allowed routine mathematical analyses of the fluid and solid mechanics of the vessels and scaffolds. For step-by-step experimental procedures, we refer the reader to published protocols (Price and Tien 2009, 2011).

To create single microfluidic channels, we form type I collagen or fibrin gels around a stainless-steel acupuncture needle (120-μm-diameter, Japanese gauge 02), and remove

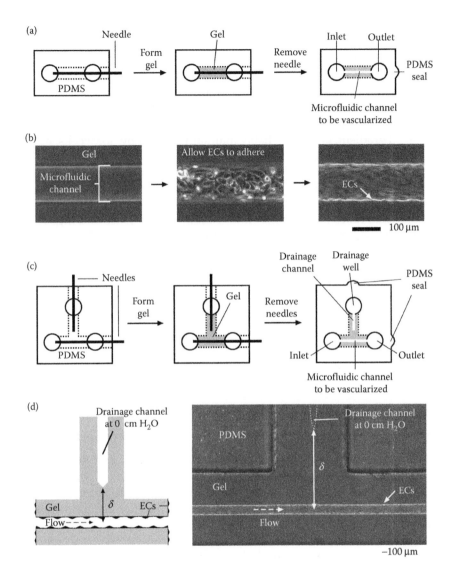

FIGURE 6.1
Strategies for forming endothelialized microfluidic scaffolds. (a) Fabrication of ECM scaffolds that contain a single microfluidic channel. The ECM hydrogel is formed around a stainless-steel needle (typically 120 µm in diameter), which is then removed to yield a microchannel that interfaces with the inlet and outlet of the PDMS chamber. The channel is then seeded with endothelial cells. (b) Endothelial cells are cultured to confluence in microchannels under constant media perfusion. (c) Fabrication of scaffolds that contain an additional drainage channel. The drainage channel is formed perpendicularly to subject the gel to atmospheric pressure. (d) Phase-contrast image of a microvessel with a drainage channel in a fibrin gel. (Adapted with permission from Wong, K. H. K. et al. 2013. *J Biomed Mater Res A*, 101, 2181–90.)

the needle after gel polymerization to leave an open channel within the gel (Figure 6.1a). The gel is directly polymerized inside a polydimethylsiloxane (PDMS) chamber, which acts as a mechanical support and allows fluidic connections across the inlet and outlet of the vessel (Figure 6.1a). We vascularize these channels by introducing a suspension of blood microvascular endothelial cells, allowing them to adhere, connecting the inlet and outlet to media reservoirs at different pressures, and culturing the seeded channels under constant perfusion. We maintain steady flow by recycling effluent media once or twice daily to the reservoir that connects to the inlet. In general, a confluent endothelial tube (Figure 6.1b) forms within 1 day. The application of pressure differences of 1–12 cm H_2O across 7- to 9-mm-long vessels generates flow rates and wall shear stresses of 0.1–3 mL/h and 1–30 dyn/cm^2, respectively (Chrobak et al. 2006, Price et al. 2010, Wong et al. 2010, Leung et al. 2012). This method precisely controls the lumenal (i.e., intravascular) pressure and its axial gradient, and can reliably maintain steady flow for weeks, but only if the endothelium remains adherent to the scaffold.

Recently, we have developed methods to control the fluid pressure of the scaffold outside the vessel (i.e., the extravascular or interstitial fluid pressure) (Wong et al. 2013). This work was inspired by the design and function of the lymphatic system *in vivo*, in which low-pressure endothelial tubes drain excess fluid from the tissue space and thus maintain the interstitial pressure at near-atmospheric pressures (Schmid-Schönbein 1990). By incorporating empty (i.e., nonvascularized) microfluidic channels that were independently held at atmospheric pressure (Figure 6.1c and d), we could independently modulate the intra- and extravascular pressures and thereby study how the transmural pressure (i.e., the difference between lumenal and interstitial pressures) affected the stability of engineered blood microvessels. Although lymphatics *in vivo* are lined by specialized (lymphatic) endothelium, the empty channels appear to serve as an effective replacement to provide drainage function in microfluidic scaffolds.

6.3 Biophysical Considerations of Scaffold Material Properties

Investigations of the vascularization of microfluidic hydrogels that are prepared under different conditions and with different structures have indicated that the physical properties of the scaffolds play an important role in determining the effectiveness of vascularization (Chrobak et al. 2006). The two principal relevant properties are the scaffold stiffness and pore size.

Depending on the subsequent degree of handling, microfluidic scaffolds require various degrees of mechanical stiffness. In our design, which encases the gel inside a PDMS chamber and requires no direct gel manipulation after polymerization, a low-density collagen gel with a shear modulus around hundreds of pascals is sufficient (Chrobak et al. 2006). If the fabrication process involves manual transfer of hydrogels, then a greater mechanical strength may be preferred to withstand deformation; for such cases, gel densities much larger than 10 mg/mL with moduli values on the order of 10–50 kPa can be advantageous (Ling et al. 2007, Nichol et al. 2010).

The physical properties of the scaffold also control the degree of cell-mediated remodeling after seeding (Figure 6.2a). We have found that collagen polymerization conditions that form soft gels (3 mg/mL, gelled at 37°C) result in irreversible vessel contraction. Depending on the softness of the gel, the degree of contraction may be so severe that perfusion desists under practically attainable pressures. The contraction appears to result from the inherent contractility of the endothelium, which can be substantial in cultured cells (Kolodney and Wysolmerski 1992). If one models the endothelium as a contractile thin film, then the inward radial stress that results from cell contraction is given by the law of Laplace as γ_{EC}/r, where γ_{EC} is the endothelial contractility and r is the

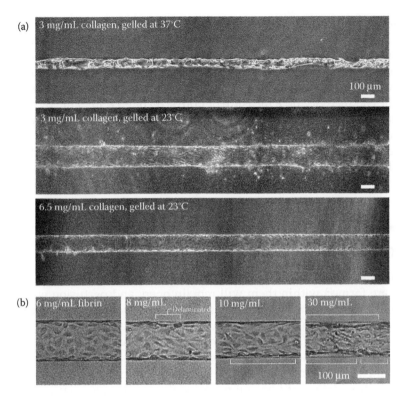

FIGURE 6.2
Effects of scaffold material properties on vessel formation. (a) In 3 mg/mL collagen gels that are polymerized at 37°C, microvessels contract by day 2 postseeding. In gels that are polymerized at 23°C, microvessels do not contract, but endothelial cells invade the surrounding collagen matrix. Invasion of endothelial cells is significantly less when collagen density is increased to 6.5 mg/mL. (b) Microvessels that are formed in fibrin gels of increasing densities (from 6 to 30 mg/mL) show a higher tendency to delaminate (i.e., detach) from the scaffold. (Adapted with permission from Chrobak, K. M., Potter, D. R., and Tien, J. 2006. *Microvasc Res*, 71, 185–96; Wong, K. H. K. et al. 2013. *J Biomed Mater Res A*, 101, 2181–90.)

contracted vascular radius. This stress leads to an inward fractional change in vessel radius of approximately $\gamma_{EC}/2Gr$, where G is the shear modulus of the scaffold (Barber 2010). That is

$$1 - \frac{r}{r_0} = \frac{\gamma_{EC}}{2Gr} \tag{6.1}$$

where r_0 is the initial (not contracted) vascular radius. This equation has no solution when

$$G < \frac{2\gamma_{EC}}{r_0} \tag{6.2}$$

Under this condition, the vessel will contract catastrophically. For the vessels and scaffolds that we typically use, $2\gamma_{EC}/r_0$ is on the order of 100 Pa, and scaffolds that are softer than this threshold cannot withstand endothelial contraction. We note that smaller vessels will require proportionally stiffer scaffolds to avoid this effect.

Conditions that form porous gels (3 mg/mL, gelled at 23°C) lead to migration of single cells into the surrounding gel (Chrobak et al. 2006); this invasion can be inhibited by increasing gel density to decrease the pore size (Figure 6.2a). Empirically optimized conditions for microfluidic collagen scaffolds are gelation at room temperature (≤25°C) and the use of gel concentrations that are ≥6.5 mg/mL, which lead to shear moduli of ~400 Pa and hydraulic conductivity (K; a measure of pore size) of ~3×10^{-8} cm^4/dyn·s (Wong et al. 2010). Although one can further stiffen scaffolds and reduce the pore size by increasing the scaffold concentration, such changes may not be desirable; in particular, we have recently found that a very low hydraulic conductivity in the scaffold can inhibit the transport of interstitial fluid and thereby destabilize vessels (see below).

6.4 How Physical Forces May Determine Vascular Stability

When scaffolds are stiff enough to resist contraction and dense enough to resist invasion, vessels that are formed within them will remain open, and perfusion will be stable, for 2–3 days. In the absence of additional stabilizing signals, however, long-term perfusion is not guaranteed. The endothelium can gradually delaminate from the microfluidic channel over the course of a week (Figure 6.2b); flow rates will likewise decrease, and often substantially. To understand the origin of this effect, and to discover signals that can promote vascular stability, we have manipulated the hydrodynamic stresses across the vessel wall directly (through variation of lumenal pressure P_{lumen} and interstitial pressure $P_{scaffold}$) or indirectly (through changes in transvascular fluid flow via alterations in endothelial permeability) (Price et al. 2010, Leung et al. 2012, Wong et al. 2010, 2013).

These experiments have pointed to a physical basis of vascular stability. In engineered microvessels, the relevant stresses involved in vascular stability include the transmural pressure $P_{TM} \equiv P_{lumen} - P_{scaffold}$, stress due to endothelial contraction γ_{EC}/r, and endothelial–scaffold adhesion σ_{adh} (Figure 6.3a). The simplest model postulates that a phenomenological force balance determines vascular stability:

$$(P_{lumen} - P_{scaffold}) + \sigma_{adh} = P_{TM} + \sigma_{adh} > \frac{\gamma_{EC}}{r} \qquad (6.3)$$

When this inequality is satisfied, the stabilizing stresses due to transmural pressure and scaffold adhesion outweigh the destabilizing stress due to contractility, and the vessel will remain adherent to the scaffold (Tien et al. 2013). In the following section, we describe experimental results in the context of this mechanical model, and provide evidence that transmural pressure is one of the predominant predictors of vascular stability. In microvascular networks *in vivo*, transmural pressure is an important driving force in transvascular filtration of fluid (Michel and Curry 1999), and it dictates microvascular perfusion by controlling the diameter and hence the fluidic resistance of capillaries (Lee and Schmid-Schönbein 1995).

6.4.1 Prediction: An Increase in Transmural Pressure Stabilizes Vascular Adhesion to the Scaffold

The mechanical model of vascular stability predicts that transmural pressure P_{TM} acts to stabilize the vessel lumen, since an increase in P_{TM} makes the inequality in Equation 6.3 more

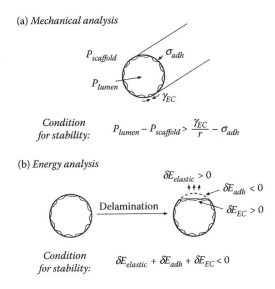

(a) *Mechanical analysis*

$P_{scaffold}$

σ_{adh}

P_{lumen}

γ_{EC}

Condition
for stability: $P_{lumen} - P_{scaffold} > \dfrac{\gamma_{EC}}{r} - \sigma_{adh}$

(b) *Energy analysis*

$\delta E_{elastic} > 0$

$\delta E_{adh} < 0$

Delamination

$\delta E_{EC} > 0$

Condition
for stability: $\delta E_{elastic} + \delta E_{adh} + \delta E_{EC} < 0$

FIGURE 6.3
Proposed models of vessel stability. (a) *Mechanical force balance:* Vessels are stable when the total stabilizing stresses—transmural pressure (i.e., lumenal pressure P_{lumen} minus scaffold pressure $P_{scaffold}$) and cell adhesion stress σ_{adh}—are greater than the destabilizing contractile stress γ_{EC}/r. (b) *Energy analysis:* Vessels are stable when the energy required to break cell–scaffold adhesion bonds $|\delta E_{adh}|$ is greater than the total energy released from elastic recoil of the scaffold $\delta E_{elastic}$ and retraction of endothelium δE_{EC}.

likely to be satisfied. In a deformable scaffold, the transmural pressure will cause the vessel to distend if $P_{TM} - \gamma_{EC}/r$ is positive. Thus, one can estimate P_{TM} by measuring the degree of radial distension of the vessel and the elastic modulus of the scaffold. Consistent with the prediction, tubes that distend outward are invariably stable over the long term (Price et al. 2010).

In contrast, we found in early experiments that endothelial tubes that were perfused under nearly zero-flow rates invariably delaminated from the scaffold over time. In this configuration, all pressures are held small so that the flow rate is very low, and the transmural pressure is close to zero. This pair of observations—stability when tubes distend, instability when they do not—is consistent with the theory that a positive transmural pressure is required for vascular stability. Moreover, since endothelial delamination occurs when P_{TM} is nearly zero, the endothelial contractile stress (γ_{EC}/r) is likely to be greater than the cell adhesion stress (σ_{adh}). Direct measurement of endothelial contractility under nearly zero-flow conditions indicates that γ_{EC}/r is equivalent to an inward radial strain of ~5% and a tensile radial stress of ~40 Pa in ~8 mg/mL collagen gels (Wong et al. 2010).

To verify that a positive transmural pressure is required to prevent endothelial delamination under normal flow conditions, we designed experiments to artificially hold $P_{scaffold}$ nearly equal to P_{lumen}, so that P_{TM} is nearly zero everywhere along the vessel. This pressure condition was achieved by placing an empty channel near and parallel to the vessel, and by subjecting both this channel and the vessel to the same perfusion pressures (Figure 6.4a). With this setup, the same axial pressure distributions will exist within the vessel lumen and the scaffold, thereby greatly reducing the transmural pressure. We found that endothelial tubes that are located adjacent to an empty channel held at a similar pressure grow to confluence, but invariably delaminate by day 3 postseeding (Price et al.

(a)

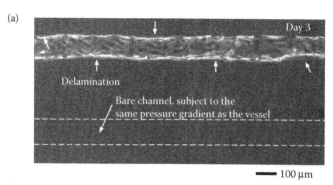

(b)

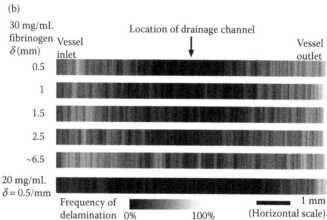

(c)

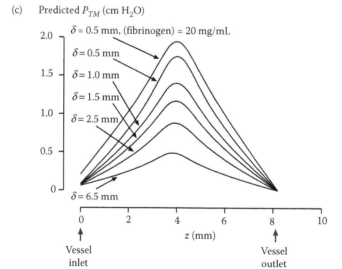

FIGURE 6.4

2010). Mechanically, this situation is similar to that of compartment syndrome, in which perfusion is impaired within an enclosed tissue (e.g., muscle surrounded by fascia) when the tissue pressure is elevated.

6.4.2 Prediction: Drainage Stabilizes Vascular Adhesion, Especially When the Scaffold Is Hydraulically Resistive

A second prediction of the mechanical model is that the hydraulic resistance of the scaffold controls the level of vascular stability. This resistance lies in series with the resistance across the vessel wall; thus, in scaffolds of higher resistance, the pressure gradients in the scaffold are higher, the pressure differences across the vessel wall (i.e., P_{TM}) are lower, and the vessel is less stable (Truslow et al. 2009). Indeed, we have experimentally found in fibrin gels of various densities that vascular stability is anticorrelated with scaffold permeability K (Figure 6.2b) (Wong et al. 2013). Vessels in fibrin gels with K above $\sim 10^{-9}$ cm^4/ dyn·s are stable, but not in gels with lower permeabilities.

In resistive scaffolds, the introduction of empty channels to directly lower the pressure within the scaffold should increase P_{TM} and hence enhance vascular stability. These artificial lymphatic-like structures serve to drain fluid from the scaffold at a small, but measurable, rate (Wong et al. 2013). By placing such drainage channels perpendicular to the vessel axis, we have shown that drainage effectively stabilizes the portions of the vessel that lie sufficiently close to the drainage channel (within ~1 mm, for a drain-to-vessel distance of ~0.5 mm in 30 mg/mL fibrin gels) (Figure 6.4b) (Wong et al. 2013). Placing the drainage channel further from the vessel wall led to a gradual loss of stabilization. Moreover, when the drain-to-vessel distance was held constant while the scaffold permeability was varied, the stabilizing effect of drainage was stronger in scaffolds of greater porosity (20 mg/mL vs. 30 mg/mL fibrin; Figure 6.4b). These results are all consistent with numerical models of the pressure distribution across the endothelium and in the scaffold (Figure 6.4c), and suggest that a large transmural pressure can be obtained by minimizing the hydraulic resistance from the vessel wall to low-pressure drainage (Truslow et al. 2009).

By comparing the experimentally determined delamination probabilities along the vessel axes with the computationally calculated transmural pressure distributions, we have obtained the first estimate of the minimum transmural pressure $\gamma_{EC}/r - \sigma_{adh}$ that is required for vascular stability (Wong et al. 2013). Given the uncertainties in the hydraulic properties of the vessel wall, we estimated the minimum required transmural pressure to be 40–140 Pa in fibrin gels. This range is remarkably consistent with the ~40 Pa estimate of endothelial contractile stress, which strengthens our claim that the delicate

---

FIGURE 6.4
Increases in transmural pressure promote vessel stability. (a) Forcing the transmural pressure to be nearly zero along the entire vessel (by subjecting the gel to the same axial pressure distribution as the vessel) leads to endothelial delamination. (b) In contrast, maintenance of a positive transmural pressure by draining the hydrogel of interstitial fluid (refer to Figure 6.1d for experimental setup) stabilizes vessels. The presence of a drainage channel (held at atmospheric pressure) in the middle of the scaffold prevents delamination in the vicinity of the channel. As the drainage channel is placed further away from the vessels (i.e., as δ increased), the stabilized region gradually shortens and eventually disappears. (c) Numerical modeling predicts that transmural pressure is highest near the middle of the vessel, consistent with experimental results. Transmural pressure decreases sharply as the drainage channel is moved away from the vessel wall. Fibrin gel concentration is 30 mg/mL unless stated otherwise. (Adapted with permission from Price, G. M. et al. 2010. *Biomaterials*, 31, 6182–9; Wong, K. H. K. et al. 2013. *J Biomed Mater Res A*, 101, 2181–90.)

balance between outward and inward radial stress at the vessel wall determines whether a vessel is stable. It also suggests that the adhesion stress σ_{adh} lies in the same range, which is consistent with measurements of the stress required to detach cultured cells from their substrata (Gallant et al. 2005).

Altogether, these data imply that the use of high-density, low-porosity scaffolds—favored by many investigators for the ease of handling and fabrication—may require a well-designed drainage system to maintain vascular stability and perfusion. Fibrin gels with concentrations above ~10 mg/mL (K below ~10^{-9} cm⁴/dyn·s) show considerable delamination in the absence of drainage, and we point out that other polymer matrices such as alginate and polyethylene glycol are even more hydraulically resistive by orders of magnitude (Comper and Zamparo 1989, Zimmermann et al. 2007).

6.4.3 Prediction: Strong Endothelial Barrier Function Promotes Vascular Stability

A third prediction of the mechanical model is that signals that promote strong endothelial barrier function will also promote stability. At first glance, the link between barrier function [typically quantified in terms of solute permeability coefficients (Michel and Curry 1999)] and physical stresses may not be obvious. The endothelial barrier, however, acts as a pathway for transport of both solutes and water; it is almost always the case that a barrier that leaks solutes will be hydraulically leaky as well (Rippe and Haraldsson 1994). When the hydraulic resistance of the vessel wall is low, the transmural pressure will also be low, all other parameters being equal. Thus, it is necessary for the vessel barrier to be strong if a large stabilizing transmural pressure is to be obtained.

In nearly all the vascular perfusion conditions that we have studied, stable vessels result only when the vessel wall restricts the movement of macromolecules of the size of serum albumin (~67 kDa). These barrier-strengthening conditions include addition of the anti-inflammatory second messenger cyclic adenosine monophosphate (AMP) (Figure 6.5a) (Wong et al. 2010), addition of polymers such as dextran and hydroxyethyl starch (Leung et al. 2012), and elevation of vascular flow rate (Figure 6.5b) (Price et al. 2010). In the absence of such signals, the endothelium becomes leaky; in particular, it displays "focal" leaks that are localized to microscale regions on the vessel wall. When enough of these leaks open simultaneously, transmural pressure can decay on the timescale of minutes, as the vascular and scaffold pressures equilibrate (Wong et al. 2010). We have observed that barrier-strengthening conditions invariably result in vessels with strong junctional expression of vascular endothelial cadherin (VE-cadherin), which establishes a structural basis for the decreased permeability (Figure 6.5b) (Price et al. 2010).

Given the disparate nature of the stabilizing conditions described in this section, it is important to consider whether they promote stability through other mechanisms, besides strengthening the endothelial barrier. For instance, cyclic AMP is known to reduce endothelial cell contractility (Morel et al. 1989), and we did find considerable cell relaxation at the highest concentrations tested (Wong et al. 2010). Reduction in contractility, however, was not able to explain the stabilizing effect of cyclic AMP at lower concentrations. Also, polymers such as dextran are known to increase the colloid osmotic pressure of solutions; so, one might expect that this increase may itself exert a stabilizing effect. We think that an effect due to osmotic pressure is unlikely, because the scaffold is already saturated with the polymers by perfusion before a confluent endothelium forms, thereby eliminating a transvascular osmotic pressure difference. In fact, these polymers induce the elevation of VE-cadherin junctional localization even in endothelial cells that are grown on plastic substrates, a culture condition in which osmotic pressure differences should be negligible

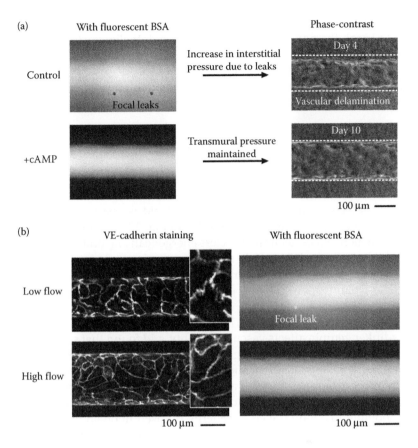

FIGURE 6.5
Conditions that prevent vascular leakage yield stable vessels. (a) Cyclic AMP, an anti-inflammatory second messenger, strengthens barrier function and maintains vessel stability. Barrier function was assayed on day 3 postseeding by taking fluorescence images of vessels perfused with Alexa Fluor 594-labeled bovine serum albumin (BSA). (b) High flow—generated with a high-pressure gradient—enhances VE-cadherin localization at cell–cell junctions and reduces leakage in engineered vessels. Vessels under high flow are typically expanded (indicative of a high transmural pressure) and are stable. (Adapted with permission from Price, G. M. et al. 2010. *Biomaterials*, 31, 6182–9; Wong, K. H. K., Truslow, J. G. and Tien, J. 2010. *Biomaterials*, 31, 4706–14.)

(Leung et al. 2012). Finally, elevation of the flow rate can lead to simultaneous increases in shear stress and lumenal pressure. We have found that it is the increase in shear stress that leads to a decrease in endothelial permeability. The increase in lumenal pressure itself can directly increase transmural pressure. Thus, an increase in flow rate appears to stabilize in two complementary ways, by a direct change in transmural pressure and by maintenance of that pressure, which is only possible when the barrier function is strong (Price et al. 2010).

6.4.4 Prediction: Scaffold Stiffening Promotes Vascular Stability Independently of Inhibition of Gel Contraction

Finally, the mechanical model predicts that stiffening of scaffolds will stabilize vessels. The physical basis of this prediction lies in the nature of the delamination process. Although so far we have expressed the analysis as a phenomenological balance of stresses in Equation 6.3, an equivalent analysis views delamination via a balance of energies

(Figure 6.3b) (Janssen et al. 2006). To detach endothelium from the scaffold, energy must be supplied to break the adhesive bonds between endothelial surface integrins and the underlying matrix proteins (i.e., $\delta E_{adh} < 0$). This energy can be supplied from the elastic recoil of the scaffold when the endothelium detaches (i.e., $\delta E_{elastic} > 0$), as well as from the release of contractile energy as the surface area of endothelium decreases (i.e., $\delta E_{EC} > 0$). In this formulation, the condition for stability becomes

$$\delta E_{elastic} + \delta E_{adh} + \delta E_{EC} < 0 \qquad (6.4)$$

When this inequality is satisfied, the energy released upon delamination is insufficient to overcome adhesion, and the vessel wall will be stable.

In principle, the elastic energy stored in the scaffold is inversely proportional to the scaffold stiffness for a given imposed stress (Beer et al. 2006). Thus, stiffening a scaffold reduces $\delta E_{elastic}$, which makes delamination less energetically favorable. Studies of endothelial mechanobiology imply that scaffold stiffening can also lead to increases in total adhesion energy (which would promote stability) (Choquet et al. 1997) and contractility (which would promote delamination) (Huynh et al. 2011); we expect these secondary effects to largely offset.

In a recent work, we have analyzed how scaffold stiffening by the small-molecular-weight cross-linker genipin affects vascular stability, and found that stiff scaffolds (moduli roughly fivefold greater than in native gels) promote stability as predicted (Chan et al. in press). Surprisingly, vessels in stiff scaffolds could withstand conditions that would normally lead to quick delamination, such as nearly zero-transmural pressure. To determine whether the energy-based mechanism described above is plausible, we have used computational modeling to obtain the elastic energy released upon endothelial delamination when transmural pressure is zero. In a scaffold of 400 Pa shear modulus (comparable to an ~8 mg/mL native collagen gel), delamination of a 400 μm^2 patch of endothelium results in a release of 1.7×10^{-6} erg of energy, nearly all of which results from elastic recoil of the scaffold; typical adhesion strengths under slow peeling of ~0.15 dyn/cm (Engler et al. 2004) yield an adhesion energy of ~6×10^{-7} erg, and we find that inequality (4) is *not* satisfied (i.e., delamination will occur). When the scaffold is stiff (shear modulus of 2 kPa), the elastic energy released upon delamination greatly decreases to 3.3×10^{-7} erg, and now, inequality (4) is satisfied (i.e., delamination is inhibited). Thus, scaffold stiffness-induced stability is both experimentally observed and computationally realistic.

6.5 Summary and Future Directions

The discussion above summarizes the evidence for a physical mechanism of vascular stability in microfluidic scaffolds. Although other, nonphysical mechanisms (e.g., perivascular proteolysis of the scaffold) could also play a role in maintaining the endothelium, the success of a purely physical theory in explaining the effects of a wide range of signals on stability suggest that these alternative mechanisms are likely to be minor. Further clarification of the proposed mechanical stabilizing signals will benefit from direct, noninvasive measurements of transmural pressure, adhesion energy or stress, and cell contractility without relying on inference from computational modeling. Such *in situ* measurements are challenging, but may become possible with the development of fluorescent reporter proteins whose optical properties are altered by local mechanical strain (Klotzsch et al. 2009, Grashoff et al. 2010).

We have presented these ideas in the context of vascularizing microfluidic collagen and fibrin scaffolds. Might similar ideas hold in other systems of vascularization? It seems likely that vascularization of other microfluidic materials, such as synthetic polymer and alginate gels, should follow the same physical principles as those that govern vascularization of ECM-based scaffolds. The relative magnitudes of the relevant stresses or energies will depend on the particular chemistry of the material (e.g., σ_{adh} in synthetic polymer gels may be close to zero without coupling of adhesive peptides), and thus, the minimum transmural pressure required for stability may vary with scaffold chemistry. Since the density of cell adhesion ligands can be tailored on synthetic polymers, this class of scaffolds may be a particularly good model system for testing the effect of cell adhesion strength on vascular stability. In any case, we believe that the mechanical model of vascular stability will hold in all microfluidic scaffolds, including those that are formed by decellularization of native tissues (Badylak et al. 2011).

It is less clear to what extent this model can explain the signals that affect vascular stability in growth factor-induced angiogenesis, which is, by far, the most popular approach in vascularization. Several differences exist between the vascularization process described here and angiogenesis, most obviously in terms of scale (50- to 100-μm-diameter tubes vs. 5- to 10-μm-diameter capillaries). In angiogenesis, the endothelial cells must create channels where none existed, and vascular regression can occur before and after initiation of perfusion through nascent sprouts (Black et al. 1998, Darland and D'Amore 2001, Shepherd et al. 2004). We suspect that the mechanical stress profile that surrounds a perfused capillary that arises by angiogenesis will depend on whether capillary sprouts grow primarily by expansion (i.e., by pushing against the surrounding matrix) or by migration into a just-formed channel. Moreover, vascular growth factors tend to activate endothelial cells and can induce hyperpermeability that may make transmural pressure less relevant as a stabilizing factor. Physical signals, such as shear stress, can stabilize capillaries through biochemical mechanisms, such as inhibition of apoptosis (Dimmeler et al. 1996). It remains to be determined whether physical signals play a major or minor *mechanical* role in stabilizing vessels that arise from angiogenesis.

Our results suggest that the ideal scaffold for microfluidic vascularization is one that is stiff and of intermediate pore size. The stiffer the scaffold, the more stable the vessels that form in it. With large pore sizes, the vascular wall becomes poorly defined as cells migrate from the channels into the scaffold; with small pore sizes, the hydraulic resistance of the scaffold destabilizes vessels. If small pore sizes are unavoidable, then addition of drainage channels to directly lower scaffold pressures is indicated. When designing scaffolds to engineer a vascularized tissue, these ideal characteristics may provide useful rules to guide the design process.

Acknowledgments

We thank John Hutchinson, Dimitrije Stamenović, and Victor Varner for insightful discussions of thin-film fracture mechanics. This work was supported by the National Heart, Lung, and Blood Institute through award HL092335.

References

Alcaraz, J., Nelson, C. M., and Bissell, M. J. 2004. Biomechanical approaches for studying integration of tissue structure and function in mammary epithelia. *J Mammary Gland Biol Neoplasia*, 9, 361–74.

Asahara, T., Murohara, T., Sullivan, A., Silver, M., van der Zee, R., Li, T., Witzenbichler, B., Schatteman, G., and Isner, J. M. 1997. Isolation of putative progenitor endothelial cells for angiogenesis. *Science*, 275, 964–7.

Au, P., Tam, J., Fukumura, D., and Jain, R. K. 2008. Bone marrow-derived mesenchymal stem cells facilitate engineering of long-lasting functional vasculature. *Blood*, 111, 4551–8.

Badylak, S. F., Taylor, D., and Uygun, K. 2011. Whole-organ tissue engineering: Decellularization and recellularization of three-dimensional matrix scaffolds. *Annu Rev Biomed Eng*, 13, 27–53.

Barber, J. R. 2010. *Elasticity*. Dordrecht; New York: Springer.

Beer, F. P., Johnston, E. R., and DeWolf, J. T. 2006. *Mechanics of Materials*. Boston: McGraw-Hill.

Black, A. F., Berthod, F., L'Heureux, N., Germain, L., and Auger, F. A. 1998. *In vitro* reconstruction of a human capillary-like network in a tissue-engineered skin equivalent. *FASEB J*, 12, 1331–40.

Bouhadir, K. H. and Mooney, D. J. 2001. Promoting angiogenesis in engineered tissues. *J Drug Target*, 9, 397–406.

Chan, K. L. S., Khankhel, A. H., Thompson, R. L., Coisman, B. J., Wong, K. H. K., Truslow, J. G., and Tien, J. Crosslinking of collagen scaffolds promotes blood and lymphatic vascular stability. *J Biomed Mater Res A*, in press.

Chen, C. S., Tan, J., and Tien, J. 2004. Mechanotransduction at cell–matrix and cell–cell contacts. *Annu Rev Biomed Eng*, 6, 275–302.

Chiu, L. L., Montgomery, M., Liang, Y., Liu, H., and Radisic, M. 2012. Perfusable branching microvessel bed for vascularization of engineered tissues. *Proc Natl Acad Sci USA*, 109, E3414–23.

Choi, N. W., Cabodi, M., Held, B., Gleghorn, J. P., Bonassar, L. J., and Stroock, A. D. 2007. Microfluidic scaffolds for tissue engineering. *Nat Mater*, 6, 908–15.

Choquet, D., Felsenfeld, D. P., and Sheetz, M. P. 1997. Extracellular matrix rigidity causes strengthening of integrin–cytoskeleton linkages. *Cell*, 88, 39–48.

Chrobak, K. M., Potter, D. R., and Tien, J. 2006. Formation of perfused, functional microvascular tubes *in vitro*. *Microvasc Res*, 71, 185–96.

Comper, W. D. and Zamparo, O. 1989. Hydraulic conductivity of polymer matrices. *Biophys Chem*, 34, 127–35.

Darland, D. C. and D'Amore, P. A. 2001. TGF beta is required for the formation of capillary-like structures in three-dimensional cocultures of 10T1/2 and endothelial cells. *Angiogenesis*, 4, 11–20.

Dimmeler, S., Haendeler, J., Rippmann, V., Nehls, M., and Zeiher, A. M. 1996. Shear stress inhibits apoptosis of human endothelial cells. *FEBS Lett*, 399, 71–4.

Discher, D. E., Janmey, P., and Wang, Y. L. 2005. Tissue cells feel and respond to the stiffness of their substrate. *Science*, 310, 1139–43.

Engler, A. J., Griffin, M. A., Sen, S., Bonnemann, C. G., Sweeney, H. L., and Discher, D. E. 2004. Myotubes differentiate optimally on substrates with tissue-like stiffness: Pathological implications for soft or stiff microenvironments. *J Cell Biol*, 166, 877–87.

Gallant, N. D., Michael, K. E., and Garcia, A. J. 2005. Cell adhesion strengthening: Contributions of adhesive area, integrin binding, and focal adhesion assembly. *Mol Biol Cell*, 16, 4329–40.

Gjorevski, N. and Nelson, C. M. 2010. The mechanics of development: Models and methods for tissue morphogenesis. *Birth Defects Res C Embryo Today*, 90, 193–202.

Golden, A. P. and Tien, J. 2007. Fabrication of microfluidic hydrogels using molded gelatin as a sacrificial element. *Lab Chip*, 7, 720–5.

Grashoff, C., Hoffman, B. D., Brenner, M. D., Zhou, R., Parsons, M., Yang, M. T., McLean, M. A. et al. 2010. Measuring mechanical tension across vinculin reveals regulation of focal adhesion dynamics. *Nature*, 466, 263–6.

Huynh, J., Nishimura, N., Rana, K., Peloquin, J. M., Califano, J. P., Montague, C. R., King, M. R., Schaffer, C. B., and Reinhart-King, C. A. 2011. Age-related intimal stiffening enhances endothelial permeability and leukocyte transmigration. *Sci Transl Med*, 3, 112ra122.

Janssen, M., Zuidema, J., and Wanhill, R. J. H. 2006. *Fracture Mechanics: Fundamentals and Applications.* Delft: VSSD.

Kaully, T., Kaufman-Francis, K., Lesman, A., and Levenberg, S. 2009. Vascularization—The conduit to viable engineered tissues. *Tissue Eng Part B Rev*, 15, 159–69.

Klotzsch, E., Smith, M. L., Kubow, K. E., Muntwyler, S., Little, W. C., Beyeler, F., Gourdon, D., Nelson, B. J., and Vogel, V. 2009. Fibronectin forms the most extensible biological fibers displaying switchable force-exposed cryptic binding sites. *Proc Natl Acad Sci USA*, 106, 18267–72.

Koike, N., Fukumura, D., Gralla, O., Au, P., Schechner, J. S., and Jain, R. K. 2004. Tissue engineering: Creation of long-lasting blood vessels. *Nature*, 428, 138–9.

Kolodney, M. S. and Wysolmerski, R. B. 1992. Isometric contraction by fibroblasts and endothelial cells in tissue culture: A quantitative study. *J Cell Biol*, 117, 73–82.

Lee, J. and Schmid-Schönbein, G. W. 1995. Biomechanics of skeletal muscle capillaries: Hemodynamic resistance, endothelial distensibility, and pseudopod formation. *Ann Biomed Eng*, 23, 226–46.

Leung, A. D., Wong, K. H. K., and Tien, J. 2012. Plasma expanders stabilize human microvessels in microfluidic scaffolds. *J Biomed Mater Res A*, 100, 1815–22.

Ling, Y., Rubin, J., Deng, Y., Huang, C., Demirci, U., Karp, J. M., and Khademhosseini, A. 2007. A cell-laden microfluidic hydrogel. *Lab Chip*, 7, 756–62.

Lovett, M., Lee, K., Edwards, A., and Kaplan, D. L. 2009. Vascularization strategies for tissue engineering. *Tissue Eng Part B Rev*, 15, 353–70.

Michel, C. C. and Curry, F. E. 1999. Microvascular permeability. *Physiol Rev*, 79, 703–61.

Miller, J. S., Stevens, K. R., Yang, M. T., Baker, B. M., Nguyen, D. H., Cohen, D. M., Toro, E. et al. 2012. Rapid casting of patterned vascular networks for perfusable engineered three-dimensional tissues. *Nat Mater*, 11, 768–74.

Morel, N. M. L., Dodge, A. B., Patton, W. F., Herman, I. M., Hechtman, H. B., and Shepro, D. 1989. Pulmonary microvascular endothelial cell contractility on silicone rubber substrate. *J Cell Physiol*, 141, 653–9.

Nichol, J. W., Koshy, S. T., Bae, H., Hwang, C. M., Yamanlar, S., and Khademhosseini, A. 2010. Cell-laden microengineered gelatin methacrylate hydrogels. *Biomaterials*, 31, 5536–44.

Price, G. M., Chu, K. K., Truslow, J. G., Tang-Schomer, M. D., Golden, A. P., Mertz, J., and Tien, J. 2008. Bonding of macromolecular hydrogels using perturbants. *J Am Chem Soc*, 130, 6664–5.

Price, G. M. and Tien, J. 2009. Subtractive methods for forming microfluidic gels of extracellular matrix proteins. In: Bhatia, S. N., and Nahmias, Y., (eds.) *Microdevices in Biology and Engineering.* Boston, MA: Artech House.

Price, G. M. and Tien, J. 2011. Methods for forming human microvascular tubes *in vitro* and measuring their macromolecular permeability. *Methods Mol Biol*, 671, 281–93.

Price, G. M., Wong, K. H. K., Truslow, J. G., Leung, A. D., Acharya, C., and Tien, J. 2010. Effect of mechanical factors on the function of engineered human blood microvessels in microfluidic collagen gels. *Biomaterials*, 31, 6182–9.

Rippe, B. and Haraldsson, B. 1994. Transport of macromolecules across microvascular walls: The two-pore theory. *Physiol Rev*, 74, 163–219.

Schmid-Schönbein, G. W. 1990. Microlymphatics and lymph flow. *Physiol Rev*, 70, 987–1028.

Shepherd, B. R., Chen, H. Y., Smith, C. M., Gruionu, G., Williams, S. K., and Hoying, J. B. 2004. Rapid perfusion and network remodeling in a microvascular construct after implantation. *Arterioscler Thromb Vasc Biol*, 24, 898–904.

Tien, J., Wong, K. H. K., and Truslow, J. G. 2013. Vascularization of microfluidic hydrogels. In: Bettinger, C. J., Borenstein, J. T., and Tao, S. L., (eds.) *Microfluidic Cell Culture Systems*, 2nd edition. pp. 205–221. Oxford, U.K: Elsevier.

Tremblay, P.-L., Hudon, V., Berthod, F., Germain, L., and Auger, F. A. 2005. Inosculation of tissue-engineered capillaries with the host's vasculature in a reconstructed skin transplanted on mice. *Am J Transplant*, 5, 1002–10.

Truslow, J. G., Price, G. M., and Tien, J. 2009. Computational design of drainage systems for vascularized scaffolds. *Biomaterials*, 30, 4435–43.

Vailhé, B., Vittet, D., and Feige, J. J. 2001. *In vitro* models of vasculogenesis and angiogenesis. *Lab Invest*, 81, 439–52.

Wong, K. H. K., Chan, J. M., Kamm, R. D., and Tien, J. 2012. Microfluidic models of vascular functions. *Annu Rev Biomed Eng*, 14, 205–30.

Wong, K. H. K., Truslow, J. G., Khankhel, A. H., Chan, K. L., and Tien, J. 2013. Artificial lymphatic drainage systems for vascularized microfluidic scaffolds. *J Biomed Mater Res A*, 101, 2181–90.

Wong, K. H. K., Truslow, J. G., and Tien, J. 2010. The role of cyclic AMP in normalizing the function of engineered human blood microvessels in microfluidic collagen gels. *Biomaterials*, 31, 4706–14.

Yeon, J. H., Ryu, H. R., Chung, M., Hu, Q. P., and Jeon, N. L. 2012. *In vitro* formation and characterization of a perfusable three-dimensional tubular capillary network in microfluidic devices. *Lab Chip*, 12, 2815–22.

Zheng, Y., Chen, J., Craven, M., Choi, N. W., Totorica, S., Diaz-Santana, A., Kermani, P. et al. 2012. *In vitro* microvessels for the study of angiogenesis and thrombosis. *Proc Natl Acad Sci USA*, 109, 9342–7.

Zimmermann, H., Wahlisch, F., Baier, C., Westhoff, M., Reuss, R., Zimmermann, D., Behringer, M. et al. 2007. Physical and biological properties of barium cross-linked alginate membranes. *Biomaterials*, 28, 1327–45.

7

Engineered In Vitro *Systems of the Microcirculation*

Monica L. Moya, Christopher C. W. Hughes, and Steven C. George

CONTENTS

7.1 Introduction

Although the microcirculation refers to the smallest blood vessels in the body, its contribution to defining microenvironmental conditions in tissues is anything but small. Serving as the intermediate between the larger vessels of the arterial and venous circulation and the tissues to which it delivers blood and nutrients, the microcirculation is responsible for meeting the metabolic demands of the tissues. Impairments or dysfunction in the microcirculation can lead to and propagate pathologies, including cardiovascular disease, cancer, and diabetes mellitus.

Recapitulating the function of the microcirculation *in vitro* is a pursuit of biologists and engineers alike, who aim to better understand the microcirculation in hopes of being able to manipulate it for therapeutic and regenerative purposes. Many of the *in vitro* platforms created to meet that need seek to imitate specific features of the *in vivo* environment. The challenge that arises in this pursuit, however, is that often there is limited understanding of what is being imitated. Only after a fundamental understanding of basic biology is achieved can more effective therapeutics be designed. Therefore, the role and challenge of these models of the microcirculation is to not simply provide models that create potential therapies but also to contribute fundamental understanding of the physiological and pathophysiological processes of the microcirculation.

This chapter will highlight and assess *in vitro* models of the microcirculation that have been developed over the past four decades. This overview will examine their potential as tools to guide the design of new therapies as well as increase our understanding of the microcirculation. While many of these models have been designed to study specific diseased states, or to examine a particular pathophysiological mechanism, this review will focus more on categorizing the basic principles and methods on which these models are built. We have organized this review into two major sections broadly categorizing these models based on their design for studying vascular function. The first part will focus on methods used to understand the microcirculations' transport and flow processes while the second part will focus on models used to investigate growth and remodeling processes of the microcirculation.

7.2 Transport and Flow

The microcirculation performs, or is characterized by, several fundamental physiological processes, which include non-Newtonian fluid mechanical properties (i.e., shear thinning), endothelial shear, platelet aggregation, leukocyte attachment and margination, selective permeability, and transport of the respiratory gases (oxygen and carbon dioxide). Our understanding of these processes has been significantly enriched by a wide range of models of the microcirculation that broadly fall into seven categories: (1) cone and plate rheometry, (2) parallel plate rheometry, (3) porous membranes, (4) glass (or other material) tubes, (5) packed columns, (6) polymeric hydrogels, and (7) microfabricated networks. Each system has advantages and disadvantages and can be successfully employed to understand features of the microcirculation depending on the objective.

7.2.1 Cone and Plate Rheometry

Blood has particulate (cells, platelets) and fluid (plasma, including fibrinogen) phases that can aggregate depending on the chemical and mechanical environments. As such, blood is a non-Newtonian fluid, displaying characteristics consistent with a Casson fluid (nonzero yield stress and shear thinning). In addition, the anatomy of the vascular tree is generally a bifurcating tree structure in which the daughter branches are shorter in length and smaller in diameter compared to the parent. This latter feature, which conforms to the Hess–Murray law, creates a wide range of fluid velocities, although shear rate at the wall is nearly constant.

These unique features of blood and the vascular system were first explored more than four decades ago using a traditional cone plate rheometer (Charm and Kurland 1969, Charm et al. 1969, Parker 1968) (Figure 7.1a). The cone and plate rheometer consists of a cone-shaped top structure that points downward and touches a stationary flat plate. As the cone spins, it creates a constant shear stress in the radial direction. The cone and plate rheometer recreates the shear and fluid velocities present in the microcirculation, and thus

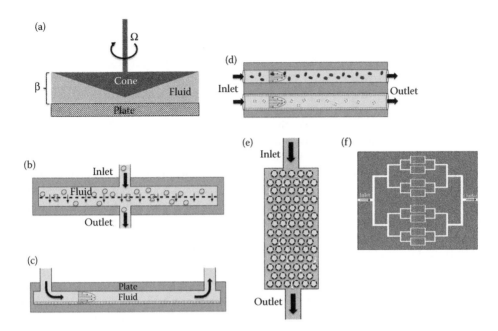

FIGURE 7.1

Schematic depiction of the major model systems used to simulate flow and mass transport features of the micro-circulation. (a) The cone and plate rheometer is characterized by a cone-shaped piece that points downward toward a flat plate, and the space between the two surfaces is filled with a fluid. An angle β characterizes the shape of the cone, and ensures that the shear stress within the fluid when the cone rotates at angular speed Ω is constant with respect to radial position. (b) A porous membrane can be used to simulate the permeability of the capillary to molecules and cells. Fluid containing a particulate phase of interest (e.g., red blood cells) can be forced by an external pressure gradient through the membrane. The rate of appearance of the particulate phase on the other side of the membrane provides an index of microcirculation permeability. (c) Two parallel plates can be separated by a fixed distance and lined by endothelial cells. A fluid can then flow between the plates (approximate laminar velocity profile depicted) to effectively mimic the controlled shear stress environ-ment in the microcirculation. (d) Tubes made from transparent materials such as glass can be embedded in a supporting material to mimic the cylindrical shape of small arterioles or capillaries. Fluid-carrying cells such as red blood cells (top tube) or white blood cells (bottom tube) can flow through the tubes (approximate lami-nar velocity profile depicted), which can also be lined by endothelial cells. (e) Small beads made from glass or dextran can be coated (circular dashed black line) with endothelial cells and packed gently into a long cylinder. Fluid mimicking plasma, lymph, or blood can flow through the pack bed to simulate the resistance to flow in the microcirculation as well as endothelial absorption functions. (f) A finely controlled network of fluidic chan-nels can be microfabricated into a polymer such as PDMS to mimic flow and resistance patterns present in the microcirculation.

constitutes one of the earliest models of the microcirculation. The cone and plate rheom-eter has been employed successfully to investigate the non-Newtonian features of blood, including values for the yield stress, and range of shear in which the viscosity decreases with increasing shear (shear thinning). In addition, the cone and plate arrangement has been used to generate numerous and sustained observations related to the role of shear and platelet aggregation (Giorgio and Hellums 1988, Levy-Shraga et al. 2006). Finally, endothelial cell response to fluid shear was initially investigated by a cone and plate rhe-ometer (Dewey et al. 1981, Davies et al. 1984). These early seminal studies demonstrated that fluid shear stress leads to cytoskeletal reorganization and alignment with flow, altered cell migration, and enhanced pinocytosis.

7.2.2 Filters and Membranes

The endothelium that lines arteries and veins, and forms the wall of the capillary in the microcirculation, provides important and selective barrier properties. For example, it has long been known that white blood cells cross the endothelium as they are transported from the blood to the interstitial space. In addition, the endothelium selectively retains larger proteins such as albumin in the blood, but lower-molecular-weight molecules diffuse across relatively freely. The selective permeability functions of the microcirculation were first mimicked using a series of porous membranes or filters (Figure 7.1b). For example, filters with pore sizes ranging from 5 to 8 μm demonstrated the impact of pH, temperature, hydrostatic pressure, osmotic pressure, cell concentration, and cell deformability on the transmembrane migration of peripheral blood leukocytes (Tanner and Scott 1976, Reinhart and Chien 1987, Eppihimer and Lipowsky 1994) and red blood cells (Reinhart and Chien 1987). Similar membranes (polycarbonate, 5 μm pore diameter) have been used to model the microcirculation of the lungs and investigate mechanisms underlying the retarded transit of neutrophils relative to red blood cells (Selby et al. 1991). While porous membranes alone have proven useful to investigate transport properties of white blood cells and red blood cells in the microcirculation, they have also proven to be a helpful mimic of the endothelial basement membrane. As such, they have proven to be a useful substrate for investigating the permeability of cultured endothelial cells to a variety of compounds and cells, including albumin (Smith and Borchardt 1989, Dull et al. 1991), dextran (Albelda et al. 1988), horseradish peroxidase (Behzadian et al. 2003), white blood cells (Giri et al. 2002, McGettrick et al. 2006, Moreland and Bailey 2006), and tumor cells (Lee et al. 2003, Sahni et al. 2009).

7.2.3 Parallel Plates

Following on the success of the cone and plate rheometer and porous membranes, were *in vitro* devices characterized by laminar flow of cell culture media between two flat parallel plates separated by a fixed distance (Figure 7.1c). Parallel plate devices share the advantages of the cone and plate rheometer and porous membranes in that they can be used to finely control fluid shear at the surface and allow transport across the plates (lined with or without endothelial cells). Thus, they can be used for similar studies of blood viscosity, white blood cell attachment, endothelial cell alignment (Levesque and Nerem 1985), endothelial cell metabolism and transport (Nollert et al. 1991), and platelet aggregation (Muggli et al. 1980, Barstad et al. 1994) within the microcirculation. In addition, the parallel plate design offers much more flexibility in altering the components of the fluid in a temporal fashion, and for visualizing dynamics in real time.

Using the inlet/outlet design, chemical or cellular components can be introduced at precise times, and the duration of exposure can be easily controlled. For example, the duration of exposure time of platelets to endothelial cells in a controlled shear environment demonstrated that platelets can adhere to specific proteins in the extracellular matrix (Sakariassen et al. 1983). Combining relatively simple microscopes and the parallel plate system made of transparent materials such as glass has facilitated real-time visualization of white blood attachment from the top (Forrester and Lackie 1984), as well as from a side view (Lei et al. 1999). The relative simplicity of the parallel plate design also stimulated the first high-throughput platform designs of the microcirculation (Low et al. 1996). More recently, the parallel plate design has been adapted to include a porous membrane and thus allow coculture studies involving endothelial cells with smooth muscle cells

(Chiu et al. 2003). More advanced designs have been able to layer parallel plate flow chambers to create models of ischemia–reperfusion injury (Lee et al. 2009).

7.2.4 Small-Diameter Tubes

A primary limitation of parallel plates is the flat geometry of the flow chamber. To overcome this limitation, while still maintaining the potential for real-time visualization, small-diameter tubes made of glass or other polymers (Figure 7.1d) have been used to mimic features of the microcirculation. Although circular in shape, the major disadvantage of this technique is a lack of wall flexibility, and in many cases, rather impermeable walls. Nonetheless, this methodology has contributed to our understanding of blood rheology, including the Farheus–Lindquist effect (Thompson et al. 1989, Kubota et al. 1996), effect of shear on endothelial cell phenotype (Eskin et al. 1984), margination of white blood cells (Bagge et al. 1983, Goldsmith and Spain 1984, Nobis et al. 1985), and oxygen transport by perfluorocarbons across permeable capillary tubes (Vaslef and Goldstick 1994).

7.2.5 Microcarrier Beads and Packed Columns

An interesting method to simulate such features of the microcirculation as endothelial cell absorption and blood coagulation is the use of microcarrier beads, either alone or placed in a long cylindrical column (Figure 7.1e). The beads can be made of glass or dextran (Cytodex©), either coated with endothelial cells or not, and then either placed in a column or a well plate. When placed in a column, cell culture media slowly introduced at the top percolates over the beads. This particular geometry dramatically increases the surface-area-to-volume ratio compared to other methods, and can also effectively mimic the resistance to flow of the entire vascular bed to a single organ such as the lung. However, the technique lacks the control of uniform or homogeneous fluid shear and easy visualization offered by parallel plates. Microcarrier beads and packed columns have been used to enhance our understanding of the formation of macromolecular complexes in the blood coagulation (DePaulis et al. 1988), endothelial cell permeability (Killackey et al. 1986, Alexander et al. 1988, Haselton and Alexander 1992, Haselton et al. 1996), and white blood cell entrapment (Haselton et al. 1996).

7.2.6 Microfabricated Networks

The application of microfabrication technology to the study of microcirculation at the turn of the twenty-first century represents a significant advance in our understanding, particularly the impact of a branching network on the microcirculation function. Microfabrication technologies were developed initially by the semiconductor industry, but methods such as plasma etching, photolithography, and polymer micromolding have proved useful for mimicking the complex structure of the microcirculation at submicron resolution (Figure 7.1f).

The earliest examples of microfabricated microvessel networks utilized a silicon wafer as the backbone, and then utilized a mask combined with a photopolymerizable polymer such as SU-8 to create a negative mold on top of the wafer. A soft biocompatible polymer such as polydimethylsiloxane (PDMS) was then placed over the mold, cured, and then peeled off to create a positive mold of the microchannel network (Borenstein et al. 2002). The molded PDMS could then be bonded to another layer of PDMS or glass to create a completely enclosed network of microchannels of diameters as small as 20 μm. This methodology represents an example of polymer micromolding and is still widely used. The network of channels can be created to precisely mimic the branching network of the microcirculation, including the drop in pressure and the desired shear rates. The channels can also be lined by endothelial cells, which have been shown to easily develop confluency in a matter of days and maintain viability for more than 2 weeks (Shin et al. 2004). Alternate polymers such as poly(glycerol sebacate) (PGS) (Fidkowski et al. 2005) and liquid elastomer RTV 615 A/B (Shevkoplyas et al. 2003) are also amendable to micromolding techniques. More recently, tissue functionality (e.g., hepatocytes) has been added to these networks (Carraro et al. 2008) demonstrating their ability to recreate organ-specific features of the microcirculation.

In general, polymer micromolding has been useful to understand the effect of complex flow patterns and shear in a vessel network on pressure loss (Borenstein et al. 2002), endothelial cell alignment (Song et al. 2005), and cellular transport (Shevkoplyas et al. 2003); however, there remain numerous disadvantages. For example, the natural variability (e.g., asymmetric branching) of the *in vivo* microcirculation has been difficult to simulate. An additional limitation of the early photolithographic and polymicromolding techniques was square channels, which do not completely replicate *in vivo* flow patterns, and endothelial cells do not coat the sharp transitions present in the corners. This has potentially been alleviated by more recent reports that present techniques to create round channels using similar micromolding techniques, and have also included traditional polystyrene in place of PDMS (Wang et al. 2007, Borenstein et al. 2010). Finally, a major limitation of polymer micromolding with synthetic polymers such as PDMS, PGS, or polystyrene is the lack of a natural extracellular matrix to encase the microvessels as well as other functional cells.

More recently, a significant advance has been the creation of microvascular networks within *in vivo* extracellular matrix proteins such as collagen and fibrin. The earliest example utilized collagen cast within a PDMS mold around a removable needle. Once the needle was removed, the resulting channel was completely surrounded by collagen, and following seeding with endothelial cells, created endothelial-lined channels within the collagen gel on the order of arterioles (50–100 µm diameter) (Chrobak et al. 2006). Later advances in this technique employed gelatin as the sacrificial polymer and demonstrated vessels as small as 6 µm diameter or that of a capillary (Golden and Tien 2007). Most recently, this method has been extended to 3D microvessel networks in which a polysaccharide (sugar) is micropatterned and used to create the sacrificial scaffold for the network using 3D printing (Lee et al. 2010, Miller et al. 2012) or melt spinning (Bellan et al. 2009).

7.3 Growth and Remodeling

While early studies of the microcirculation focused on the understanding of fundamental physiological processes, the past three decades have been marked by increased interest in forming or guiding the development of living dynamic vascular networks. An understanding of growth and remodeling processes such as migration, proliferation, vessel sprouting, tubulogenesis, stabilization, and maturation is required to effectively develop strategies for vascularization. This section will focus on the *in vitro* models that have contributed and continue to add to our understanding of these processes. These models will be divided into the following categories: (1) 2D culture (cells grown on various substrates including patterned substrates); (2) 3D culture (static culture of cells within a scaffold); and (3) dynamic 3D culture (3D bioreactor cultures).

7.3.1 Two-Dimensional Cultures

7.3.1.1 Substrates and Coatings

Two-dimensional cultures of cells provided some of the earliest studies of the microcirculation and its growth processes. Among the first of the 2D cultures were cultivation of microvascular endothelial cells on plastic with gelatin and collagen coatings (Figure 7.2a) (Folkman and Haudenschild 1980, Montesano et al. 1983, Ingber and Folkman 1989). These early *in vitro* models provided a simple approach for investigating important factors that stimulate endothelial cell proliferation and rudimentary capillary assembly. This method of growing cells on 2D surfaces continues to be a useful tool for studying singular process such as migration, proliferation, and endothelial cell differentiation. Typically, in these models, a singular vascular cell type (e.g., endothelial cells or mural cells) is investigated. The 2D surfaces that are used to stimulate the interaction of the cells with the extracellular matrix range from coatings of basement proteins such as fibronectin, collagen IV or I, or matrigel to seeding cells on gels of fibrin (Vailhe et al. 1998), collagen (Sieminski et al. 2004, Hong and Stegemann 2008, Francis-Sedlak et al. 2010), or tissue-derived matrix.

7.3.1.2 Patterned Substrates

A more recent variation of this type of 2D *in vitro* model is growing cells on patterned substrates (Figure 7.2b). Adhesive patterns of extracellular matrix proteins such as fibronectin or other biological molecules are patterned on a rigid surface such as silicon or gold (Chen et al. 1998). This approach allows for vascular structures to be patterned in a precise location or in the form of a vascular structure rather than relying on the spontaneous self-organization of cells. Using this technique, growth, apoptosis, migration, and proliferation of vascular cells can be examined with regard to the influence of size and geometry of the patterned substrate.

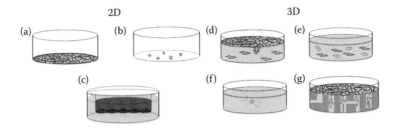

FIGURE 7.2
Schematic depiction of the major model systems used to study growth and remodeling processes of the microcirculation. (a) 2D substrates and coatings can be used to simulate the interaction of cells with extracellular matrix. (b) Cells or vascular structures can be patterned in a precise location or shape using adhesive patterns of extracellular matrix proteins. (c) A Boyden chamber consists of a cell culture insert placed into the well of a cell culture plate. Migration or invasion from cells seeded in the top of the insert can be stimulated by placing chemoattractant in the well below. (d) Assays where 2D cultured cells invade a surrounding 3D matrix allow for visualization of angiogenesis-like processes such as sprouting. (e) Endothelial cells dispersed in a 3D matrix either as a single cell type or mixed with stromal cells self-assemble into tubes. This process resembles the vascularization process of vasculogenesis. (f) A mass of cells embedded in a 3D scaffold, either as an aggregate or a cell-covered bead, can sprout radially and form lumenized capillary networks. Sprouting from a focal point allows for easier quantification of forming capillaries. (g) Patterned 3D scaffolds consist of scaffolds with predefined pathways for cell and vascular growth.

7.3.1.3 Boyden Chamber

Another commonly used variation of 2D *in vitro* models is the Boyden chamber assay. This assay aims to recapitulate the migration of endothelial cells from existing blood vessels into surrounding tissue areas during angiogenesis. For this assay, endothelial cells are grown as a monolayer on a cell-permeable membrane (8+ μm pore size) sometimes coated with extracellular matrix, a Transwell insert inside the wells of a multiwell plate (Figure 7.2c). Cell medium is placed above the monolayer while medium with chemoattractants or test agent is placed below the membrane. This assay is commonly used to test the migratory response of endothelial cells to angiogenic inducers or inhibitors. A major advantage of this assay is that it requires low levels of angiogenic stimulus to induce migration of endothelial cells into the second chamber.

7.3.2 Three-Dimensional Cultures

Since the early studies using 2D models, interest has been growing in 3D cultures as most cells in the body are in a 3D environment that is not recapitulated at all in monolayer, 2D cultures. Migration, proliferation, endothelial cell differentiation, tubulogenesis, branching, and stabilization are all crucial steps in the formation of the microvasculature, both embryologically and in the adult during both normal and diseased processes such as would healing and tumorigenesis. 3D models of the microcirculation are now becoming important tools for screening and developing potential therapies associated with the microvasculature.

7.3.2.1 Cells Dispersed in 3D Scaffolds

Many of the early studies with substrate coatings led to a growing awareness of the importance of cell–ECM interactions. Cells not only respond to the proteins and soluble factors associated with the ECM but they also respond to biomechanical cues that may serve to regulate capillary development. *In vitro* models of cells within a scaffold aim to recapitulate this environment. Initial variations to mimic the development of vessel networks in a 3D environment consisted of seeding endothelial cells sandwiched between gels (Montesano et al. 1983), under gels, or on top of malleable gels made from fibrin (Pepper et al. 1990, Vailhe et al. 1998, Bayless et al. 2000), collagen (Kubota et al. 1988, Davis et al. 2000), or Matrigel. Similar to *in vivo* angiogenesis, confluent monolayers of endothelial cells are stimulated to invade into the surrounding 3D scaffold by the addition of growth factors or other angiogenic factors (Montesano and Orci 1985). These 2D/3D assays allow for investigation of several stages in angiogenesis (Figure 7.2d) specifically allowing for the visualization of endothelial cell invasion and morphogenesis into 3D matrices.

The addition of exogenous factors for coaxing single-cell cultures to form tube structures was especially important with assays in which endothelial cells were dispersed within 3D scaffolds (Sieminski et al. 2005). Similar to the 2D cultures, these types of assay relied on the self-assembly of endothelial cells into tubes, a process that resembles the vascularization process of vaculogenesis rather than sprouting angiogenesis, where endothelial cells suspended as single cells form lumens and tubes through the coalescence of vacuoles (Davis and Camarillo 1996). Although these single-cell 3D models required addition of exogenous factors such as TGFβ (Madri et al. 1988), VEGF or bFGF (Yang et al. 1999), or HGF/SF (Lafleur et al. 2002), they allowed for the study of capillary morphogenesis in a controlled manner and leading to successful identification of many signaling components involved in tube formation (Sacharidou et al. 2012). Such systems are amenable for studying cell proliferation and migration, and also provide the ability to study remodeling processes such as matrix degradation (Vailhe et al. 1998, Sacharidou et al. 2010).

While dispersing cells in a scaffold provides cell–ECM interaction, another feature of the microcirculation microenvironment is cell–cell interaction. A shift toward 3D culture of multicellular systems allowed for this interaction to be studied, and produced systems that do not require exogenous factors. In the 2D/3D models, rather than being cultured on top of acellular gels, endothelial cells are cultured on top of gels with dispersed stromal cells. This not only creates a more *in vivo*-like environment, where the endothelial cells are responding to cues on the basal side rather than medium (and growth/morphogenic factors) placed on their apical side, but also allows for the creation of an environment where the cells are responding to paracrine soluble factors from stromal cells (Montesano et al. 1993) (Kuzuya and Kinsella 1994, Tille and Pepper 2002).

Additional cell types were also eventually added to the model of dispersed cells in 3D scaffold (Figure 7.2e). Endothelial cells can be mixed with fibroblasts or other stromal cells in a 3D scaffold. Like the monoculture dispersed model, capillary networks self-assemble but without addition of exogenous factors. In addition to providing support for growth, stromal cells have been found to wrap around endothelial cells and thus take on a pericyte-like behavior allowing for this type of cell–cell interaction behavior to be observed (Darland and D'Amore 2001, Chen et al. 2010). One benefit of these models is that their functionality can be tested *in vivo* (Sieminski et al. 2002, Koike et al. 2004, Levenberg et al. 2005, Chen et al. 2009). This offers the advantage of being able to examine not only their functionality *in vivo* but also to observe, and thus understand, the process of anastomosis (Chen et al. 2009, White et al. 2012). To further examine the cell–matrix interaction some of these models have used other matrices that better approximate the mechanical and physical properties of tissue *in vivo* (Levenberg et al. 2005).

7.3.2.2 Cell Aggregates or Cells on Beads

One limitation of *in vitro* models with dispersed cells in 3D scaffolds is the difficulty in quantifying the formation of capillary-like structures. Models with cell aggregates or cell-covered beads embedded in scaffolds provide an easier method for observing the radial formation and sprouting of capillary-like structures (Figure 7.2f). Interestingly, while the culturing of single endothelial cells dispersed in a matrix requires the addition of exogenous factors for survival, cells in aggregate form do not (Korff and Augustin 1998). Although these cells survive, they still require the addition of angiogeneic factors to stimulate sprouting. For these models, endothelial cells are coated onto Cytodex beads (Nehls and Drenckhahn 1995) or embedded as a mass of cells (Pepper et al. 1991, Korff and Augustin 1998) and are allowed to sprout, forming lumenized capillary-like structures that grow out from these spheroids into the surrounding scaffold. In some variations of these models, multiple cell-covered beads or aggregates are placed within a scaffold whereas in other variations a single aggregate is placed within a gel (Vernon and Sage 1999). The use of multiple beads or aggregates has the added advantage of allowing for the study of anastomosis between sprouts from neighboring beads or aggregates. These models are well suited for studying different steps of the angiogenesis process in response to various stimuli, including degradation of basement membrane, proliferation, sprouting of endothelial cells, and branching. Endothelial migration or sprouting is induced by various factors, including FGF-1 (Uriel et al. 2006, Moya et al. 2010) and VEGF (Korff and Augustin 1999, Xue and Greisler 2002). Addition of a stromal cell in these models as coembedded fibroblast-coated Cytodex beads, cocultured spheriods (i.e., aggegates made up of both endothelial cells and stromal cells) (Korff et al. 2001) seeded on top of the gel (Nakatsu et al. 2003), or disperesed within the gel (Ghajar et al. 2008) allows for the investigation of not only the cross talk between endothelial cells and the stromal cells but also allows for the study of vessel maturity and stability (Newman et al. 2011, 2013).

7.3.2.3 Patterned 3D Scaffolds

Like its 2D predecessor, patterning in 3D scaffolds allows for the precise control of vascular-like structure assembly (Figure 7.2g). The creation of engineered scaffolds aims to mimic the complexity and microarchitecture of the tissues. Other *in vitro* models of angiogenesis or vasculogenesis allow for cell-guided formation of vessel structures that are often randomly structured and do not recapitulate vessel hierarchy. Although patterning scaffolds allows for the spatial arrangement of endothelial tubulogenesis within 3D ECM, many of these models do not provide the flexibility of allowing the vessel network to adapt or remodel in response to an agonist or to flow.

For these models, various methods are used to create patterned structures of adhesion proteins or growth factors into either natural scaffolds such as collagen (Oh et al. in press), Matrigel (Nahmias et al. 2005), or synthetic scaffolds. These methods include 3D laser printing, microfrabrication technologies, and photopolymerizable chemistries of polymers such as PEG-based hydrogels (Chiu et al. 2009). These models serve to help understand the interaction between cells and their scaffold (Tan and Desai 2003), tubulogenesis (Raghavan et al. 2010), and cell migration along predefined pathways (Lee et al. 2008).

7.3.3 Dynamic 3D Cultures

Cells are responsive to their microenviroment sensing both mechnical and biochemical cues. While growing cells in static 3D cultures has allowed for the investigation of some mechnical and biochemical cues, these static cultures do not recapitulate the dynamic forces that drive morphogenesis *in vivo*. Bulk fluid flow can generate gradients of chemokines as well as exert forces directly on cells, or indirectly by exerting forces on the scaffolds. Dynamic 3D cultures not only help elucidate transport and flow processes (see the previous section) but are also important for understanding growth and remodeling processes in response to chemokine gradients or intersitial flow.

Early work that introduced bulk perfusion into gels demonstrated significant influence on capillary morphogenesis, cell migration, and remodeling (Ng and Swartz 2003, Helm et al. 2005, 2007). More recently, the use of microfabricated networks (see the previous section) has increased the number of dynamic 3D cultures (Wong et al. 2012). Microfabricated networks add dynamic components to 3D cell culture by typically circulating media through a hydrogel through the use of microfabricated channels. Various designs in the microfluidic network allow for numerous patterns and methods to control mass transport gradients. This percision is advantageous as it allows for fine spatial and temporal control. In some models, the design is derived from static culture but with the added complexity and sophistication of mass transport. For example, one such model uses the endothelial cell-coated Cytodex beads to examine sprouting in a microfluidic device under VEGF gradients (Shamloo et al. 2012). Like the static model, this dynamic model can examine tip sprouting, migration, and branching; however, it has the added advantage that this model can examine sprouting navigation dynamics in response to VEGF gradients.

These dynamic 3D cultures allow us to investigate all the endpoints that are possible with static cultures, but with more complexity and physiological accuracy as they also combine transport and flow processes with growth and remodeling processes. These models can be used to look at angiogenic sprouting into a 3D scaffold, but unlike previous static models, this sprouting process involves perfused native-like vessel structures (i.e., channels lined with endothelial cells) (Song et al. 2012b, Zheng et al. 2012) that can respond to flow. In addition, the sprouting is in response to not just bolus addition of growth factors but to *in vivo*-like gradients of solutes and growth factors. Some of these models have also been designed to specicifically demonstrate the vascular response to other cell types such as tumor cells (Jeon et al. 2013) or even encapsulated cells (Kim et al. 2012). More uniquely, these dynamic cultures allow for examining and understanding the development of vessels into continuous networks in response to dynamic cues such as pressure and interstitial flow (Vickerman and Kamm 2012, Hsu et al. 2013). Finally, the true utility of these models is being realized, as they not only capture the growth of vascular networks but have now been shown to support flow. The further development of perfusable microvessels *in vitro*, capable of supporting the growth of surrounding tissues, will be a major step forward in the now rapidly evolving world of tissue engineering (Yeon et al. 2012, Song et al. 2012a, Moya et al. 2013).

7.4 Conclusions

Despite the numerous *in vitro* models developed over the years, an epitome of the microcirculation is not currently available. A lack of a standardized assay provides a challenge when comparing findings from *in vitro* assays across groups. Variations in cell origins, substrate or matrix material, and growth media all need to be considered when attempting to extrapolate *in vitro* observations to an *in vivo* setting. Research in developing more *in vivo*-inspired *in vitro* models will continue to enhance our understanding of the microcirculation and provide opportunities for new therapies.

References

Albelda, S. M., Sampson, P. M., Haselton, F. R., Mcniff, J. M., Mueller, S. N., Williams, S. K., Fishman, A. P., and Levine, E. M. 1988. Permeability characteristics of cultured endothelial cell monolayers. *J Appl Physiol*, 64, 308–22.

Alexander, J. S., Hechtman, H. B., and Shepro, D. 1988. Phalloidin enhances endothelial barrier function and reduces inflammatory permeability *in vitro*. *Microvasc Res*, 35, 308–15.

Bagge, U., Blixt, A., and Strid, K. G. 1983. The initiation of post-capillary margination of leukocytes: Studies *in vitro* on the influence of erythrocyte concentration and flow velocity. *Int J Microcirc Clin Exp*, 2, 215–27.

Barstad, R. M., Roald, H. E., Cui, Y., Turitto, V. T., and Sakariassen, K. S. 1994. A perfusion chamber developed to investigate thrombus formation and shear profiles in flowing native human blood at the apex of well-defined stenoses. *Arterioscler Thromb*, 14, 1984–91.

Bayless, K. J., Salazar, R., and Davis, G. E. 2000. RGD-dependent vacuolation and lumen formation observed during endothelial cell morphogenesis in three-dimensional fibrin matrices involves the alpha(v)beta(3) and alpha(5)beta(1) integrins. *Am J Pathol*, 156, 1673–83.

Behzadian, M. A., Windsor, L. J., Ghaly, N., Liou, G., Tsai, N. T., and Caldwell, R. B. 2003. VEGF-induced paracellular permeability in cultured endothelial cells involves urokinase and its receptor. *FASEB J*, 17, 752–4.

Bellan, L., Singh, S., Henderson, P., Porri, T., Craighead, H., and Spector, J. 2009. Fabrication of an artificial 3-dimensional vascular network using sacrificial sugar structures. *Soft Matter*, 5, 1354–7.

Borenstein, J., Teraj, H., King, K., Weinberg, E., Kaazempur-Mofrad, M., and Vacanti, J. 2002. Microfabrication technology for vascularized tissue engineering. *Biomed Microdevices*, 4, 167–75.

Borenstein, J. T., Tupper, M. M., Mack, P. J., Weinberg, E. J., Khalil, A. S., Hsiao, J., and Garcia-Cardena, G. 2010. Functional endothelialized microvascular networks with circular cross-sections in a tissue culture substrate. *Biomed Microdevices*, 12, 71–9.

Carraro, A., Hsu, W. M., Kulig, K. M., Cheung, W. S., Miller, M. L., Weinberg, E. J., Swart, E. F. et al. 2008. In vitro analysis of a hepatic device with intrinsic microvascular-based channels. *Biomed Microdevices*, 10, 795–805.

Charm, S. E. and Kurland, G. S. 1969. A comparison of couette, cone and plate and capillary tube viscometry for blood. *Bibl Anat*, 10, 85–91.

Charm, S. E., Kurland, G. S., and Schwartz, M. 1969. Absence of transition in viscosity of human blood between shear rates of 20 and 100 sec^{-1}. *J Appl Physiol*, 26, 389–92.

Chen, C. S., Mrksich, M., Huang, S., Whitesides, G. M., and Ingber, D. E. 1998. Micropatterned surfaces for control of cell shape, position, and function. *Biotechnol Prog*, 14, 356–63.

Chen, X., Aledia, A. S., Ghajar, C. M., Griffith, C. K., Putnam, A. J., Hughes, C. C. W., and George, S. C. 2009. Prevascularization of a fibrin-based tissue construct accelerates the formation of functional anastomosis with host vasculature. *Tissue Eng Part A*, 15, 1363–71.

Chen, X., Aledia, A. S., Popson, S. A., Him, L., Hughes, C. C. W., and George, S. C. 2010. Rapid anastomosis of endothelial progenitor cell-derived vessels with host vasculature is promoted by a high density of cotransplanted fibroblasts. *Tissue Eng Part A*, 16, 585–94.

Chiu, J. J., Chen, L. J., Lee, P. L., Lee, C. I., Lo, L. W., Usami, S., and Chien, S. 2003. Shear stress inhibits adhesion molecule expression in vascular endothelial cells induced by coculture with smooth muscle cells. *Blood*, 101, 2667–74.

Chiu, Y.-C., Larson, J. C., Perez-Luna, V. H., and Brey, E. M. 2009. Formation of microchannels in poly(ethylene glycol) hydrogels by selective degradation of patterned microstructures. *Chem Mater*, 21, 1677–82.

Chrobak, K. M., Potter, D. R., and Tien, J. 2006. Formation of perfused, functional microvascular tubes in vitro. *Microvasc Res*, 71, 185–96.

Darland, D. C. and D'Amore, P. A. 2001. TGF beta is required for the formation of capillary-like structures in three-dimensional cocultures of 10T1/2 and endothelial cells. *Angiogenesis*, 4, 11–20.

Davies, P. F., Dewey, C. F., Jr., Bussolari, S. R., Gordon, E. J., and Gimbrone, M. A., Jr. 1984. Influence of hemodynamic forces on vascular endothelial function. In vitro studies of shear stress and pinocytosis in bovine aortic cells. *J Clin Invest*, 73, 1121–9.

Davis, G. E., Black, S. M., and Bayless, K. J. 2000. Capillary morphogenesis during human endothelial cell invasion of three-dimensional collagen matrices. *In Vitro Cell Dev Anim*, 36, 513–9.

Davis, G. E. and Camarillo, C. W. 1996. An alpha 2 beta 1 integrin-dependent pinocytic mechanism involving intracellular vacuole formation and coalescence regulates capillary lumen and tube formation in three-dimensional collagen matrix. *Exp Cell Res*, 224, 39–51.

Depaulis, R., Mohammad, S. F., Chiariello, L., Morea, M., and Olsen, D. B. 1988. A pulmonary mock circulation model for a better understanding of protamine reversal of heparin. *ASAIO Trans*, 34, 367–70.

Dewey, C. F., J. R., Bussolari, S. R., Gimbrone, M. A., Jr., and Davies, P. F. 1981. The dynamic response of vascular endothelial cells to fluid shear stress. *J Biomech Eng*, 103, 177–85.

Dull, R. O., Jo, H., Sill, H., Hollis, T. M., and Tarbell, J. M. 1991. The effect of varying albumin concentration and hydrostatic pressure on hydraulic conductivity and albumin permeability of cultured endothelial monolayers. *Microvasc Res*, 41, 390–407.

Eppihimer, M. J. and Lipowsky, H. H. 1994. The mean filtration pressure of leukocyte suspensions and its relation to the passage of leukocytes through nuclepore filters and capillary networks. *Microcirculation*, 1, 237–50.

Eskin, S. G., Ives, C. L., Mcintire, L. V., and Navarro, L. T. 1984. Response of cultured endothelial cells to steady flow. *Microvasc Res*, 28, 87–94.

Fidkowski, C., Kaazempur-Mofrad, M. R., Borenstein, J., Vacanti, J. P., Langer, R., and Wang, Y. 2005. Endothelialized microvasculature based on a biodegradable elastomer. *Tissue Eng*, 11, 302–9.

Folkman, J. and Haudenschild, C. 1980. Angiogenesis *in vitro*. *Nature*, 288, 551–6.

Forrester, J. V. and Lackie, J. M. 1984. Adhesion of neutrophil leucocytes under conditions of flow. *J Cell Sci*, 70, 93–110.

Francis-Sedlak, M. E., Moya, M. L., Huang, J. J., Lucas, S. A., Chandrasekharan, N., Larson, J. C., Cheng, M. H., and Brey, E. M. 2010. Collagen glycation alters neovascularization *in vitro* and *in vivo*. *Microvasc Res*, 80, 3–9.

Ghajar, C. M., Chen, X., Harris, J. W., Suresh, V., Hughes, C. C. W, Jeon, N. L., Putnam, A. J., and George, S. C. 2008. The effect of matrix density on the regulation of 3-D capillary morphogenesis. *Biophys J*, 94, 1930–41.

Giorgio, T. D. and Hellums, J. D. 1988. A cone and plate viscometer for the continuous measurement of blood platelet activation. *Biorheology*, 25, 605–24.

Giri, R., Selvaraj, S., Miller, C. A., Hofman, F., Yan, S. D., Stern, D., Zlokovic, B. V., and Kalra, V. K. 2002. Effect of endothelial cell polarity on beta-amyloid-induced migration of monocytes across normal and AD endothelium. *Am J Physiol Cell Physiol*, 283, C895–904.

Golden, A. P. and Tien, J. 2007. Fabrication of microfluidic hydrogels using molded gelatin as a sacrificial element. *Lab Chip*, 7, 720–5.

Goldsmith, H. L. and Spain, S. 1984. Margination of leukocytes in blood flow through small tubes. *Microvasc Res*, 27, 204–22.

Haselton, F. R. and Alexander, J. S. 1992. Platelets and a platelet-released factor enhance endothelial barrier. *Am J Physiol*, 263, L670–8.

Haselton, F. R., Woodall, J. H., and Alexander, J. S. 1996. Neutrophil-endothelial interactions in a cell-column model of the microvasculature: Effects of fmlp. *Microcirculation*, 3, 329–42.

Helm, C. L., Fleury, M. E., Zisch, A. H., Boschetti, F., and Swartz, M. A. 2005. Synergy between interstitial flow and VEGF directs capillary morphogenesis *in vitro* through a gradient amplification mechanism. *PNAS*, 102, 15779–84.

Helm, C. L., Zisch, A., and Swartz, M. A. 2007. Engineered blood and lymphatic capillaries in 3-D VEGF-fibrin-collagen matrices with interstitial flow. *Biotechnol Bioeng*, 96, 167–76.

Hong, H. and Stegemann, J. P. 2008. 2D and 3D collagen and fibrin biopolymers promote specific ECM and integrin gene expression by vascular smooth muscle cells. *J Biomater Sci Polym Ed*, 19, 1279–93.

Hsu, Y. H., Moya, M. L., Abiri, P., Hughes, C. C. W, George, S. C., and Lee, A. P. 2013. Full range physiological mass transport control in 3D tissue cultures. *Lab Chip*, 13, 81–9.

Ingber, D. E. and Folkman, J. 1989. Mechanochemical switching between growth and differentiation during fibroblast growth factor-stimulated angiogenesis *in vitro*: Role of extracellular matrix. *J Cell Biol*, 109, 317–30.

Jeon, J. S., Zervantonakis, I. K., Chung, S., Kamm, R. D., and Charest, J. L. 2013. *In vitro* model of tumor cell extravasation. *PloS One*, 8, e56910.

Killackey, J. J., Johnston, M. G., and Movat, H. Z. 1986. Increased permeability of microcarrier-cultured endothelial monolayers in response to histamine and thrombin. A model for the *in vitro* study of increased vasopermeability. *Am J Pathol*, 122, 50–61.

Kim, C., Chung, S., Yuchun, L., Kim, M. C., Chan, J. K., Asada, H. H., and Kamm, R. D. 2012. *In vitro* angiogenesis assay for the study of cell-encapsulation therapy. *Lab Chip*, 12, 2942–50.

Koike, N., Fukumura, D., Gralla, O., Au, P., Schechner, J. S., and Jain, R. K. 2004. Tissue engineering: Creation of long-lasting blood vessels. *Nature*, 428, 138–9.

Korff, T. and Augustin, H. G. 1998. Integration of endothelial cells in multicellular spheroids prevents apoptosis and induces differentiation. *J Cell Biol*, 143, 1341–52.

Korff, T. and Augustin, H. G. 1999. Tensional forces in fibrillar extracellular matrices control directional capillary sprouting. *J Cell Sci*, 112 (Pt 19), 3249–58.

Korff, T., Kimmina, S., Martiny-Baron, G., and Augustin, H. G. 2001. Blood vessel maturation in a 3-dimensional spheroidal coculture model: Direct contact with smooth muscle cells regulates endothelial cell quiescence and abrogates VEGF responsiveness. *FASEB J*, 15, 447–57.

Kubota, K., Tamura, J., Shirakura, T., Kimura, M., Yamanaka, K., Isozaki, T., and Nishio, I. 1996. The behaviour of red cells in narrow tubes *in vitro* as a model of the microcirculation. *Br J Haematol*, 94, 266–72.

Kubota, Y., Kleinman, H. K., Martin, G. R., and Lawley, T. J. 1988. Role of laminin and basement-membrane in the morphological-differentiation of human-endothelial cells into capillary-like structures. *J Cell Biol*, 107, 1589–98.

Kuzuya, M. and Kinsella, J. L. 1994. Induction of endothelial cell differentiation *in vitro* by fibroblast-derived soluble factors. *Exp Cell Res*, 215, 310–18.

Lafleur, M. A., Handsley, M. M., Knauper, V., Murphy, G., and Edwards, D. R. 2002. Endothelial tubulogenesis within fibrin gels specifically requires the activity of membrane-type-matrix metalloproteinases (MT-MMPs). *J Cell Sci*, 115, 3427–38.

Lee, S.-H., Moon, J. J., and West, J. L. 2008. Three-dimensional micropatterning of bioactive hydrogels via two-photon laser scanning photolithography for guided 3D cell migration. *Biomaterials*, 29, 2962–8.

Lee, T. H., Avraham, H. K., Jiang, S., and Avraham, S. 2003. Vascular endothelial growth factor modulates the transendothelial migration of MDA-MB-231 breast cancer cells through regulation of brain microvascular endothelial cell permeability. *J Biol Chem*, 278, 5277–84.

Lee, W., Lee, V., Polio, S., Keegan, P., Lee, J. H., Fischer, K., Park, J. K., and Yoo, S. S. 2010. On-demand three-dimensional freeform fabrication of multi-layered hydrogel scaffold with fluidic channels. *Biotechnol Bioeng*, 105, 1178–86.

Lee, W. H., Kang, S., Vlachos, P. P., and Lee, Y. W. 2009. A novel *in vitro* ischemia/reperfusion injury model. *Arch Pharm Res*, 32, 421–9.

Lei, X., Lawrence, M. B., and Dong, C. 1999. Influence of cell deformation on leukocyte rolling adhesion in shear flow. *J Biomech Eng*, 121, 636–43.

Levenberg, S., Rouwkema, J., Macdonald, M., Garfein, E. S., Kohane, D. S., Darland, D. C., Marini, R. et al. 2005. Engineering vascularized skeletal muscle tissue. *Nat Biotechnol*, 23, 879–84.

Levesque, M. J. and Nerem, R. M. 1985. The elongation and orientation of cultured endothelial cells in response to shear stress. *J Biomech Eng*, 107, 341–7.

Levy-Shraga, Y., Maayan-Metzger, A., Lubetsky, A., Shenkman, B., Kuint, J., Martinowitz, U., and Kenet, G. 2006. Platelet function of newborns as tested by cone and plate(let) analyzer correlates with gestational Age. *Acta Haematol*, 115, 152–6.

Low, J., Kellner, D., and Schuette, W. 1996. An automated high capacity data capture and analysis system for the *in vitro* assessment of leukocyte adhesion under shear-stress conditions. *J Immunol Methods*, 194, 59–70.

Madri, J. A., Pratt, B. M., and Tucker, A. M. 1988. Phenotypic modulation of endothelial-cells by transforming growth factor-beta depends upon the composition and organization of the extracellular-matrix. *J Cell Biol*, 106, 1375–84.

Mcgettrick, H. M., Lord, J. M., Wang, K. Q., Rainger, G. E., Buckley, C. D., and Nash, G. B. 2006. Chemokine- and adhesion-dependent survival of neutrophils after transmigration through cytokine-stimulated endothelium. *J Leukoc Biol*, 79, 779–88.

Miller, J. S., Stevens, K. R., Yang, M. T., Baker, B. M., Nguyen, D. H., Cohen, D. M., Toro, E. et al. 2012. Rapid casting of patterned vascular networks for perfusable engineered three-dimensional tissues. *Nat Mater*, 11, 768–74.

Montesano, R. and Orci, L. 1985. Tumor-promoting phorbol esters induce angiogenesis *in vitro*. *Cell*, 42, 469–77.

Montesano, R., Orci, L., and Vassalli, P. 1983. *In vitro* rapid organization of endothelial cells into capillary-like networks is promoted by collagen matrices. *J Cell Biol*, 97, 1648–52.

Montesano, R., Pepper, M. S., and Orci, L. 1993. Paracrine induction of angiogenesis *in vitro* by Swiss 3T3 fibroblasts. *J Cell Sci*, 105, 1013–24.

Moreland, J. G. and Bailey, G. 2006. Neutrophil transendothelial migration *in vitro* to Streptococcus pneumoniae is pneumolysin dependent. *Am J Physiol Lung Cell Mol Physiol*, 290, L833–40.

Moya, M. L., Cheng, M. H., Huang, J. J., Francis-Sedlak, M. E., Kao, S. W., Opara, E. C., and Brey, E. M. 2010. The effect of FGF-1 loaded alginate microbeads on neovascularization and adipogenesis in a vascular pedicle model of adipose tissue engineering. *Biomaterials*, 31, 2816–26.

Moya, M. L., Hsu, Y. H., Lee, A. P., Hughes, C. C. W., and George, S. C. 2013. *In vitro* perfused human capillary networks. *Tissue Eng Part C Methods*, 19(9), 730–7.

Muggli, R., Baumgartner, H. R., Tschopp, T. B., and Keller, H. 1980. Automated microdensitometry and protein assays as a measure for platelet adhesion and aggregation on collagen-coated slides under controlled flow conditions. *J Lab Clin Med*, 95, 195–207.

Nahmias, Y., Schwartz, R. E., Verfaillie, C. M., and Odde, D. J. 2005. Laser-guided direct writing for three-dimensional tissue engineering. *Biotechnol Bioeng*, 92, 129–36.

Nakatsu, M. N., Sainson, R. C. A., Aoto, J. N., Taylor, K. L., Aitkenhead, M., Perez-Del-Pulgar, S. A., Carpenter, P. M., and Hughes, C. C. W. 2003. Angiogenic sprouting and capillary lumen formation modeled by human umbilical vein endothelial cells (HUVEC) in fibrin gels: The role of fibroblasts and angiopoietin-1. *Microvasc Res*, 66, 102–12.

Nehls, V. and Drenckhahn, D. 1995. A novel, microcarrier-based *in vitro* assay for rapid and reliable quantification of three-dimensional cell migration and angiogenesis. *Microvasc Res*, 50, 311–22.

Newman, A. C., Chou, W., Welch-Reardon, K. M., Fong, A. H., Popson, S. A., Phan, D. T., Sandoval, D. R., Nguyen, D. P., Gershon, P. D., and Hughes, C. C. W. 2013. Analysis of stromal cell secretomes reveals a critical role for stromal cell-derived hepatocyte growth factor and fibronectin in angiogenesis. *Arteriosc Thromb Vasc Biol*, 33, 513–22.

Newman, A. C., Nakatsu, M. N., Chou, W., Gershon, P. D., and Hughes, C. C. W. 2011. The requirement for fibroblasts in angiogenesis: Fibroblast-derived matrix proteins are essential for endothelial cell lumen formation. *Mol Biol Cell*, 22, 3791–800.

Ng, C. P. and Swartz, M. A. 2003. Fibroblast alignment under interstitial fluid flow using a novel 3-D tissue culture model. *Am J Physiol Heart Circ Physiol*, 284, H1771–7.

Nobis, U., Pries, A. R., Cokelet, G. R., and Gaehtgens, P. 1985. Radial distribution of white cells during blood flow in small tubes. *Microvasc Res*, 29, 295–304.

Nollert, M. U., Diamond, S. L., and Mcintire, L. V. 1991. Hydrodynamic shear stress and mass transport modulation of endothelial cell metabolism. *Biotechnol Bioeng*, 38, 588–602.

Oh, H. H., Lu, H., Kawazoe, N., and Chen, G. In press. Spatially guided angiogenesis by three-dimensional collagen scaffolds micropatterned with vascular endothelial growth factor. *J Biomater Sci Polym Ed*.

Parker, D. J. 1968. The effect of operations of moderate severity on the rheological properties of blood as measured by a rotating cone and plate microviscometer. *Br J Surg*, 55, 857–8.

Pepper, M. S., Belin, D., Montesano, R., Orci, L., and Vassalli, J. D. 1990. Transforming growth factor-beta-1 modulates basic fibroblast growth-factor induced proteolytic and angiogenic properties of endothelial-cells *in vitro*. *J Cell Biol*, 111, 743–55.

Pepper, M. S., Montesano, R., Vassalli, J. D., and Orci, L. 1991. Chondrocytes inhibit endothelial sprout formation *in vitro*—Evidence for involvement of a transforming growth-factor-beta. *J Cell Physiol*, 146, 170–9.

Raghavan, S., Nelson, C. M., Baranski, J. D., Lim, E., and Chen, C. S. 2010. Geometrically controlled endothelial tubulogenesis in micropatterned gels. *Tissue Eng Part A*, 16, 2255–63.

Reinhart, W. H. and Chien, S. 1987. The time course of filtration test as a model for microvascular plugging by white cells and hardened red cells. *Microvasc Res*, 34, 1–12.

Sacharidou, A., Koh, W., Stratman, A. N., Mayo, A. M., Fisher, K. E., and Davis, G. E. 2010. Endothelial lumen signaling complexes control 3D matrix-specific tubulogenesis through interdependent Cdc42- and MT1-MMP-mediated events. *Blood*, 115, 5259–69.

Sacharidou, A., Stratman, A. N., and Davis, G. E. 2012. Molecular mechanisms controlling vascular lumen formation in three-dimensional extracellular mMatrices. *Cells Tissues Organs*, 195, 122–43.

Sahni, A., Arevalo, M. T., Sahni, S. K., and Simpson-Haidaris, P. J. 2009. The VE-cadherin binding domain of fibrinogen induces endothelial barrier permeability and enhances transendothelial migration of malignant breast epithelial cells. *Int J Cancer*, 125, 577–84.

Sakariassen, K. S., Aarts, P. A., De Groot, P. G., Houdijk, W. P., and Sixma, J. J. 1983. A perfusion chamber developed to investigate platelet interaction in flowing blood with human vessel wall cells, their extracellular matrix, and purified components. *J Lab Clin Med*, 102, 522–35.

Selby, C., Drost, E., Wraith, P. K., and Macnee, W. 1991. *In vivo* neutrophil sequestration within lungs of humans is determined by *in vitro* filterability. *J Appl Physiol*, 71, 1996–2003.

Shamloo, A., Xu, H., and Heilshorn, S. 2012. Mechanisms of vascular endothelial growth factor-induced pathfinding by endothelial sprouts in biomaterials. *Tissue Eng Part A*, 18, 320–30.

Shevkoplyas, S. S., Gifford, S. C., Yoshida, T., and Bitensky, M. W. 2003. Prototype of an *in vitro* model of the microcirculation. *Microvasc Res*, 65, 132–6.

Shin, M., Matsuda, K., Ishii, O., Terai, H., Kaazempur-Mofrad, M., Borenstein, J., Detmar, M., and Vacanti, J. P. 2004. Endothelialized networks with a vascular geometry in microfabricated poly(dimethyl siloxane). *Biomed Microdevices*, 6, 269–78.

Sieminski, A. L., Hebbel, R. P., and Gooch, K. J. 2004. The relative magnitudes of endothelial force generation and matrix stiffness modulate capillary morphogenesis *in vitro*. *Exp Cell Res*, 297, 574–84.

Sieminski, A. L., Hebbel, R. P., and Gooch, K. J. 2005. Improved microvascular network *in vitro* by human blood outgrowth endothelial cells relative to vessel-derived endothelial cells. *Tissue Eng*, 11, 1332–45.

Sieminski, A. L., Padera, R. F., Blunk, T., and Gooch, K. J. 2002. Systemic delivery of human growth hormone using genetically modified tissue-engineered microvascular networks: Prolonged delivery and endothelial survival with inclusion of nonendothelial cells. *Tissue Eng*, 8, 1057–69.

Smith, K. R. and Borchardt, R. T. 1989. Permeability and mechanism of albumin, cationized albumin, and glycosylated albumin transcellular transport across monolayers of cultured bovine brain capillary endothelial cells. *Pharm Res*, 6, 466–73.

Song, J. W., Bazou, D., and Munn, L. L. 2012a. Anastomosis of endothelial sprouts forms new vessels in a tissue analogue of angiogenesis. *Integr Biol Quant Biosci Nano Macro*, 4, 857–62.

Song, J. W., Daubriac, J., Tse, J. M., Bazou, D., and Munn, L. L. 2012b. RhoA mediates flow-induced endothelial sprouting in a 3-D tissue analogue of angiogenesis. *Lab Chip*, 12, 5000–6.

Song, J. W., Gu, W., Futai, N., Warner, K. A., Nor, J. E., and Takayama, S. 2005. Computer-controlled microcirculatory support system for endothelial cell culture and shearing. *Anal Chem*, 77, 3993–9.

Tan, W. and Desai, T. A. 2003. Microfluidic patterning of cells in extracellular matrix biopolymers: Effects of channel size, cell type, and matrix composition on pattern integrity. *Tissue Eng*, 9, 255–67.

Tanner, L. M. and Scott, R. B. 1976. A filtration model for study of leukocyte transit in the microcirculation. *Am J Hematol*, 1, 293–305.

Thompson, T. N., La Celle, P. L., and Cokelet, G. R. 1989. Perturbation of red blood cell flow in small tubes by white blood cells. *Pflugers Arch*, 413, 372–7.

Tille, J. C. and Pepper, M. S. 2002. Mesenchymal cells potentiate vascular endothelial growth factor-induced angiogenesis *in vitro*. *Exp Cell Res*, 280, 179–91.

Uriel, S., Brey, E. M., and Greisler, H. P. 2006. Sustained low levels of fibroblast growth factor-1 promote persistent microvascular network formation. *Am J Surg*, 192, 604–9.

Vailhe, B., Lecomte, M., Wiernsperger, N., and Tranqui, L. 1998. The formation of tubular structures by endothelial cells is under the control of fibrinolysis and mechanical factors. *Angiogenesis*, 2, 331–44.

Vaslef, S. N. and Goldstick, T. K. 1994. Enhanced oxygen delivery induced by perfluorocarbon emulsions in capillary tube oxygenators. *ASAIO J*, 40, M643–8.

Vernon, R. B. and Sage, E. H. 1999. A Novel, Quantitative model for study of endothelial cell migration and sprout formation within three-dimensional collagen matrices. *Microvasc Res*, 57, 118–33.

Vickerman, V. and Kamm, R. D. 2012. Mechanism of a flow-gated angiogenesis switch: Early signaling events at cell-matrix and cell-cell junctions. *Integr Biol*, 4, 863–74.

Wang, G. J., Ho, K. H., Hsu, S. H., and Wang, K. P. 2007. Microvessel scaffold with circular microchannels by photoresist melting. *Biomed Microdevices*, 9, 657–63.

White, S. M., Hingorani, R., Arora, R. P., Hughes, C. C. W, George, S. C., and Choi, B. 2012. Longitudinal *in vivo* imaging to assess blood flow and oxygenation in implantable engineered tissues. *Tissue Eng Part C Methods*, 18, 697–709.

Wong, K. H., Chan, J. M., Kamm, R. D., and Tien, J. 2012. Microfluidic models of vascular functions. *Annu. Rev. Biomed. Eng*, 14, 205–30.

Xue, L. and Greisler, H. P. 2002. Angiogenic effect of fibroblast growth factor-1 and vascular endothelial growth factor and their synergism in a novel *in vitro* quantitative fibrin-based 3-dimensional angiogenesis system. *Surgery*, 132, 259–67.

Yang, S., Graham, J., Kahn, J. W., Schwartz, E. A., and Gerritsen, M. E. 1999. Functional roles for Pecam-1 (CD31) and VE-cadherin (CD144) in tube assembly and lumen formation in three-dimensional collagen gels. *Am J Pathol*, 155, 887–95.

Yeon, J. H., Ryu, H. R., Chung, M., Hu, Q. P., and Jeon, N. L. 2012. *In vitro* formation and characterization of a perfusable three-dimensional tubular capillary network in microfluidic devices. *Lab Chip*, 12, 2815–22.

Zheng, Y., Chen, J., Craven, M., Choi, N. W., Totorica, S., Diaz-Santana, A., Kermani, P. et al. 2012. *In vitro* microvessels for the study of angiogenesis and thrombosis. *PNAS*, 109, 9342–7.

8

Microfabrication of Three-Dimensional Vascular Structures

Xin Zhao, Šeila Selimović, Gulden Camci-Unal, Mehmet R. Dokmeci, Lara Yildirimer, Nasim Annabi, and Ali Khademhosseini

CONTENTS

8.1 Introduction

Tissue engineering is a rapidly growing research area aiming to repair damaged organs or tissues by utilizing an interdisciplinary approach at the interface between medicine, physical sciences, and engineering (Langer and Vacanti, 1993). The functionality of engineered tissues highly depends on the vascularization of the resulting constructs because most of the vital organs are highly vascularized in the native environment. For example, heart, kidney, liver, and lung possess complex microarchitectural units surrounded by a dense network of blood vessels. The vasculature in the body provides the tissues with oxygen and nutrients, as well as facilitates the removal of metabolic waste and unwanted metabolites (Du et al., 2008). The engineered tissue units, if not prevascularized, are likely to experience substantial problems due to diffusion-based limitations. Although avascular tissues, such as bladder, heart valve, cartilage, or skin, have previously been successfully engineered, currently, there is a tremendous need for developing strategies to generate vascularized tissues and organs (Ashiku et al., 1997, Atala et al., 2006, Auger et al., 2009, Cebotari et al., 2010).

One way to mimic the native tissue complexity and architecture to generate vascularized constructs is to utilize microfabrication approaches. These technologies have emerged as powerful approaches for the fabrication of microvascular geometries inside biomaterials. To precisely control cellular microenvironments, various microfabrication techniques, including micromolding, photolithography, and rapid prototyping (RP), have been used to enable the generation of shape, size, and geometrically controlled tissue constructs. For example, micromolding techniques have been previously used to fabricate bifurcating capillary channels using poly(glycerolsebacate) (PGS) (Fidkowski et al., 2005). Endothelial cells (ECs) were successfully cultured in this system for up to 2 weeks. In addition, photolithography has been used to generate engineered vascular units, for example, ECs adhered, spread, and proliferated on methacrylated gelatin (GelMA) hydrogels that had been micropatterned using photomasks (Nichol et al., 2010). Furthermore, RP has been used to print microvascular networks utilizing alginate hydrogels and it was found that the resultant networks could be lined with ECs and perfused with blood under specific circumstances (Cabodi et al., 2005, Miller et al., 2012).

Nevertheless, a major limitation of current regenerative models for engineering large tissues is the requirement of extensive time scales for the generation of capillary structures by implanted cells or formation of angiogenesis from existing vessels. As a result, low cellular viability and dysfunction of the transplanted engineered tissues are commonly encountered. It is therefore suggested that prevascularized scaffolds with perfusion ability may be used to address the challenges associated with the transport of nutrients and oxygen. Advanced microfabrication techniques adapted for this purpose have yielded microfluidic systems for vascularization (Bettinger and Borenstein, 2010). Microfluidics is defined as the manipulation of liquids and suspensions in confined volumes on the scale of 1 to several hundred microns. Microfluidic systems can provide excellent control over environmental parameters on the same or similar-size scales to cells, and thus are considered to be useful tools for cell culture studies. Additionally, these systems with vascular geometries can be incorporated with biomolecules to induce cell adhesion of ECs and further neovessel formation. For example, Kim et al. (2013) used micromolding techniques to create microfluidic devices made of fibrin hydrogel with one central channel encapsulated with human umbilical vein endothelial cells (HUVECs) and two adjacent channels on either side of the central channel encapsulated with human normal lung fibroblasts. It was found that vascular networks resembling primary plexus spontaneously emerged from the HUVECs in the central channel to the human normal lung fibroblasts in flanking channels (Kim et al., 2013). It is therefore anticipated that microfabrication techniques could enable the production of tissue constructs with controlled properties to engineer vascularized organs in the future.

In this chapter, we provide a brief introduction of vascular structure and then discuss a number of different microfabrication approaches to fabricate vascularized tissue constructs with corresponding examples. Finally, biomaterials currently investigated for vascularized tissue engineering scaffolds are summarized and additional strategies for enhancing vasculature are discussed.

8.2 Vascular Structures

8.2.1 Structure of Blood Vessels

Blood vessels are a key component of the body's circulatory system. Their tubular network serves to continuously deliver blood from the heart to the rest of the body. Blood vessels are mainly classified into five types: arteries, arterioles, capillaries, veins, and venules. During a normal circulatory cycle, arteries carry blood from the heart to the smaller arterioles and further into the capillary network where the exchange of substances such as oxygen, water, carbon dioxide, and other chemicals occurs between the blood and extracellular fluid to maintain sufficiently oxygenated tissues while removing waste products. Finally, deoxygenated blood is returned to the heart via the venous network (Cliff, 1976a,b). The following sections give a brief introduction of different types of blood vessels.

8.2.1.1 Arteries and Arterioles

Arteries consist of three layers (from outermost to innermost: *tunica adventitia, tunica media,* and *tunica intima*) (Figure 8.1a). The outer *tunica adventitia* is based on collagenous connective tissue. The middle *tunica media* is composed of a mixture of smooth muscle cells

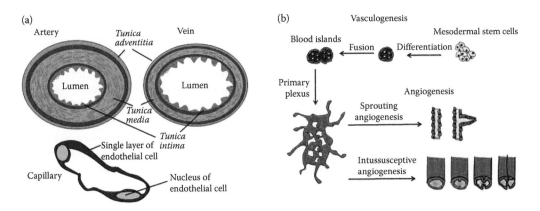

FIGURE 8.1
(a) Schematic representation of arterial, venous, and capillary structures. Arteries and veins consist of three layers, including *tunica intima, tunica media,* and *tunica adventitia,* whereas capillaries consist of a single layer of endothelial cells (ECs). (b) Illustration of vasculogenesis and angiogenesis. Vasculogenesis involves differentiation from mesodermal stem cells to angioblasts, which migrate to form blood islands, differentiation from angioblasts to ECs, and formation of a primitive vascular plexus. Angiogenesis involves the remodeling and growing processes of the preexisting vessels. It can be divided into sprouting angiogenesis, which involves EC proliferation and sprouting from an existing vessel, and intussusceptive angiogenesis, which involves splitting by the formation of transluminal pillars and growing of vessels. (Inspired by Adair, T. H. and Montani, J. P. 2010. Overview of angiogenesis. *Angiogenesis.* 1–8. San Rafael, CA: Morgan and Claypool Life Sciences.)

and elastic fibers, which are arranged in a circular manner. The inner *tunica intima* consists of a layer of squamous cells called endothelium that lines all blood vessels and the endocardium (Semenza, 2007). The basement membrane is a subendothelial connective tissue layer that mechanically supports the endothelium. The arteries' major function is to transport oxygenated blood from the heart into the body and to regulate the blood flow. Arterioles are the smaller branches of arteries that subdivide into even smaller vessels, eventually ending in a dense network of capillaries.

8.2.1.2 Veins and Venules

Veins have the same three-layered structure as arteries (Figure 8.1a). The main difference lies in their respective thicknesses: in veins, the wall of the middle *tunica media* is thinner than that of the arteries due to the reduced muscle component. Furthermore, veins have a larger lumen diameter accommodating a larger volume of blood within them. Valves within the luminal aspect of veins prevent backflow of blood, thus enabling active propagation of deoxygenated blood back toward the heart (Cliff, 1976a,b). Venules, or small veins, serve as a conduit between the tiny capillaries and the larger-diameter veins. Multiple venules usually merge to form one larger vein (Cliff, 1976a,b).

8.2.1.3 Capillaries

Capillaries consist of a single layer of squamous ECs and a basement membrane, allowing for the efficient exchanges of nutrients and waste (Figure 8.1a). Their diameter ranges from approximately 5 to 20 µm. Their fine networks penetrate all tissues and organs to provide oxygen and collect waste. Deoxygenated blood leaves the capillary network via venules into veins to ultimately rejoin the heart for yet another transportation cycle (Cliff, 1976a,b).

8.2.2 Overview of Blood Vessel Formation

To generate energy through the metabolic process, a continuous supply of oxygen is necessary for all tissues of the body. Oxygen is transported to these tissues via capillaries. Increasing the number of capillaries can improve tissue oxygenation, thus enhancing energy production, whereas reduction in capillaries may result in hypoxia and even anoxia (a total depletion in the level of oxygen, an extreme form of hypoxia) in the tissues, which may eventually result in reduced field of vision, sleepiness, or even brain damage. Therefore, for sufficient oxygen transport, cells in metabolically active tissues (e.g., liver, heart, and brain) must be located within a few hundred microns from a capillary, which are formed by a process called angiogenesis (see below for details) (Adair and Montani, 2010).

8.2.2.1 Origin of Blood Vessels

Vasculogenesis is the first step in *de novo* blood vessel formation. Mesodermal stem cells first differentiate into angioblasts, which with the aid of growth factors (GF, e.g., vascular endothelial growth factor, VEGF) migrate to form blood islands (small foci of hemangioblasts) and, over time, are capable of differentiating into ECs (Figure 8.1b). Proliferation and further migration of ECs result in the nascent endothelial tube formation and eventually the formation of blood vessels (Risau, 1995, 1997, Schmidt et al., 2007). Normally, vasculogenesis occurs in intra- and extraembryonic tissues. However, sometimes the circulating endothelial progenitor cells may also result in vasculogenesis in the adult organism. Vasculogenesis is a dynamic process that involves intracellular and cell–matrix interactions directed spatially and temporally by GFs (Coffin, 1991, Risau, 1995, Schmidt et al., 2007).

8.2.2.2 Angiogenic Process

The process of angiogenesis, that is, new blood vessel formation from preexisting ones, occurs throughout life, in health and disease states. It can be divided into two distinct subgroups, sprouting and splitting angiogenesis (Figure 8.1b). Both types require a finely tuned interplay between several molecules, including acidic and basic fibroblast growth factors (aFGF and bFGF, respectively), VEGF, angiopoietins 1 and 2 (Ang1 and Ang2), and matrix metalloproteinases (MMPs). Sprouting angiogenesis is a multistep process that requires the binding of angiogenic GFs to EC receptors located within preexisting vessels and the activation of such cell-signaling receptors. These, in turn, stimulate the release of proteases from ECs, which degrade the surrounding basement membrane. Via this route, newly proliferating ECs can then escape from their parent vessel and proliferate and migrate to form new EC tubes. These sprouts continue to migrate toward the source of angiogenic stimulation and connect neighboring vessels to close gaps in vasculatures (Chien, 2007). Sprouting angiogenesis is dramatically different from splitting (intussusceptive) angiogenesis since it forms entirely new vessels instead of splitting existing ones.

Splitting angiogenesis comprises of distinct steps to form two vessel lumina from an existing one. Initially, two opposing vessel walls extend into the lumen of existing blood vessels to establish a zone of contact in which resident ECs undergo a process of junctional reorganization. Then the newly formed endothelial bilayer and basement membrane are perforated to allow for GF exchange. Afterwards, fibroblasts migrate into the site of perforation and lay down extracellular matrix (ECM), and eventually, a tissue pillar is formed. The process of splitting angiogenesis is faster and more efficient compared to sprouting angiogenesis since it merely requires the reorganization of ECs rather than proliferation or migration. During embryogenesis where growth is fast but resources are limited, this form of rapid blood vessel formation is exploited (Djonov et al., 2002, Djonov et al., 2003, Kurz et al., 2003). However, intussusception mainly results in new capillaries where capillaries already exist.

8.3 Microfabrication Technologies

For tissue engineering techniques and technologies to translate to clinical treatments, it is necessary to develop macroscale functional (and therefore also vascularized) tissues in the laboratory. However, it has been challenging to vascularize *in vitro* engineered tissues on a large scale and to allow for efficient transport of metabolites to all contained cells. Namely, current research shows that cells that are far away (more than a few hundred microns) from a vascular channel do not receive sufficient oxygen and nutrients. This may lead to cell death and potentially even necrosis of the affected part of the tissue. Thus, it is prudent to design and fabricate a dense network of vascular channels into a tissue construct. This may be done by using microfabrication techniques, such as micromolding, photolithography, and RP, all of which are capable of generating features on the scale of micrometers or nanometers. It has been found that the microfabricated structures containing cells can aid in the vasculogenesis process (Nichol et al., 2010, Koroleva et al., 2012, Miller et al., 2012). Hence, microfabrication techniques provide a broad set of tools for probing and manipulating cell behavior by enabling the generation of cell-scale features in devices and enabling precise control over the cellular microenvironment (Teruo, 2002, Park et al., 2007, Gauvin et al., 2012). In addition, microscale techniques introduced in this section can be used to generate macroscopic constructs that are easy to handle and compatible with traditional laboratory analysis techniques.

8.3.1 Micromolding

Micromolding is one of the most widely used strategies for generating micropatterns in three-dimensional (3D) tissue engineering constructs. In this technique, the negative shape of a preformed master pattern is transferred onto a liquid prepolymer through direct contact, during which the prepolymer undergoes cross-linking (Figure 8.2a). More specifically, the prepolymer solution is dispensed onto a patterned master and covered with a glass slide, cross-linked, and finally peeled off for further use. In some variations of this approach, elastomers such as polydimethylsiloxane (PDMS) are used as masters to mold biomaterials, although the master can be made out of a solid material (glass, silicon, poly (methyl methacrylate) (PMMA)) or a hydrogel such as polyethylene glycol (PEG). The cross-linking of the prepolymer can be carried out by UV irradiation and other strategies, including thermal and ionic methods (Du et al., 2008, Tekin et al., 2011).

Since micromolding is a quick and easy-to-use experimental method, examples utilizing micromolding are widely available. For example, Koroleva et al. (2012) fabricated 3D fibrin scaffolds with interconnected pores and controllable pore sizes using a combination of two-photon polymerization (2PP) and micromolding techniques. In the paper, first, the master structures (a photocurable acrylate-based polymer, E-Shell 300) were fabricated by 2PP. This 2PP-based structure was subsequently used as a mold for a PDMS pattern. The resulting PDMS device was then coated with fibrinogen, followed by thrombin, and the

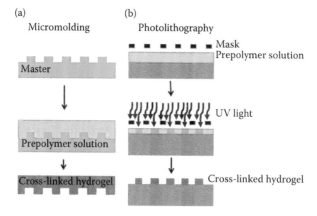

FIGURE 8.2
Schematic of procedures of (a) micromolding and (b) photolithography. (a) In micromolding, a prepolymer solution is added on top of a solid master and cross-linked. The resulting gel is removed to generate a structure with the imprint of the master. (b) Photolithography involves coating a substrate with a liquid prepolymer and exposing it to UV light through a patterned mask. Only the prepolymer exposed to the light is cross-linked and the remaining polymer solution can be washed off.

full fibrin scaffold removed from the PDMS device after curing. Scanning electron microscope and optical microscope images verified the porous fibrin network. The presence of this network allowed ECs to line and spread along the fibrin structure, while ECs encased inside dense fibrin blocks were oriented isotropically (Koroleva et al., 2012). This observation served as evidence that complex tissue structures could be engineered from natural proteins using the 2PP-micromolding approach.

Micromolding has also been applied to induce blood vessel formation by spatially patterning ECs to certain parts of collagen gels to encourage formation of tubules (Miller et al., 2012). Using this approach, hollow 1-cm-long tubes consisting of ECs encapsulated in collagen could be visualized within 48 h of cell seeding. These tubes had cell–cell junctions that were similar to the early-stage vascular development. Furthermore, the size of the lumen was found to be dependent on both microgel size and the concentration of collagen. Interestingly, time-lapse videos of the tubulogenesis process showed evidence of EC coalescence or fusion as well as structural changes in the organization of collagen fibers. This tissue engineering approach could potentially be applied for the formation of more complex blood vessel structures, while at the same time enabling fundamental studies of endothelial tubulogenesis (Miller et al., 2012).

Another example of micromolding being used to generate vascularized structures was recently demonstrated by He et al. (2013). The authors molded a silk fibroin precursor solution using a PDMS master in a mechanical, rapid micromolding setup. Several micromolded layers of silk fibroin were solidified using sacrificial layer of ice and stacked layer by layer, thereby generating a complex 3D structure. Namely, the low temperature of the icy layer helped polymerize the micromolded bioprotein solution and bond two polymer layers to each other. Upon warming, the sacrificial ice melted and was absorbed into the porous scaffold. Repetitive execution of this technique enabled the authors to generate cylindrical, porous vascular networks of silk fibroin gel. The generated scaffolds are particularly noteworthy due to the presence of a central vein—a central perfusion channel inside the 3D structure—as well as multiple peripheral veins that helped distribute the medium throughout the scaffold. The thickness of the scaffold at any given point was maximally 500 μm, such that encapsulated HUVECs could be sufficiently close to the vascular channels to receive adequate amounts of nutrients and survive with high viability values.

While micromolding can be used to generate a variety of microscale structures, it is limited to features whose width and height dimensions are comparable to each other. When the feature height greatly exceeds its width (e.g., in a 5:1 ratio), the generated mold cannot be easily removed from the master, which tends to lead to defects. Techniques that do not require masters, such as photolithography, do not present such problems (Selimović et al., 2012).

8.3.2 Photolithography

Photolithography relies on light (usually ultraviolet or UV wavelengths) to transfer a pattern from a transparency or chrome mask onto a liquid light-sensitive material via polymerizing or cross-linking (Figure 8.2b) (Fisher et al., 2001, Dendukuri et al., 2007, Bahney et al., 2011). In bioengineering applications, these light-sensitive materials are often polymers with acrylate or methacrylate group (e.g., PEG diacrylate, PEGDA, methacrylated hyaluronic acid (MeHA), and GelMA (Nichol et al., 2010, Hutson et al., 2011)). When exposed to light, the macromers polymerize, while the unexposed material remains liquid and can be easily removed. The cross-linking process is most commonly induced by free radical-based chemical reactions between the photoinitiator and the photosensitive material (Fisher et al., 2001).

Photolithography has been widely used in generating microscale structures for bioengineering applications. For example, Nemir et al. (2010) has used photolithography to endow PEGDA hydrogels of various polymer chain lengths with tunable patterning and mechanical properties. In the paper, first, a 20-kDa PEGDA base hydrogel was cross-linked under UV light for 45 s, then soaked in a 3.4-kDa PEGDA solution and cross-linked again under UV for 1 min. During the second cross-linking step, a photomask was used to restrict cross-linking to patterned areas. The uncross-linked hydrogel was washed off using HEPES-buffered saline (Figure 8.3). The hydrogels were patterned to fabricate anisotropic structures and have exhibited different modulus with different strip width and spacing. These hydrogels could potentially be used in studies of cell behavior in response to substrate stiffness, in both two and three dimensions, and they could also be used as scaffolds in tissue engineering applications.

In another instance, a photopolymerization approach was used to generate GelMA-based tissue structures (Nichol et al., 2010). Namely, the authors patterned GelMA prepolymer into microfibers that contained cells or were surface-coated with cells. In either case, cells were shown to readily attach to, proliferate, and elongate on such GelMA platforms, particularly when the gel concentration or the degree of methacrylation was optimized to yield the desired elasticity and swelling characteristics. In addition, such GelMA-based structures could be fabricated with high resolution to yield perfusable microscale structures, for example, channels, for *in vitro* cell studies (Nichol et al., 2010).

Major advantages of photolithography for generating microscale structures lie in its simple technical approach and compatibility with a wide range of hydrogels and elastomers. The features fabricated using this technique enable studies of tissue morphogenesis, which describes cell behavior corresponding to microarchitectural cues. It should be noted that the total UV energy delivered to the cells as well as the concentration of the photoinitiator can be damaging to cells, and must therefore be optimized and well controlled. Similarly, DNA damage can occur due to UV exposure, although the effect may

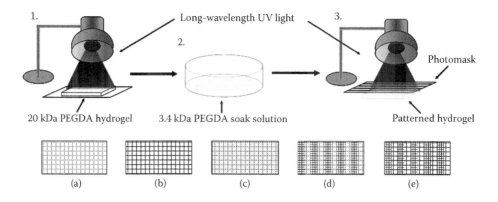

FIGURE 8.3
Schematic of fabrication of a PEGDA hydrogel with different strip width and spacing using photolithography. A high-molecular-weight PEGDA prepolymer solution is UV-cross-linked, then soaked overnight in a low-molecular-weight PEGDA solution (1,2, and a–c) and finally cross-linked a second time through a photomask (3, d) and washed (e). (Reprinted with permission from Nemir, S., Hayenga, H. N. and West, J. L. 2010. PEGDA Hydrogels with patterned elasticity: Novel tools for the study of cell response to substrate rigidity. *Biotechnol. Bioeng.* 105: 636–44.)

be minimized by controlling the concentration of acrylate or methacrylate groups of the polymer (Gao et al., 2010).

8.3.3 Rapid Prototyping

RP-based methods can potentially be used to produce 3D constructs. RP is an automatic additive manufacturing technique, which joins materials, usually layer upon layer, to make objects from 3D model data using computer-aided design (CAD) software. Contrary to traditional prototyping methods, which may require several hours or even days, RP methods can be optimized to engineer tissue constructs within an hour or two. In case of RP, a robotic stage and a printing head, for example, a glass needle, are used to deposit biomaterials with or without cells in a layer-by-layer manner in a predetermined pattern (Figure 8.4) (Mironov et al., 2007). Features that can be printed range from tens of micrometer to millimeter and even centimeter in size.

One recent example featured an RP-generated 3D carbohydrate glass network, which was used as a sacrificial substrate for cell-containing tissue structures. Specifically, the lattice was submerged in ECM together with cells. The resulting structures contained a network of perfusable tubular channels, lined with ECs, and were robust enough to withstand high pulsatile flows, while the original carbohydrate network was dissolved (Miller et al., 2012). Moreover, in this vascular casting technique, the lattice geometry as well as the biological components (cells and ECM) could be independently chosen, allowing for a range of cross-linking methods (Miller et al., 2012).

Another example used RP technique to generate porous microscale scaffolds, for the purpose of engineering blood vessel-like structures in a rat liver tissue (Figure 8.5) (Liu et al., 2012). In this approach, a PDMS master was fabricated using RP and micromolding techniques, onto which a silk fibroin/gelatin solution was dispensed and freeze-dried to form a planar microchannel network. The network was seeded with primary rat hepatocytes and subsequently rolled to generate a 3D construct, which was confirmed using imaging and computer simulations. The 3D network proved to aid cell adhesion and proliferation, when compared to a 2D control platform, which made it potentially a suitable candidate for the fabrication of implantable tissues (Liu et al., 2012).

Generating physical objects predetermined from CAD models represents a promising design and fabrication strategy for the establishment of complex organ vascular systems (Xu et al., 2008). RP has been successfully applied to generate bladder tissue, functional

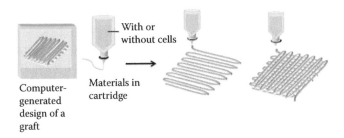

FIGURE 8.4
Schematic of procedures of rapid prototyping (RP). Materials in cartridge with or without cells are deposited through a printing head layer by layer into a pattern predetermined using computer-generated design. Once deposited, materials may solidify spontaneously, upon contact with the substrate layer, or when exposed to UV light.

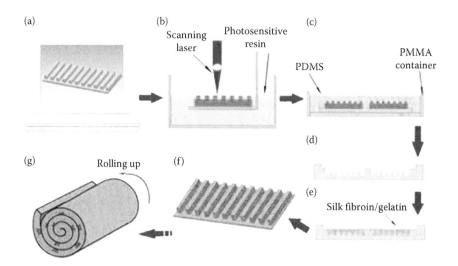

FIGURE 8.5
Schematic of fabrication of a silk fibroin/gelatin hydrogel scaffold with blood vessel-like structures using RP and micromolding. A computer-generated design (a) is used to fabricate a prototype using RP (b), which is then cast in PDMS (c–d). A silk fibroin/gelatin solution is dispensed (e), freeze-dried (f) to form a planar scaffold, which is then rolled to form the final product (g). (Reprinted with permission from Liu, Y. X. et al. 2012. The fabrication and cell culture of three-dimensional rolled scaffolds with complex micro-architectures. *Biofabrication* 4: 015004.)

heart tissues using embryonic endothelial and cardiac cells, a model of human skin based on fibroblasts and keratinocytes in collagen, and also vascular channels based on aortic smooth muscle cells (Oberpenning et al., 1999, Jakab et al., 2008, Norotte et al., 2009). The applications of CAD-assisted fabrication methods could be further expanded if they were combined with other micro- or nano-scale feature-generation techniques, including electrospinning, electrospraying, and self-assembly.

8.4 Generating 3D Vascularized Tissue Constructs Using Microfluidic Systems

Microfluidics is typically defined as the manipulation of liquids and suspensions in confined volumes on the scale of 1 to several hundred microns. Microfluidic technologies have proved especially helpful for cell and tissue culture studies, as they provide excellent control over multiple cell microenvironmental parameters on the same or similar size scales. These include, among others, perfusion of the cell and tissue samples with nutrients and oxygen over time, the rate of media flow and the magnitude of shear stress acting on the cells, formation of temporal and spatial concentration gradients of soluble factors, and externally controlled signals such as temperature or electrical impulses (Yi et al., 2006, Warrick et al., 2008, Barbulovic-Nad et al., 2010, Kim et al., 2010). In the early years of microfluidics, Borenstein and colleagues were one of the first groups to apply microfluidic approaches to vascularized tissue engineering scaffolds. For example, they generated polymeric constructs, both from PDMS and from biodegradable poly(DL-lactide-*co*-glycolide) (PLGA), with (vascular) channel resolution of up to 1 μm, mimicking the sizes and structure of capillaries inside living tissues (Borenstein et al., 2002). In addition, they successfully cultured ECs for up to 4 weeks inside these constructs, using peristaltic flow of culture media.

Subsequent research centered on generating vascular channels inside hydrogel structures, for example, in the work by Ling et al. (2007). In Ling's work, perfusable vascular channels were introduced into agarose structures by first micromolding two separate layers of agarose gel and then bonding them to each other, such that an empty channel appeared surrounded by the gel. Murine hepatocytes were encapsulated in the gel and the construct perfused with nutrients through the central channel. The cell viability was assessed as a function of distance from the nutrient source, that is, from the central channel. It was found that the viability decreased with increasing distance from the perfusion channel (Ling et al., 2007).

In more recent studies, microfluidic networks have been built into biomaterials using micromolding, photolithography, and/or RP to support 3D cell culture and vascular ingrowth. For example, a hydrogel-based microfluidic platform was generated using micromolding that was capable of delivering multiple microenvironmental cues to cells and to monitor the responses of the 3D cell culture (Vickerman et al., 2008). Here, the main device material was PDMS: two microfluidic channels were fabricated in PDMS and connected with a gel-filled central channel. The channels served as perfusion pathways and the directly injected gel contained encapsulated cells. This type of device structure was used to study cell migration, and in particular of sprouting in angiogenesis research. Among the microenvironmental cues that were tested in the study were solute concentration gradients, shear stress, and perfusion flow through the gel, leading to observation of tubular cell (adult dermal microvascular endothelial cells, HMVEC-ad) structures within a week of culture. This microfluidic platform also enabled visualization of filopodial projection and retraction, cell migration and division, and even lumen formation (Vickerman et al., 2008).

Another microscale platform that was fabricated using micromolding and photolithography was presented by Cuchiara et al. (2010). They fabricated multilayered constructs of PEGDA channel networks surrounded by independent PDMS walls, in which the hydrogel concentration could be adjusted to tune the diffusivity of solutes within the network (see Figure 8.6 for details). The PEGDA channel network served as a perfusion system and was shown to improve the viability of encapsulated cells compared to static control systems. This was the case even for cells that were comparatively far away from the perfusion channels (up to 1 mm away). As a consequence, this particular platform may be useful for studies of hydrogels as *in vitro* diagnostic materials and potentially for the development of therapeutics in regenerative medicine (Cuchiara et al., 2010).

In addition, Yeon et al. (2012) described the *in vitro* formation and characterization of perfusable capillary networks made of HUVECs in microfluidic devices fabricated using micromolding and RP. In order to fabricate the device, the master molds were produced by patterning 100-μm-thick negative SU8 photoresist on silicon wafers, then molding it in PDMS through heat-activated curing. The PDMS device contained microports for the delivery of fluids and was oxygen–plasma bonded to a rigid glass substrate. The main microfluidic structures on-chip were two 1-mm-wide and 100-μm-high channels, connected with eight 50–200-μm-wide bridging channels (Figure 8.7a). All of the channels were interconnected, allowing media, nutrients, and cells to be transported through the system. To generate connected blood vessels, the microfluidic devices were completely filled with fibrin gel and subsequently processed to generate gel structures only inside the bridge channels (Figure 8.7b). Following gel formation, HUVECs were coated along

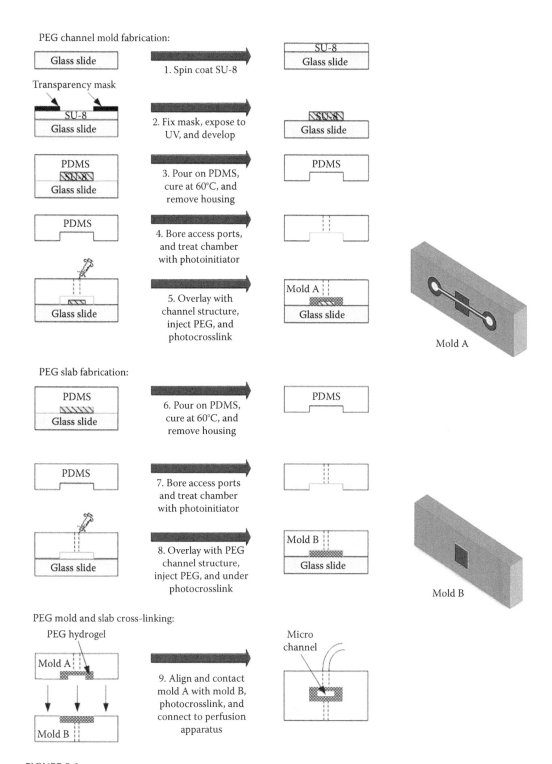

PEG channel mold fabrication:

Glass slide

1. Spin coat SU-8

SU-8
Glass slide

Transparency mask

SU-8
Glass slide

2. Fix mask, expose to UV, and develop

SU-8
Glass slide

PDMS
SU-8
Glass slide

3. Pour on PDMS, cure at 60°C, and remove housing

PDMS

PDMS

4. Bore access ports, and treat chamber with photoinitiator

Glass slide

5. Overlay with channel structure, inject PEG, and photocrosslink

Mold A
Glass slide

Mold A

PEG slab fabrication:

PDMS
Glass slide

6. Pour on PDMS, cure at 60°C, and remove housing

PDMS

PDMS

7. Bore access ports and treat chamber with photoinitiator

Glass slide

8. Overlay with PEG channel structure, inject PEG, and under photocrosslink

Mold B
Glass slide

Mold B

PEG mold and slab cross-linking:

PEG hydrogel

Mold A

9. Align and contact mold A with mold B, photocrosslink, and connect to perfusion apparatus

Micro channel

Mold B

FIGURE 8.6

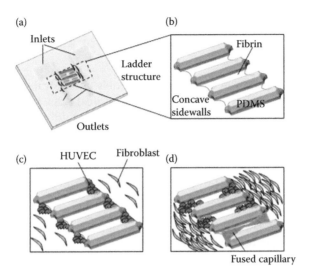

FIGURE 8.7
Illustration of capillary formation inside a microfluidic device. (a) Schematic representation of a microfluidic device consisting of two main channels and connected by a series of small parallel channels. (b). Enlarged diagram of ladder structures filled with fibrin gel forming concave sidewalls. (c) Loaded and attached HUVECs on the concave sidewalls. Fibroblasts loaded through the two main inlet channels after HUVECs attachment on the fibrin gel. (d) Self-organized capillary networks made by HUVECs grown from both sides. (Inspired by Yeon, J. H. et al. 2012. *In vitro* formation and characterization of a perfusable three-dimensional tubular capillary network in microfluidic devices. *Lab Chip* 12: 2815–22.)

the gel walls, on opposite ends of the patterned 3D fibrin gel (Figure 8.7c). After 3–4 days, HUVECs migrate into the fibrin gel from the opposite ends fused with each other, spontaneously forming a connected vessel that expressed tight junction proteins (e.g., ZO-1), which were characteristic of postcapillary venules (Figure 8.7d). With ready access to a perfusable capillary network, the hollow shape of the generated vessels was demonstrated by flowing suspensions of red blood cells (RBCs) and beads through them. The results were reproducible and could be performed in a parallel manner (nine devices per well plate). Additionally, compatibility with high-resolution live-cell microscopy and the possibility of incorporating other cell types has made this a unique experimental platform for investigating basic and applied aspects of angiogenesis and vascular biology (Yeon et al., 2012).

FIGURE 8.6
Fabrication schematics of generating PDMS/PEGDA microfluidic networks. To fabricate PEG channel mold (mold A), SU-8 is first coated on glass slide (step 1) and cross-linked using photomask (step 2). PDMS is then cast on the resultant SU-8 layer (step 3). Access ports are bored into the PDMS housing, which is then treated with photoinitiator solution (step 4). The resultant PDMS housing is overlayed with the glass slide coated with a thinner SU-8 layer produced in the same manner as steps 1–2. PEG solution is then injected through the access ports and cross-linked (step 5). To fabricate PEG slab (mold B), PDMS is first cast on glass slide coated with SU-8 layer (step 6), into which access ports are bored and treated with photoinitiator (step 7). PEG solution is then injected through the access ports and cross-linked in the chamber formed via overlaying the PDMS housing and a glass slide (step 8). Molds A and B are then aligned and photocross-linked to obtain a perfusable microchannel (step 9). (Reprinted with permission from Cuchiara, M. P. et al. 2010. Multilayer microfluidic PEGDA hydrogels. *Biomaterials* 31: 5491–7.)

8.5 Application of Biomaterials, Growth Factors, and Cell Coculture for Vascular Structure Development

In vivo blood vessel formation is the result of a finely orchestrated interaction between cells, ECM, stimulatory/inhibitory factors, and mechanical cues. It is, therefore, suggested that an exact recreation of naturally occurring events should result in the formation of stable and functional blood vessels capable of nutrient and waste exchange. The following sections discuss the vascularization strategies using microfluidic systems, including application of biomaterials, growth factors, and cell coculture.

8.5.1 Biomaterials for Guiding Vessel Growth

One of the key factors for successful fabrication of *in vitro* vasculature is the biomaterials used for the scaffolds, which will not only determine the cellular behaviors but also affect the fabrication feasibility and strategies. Both natural and synthetic materials have been investigated as scaffolds to guide the development of neovessels from host or transplanted cells. Ideally, these biomaterials should enable the formation of new vessels with minimal immune response, induce vessels of design similar to native ones, and provide sufficient mechanical support to surrounding tissues and biodegrade after vessel induction (Saik et al., 2012).

8.5.1.1 Natural Materials

Natural materials, including collagen and its gelatin derivatives, fibrin, alginate, and hyaluronic acid (HA), have been widely used as base materials for microfluidic devices for guiding vessel growth. Collagen and its gelatin derivatives are considered gold standard tissue engineering materials due to their excellent biocompatibility, support of cell attachment and proliferation, and *in vivo* biodegradability. Cellular behavior and the extent of vascular network formation in collagen/gelatin scaffolds can further be modulated by changing the degree of methacrylation or acrylation (Chen et al., 2012). Additionally, their mechanical properties can be tuned and enhanced by adding chemical cross-linkers (Drury and Mooney, 2003, de Moraes et al., 2012). For example, Nichol et al. (2010) used GelMA, a photopolymerizable hydrogel made of modified natural ECM components, to seed HUVECs into preformed microchannels and cocultured 3T3 cells to demonstrate the ability to support cellular adhesion, proliferation, and perfusability of GelMA microchannels. Other groups exploited the structural and biochemical advantages of fibrin to generate prevascularized scaffolds with well-formed capillaries, which accelerated anastomosis (connections of blood vessels) with the host vasculature upon implantation, and promoted cellular activity consistent with tissue remodeling (Patel and Mikos, 2004, Chen et al., 2009). Alginate, a hydrophilic and linear polysaccharide, also demonstrated proangiogenic properties when modified with different functional groups. This property was exploited by Shin et al. (2008) who successfully developed microfluidic chip-based micro-alginate tubes, which supported EC survival throughout a 5-day culture period. HA has a relatively stronger mechanical property to maintain microchannel integrity compared to other natural materials but has a lower affinity to bind cells (Brigham et al., 2009, Nichol et al., 2010). In order to combine relatively better mechanical stability of HA and cytocompatibility of collagen, Jeong et al. (2011) synthesized a collagen/HA hybrid microfluidic device capable of supporting cell attachment onto the channel walls and proliferation into near-confluent layers. However, despite favorable biological advantages, natural bio-materials lack the ability to create lasting microchannels due to insufficient mechanical robustness.

8.5.1.2 Synthetic Materials

Synthetic polymers are also utilized for fabricating scaffolds due to the larger range of material parameters, such as mechanical properties, degradation rates, formulation and processing feasibility, and drug delivery profiles. By changing the chemical structures, molecular weights, and cross-linking degree of the synthetic polymers, their physicochemical and biological properties could be finely tuned, thus providing a variety of options for tissue engineering applications (Saik et al., 2012). There are hundreds of types of synthetic polymers that have been used for vascular tissue engineering applications and some of the most highly researched synthetic biomaterials include nondegradable PEG, degradable PLGA, and PGS.

PEG is approved by the Food and Drug Administration (FDA) for certain applications and can be cross-linked via light or other chemical approaches after introducing diacrylate groups to the chain ends (e.g., poly (ethylene glycol) diacrylate (PEGDA)). PEG is hydrophilic and resists cell adhesion, which can be changed via modifying PEG with bioactive ligands such as ECM-derived Arg–Gly–Asp–Ser (RGDS) to create biofunctionality, including specific cell adhesion (DeLong et al., 2005). It was found that acrylate–PEG–RGDS could be lined with ECs and perfused with blood under high-pressure pulsatile flow (Miller et al., 2012). Contrary to PEG, PLGA is a biodegradable and hydrophobic polymer. PLGA films with integrated micrometer-scale features were stacked and fusion glued to construct fully multilayer microfluidic networks. It has been suggested that integration with cells or drug compounds could enable the use of these devices in therapeutic applications (King et al., 2004). In addition, PGS with microfabricated capillary networks was found to be able to endothelialize under flow conditions, suggesting that these approaches may lead to tissue-engineered microvasculature that is critical for engineering vital organs (Fidkowski et al., 2005). Other synthetic polymers, including polycaprolactone and polyhydroxybutyrate, have also been utilized for vascular tissue engineering applications. Alternatively, strategies that rely on the application of growth factors and specialized cells to promote angiogenesis can be used to support long-term cell and tissue survival (Gauvin et al., 2012).

8.5.2 Vascularization Enhancement by Growth Factors

In addition to biomaterials, there are several essential factors contributing to a well-established capillary-like network. For instance, the addition of angiogenic growth factors such as VEGF to the 3D matrix has been proved to enhance capillary-like tubular structure formation from EC lines by inducing vasculogenesis and angiogenesis (Kaully et al., 2009). This has been demonstrated in a study by Song et al. (2012). It was found that exposure of HUVECs encapsulated respectively in two parallel microfluidic channels within a collagenous matrix with 50 ng/mL VEGF in culture medium caused the HUVECs to abandon the preexisting vessel wall (formed by the seeded HUVECs) and sprout into the 3D matrix, leading to anastomosis of the endothelial sprouts. However, no anastomoses were observed in the preexisting vessel without VEGF stimulation (Song et al., 2012).

Angiogenesis can also be induced and promoted using GF gradients. For example, Barkefors et al. (2008) investigated the effect of VEGF gradients on EC migration. A microfluidic device with three inlets was designed to generate gradient concentration by three parallel fluid streams. The VEGF gradient was tunable by adjusting the solution flow rates. Their results showed that the ECs obviously migrated toward a high concentration of VEGF. These results also indicated that the steep gradients induced faster cell migration from 0 to 50 ng/mL, but no obvious migration from 50–100 ng/mL because of receptor saturation (Barkefors et al., 2008).

8.5.3 Vascularization Enhancement by Cell Coculture

Angiogenesis is a complex process that results from a sequential set of events modulated by cell–cell interactions. It relies on ECs or progenitor cells that can differentiate into ECs and is found to be enhanced by the coculture of different cell types. For example, Trkov et al. (2010) had used micropatterned 3D fibrin hydrogel to localize hydrogel-encapsulated HUVECs and mesenchymal stem cells (MSCs) within separate channels with spacing at 500 μm. It was found that MSCs migrated toward HUVECs and supported the formation of mature and stable vascular networks over a 2-week period, whereas tube-like structures formed by HUVECs alone resulted in their regression over this time period. Sudo et al. (2009) additionally confirmed the advantages of coculture systems in the development of 3D vascular structures. A microfluidic platform seeded with HUVECs was shown to form 3D capillary-like structures when grown in coculture with hepatocytes whereas no capillary formation was observed when HUVECs were cultured alone.

8.6 Conclusions and Future Directions

In this chapter, we provide an overview of the fabrication strategies for vascular networks integrated into 3D tissue constructs. While building thick tissue constructs, a perfusable vasculature is crucial as it provides nutrients and oxygen to the cells and removes waste and other metabolic by-products. An ideal vascular network should be easy to fabricate, combine into tissue scaffolds, preserve the mechanical properties of the constructs, and be cytocompatible and perfusable. Depending on the tissue engineering and regeneration medicine applications, different scaffold materials can be combined with different GFs and/or cell types to fabricate scaffolds with vasculature.

The concept of incorporating microfluidics into the field of vascular tissue engineering has been steadily moving forward (Borenstein et al., 2002, King et al., 2004, Fidkowski et al., 2005, Stroock and Cabodi, 2006, Nichol et al., 2010). The process of generating 3D vascularized tissue constructs using microfluidic systems can be broken down into (1) synthesis of a microfluidic network within a biocompatible material using microfabrication techniques, including micromolding, photolithography, and/or RP, (2) exposure of the fabricated microchannels to an EC suspension and upon perfusion of nutrients and oxygen, culture of these cells until confluency is reached over the inner surface of the channels, and (3) upon implantation, anastomosis of resultant vessels in a grafted microfluidic scaffold to vessels in a recipient's tissue bed, enabling immediate perfusion of the scaffold with blood.

To create ideal 3D tissue constructs, the aforementioned properties need to be considered early in the design stage. Among several competing factors that need to be considered in designing vascularized scaffolds are first, the tendency to lose some mechanical stability (e.g., compressive modulus) when increasing the density of the perfusable networks.

Second, depending on the type of tissue, different perfusion rates must be utilized to preserve the cells inside. Furthermore, the behavior of ECs under different flow conditions, their stability, and attachment/proliferation behavior need to be carefully evaluated while generating cell-containing, perfusable scaffolds.

The future direction is to create large, thick vascularized tissue constructs, with the ultimate goal of generating organs for transplantation. Several challenges need to be resolved prior to translating the discoveries of this field into clinical practice. One concern is the need for standardization; the advances in microfabrication technologies have produced a plethora of microchannel-laden scaffolds embedded with cells. Prior to use, metrics need to be in place to evaluate these methods and develop one that fulfills the needs of the intended organ application.

Another concern that arises alongside the increasing popularity of microfabrication approaches is that these methods are essentially two dimensional in nature. More work is needed to integrate these approaches into 3D scaffolds. For example, inventive, yet simple approaches to create channels in a 3D multilayer format and to connect these in a reliable and stable manner are open areas for research. The ability to create scalable channels with different dimensions (from micrometer to millimeter) that connect multiple layers in all three spatial directions is one of the key challenges in the field and requires further research.

Another area of interest is the development of real-time monitoring of perfusable organ constructs. The rapid advances in micro- and nanoscale sensors have led to novel commercial applications. The integration of these sensors into vascularized scaffolds for monitoring cellular viability and metabolic activity in real time is another area wide open for future investigations. Finally, the rapid development of new functional materials and new formulation techniques may bring revolutionary changes to this area.

Overall, the field of vascularized scaffolds has seen formidable progress over the years, yet there is still more to be done. Although the final product goal is evident—a fully vascularized perfusable organ—there are several different approaches that most likely will include a hybrid approach combining parts from microfabrication/microfluidics, surface patterning, and RP to achieve the final goal.

References

Adair, T. H. and Montani, J. P. 2010. Overview of angiogenesis. *Angiogenesis*. 1–8. San Rafael, CA: Morgan and Claypool Life Sciences.

Ashiku, S. K., Randolph, M. A., and Vacanti, C. A. 1997. Tissue engineered cartilage. *Mater. Sci. Forum* 250: 129–50.

Atala, A., Bauer, S. B., Soker, S., Yoo, J. J. and Retik, A. B. 2006. Tissue-engineered autologous bladders for patients needing cystoplasty. *Lancet* 367: 1241–6.

Auger, F. A., Lacroix, D. and Germain, L. 2009. Skin substitutes and wound healing. *Skin Pharmacol. Phys.* 22: 94–102.

Bahney, C. S., Lujan, T. J., Hsu, C. W., Bottlang, M., West, J. L. and Johnstone, B. 2011. Visible light photoinitiation of mesenchymal stem cell-laden bioresponsive hydrogels. *Eur. Cells Mater.* 22: 43–55.

Barbulovic-Nad, I., Au, S. H. and Wheeler, A. R. 2010. A microfluidic platform for complete mammalian cell culture. *Lab Chip* 10: 1536–42.

Barkefors, I., Le Jan, S., Jakobsson et al. 2008. Endothelial cell migration in stable gradients of vascular endothelial growth factor A and fibroblast growth factor 2: Effects on chemotaxis and chemokinesis. *J. Biol. Chem.* 283: 13905–12.

Bettinger, C. J. and Borenstein, J. T. 2010. Biomaterials-based microfluidics for engineered tissue constructs. *Soft Matter* 6: 4999–5015.

Borenstein, J. T., Terai, H., King, K. R., Weinberg, E. J., Kaazempur-Mofrad, M. R., and Vacanti, J. P. 2002. Microfabrication technology for vascularized tissue engineering. *Biomed. Microdevices* 4: 167–75.

Brigham, M. D., Bick, A., Lo, E., Bendali, A., Burdick, J. A. and Khademhosseini A. 2009. Mechanically robust and bioadhesive collagen and photocrosslinkable hyaluronic acid semi-interpenetrating networks. *Tissue Eng. Pt. A*. 15: 1645–53.

Cabodi, M., Choi, N. W., Gleghorn, J. P., Lee, C. S. D., Bonassar, L. J. and Stroock, A. D. 2005. A microfluidic biomaterial. *J. Am. Chem. Soc.* 127: 13788–9.

Cebotari, S., Tudorache, I., Schilling, T. and Haverich, A. 2010. Heart valve and myocardial tissue engineering. *Herz* 35: 334–40.

Chen, X. F., Aledia, A. S., Ghajar, C. M. et al. 2009. Prevascularization of a fibrin-based tissue construct accelerates the formation of functional anastomosis with host vasculature. *Tissue Eng. Pt. A* 15: 1363–71.

Chen, Y. C., Lin, R. Z., Qi, H. et al. 2012. Functional human vascular network generated in photocrosslinkable gelatin methacrylate hydrogels. *Adv. Funct. Mater.* 22: 2027–39.

Chien, S. 2007. Mechanotransduction and endothelial cell homeostasis: The wisdom of the cell. *Am. J. Physiol. Heart Circ. Physiol.* 292: H1209–24.: 17098825.

Cliff, W. J. 1976a. The extra-endothelial cells of blood vessel walls. *Blood Vessels.* 68–96. Cambridge: Cambridge University Press.

Cliff, W. J. 1976b. Vessels as functional units. *Blood Vessels.* 141–56. Cambridge: Cambridge University Press.

Coffin, J. D. and Poole, T. J. 1991. Endothelial cell origin and migration in embryonic heart and cranial blood vessel development. *Anat. Rec.* 231: 383–95.

Cuchiara, M. P., Allen, A. C. B., Chen, T. M., Miller, J. S. and West, J. L. 2010. Multilayer microfluidic PEGDA hydrogels. *Biomaterials* 31: 5491–7.

De Moraes, M. A., Paternotte, E., Mantovani, D. and Beppu, M. M. 2012. Mechanical and biological performances of new scaffolds made of collagen hydrogels and fibroin microfibers for vascular tissue engineering. *Macromol. Biosci.* 12: 1253–64.

DeLong, S. A., Moon, J. J. and West, J. L. 2005. Covalently immobilized gradients of bFGF on hydrogel scaffolds for directed cell migration. *Biomaterials* 26: 3227–34.

Dendukuri, D., Gu, S. S., Pregibon, D. C., Hatton, T. A. and Doyle, P. S. 2007. Stop-flow lithography in a microfluidic device. *Lab Chip* 7: 818–28.

Djonov, V. G., Baum, O. and Burri, P. H. 2003. Vascular remodeling by intussusceptive angiogenesis. *Cell Tissue Res.* 314: 107–17.

Djonov, V. G., Kurz, H. and Burri, P. H. 2002. Optimality in the developing vascular system: Branching remodeling by means of intussusception as an efficient adaptation mechanism. *Dev. Dyn.* 224: 391–402.

Drury, J. L. and Mooney, D. J. 2003. Hydrogels for tissue engineering: Scaffold design variables and applications. *Biomaterials* 24: 4337–51.

Du, Y., Cropek, D., Mofrad, M. R. K., Weinberg, E. J., Khademhosseini, A. and Borenstein, J. 2008. Microfluidic systems for engineering vascularized tissue constructs. In *Microfluidics for Biological Applications*, ed. Tian, W.-C. and Finehout, E. New York: Springer.

Fidkowski, C., Kaazempur-Mofrad, M. R., Borenstein, J., Vacanti, J. P., Langer, R. and Wang, Y. 2005. Endothelialized microvasculature based on a biodegradable elastomer. *Tissue Eng.* 11: 302–9.

Fisher, J. P., Dean, D., Engel, P. S. and Mikos, A. G. 2001. Photoinitiated polymerization of biomaterials. *Ann. Rev. Mater. Res.* 31: 171–81.

Gao, X., Zhou, Y., Ma, G., Shi, S., Yang, D., Lu, F., Nie, J. 2010. A water-soluble photocrosslinkable chitosan derivative prepared by Michael-addition reaction as a precursor for injectable hydrogel. *Carbohydr. Polym.* 79: 507–12.

Gauvin, R., Parenteau-Bareil, R., Dokmeci, M. R., Merryman, W. D. and Khademhosseini, A. 2012. Hydrogels and microtechnologies for engineering the cellular microenvironment. *Wiley Interdiscip. Rev. Nanomed. Nanobiotechnol.* 4: 235–46.

He, J., Wang, Y., Liu, Y., Li, D. and Jin, Z. 2013. Layer-by-layer micromolding of natural biopolymer scaffolds with intrinsic microfluidic networks. *Biofabrication* 5: 025002.

Hutson, C. B., Nichol, J. W., Aubin, H. et al. 2011. Synthesis and characterization of tunable poly-(ethylene glycol): Gelatin methacrylate composite hydrogels. *Tissue Eng. Pt. A* 17: 1713–23.

Jakab, K., Norotte, C., Damon, B. et al. 2008. Tissue engineering by self-assembly of cells printed into topologically defined structures. *Tissue Eng. Pt. A* 14: 413–21.

Jeong, G. S., Kwon, G. H., Kang, A. R. et al. 2011. Microfluidic assay of endothelial cell migration in 3D interpenetrating polymer semi-network HA-Collagen hydrogel. *Biomed. Microdevices.* 13: 717–23.

Kaully, T., Kaufman-Francis, K., Lesman, A. and Levenberg, S. 2009. Vascularization-the conduit to viable engineered tissues. *Tissue Eng. Pt. B-Rev.* 15: 159–69.

Kim, S., Kim, H. J. and Jeon, N. L. 2010. Biological applications of microfluidic gradient devices. *Integr. Biol.* 2: 584–603.

Kim, S., Lee, H., Chung, M. and Jeon, N. L. 2013. Engineering of functional, perfusable 3D microvascular networks on a chip. *Lab Chip* 19:13: 1489–500.

King, K. R., Wang, C. C. J., Kaazempur-Mofrad, M. R., Vacanti, J. P. and Borenstein, J. T. 2004. Biodegradable microfluidics. *Adv. Mater.* 16: 2007–12.

Koroleva, A., Gittard, S., Schlie, S., Deiwick, A., Jockenhoevel, S. and Chichkov, B. 2012. Fabrication of fibrin scaffolds with controlled microscale architecture by a two-photon polymerization-micromolding technique. *Biofabrication* 4: 015001.

Kurz, H., Burri, P. H. and Djonov, V. G. 2003. Angiogenesis and vascular remodeling by intussusception: From form to function. *News Physiol. Sci.* 18: 65–70.

Langer, R. and Vacanti, J. P. 1993. Tissue engineering. *Science* 260: 920–6.

Ling, Y., Rubin, J., Deng, Y. et al. 2007. A cell-laden microfluidic hydrogel. *Lab Chip* 7: 756–62.

Liu, Y. X., Li, X., Qu, X. L. et al. 2012. The fabrication and cell culture of three-dimensional rolled scaffolds with complex micro-architectures. *Biofabrication* 4: 015004.

Miller, J. S., Stevens, K. R., Yang, M. T. et al. 2012. Rapid casting of patterned vascular networks for perfusable engineered three-dimensional tissues. *Nat. Mater.* 11: 768–74.

Mironov, V., Prestwich, G. and Forgacs, G. 2007. Bioprinting living structures. *J. Mater. Chem.* 17: 2054–60.

Nemir, S., Hayenga, H. N. and West, J. L. 2010. PEGDA Hydrogels with patterned elasticity: Novel tools for the study of cell response to substrate rigidity. *Biotechnol. Bioeng.* 105: 636–44.

Nichol, J. W., Koshy, S. T., Bae, H., Hwang, C. M., Yamanlar, S. and Khademhosseini, A. 2010. Cell-laden microengineered gelatin methacrylate hydrogels. *Biomaterials* 31: 5536–44.

Norotte, C., Marga, F. S., Niklason, L. E. and Forgacs, G. 2009. Scaffold-free vascular tissue engineering using bioprinting. *Biomaterials* 30: 5910–7.

Oberpenning, F., Meng, J., Yoo, J. J. and Atala, A. 1999. *De novo* reconstitution of a functional mammalian urinary bladder by tissue engineering. *Nat. Biotechnol.* 17: 149–55.

Park, H., Cannizzaro, C., Vunjak-Novakovic, G., Langer, R., Vacanti, C. A. and Farokhzad, O. C. 2007. Nanofabrication and microfabrication of functional materials for tissue engineering. *Tissue Eng* 13: 1867–77.

Patel, Z. S. and Mikos, A. G. 2004. Angiogenesis with biomaterial-based drug- and cell-delivery systems. *J. Biomat. Sci.-Polym. E.* 15: 701–26.

Risau, W. 1995. Differentiation of endothelium. *FASEB J.* 9: 926–33.

Risau, W. 1997. Mechanisms of angiogenesis. *Nature* 386: 671–4.

Saik, J. E., Mchale, M. K. and West, J. L. 2012. Biofunctional materials for directing vascular development. *Curr. Vasc. Pharmacol.* 10: 331–41.

Schmidt, A., Brixius, K. and Bloch, W. 2007. Endothelial precursor cell migration during vasculogenesis. *Circ. Res.* 101: 125–36.

Selimović, Š., Oh, J., Bae, H., Dokmeci, M. and Khademhosseini, A. 2012. Microscale strategies for generating cell-encapsulating hydrogels. *Polymers* 4: 1554–79.

Semenza, G. L. 2007. Vasculogenesis, angiogenesis, and arteriogenesis: Mechanisms of blood vessel formation and remodeling. *J. Cell Biochem.* 102: 840–7.

Shin, S. J., Lee, K. H. and Lee, S. H. 2008. Development of microfluidic chip-based alginate microtube for angiogenesis. In *Proceedings of Twelfth International Conference on Miniaturized Systems for Chemistry and Life Sciences*. San Diego, CA, October 12–16, 2008.

Song, J. W., Bazou, D. and Munn, L. L. 2012. Anastomosis of endothelial sprouts forms new vessels in a tissue analogue of angiogenesis. *Integr Biol (Camb)*. 4: 857–62.

Stroock, A. D. and Cabodi M. 2006. Microfluidic biomaterials. *MRS Bulletin* 31: 114–9.

Sudo, R., Chung, S., Zervantonakis, I. K. et al. 2009. Transport-mediated angiogenesis in 3D epithelial coculture. *FASEB J*. 23: 2155–64.

Tekin, H., Ozaydin-Ince, G., Tsinman, T. et al. 2011. Responsive microgrooves for the formation of harvestable tissue constructs. *Langmuir* 27: 5671–9.

Teruo, F. 2002. PDMS-based microfluidic devices for biomedical applications. *Microelectron. Eng.* 61–62: 907–14.

Trkov, S., Eng, G., DiLiddo, R., Parnigotto, P. P. and Vunjak-Novakovic, G. 2010. Micropatterned three-dimensional hydrogel system to study human endothelial-mesenchymal stem cell interactions. *J Tissue Eng Regen Med*. 4: 205–15.

Vickerman, V., Blundo, J., Chung, S. and Kamm, R. 2008. Design, fabrication and implementation of a novel multi-parameter control microfluidic platform for three-dimensional cell culture and real-time imaging. *Lab Chip* 8: 1468–77.

Warrick, J. W., Murphy, W. L. and Beebe, D. J. 2008. Screening the cellular microenvironment: A role for microfluidics. *IEEE Rev. Biomed. Eng.* 1: 75–93.

Xu, W., Wang, X. H., Yan, Y. N. and Zhang, R. J. 2008. Rapid prototyping of polyurethane for the creation of vascular systems. *J. Bioact. Compat. Pol.* 23: 103–14.

Yeon, J. H., Ryu, H. R., Chung, M., Hu, Q. P. and Jeon, N. L. 2012. *In vitro* formation and characterization of a perfusable three-dimensional tubular capillary network in microfluidic devices. *Lab Chip* 12: 2815–22.

Yi, C., Li, C.-W., Ji, S. and Yang, M. 2006. Microfluidics technology for manipulation and analysis of biological cells. *Anal. Chim. Acta* 560: 1–23.

9

Gradient Scaffolds for Vascularized Tissue Formation

Michael V. Turturro and Georgia Papavasiliou

CONTENTS

9.1 Introduction

The clinical success of tissue-engineered scaffolds used for the replacement and reconstruction of damaged and/or diseased tissues and organs requires the formation of an extensive and stable microvascular network. In the absence of a perfusable network capable of mediating adequate mass transport of oxygen and nutrients, and removal of waste products, tissue-engineered constructs are unable to support cell and tissue viability throughout the scaffold. The ability to induce and maintain neovascularization (new blood vessel formation) in scaffolds presents unique challenges in tissue engineering due to the complexity and volume of the tissues targeted (Papavasiliou et al. 2010).

The process of neovascularization is dependent on coordinated interactions of multiple cell types with signals provided by the extracellular matrix (ECM) (Francis et al. 2008, Chan et al. 2011). The induction phase of neovascularization, as defined by endothelial cell (EC) polarization and directional sprouting, is highly dependent on cellular response to gradients of diffusible and matrix-bound growth factors, mechanical and physical properties of the ECM, as well as gradients of interstitial flow. Studies have shown that during angiogenesis, vascular ECs migrate in response to gradients of soluble chemoattractants (chemotaxis), gradients of immobilized ECM ligands (haptotaxis), as well as gradients of mechanical forces and matrix properties (mechanotaxis). Guided cellular response and vascular sprout formation is also dependent on different gradient characteristics, including gradient orientation, steepness, and magnitude, which play a critical role in regulating EC polarization (Zeng et al. 2007), guiding sprout formation (MacGabhann et al. 2007), and perivascular cell recruitment (Abramsson et al. 2003). In order to recreate a microenvironment that mimics the coordinated cell–ECM responses to gradients that occur during neovascularization, a variety of biomaterial approaches have been used to embed gradients of angiogenic factors, ECM biosignals, as well as gradients of physical and mechanical material properties in natural and synthetic scaffolds. Regardless of the approach, optimization of the spatial presentation of multiple biofunctional and material cues is required to allow for rapid vascular ingrowth and vascularization of the entire tissue-engineered construct. Gradient scaffolds are defined as those that contain anisotropy in chemical and biochemical composition, or physical and mechanical properties (Seidi et al. 2011). This chapter begins with a description of neovascularization with an emphasis placed on the role that particular gradients play during this process. Specific examples from the literature focusing on three-dimensional (3D) cell response to chemotactic, haptotactic, and mechanotactic gradients for stimulation of neovascularization within scaffolds will be presented and the fabrication techniques used to create these gradients will be described. This chapter concludes with a summary of future directions and recommendations for stimulating neovascularization within gradient scaffolds.

9.2 Biology of Neovascularization and Endothelial Cell Response to Gradients

Neovascularization arises either through an assembly process (vasculogenesis) or via the coordinated expansion of a preexisting capillary network (angiogenesis) (Geudens and Gerhardt 2011). Vasculogenesis relies on the formation of blood vessels from EC progenitors while angiogenesis involves a variety of coordinated morphogeneic events during which preexisting quiescent ECs sprout, branch, form lumens, anastomose, and remodel to form functional networks (Swift and Weinstein 2009). Regardless of the mechanism, the process of neovascularization entails multiple and consecutive steps all of which require coordinated interactions between multiple cell types and the ECM: (1) proteolytic degradation of the basement membrane, (2) directional cell migration through the ECM toward an angiogenic stimulus, (3) proliferation, (4) lumen formation, (5) maturation, (6) and return to quiescence (Nakatsu et al. 2003).

During the initiation stage of angiogenesis, a quiescent EC vessel monolayer results in vessel permeability in response to a stimulus such as hypoxia or an angiogenesis-promoting signal such as vascular endothelial growth factor (VEGF) (Germain et al. 2010). VEGF stimulates ECs to adopt a proteolytic phenotype and to break down the basement membrane and surrounding ECM, thereby releasing ECM components and sequestered VEGF through the action of matrix metalloproteinases (MMPs) (Chung and Ferrara 2010), allowing for cellular infiltration into the ECM. As the intercellular junctions in the vessel sprouting process are disrupted, increases in interstitial flow sweep VEGF and protease away from the origin of the sprout, thereby contributing to the established VEGF gradient (Lelkes et al. 1998). Upon induction of angiogenic sprouting, ECs exposed to angiogenic stimuli differentiate to enable migration, proliferation, lumen formation, and vessel assembly. Specifically, the

EC differentiates into a specialized tip cell that emerges from the existing blood vessel, becoming the leading cell of the sprouting vessel (Siemerink et al. 2013). Tip cell migration is mediated by extracellular gradients of VEGF that guide the growing sprout toward the angiogenic stimulus (Figure 9.1). *In vivo* studies in the postnatal retina have shown that the formation of vascular sprouts from preexisting quiescent vessels involves the emergence of tip cells exposed at the leading front of the sprout that extend long filopodial protrusions in a polarized morphology expressing receptors that are capable of sensing cell guidance cues and gradients of VEGF (Gerhardt et al. 2003). The established VEGF gradients are required for proper tip cell polarization, directed filopodial extension, and migration. While tip cell migration has been shown to depend on the VEGF gradient, tip cell proliferation is regulated by the actual VEGF concentration (Gerhardt et al. 2003).

ECs directly following the migrating tip cells differentiate into stalk cells that proliferate to elongate the sprout and generate the blood vessel lumen (Figure 9.1). Tip cell migration can occur without stalk cell proliferation and vice versa; however, the regulated balance of these two processes is required for the establishment of adequately shaped nascent sprouts (Ruhrberg et al. 2002, Gerhardt et al. 2003, Ruhrberg 2003). *In vivo*, this balance is achieved when the correct relationship between the VEGF gradient and VEGF concentration is established (Gerhardt and Betsholtz 2005). The interaction of the tip cell with the surrounding matrix pulls the growing sprout forward in the direction of growth (Ausprunk and Folkman 1977). Tip cells and stalk cells result in the formation of vascular

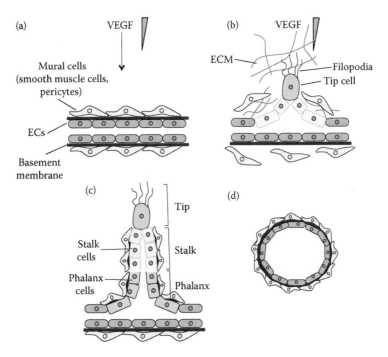

FIGURE 9.1
Neovascularization is mediated by cell–ECM interactions and gradients of stimuli. (a) Quiescent EC monolayer exposed to gradients of VEGF. (b) Mural cell detachment and basement membrane degradation due to response to a gradient. Tip cells at the leading edge of the sprout extend filapodial protrusions and migrate in response to gradients of VEGF. (c) Cell matrix interactions mediate EC tip cell migration and proliferation followed by stalk cell proliferation. ECs behind stalk cells differentiate into phalanx cells. (d) Cross section of a quiescent vessel stabilized by mural cells.

sprouts that grow toward the VEGF gradient and in response to biochemical, mechanical, and physical cues provided by surrounding cells and the ECM (Siemerink et al. 2013). ECs behind stalk cells differentiate into phalanx cells that align to form an inner cell monolayer of the new blood vessel, which elongates to create the vessel lumen by merging pinocytic vacuoles (Figure 9.1). Stalk cells and phalanx cells express tight junctions and associate with supporting mural cells (vascular smooth muscle cells or pericytes) to stabilize the neovessels formed (Figure 9.1). While gradients of VEGF are responsible for the initiation of neovascularization, gradients of other growth factors play a critical role in vessel maturation. Platelet-derived growth factor (PDGF) gradients stimulate chemotaxis of pericytes to nascent vessels (Hellström et al. 1999) and enable pericyte recruitment from distant tissues (Rajantie et al. 2004). Finally, tip cells from multiple sprouts undergo anastomosis to form a fully perfused vascular network, which recent studies suggest may be mediated by tissue-resident macrophages (Geudens and Gerhardt 2011). Blood flow to the newly vascularized area results in a cessation of migratory and proliferative activity of ECs, a quiescent EC phenotype, increases in local oxygen levels, and decreased VEGF levels eventually terminating the angiogenic cycle (Chung and Ferrara 2010).

The establishment of both chemotactic and haptotactic gradients of growth factors and other ECM constituents is critical for guided cell behavior and neovascularization. The range of action of these factors is dependent on their (1) diffusivity through the ECM, (2) secreted levels, (3) extracellular stability, and (4) propensity to adhere to ECM components. The restricted diffusion of growth factors in the ECM results in the formation of local chemotactic and haptotactic growth factor gradients. In addition, proteolysis of the ECM results in the creation of haptotactic gradients of growth factors and ECM components, which cells use to move in a directed fashion (Perumpanani et al. 1998). While *in vivo* cell response to chemotactic gradients has been demonstrated (Gerhardt et al. 2003), limited *in vivo* studies exist confirming the existence of haptotactic signals due to the associated difficulties of measuring gradients in tissues (Gritli-Linde et al. 2001, Ruhrberg et al. 2002). Among the growth factors of the VEGF family, VEGF-A has been shown to be a fundamental regulator of neovascularization acting in concert with a variety of other growth factors such as acidic (aFGF) and basic fibroblast growth factors (bFGF), angiopoietins (ANG-1, ANG-2), transforming growth factor beta 1 (TGF-β1), PDGF, and sphingosine-1-phosphate (S1P) (Wong et al. 2012). Gradients of VEGF-A are established *in vivo* through spatial variations in synthesis and secretion rates as well as through distinct VEGF-A isoforms with different affinities for heparan sulfate proteoglycans. Although all VEGF-A isoforms appear to induce the formation of filopodia on tip cells, the effects on filopodia morphology, vascular patterning, and guided tip cell migration are isoform specific (Gerhardt and Betsholtz 2005). Specifically, VEGF-A isoforms differ by the presence or absence of heparin-binding domains that confer their ability to exist in either soluble or bound form, or both. These different isoforms are distributed differently and differentially in the local microenvironment of a VEGF-secreting cell. As confirmed by *in vitro* cell culture experiments, the short isoform VEGF 120/121 (120 in mice and 121 in humans) lacks heparin-binding domains and is freely diffusible. The longer isoforms, VEGF 188/189 and VEGF 164/165, enable binding to heparan sulfate proteoglycans. While VEGF 188/189 contains two heparin-binding domains and is only found on the cell surface or bound to the ECM, VEGF164/165 lacks one of the two binding domains found in VEGF 188/189 and possesses intermediate properties allowing it to exist in both soluble and bound form (Park et al. 1993). *In vivo* studies by Ruhrberg et al. (2002) have demonstrated that the relative ratio of the various VEGF-A isoforms as well as the VEGF-A concentration gradient in the extracellular space is critical for regulating vascular patterning. Mouse embryos that only expressed the non-heparin-binding form VEGF 120 exhibited decreases in branch formation, tip cell formation and length, and increased stalk cell proliferation, resulting in vessels with enlarged diameters. In contrast, animals expressing the VEGF 188 isoform exhibited an opposing phenotype with a highly branched network consisting of long and thin vessels leading to ectopic branching. These defects were not present in mice engineered to express both the VEGF 120 and VEGF 188 isoforms. Examination of VEGF expression and localization demonstrated that the diffusible form (VEGF 120) reached the endothelium over large distances, stimulating continued proliferation of ECs resulting in the observed increase in vessel diameter, while the matrix-bound form, VEGF 188, produced steep gradients that caused excessive branching. Furthermore, alterations in the VEGF 120 gradient in mice solely expressing VEGF 120 resulted in shallow gradients that correlated with decreased filopodial extension and tip cell migration and in perturbed vessel branching. Mice producing only VEGF 164 (the heparin-binding isoform that is also diffusible) developed normal vessel networks that were indistinguishable from those found in wild types; however, alterations in the VEGF 164 gradient led to similar observations as those found in mice expressing VEGF 120 alone. These results illustrate that guided EC migration and stable neovascularization is highly dependent on gradient steepness and magnitude as well as on the established equilibrium between matrix-bound (haptotactic) and soluble growth factor (chemotactic) gradients in the 3D local microenvironment.

9.3 Scaffold Gradients for Neovascularization of Engineered Tissues

A variety of studies have explored the role of different types of gradients embedded into scaffolds to recapitulate the guided process of vascular cell response that occurs during neovascularization. While gradients of soluble (chemotactic) and immobilized (haptotactic) signals as well as gradients of physical and mechanical biomaterial properties (mechanotactic) have been incorporated into 3D scaffolds, earlier efforts focused on two-dimensional (2D) cell response (on the surface of the biomaterial) to gradients, which is not representative of 3D vascular response that occurs *in vivo*. During neovascularization, ECs lining blood vessels initially experience a 2D environment of the luminal surface followed by cell migration into the 3D ECM (Shamloo et al. 2012). Owing to the morphological and functional differences exhibited by cells in 2D versus 3D environments, recent efforts in the field of biomaterials and tissue engineering have focused on designing scaffolds and *in vitro* cell culture systems that allow for 3D cellular invasion and neovascularization. This section will present specific examples of studies from the literature that have designed scaffolds to contain chemotactic, haptotactic, mechanotactic, and combined gradients that stimulate guided 3D EC invasion and vascular sprouting within scaffolds. In each case, the fabrication technique used to create these gradients will be described.

9.3.1 Chemotactic Gradient Scaffold Systems

Chemotaxis involves directed cell migration in response to gradients of soluble chemoattractants. The creation of soluble growth factor gradients within scaffolds for the promotion of EC-guided migration, directional sprouting, and neovascularization has been achieved using microfluidic devices, through the spatial localization of growth factors in multilayered scaffolds, and through the confinement of growth factor-loaded particle carriers in distinct scaffold regions to form protein-releasing "depots" (Peret and Murphy 2008).

9.3.1.1 Microfluidic Devices

The use of microfluidic devices allows for the creation of physiologically relevant cell culture models to be combined with 3D scaffolds in order to reproduce many of the features of the *in vivo* vascular microenvironment (Wong et al. 2012). The small dimensions of these systems (<1 mm) enable observation of growth factor distribution as well as individual cell responses within 3D scaffolds in real time when coupled with image acquisition. Laminar flows with inherently low Reynolds numbers (Re < 10) allow for the creation of stable gradients of soluble growth factors as well as culturing of cells in low-shear environments for mechanistic studies of sprouting morphogenesis (Benitez and Heilshorn 2013). Microfluidic devices also enable control of gradient shape, slope, and magnitude, signaling distance, and timing of growth factor delivery through variations in input and output flow rates and channel geometry. While the scaffolds themselves are not created to contain the desired gradient, as will be described later, these systems enable the diffusion of chemotactic gradients through the scaffold. This is achieved using convective flow platforms that incorporate a barrier perpendicular to flow in the form of microcapillaries (Keenan et al. 2006) that continuously maintain soluble signal concentrations in source and sink flow channels (Nguyen et al. 2011). Typically, the soluble factors diffuse through microcapillary channels that connect reagent channels to the scaffold cell culture chamber inhibiting convective transport in this region (Figure 9.2). The stability of the gradient requires that the characteristic time for diffusion across the chamber be less than that for convection in the source and sink channels (Benitez and Heilshorn 2013).

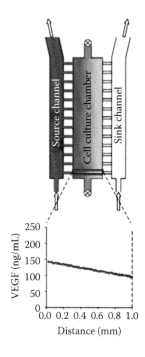

FIGURE 9.2
Microfluidic device used to create gradients of VEGF within a scaffold using perpendicular capillary extensions from source and sink channels. (From Shamloo A and Heilshorn SC. 2010. Matrix density mediates polarization and lumen formation of endothelial sprouts in VEGF gradients. *Lab on a Chip* 10: 3061–3068. Copyright 2010. Reproduced by permission of The Royal Society of Chemistry.)

It is well established that ECM mechanical stiffness mediates EC migration and capillary sprout formation (Sieminski et al. 2004, Ghajar et al. 2008). To this end, Shamloo and coworkers designed a microfluidic system to investigate the interplay between uniform scaffold stiffness and soluble VEGF gradients on the 3D sprouting morphogenesis of human microvascular ECs (Shamloo and Heilshorn 2010). Collagen fibronectin scaffolds with increasing moduli ranging from 7 to 700 Pa, and consequently altered diffusivity, resulted in increases in the time required to achieve an identical equilibrium VEGF concentration (~125 ng mL^{-1}) and a steady-state VEGF gradient (~50 ng mL^{-1} mm^{-1}). The time to reach the steady-state gradient was within 2 h, much shorter than the experimental time scale for sprouting morphogenesis (1–4 days). This study showed that the ability of the VEGF gradient and concentration to regulate stable sprout formation was mediated by scaffold stiffness with EC behavior changing from migratory to proliferative with increases in stiffness. ECs exposed to the VEGF gradient resulted in uncoordinated individual chemotactic cell migration within the least stiff scaffolds, while at intermediate stiffness, cells coordinated their chemotaxis and proliferation resulting in stable sprout formation. Scaffolds with the highest stiffness formed short and thick stumps that did not migrate into the surrounding matrix. Among the stiffness range that permitted sprout formation, sprouts within higher-density matrices were more likely to polarize toward the VEGF gradient and demonstrated more stable lumen formation as compared to scaffolds of lower stiffness. These results demonstrate that isotropic scaffold stiffness mediates VEGF-induced chemotactic cell migration, polarization, and lumen formation through the regulated balance of migration and proliferation. This study was later extended to investigate the effects of variations in the maximum VEGF concentration as well as the slope of the VEGF gradient on EC sprouting morphogenesis (Shamloo et al. 2012). *In vitro* studies showed that collagen–fibronectin scaffolds of higher stiffness required steeper gradients and higher VEGF concentration to initiate proper EC sprout polarization. In scaffolds of lower stiffness, steep gradients were found to induce significant sprout turning for proper navigation. This example illustrates how identical VEGF gradients in 3D scaffolds can induce different types of coordinated cellular responses mediated by the stiffness of the biomaterial. Therefore, the resulting growth factor concentration profiles and established gradients required to stimulate 3D neovascularization are highly dependent on scaffold properties that may differ based on the particular tissue engineering application.

While the previous example illustrates the importance of mechanical and/or physical scaffold properties in regulating vascular sprouting during neovascularization, other biophysical stimuli, including shear stress and interstitial flow, are also key regulators of this process (Vickerman et al. 2008, Jeong et al. 2011). Kamm and coworkers created a microfluidic device capable of controlling surface shear stress, interstitial flow, and gradients of S1P through 3D collagen scaffolds (Vickerman et al. 2008). In this study, EC monolayers cultured on scaffolds for several days with VEGF-enriched media and soluble gradients of S1P resulted in membrane protrusion, filopodial projection, and lumen formation as well as in the creation of complex multicellular capillary structures with the scaffold. This platform is particularly useful for studies involving ECM remodeling during neovascularization in response to gradients since proteolytic degradation and synthesis of ECM are inherently linked to interstitial flow (Benitez and Heilshorn 2013). The microfluidic device was later modified to accommodate cocultures of ECs and smooth muscle precursor cells (10T 1/2) to investigate perivascular–EC communication and migration within a 3D collagen scaffold as a function of VEGF gradients and scaffold stiffness (Chung et al. 2009). The modified device was designed to contain three independent flow channels each separated by a collagen scaffold filled through other microchannels. ECs were grown in the central flow channel and the stimulus (second cell type or soluble growth factor) was applied to one of the exterior flow channels. This microfluidic platform offers the advantage for investigating 2D and 3D EC migration response to soluble gradients in coculture environments and conditions that are highly representative of the multicellular process of neovascularization that occurs *in vivo*. EC monolayers formed on the surface of the scaffolds acted as a barrier to diffusion resulting in sudden and steep gradients in VEGF concentration established at the monolayer, similar to the situation that occurs *in vivo* near existing EC-enclosed capillaries. Cells preferentially migrated from the EC monolayer into the scaffold up the VEGF gradient. Furthermore, the presence of 10T 1/2 cells was found to have a stabilizing influence on ECs. Later studies using this microfluidic platform generated chemotactic gradients of multiple growth factors (VEGF and Ang-1) to investigate chemotactic EC sprouting within 3D hydrogels (Shin et al. 2011). The presence of the VEGF gradient alone was capable of inducing the formation of tip cells within the scaffold; however, separation of tip cells from initially formed stalk cells that eventually regressed was observed in scaffolds in the presence of VEGF gradients alone. When both VEGF and Ang-1 gradients were present, the activated tip cells were capable of remaining attached to the collectively migrating stalk cells confirming that Ang-1 stabilizes tip–stalk cell connections, illustrating the important role of gradients of stabilizing factors in vessel maturation.

9.3.1.2 Localized Controlled Release in Scaffolds

While microfluidic systems provide excellent platforms for investigating the role of soluble gradients on neovascularization within scaffolds, these systems are designed for *in vitro* experiments rather than for *in vivo* use. An alternative approach involves utilizing scaffolds that locally confine the growth factor in distinct regions, enabling its spatiotemporal release. The characteristics of the gradient (magnitude, slope, duration) in these systems can be controlled through a variety of material properties (scaffold pore size, mesh size, stiffness, crosslink density, and degradation rate) as well as through particle carrier loading and concentration. Mooney and coworkers designed porous two-layered poly(lactide-*co*-glycolide) (PLG) scaffolds capable of delivering spatial gradients of soluble VEGF with differential loading of the growth factor in each scaffold layer (Chen et al. 2007a). The scaffolds were designed using a mathematical model of VEGF distribution in ischemic mouse hindlimbs as a guide. The computational model described the VEGF profile due to diffusion, release, and VEGF degradation in each scaffold layer. Model parameters, including the VEGF diffusion coefficient, half-life, and cellular uptake, were determined from a series of *in vitro* experimental studies. The model predicted the presence of a steep VEGF gradient (40 ng mL^{-1} mm^{-1}) along the tissue scaffold interface, a decrease in gradient slope with increasing distance from the scaffold, and maintenance of the steady-state VEGF gradient in tissue up to 1 mm away from the scaffold surface. Based on model predictions, scaffolds were implanted in mouse models of hindlimb ischemia to assess the role of spatial VEGF delivery on neovascularization and restoration of perfusion. *In vivo* studies showed that scaffolds providing spatial delivery of VEGF led to extensive vessel networks as compared to control scaffolds with no VEGF and uniformly distributed VEGF of the same total content. While implantation of scaffolds with uniform VEGF resulted in enhanced hindlimb blood flow levels above those of control scaffolds lacking VEGF, scaffolds that spatially delivered VEGF rapidly restored blood flow to normal levels after 2 weeks postimplantation. This study provides an excellent example of the importance of integrating *in silico*, *in vitro*, and *in vivo* quantitative studies to effectively design scaffolds with the appropriate growth factor gradient characteristics for therapeutic neovascularization.

As previously mentioned, the process of neovascularization involves the coordinated spatial and temporal presentation of multiple growth factors. Owing to the important roles of VEGF gradients in stimulating vessel sprouting as well as gradients of other growth factors (e.g., Ang-1 and PDGF-BB) involved in the later stage of vessel stabilization, Mooney and coworkers designed porous, spatially compartmentalized, bilayered PLG scaffolds that allowed for the sequential delivery of VEGF and PDGF (Chen et al. 2007b). The inclusion of these growth factors and growth factor combinations within a single scaffold was used to investigate whether the local delivery and spatial segregation of VEGF and PDGF combinations was capable of controlling vascular patterning (i.e., vessel density and maturity) in ischemic sites over prolonged time periods *in vivo*. In the fabricated scaffolds, VEGF delivery preceded PDGF delivery kinetics by loading VEGF in the scaffold pores and by physically entrapping PDGF-loaded PLG microspheres into the scaffolds enabling for its sustained and delayed release. This scaffold delivery system was designed to contain two distinct layers; one that locally presented VEGF alone in one spatial region and a second layer that sequentially delivered VEGF and PDGF in an adjacent region. *In vivo* results demonstrated that vascular patterning was guided in the spatially segregated VEGF and PDGF scaffold delivery system. In scaffold compartments delivering VEGF alone, a high density of small and immature blood vessels was observed, while scaffold compartments that allowed for sequential delivery of VEGF followed by PDGF demonstrated slight increases in vascular density, but significantly increased vessel size and maturity. This study illustrates the importance of controlling the delivery and timing of multiple growth factors for the regulation of neovascularization.

The delivery of growth factors from particle carriers has been shown to protect growth factors from rapid degradation, prevent their undesirable systemic effects and toxicity due to nonspecific distribution and accumulation *in vivo*, and allow for prolonged and sustained growth factor release (Tayalia and Mooney 2009, Zhang and Uludağ 2009). Guo et al. (2012) encapsulated bFGF-loaded PLG microspheres in fibrous polycaprolactone (PCL) scaffolds to present bFbF in a gradient. These gradients were created by simultaneously electrospinning PCL fibers, encapsulating a gradient amount of bFGF onto microspheres and electrospraying the bFGF-loaded microspheres within the PCL fibers. This approach resulted in scaffolds with homogeneously embedded microspheres with the loading amount of bFGF gradually increasing depth-wise into the fibrous scaffold by varying the loading rate for the bFGF solution while keeping the loading rate of the PLGA solution constant. The sustained release of bFGF from the microspheres created a concentration gradient of bFGF with the scaffold. *In vivo* studies in subcutaneous implant models demonstrated that bFGF gradients significantly promoted cell invasion into the scaffolds with a high density of mature blood vessels 10 days post implantation. Vessel density was also found to nearly double with increased bFGF gradient steepness as compared to scaffolds of lower gradient steepness illustrating the importance of the embedded gradient slope in controlling neovascularization.

9.3.2 Haptotactic Gradients

Haptotaxis involves directed cellular migration toward an immobilized stimulus such as a growth factor, adhesion sequence, or ECM signal. *In vitro* assays of vascular cell migration on the surfaces of 2D rigid substrates (Liu et al. 2007, Smith et al. 2009, Cai et al. 2009) and on 3D scaffolds (DeLong et al. 2005b) have demonstrated increased cell speed in response to gradients of immobilized growth factors as well as to combinations of immobilized growth factors and adhesion proteins (Liu et al. 2007) suggesting that these gradients may lead to more rapid neovascularization of engineered tissues.

Hydrogel scaffolds provide environments for investigating the role of 3D cell invasion and neovascularization in response to embedded gradients. Multiphoton patterning techniques have been used to immobilize a series of concentration gradients of VEGF 165 in agarose hydrogels modified to contain the cell adhesion peptide sequence arginine-glycine-aspartic Acid (RGD). (Aizawa et al. 2010). As an example, VEGF gradients were immobilized in defined scaffold volumes by chemically modifying agarose with coumarin-protected photolabile cysteine groups, which upon excitation to UV or pulsed infrared laser yielded agarose sulfide that reacted with the maleimide-modified VEGF via Michael-type addition. This technique enabled the creation of scaffolds with a series of vertical linear concentration gradients of fluorescently tagged VEGF with varying slopes equivalent to 0.99, 1.65, and 2.48 ng mL^{-1} μm^{-1} tuned through variations in laser scanning number and region. To study cell response to the immobilized gradient and its steepness, EC aggregates were seeded on top of the patterned agarose hydrogels. Hydrogels containing 0.99 and 1.65 ng mL^{-1} μm^{-1} gradients of immobilized VEGF that were seeded with cells resulted in tubule-like structures that penetrated into the scaffolds over 200 μm, while scaffolds embedded with the steepest gradient (2.48 ng mL^{-1} μm^{-1}) showed no evidence of tubule-like structures and limited sprout formation. Scaffolds with no VEGF and homogeneously immobilized VEGF demonstrated limited depth of cell invasion (<30–50 μm) regardless of the immobilized VEGF concentration. To explore the effect of VEGF concentration on the elongation of the tubular structures formed, a variety of hydrogels were synthesized to contain the same immobilized gradient of VEGF (1.8 ng mL^{-1} μm^{-1}) but with varying initial VEGF concentration. It was found that ECs were unable to migrate within gradient scaffolds containing VEGF concentrations above 600 ng mL^{-1}. This study suggested that gradients that are too steep may lead to saturation of VEGFR2 receptors on EC tip cells, limiting their guidance response and that critical values of VEGF concentrations exist above which VEGFR2 receptors on stalk cells become saturated, inhibiting proliferation and migration into the scaffolds. This work demonstrates the importance of the presence of the gradient in guided EC invasion within a 3D scaffold as well as the significance of immobilized growth factor gradient steepness and magnitude in regulating this process.

A key challenge with tissue-engineered constructs is overcoming cell death in the scaffold interior due to lack of oxygen and nutrients in this region. Biomaterial strategies have attempted to overcome this limitation by immobilizing gradients of growth factors that span from the scaffold center to the periphery in order to guide cellular infiltration and neovascularization into the construct upon implantation. As an example, immobilized growth factor gradients of VEGF 165 in porous fibrous collagen scaffolds were created opposite the oxygen gradient to enable cell guidance into the scaffold interior (Odedra et al. 2011). The immobilized VEGF gradients were covalently attached to preformed collagen scaffolds by activating VEGF with EDC/NHS to allow for binding to carboxyl groups on the scaffold followed by injection of the protein solution to the preformed construct. This was achieved using a mounted syringe enabling VEGF penetration in the center of the scaffold and its subsequent diffusion outwards in the radial direction (point source method). This gradient generation approach allowed for the formation of millimeter-scale gradients in prefabricated scaffolds without the need to polymerize and/or fabricate the scaffold simultaneously with gradient establishment. Consistent with previous studies (Aizawa et al. 2010), scaffolds that contained immobilized VEGF concentration gradients of ~2 ng mL^{-1} mm^{-1} promoted significant increases in the number of ECs that invaded the scaffold core as compared to scaffolds containing uniformly immobilized VEGF of similar total VEGF concentration as well as to VEGF-free controls. It was also noted that all scaffolds yielded the same total number of infiltrated cells with higher cell density observed in the central region of the gradient scaffolds. This suggested that the observed increase in cell number in the scaffold core was not a result of increased cellular proliferation in response to the higher VEGF concentration, but due to guided EC migration in response to the haptotactic gradient.

In addition to the importance of growth factor gradients in stimulating directed cell migration and neovascularization, gradients of immobilized adhesion proteins (Cai et al. 2009, Smith et al. 2009) and peptides (DeLong et al. 2005a, Guarnieri et al. 2008, 2010, Turturro and Papavasiliou 2012), as well as other ECM components (Borselli et al. 2007) have been shown to guide cell behavior. Two-dimensional studies of cells seeded on the surfaces of scaffolds and rigid substrates have shown that gradients of adhesion ligands of the RGD peptide sequence stimulate preferential cell adhesion toward regions of increased ligand density (Burdick et al. 2004, He et al. 2010), and increased cell speed as compared to surfaces immobilized with uniform density (DeLong et al. 2005a, Guarnieri et al. 2008, 2010). Quantitative experimental analyses of EC migration on gold surfaces immobilized with fibronectin have indicated increases in cell drift speed in the presence of the gradient as compared to uniformly immobilized substrates (Smith et al. 2004) as well as increased frequency of discrete cellular motion in the gradient direction with increases in gradient slope and EC cell polarization (Smith et al. 2006). Furthermore, combined gradients of adhesion proteins and growth factors on rigid substrates have shown significant increases in directional EC motility as compared to individual gradients of either protein (Liu et al. 2007). This suggests that combinations of haptotactic protein gradients influence EC behavior and neovascularization.

Recent efforts have focused on translating haptotactic gradients of adhesive cues into scaffolds to investigate their effect on 3D cellular response (Hahn et al. 2006). Gradients of immobilized adhesion peptides of tyrosine-arginine-glycine-aspartic acid-serine (YRGDS) were created depth-wise within fibrous methacrylated hyaluronic acid (MeHA) scaffolds using a combination of electrospinning and photopolymerization (Sundararaghavan and Burdick 2012). Adhesive gradients were created through controlled delivery of flow rates using two programmable syringe pumps that contained distinct precursor solutions, one with decreasing flow rate (0 mM RGD, Polymer 1) and the other with increasing flow rate (with RGD, Polymer 2) prior to meeting at a t-junction and ejection through a charged needle for electrospinning and fiber collection on a rotating mandrel (Figure 9.3). This was followed by photopolymerization and crosslinking to lock in depth-wise gradients of YRGDS within the scaffold. To investigate the role of gradient orientation on 3D cellular infiltration, chick aortic arch explants were seeded on anisotropic scaffolds with the gradient either oriented away (low to high RGD) or adjacent to the explants (reversed gradient of high to low RGD) with control scaffolds containing uniform low and uniform high RGD densities similar to surface values of the gradient scaffolds. Significantly increased cellular infiltration occurred into scaffolds toward regions of increasing adhesivity (low to high RGD) as compared to the other scaffold groups. The maximum infiltration depth, defined as the furthest that distance that any cell could infiltrate the scaffold, was found to be highest in low–high RGD gradient scaffolds as well as in the high uniform RGD control scaffolds. Comparisons of maximum cellular infiltration between scaffolds that contained equivalent surface adhesivity demonstrated that infiltration increased from $34 \pm 17 \ \mu m$ in the uniform low RGD scaffolds to $187 \pm 47 \ \mu m$ in the low-to-high RGD gradient scaffolds. These results indicate that the orientation of the gradient as well as the surface concentration of embedded biosignals at the cell–scaffold interface are critical parameters that influence cellular invasion and neovascularization of implantable constructs.

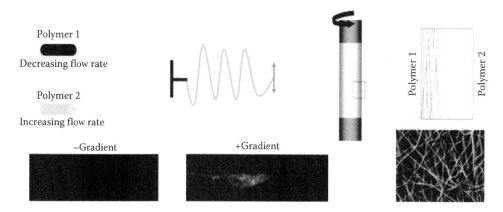

FIGURE 9.3
Schematic of the electrospinning photopolymerization process used to create depth-wise gradients of YRGDS within fibrous methacrylated hyaluronic acid scaffolds. (Reproduced with permission from Sundararaghavan HG and Burdick JA. 2012. Gradients with depth in electrospun fibrous scaffolds for directed cell behavior. *Biomacromolecules* 12: 2344–2350. Copyright 2011. American Chemical Society.)

Gradients of other ECM components, such as glycosaminoglycans, have also been shown to guide neovascularization. Scaffolds composed of semi-interpenetrated networks of collagen and hyaluronic acid (HA) showed that gradients of 0.5 mg mL^{-1} mm^{-1} in HA modulate vascular sprouting of EC-embedded spheroids *in vitro* (Borselli et al. 2007). Spheroids encapsulated in scaffolds that contained sigmoidal-shaped HA gradients showed preferential sprouting in the direction downstream of the HA gradient while aggregates embedded in scaffolds containing symmetric HA gradients (a collagen matrix on each side of the collagen HA gradient scaffold) resulted in a bidirectional response with sprout direction forced toward both sides of the collagen-rich matrices. Furthermore, reduced sprouting in the direction of the gradient led to an increase in sprout length within the collagen-rich regions. This study demonstrates how specific haptotactic gradients may inhibit and simultaneously direct neovascularization toward a particular direction.

9.3.3 Mechanotactic Gradients

Mechanical forces in the ECM play a critical role in directing EC behavior and tissue morphogenesis (Califano and Reinhart-King 2010, Kim and Peyton 2012). These forces are transferred between cells through cadherins or between cells and the matrix through focal adhesions and the actin cytoskeleton (Kim and Peyton 2012). Changes in the mechanical properties of the ECM are translated into functional changes in cell viability, morphology, spreading, and migration. Mechanotaxis refers to cell response toward a mechanical stimulus (e.g., fluid shear stress, pressure gradient, stiffness, and/or porosity gradient).

9.3.3.1 Durotaxis

A subcategory of mechanotaxis is durotaxis whereby cells respond to gradients of matrix density or stiffness. The first evidence of stiffness-directed cell motility was reported by Lo et al. (2000). This study demonstrated the preferential cell movement of fibroblasts toward stiffer regions on polyacrylamide hydrogel surfaces with covalently attached type I collagen. Stiffness gradients were achieved by creating a 2D substrate that contained an interface between a soft side and a stiff side. Fibroblasts were then seeded on these surfaces in a manner that enabled cell migration toward the transition zone from either region. Cells that approached the transition region from the compliant side were capable of migrating across the boundary while those approaching from the stiff side retracted as they reached the boundary. Following this study, quantitative differences of cell response to stiffness gradients on 2D biomaterial surfaces have been reported for a variety of cell types (Saez et al. 2007, He et al. 2010, Tse and Engler 2011, Diederich et al. 2013).

While stiffness gradients have been created in 3D hydrogel biomaterials, cellular response to durotactic cues has been primarily limited to 2D (Saez et al. 2007, He et al. 2010, Nemir et al. 2010, Tse and Engler 2011, Diederich et al. 2013). Furthermore, limited studies exist involving 3D cell response to stiffness gradients embedded in natural (Hadjipanayi et al. 2009, Sundararaghavan et al. 2009) and synthetic materials (Kloxin et al. 2010). The use of natural materials makes it difficult to isolate stiffness cues from adhesive and proteolytic cues in order to assess the mechanisms by which cells are capable of responding to durotactic gradients in a 3D microenvironment. Synthetic biocompatible crosslinked hydrogels provide an inert environment allowing for ease of manipulation of physical (crosslink density, mesh size, and swelling), mechanical (or stiffness), and degradative properties of the scaffold through alterations in macromer chemistry and polymerization conditions (Elbert and Hubbell 2001, Bryant and Anseth 2003, Turturro et al. 2013b). Among the classes of synthetic biomaterials, crosslinked hydrogels of poly(ethylene glycol) (PEG) have been extensively utilized as scaffolds in tissue engineering. The intrinsic resistance of these hydrogels to nonspecific cell adhesion and protein adsorption allows for selective incorporation of key signals of the ECM to be embedded into their crosslinked network enabling controlled studies of cell–biomaterial interactions (Hern and Hubbell 1998).

Free-radical polymerization chemistries resulting in crosslinking have been combined with gradient fabrication techniques to spatially embed signals and material properties in PEG-based scaffolds (Burdick et al. 2004, DeLong et al. 2005a,b, Guarnieri et al. 2008, 2010, Lee et al. 2008, Nemir et al. 2010, Turturro and Papavasiliou, 2012, Turturro et al. 2013a). Commonly employed methods used to create spatial variations in embedded biofunctionality or crosslink density/modulus in these scaffolds include the use of gradient makers (DeLong et al. 2005a,b, Nemir et al. 2010), microfluidic techniques (Burdick et al. 2002, Guarnieri et al. 2008, 2010, He et al. 2010) noncontact photolithography using sliding (Kloxin et al. 2010) or patterned photomasks (Nemir et al. 2010), and micropatterning techniques using two-photon confocal-based laser scanning lithography (Hahn et al. 2005, 2006, Lee et al. 2008) (Figure 9.4). Commercially available gradient makers rely on feeding prepolymer solutions with unique composition into two separate chambers that are mixed using a control valve that is manually adjusted and centered between the chambers. The mixed precursor solution enters an exit steam that relies on gravity flow to lock in the gradient via subsequent photopolymerization. Microfluidic devices are composed of a series of interconnected microchannels to combine the feed streams and simultaneously create multiple layers each with a unique composition that can be then photocrosslinked to lock in the hydrogel gradient. Two-photon laser scanning lithography utilizes the absorption of finely focused photons resulting in two-photon excitation within microscale focal volumes (Lee et al. 2008). Focal excitation occurs along the laser focal point allowing for photoreactive processes such as photopolymerization to create microscale gradients (Hahn et al. 2006).

Recently, PEG scaffolds with spatially controlled centimeter gradients in crosslink density or modulus were engineered using a novel gradient generation technique involving

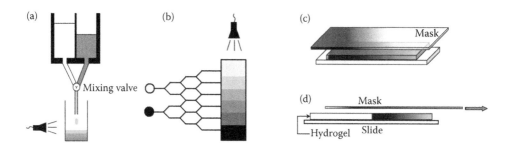

FIGURE 9.4
Common methods used to create gradients within PEG hydrogel scaffolds. (a) Gradient maker, (b) microfluidic device with a series of interconnected channels combining feed streams, (c) noncontact photolithography involving the use of photocrosslinking through gradient patterned photomasks, and (d) use of an opaque sliding photomask over a prepolymer solution or a preformed sample to create an irradiation gradient within the scaffold.

passive pumping, or surface-energy-driven fluid flow, by means of microfluidic tubeless flow mechanisms (He et al. 2010). The gradients were formed by injecting PEG precursors into microchannels that transported down the length of the channels via passive pumping while evaporation-driven backward flow established a gradient in PEG diacrylate (PEGDA) concentration. This was then followed by photopolymerization and crosslinking to lock in the stiffness gradient in the scaffold. ECs seeded on the surface of PEGDA hydrogels transitioned from a round to well-spread morphology with increasing gradient stiffness consistent with previous findings of vascular smooth muscle cell morphology on gradient substrates (Isenberg et al. 2009).

Limited studies have focused on investigating 3D cell behavior in response to gradients of crosslink density in synthetic scaffolds. To achieve 3D prolonged scaffold cell survival, migration and invasion, PEG-based hydrogels must be modified with degradable moieties within network crosslinks. Incorporation of degradable domains into the crosslinked network also enables temporal changes of scaffold physical and mechanical properties similar to ECM changes that occur *in vivo*. PEG hydrogels can be rendered degradable by incorporating hydrolytically (e.g., poly(lactic acid)-*b*-PEG-*b*-poly(lactic acid) (PLA-*b*-PEG-*b*-PLA)) (Bryant and Anseth 2003) or proteolytically (Mann et al. 2001, Zisch et al. 2003, Seliktar et al. 2004, Moon et al. 2007, 2010, Sokic and Papavasiliou 2012a,b, Turturro et al. 2013a) degradable blocks within their crosslinks. In the case of hydrolytic degradation, predictable changes in degradation profiles can be achieved that cannot be altered post-gel formation. For enzymatically degradable scaffolds, the degradation rate is not fixed upon gelation but dictated locally by cell-secreted enzymes. Using an alternative approach, Kloxin et al. (2010) created photodegradable diacrylated cell-laden PEG hydrogels which allow for real-time external manipulation of spatial and temporal changes in scaffold crosslink density. This was achieved using diacrylated PEG crosslinkers that contained a nitrobenzyl ether photolabile derivative between the terminal acrylate groups that is susceptible to cleavage upon exposure to 365 nm light enabling postgelation modification. After scaffold fabrication, flood irradiation was used to attenuate the light absorbed by the photolabile groups, resulting in the formation of gradients in crosslink density. The intensity profile within the gel was set by the irradiation wavelength and photolabile group concentration, which dictated the degradation rate for tuning hydrogel crosslink density in space and time. The resulting degradation-induced crosslink density gradient scaffolds were used to investigate how predictable changes influence the 3D morphology of encapsulated human mesenchymal stem cells (hMSCs). Increased cell spreading was found to occur within regions of lowest crosslink density in these scaffolds. This unique biomaterials approach can be readily applied to investigate the influence of dynamic changes of crosslink density gradients on 3D neovascularization of engineered tissues.

9.3.3.2 Porosity Gradients

The creation of gradients in scaffolds has also been achieved using microsphere-based scaffold fabrication techniques (Singh et al. 2008, Roam et al. 2010, Scott et al. 2010, Mohan et al. 2011). The use of different types of microspheres provides a modular strategy to produce scaffolds with heterogeneous macroporosity, microarchitecture, and spatial organization of embedded biofunctionality (Roam et al. 2010, Scott et al. 2010). Elbert and coworkers exploited the use of PEG microspheres of different densities (buoyancies) that self-assembled into gradients upon centrifugation (Roam et al. 2010). In this study, microsphere size and density were controlled by varying the temperature of phase separation and the reaction time of the PEG above the cloud point. Scaffold gradients were then produced by exploiting the density differences of the microspheres to create microsphere gradients upon centrifugation (Figure 9.5) followed by their simultaneous crosslinking to form the gradient scaffold. Scaffolds with sharp interfaces or gradual transitions with up to five tiers of different microsphere types were created (Roam et al. 2010). This microsphere-based approach was extended to fabricate scaffolds with homogeneous presentation of the RGD cell adhesion ligand and gradients in scaffold porosity ranging from highly to less porous layers (Scott et al. 2010). By varying the buoyancy of the porogenic microspheres, crosslinking the microspheres to form the scaffold, followed by microsphere dissolution, gradients in porosity were created within the scaffolds. *In vitro* studies demonstrated EC infiltration within highly porous regions with continued migration toward the less porous region. This study provides an example of the importance of gradients in scaffold architecture for guiding 3D EC behavior and neovascularization. This modular approach allows for the assembly of multiple types of microsphere gradients (in the presence of cells) with a diverse set of properties for the creation of complex scaffolds

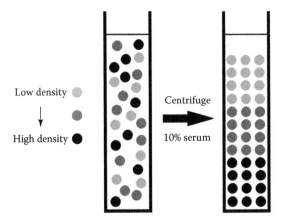

FIGURE 9.5
Schematic of modular microsphere-based scaffold approach used by Roam JL et al. (2010) to create gradient scaffolds. Gradients of PEG microspheres with different densities are produced and centrifuged to create microsphere gradients followed by crosslinking to form different gradients within scaffolds.

with spatial and temporal architectures, mechanical properties, and bioactivity for the promotion of scaffold neovascularization.

9.3.3.3 *Protease and Matrix-Bound Gradients Induced by Interstitial Flow*

While scaffold stiffness and porosity play a critical role in regulating cell behavior and neotissue formation, dynamic stresses, present in all living tissues, are also significant factors. In particular, dynamic stresses that drive slow flows ($\sim\mu m\ s^{-1}$), or interstitial flows, through the ECM have been shown to influence neovascularization (Helm et al. 2005, Vickerman et al. 2008). Interstitial flow refers to fluid flow around an interstitial cell or a cell attached to the ECM. This type of flow differs from open-channel flow within a blood vessel in that it is associated with lower velocities due to the high flow resistance of the ECM, it is transported around the cell matrix interface in three dimensions rather than on the apical side of an EC monolayer, and it results in the formation of matrix-bound gradients within the ECM (Rutkowski and Swartz 2007). *In vitro* studies have shown that interstitial flow acts synergistically with VEGF to create 3D gradients that direct capillary morphogenesis (Helm et al. 2005). Fibrin scaffolds containing VEGF modified to be released proteolytically were embedded with ECs. The presence of interstitial flow created protease gradients that skewed the release of matrix-bound VEGF that was subsequently capable of convecting and diffusing within the scaffolds, leading to the established VEGF gradient. More importantly, directed 3D capillary morphogenesis was only observed with the combination of VEGF and interstitial flow. This study provides insight on how dynamical stresses contribute to the formation of protease and haptotactic gradients in 3D scaffold environments that guide neovascularization.

9.3.4 Combined Scaffold Gradients

The studies described above have focused on creating scaffolds or utilizing scaffold systems that contain gradients of embedded biofunctionality or mechanical properties; however, multiple types of gradients influence neovascularization. In a recent study, proteolytically degradable crosslinked hydrogel scaffolds of PEG diacrylate (PEGDA) were engineered with combined haptotactic (RGD- and MMP-sensitive crosslinks) and durotactic gradients, resulting in directed 3D neovascularization *in vitro* in the absence of growth factors (Turturro et al. 2013a). These gradients were created using a novel polymerization technique: perfusion-based frontal photopolymerization (PBFP) (Turturro and Papavasiliou 2012), which is classified as a frontal polymerization process (Pojman et al. 1996). Frontal polymerization has been reported in the literature since the 1970s and is defined as a polymerization that proceeds subsequent to an initial ignition (chemical or physical) as a self-sustained propagating front (Chechilo et al. 1972). Free-radical polymerization is considered among the most suitable chemistries for driving this process since it involves reactions that are rapid, highly exothermic with a high activation energy based on the type of initiator utilized (Pojman 2012). Typically, traveling fronts based on free-radical polymerization have been shown to require an initial energy source to produce radicals that initiate the polymerization followed by the conversion of monomer to polymer to induce a self-propagating reaction with no further energy supply required to sustain the front. Three modes of frontal polymerization have been reported in the literature: thermal frontal polymerization (TFP), photofrontal polymerization (PFP), and isothermal frontal polymerization (ISP). TFP is the most commonly employed mode of frontal polymerization involving a localized reaction zone that propagates due to the coupling of thermal transport and the Arrhenius temperature dependence of the reaction rate during an exothermic reaction (Pojman et al. 1996). In PFP, a continuous flux of radiation (e.g., UV light) in a localized zone is used to create and sustain the propagating front as a result of the temperature rise from the exothermicity of the reaction induced by varying light intensity and wavelength (Nason et al. 2005). ISP involves a localized polymerization that propagates due to the Trommsdorf or gel effect that occurs when monomer and initiator diffuse into a polymer seed dissolving its uppermost layer creating a viscous region, which polymerizes and propagates throughout the reaction vessel under isothermal conditions (Lewis et al. 2005).

The use of PBFP to create gradients in crosslinked hydrogel scaffolds is distinct from the frontal polymerization processes described above. Specifically, PBFP can be classified as a type of isothermal frontal photopolymerization resulting from controlled and scheduled delivery of a visible light photoinitiator, eosin Y, during photopolymerization (Turturro and Papavasiliou 2012). The photoinitiator is localized in a reservoir that is separated from the precursor solution. During polymerization, the photoinitiator is perfused through a glass frit filter disk using a programmable syringe pump to schedule its delivery (Figure 9.6a). Although the photoinitiator is freely soluble in aqueous solution, density differences between the photoinitiator and the precursor prevent instantaneous miscibility, resulting in density fingering (Shevtsova et al. 2006). Buoyant photoinitiator rises to the surface

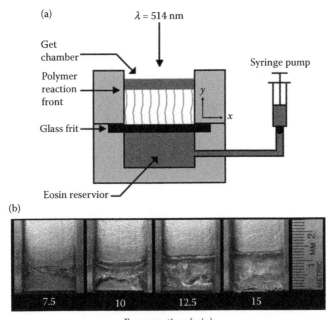

FIGURE 9.6
(a) Schematic representation of the perfusion-based frontal polymerization (PBFP) process used to create PEG scaffolds with embedded gradients. The hydrogel front is sustained by the continual perfusion of buoyant eosin Y and exposure to visible light ($\lambda = 514$ nm). (b) Series of hydrogel snapshots produced using PBFP at various polymerization times. (From Turturro MV et al. 2013c. Kinetic investigation of poly(ethylene glycol) hydrogel formation via perfusion-based frontal photopolymerization: Influence of free-radical polymerization conditions on frontal velocity and swelling gradients. *Macromolecular Reaction Engineering* 7: 107–115. Copyright 2013 Wiley-VCH Verlag GmbH and Co. KGaA. Reproduced with permission.)

and accumulates in a confined region, thus initiating polymerization in a localized reaction zone upon exposure to visible light ($\lambda = 514$ nm) as shown in Figure 9.6. After the top hydrogel layer has polymerized, additional photoinitiator is trapped below the gel causing a descending polymer front that leads to the progressive growth and expansion of a hydrogel (Figure 9.6b). The speed of the propagating front is controlled through variations in polymerization conditions and photoinitiator perfusion rate, which can be used to tune the slope of the crosslink density or stiffness gradient as well as gradients in swelling (Turturro and Papavasiliou 2012, Turturro et al. 2013c).

Since the spatial presentation of immobilized ECM cues as well as matrix properties play an important role in neovascularization, proteolytically degradable PEGDA hydrogels were engineered to contain gradients in elastic modulus, immobilized RGD, and MMP-sensitive peptide sequences (VPMS ↓ MRGG; ↓ denotes cleavage point) between crosslinks using PBFP (Figure 9.7) (Turturro et al. 2013a). The scaffolds were designed to degrade by MMP-mediated proteolytic mechanisms due to the fact that MMPs are highly expressed during neovascularization (Moses 1997, Davis and Senger 2005). By scheduling the delivery of the photoinitiator in the perfusion system, scaffolds were formed with an elastic modulus that decreased 80.4% from 3.17 to 0.62 kPa (250 Pa mm^{-1}), immobilized concentration of YRGDS that ranged from 152.2 to 66.7 mM (8.6 mM mm^{-1}), and with spatial variations in hydrogel degradation times from 4 to 12 h (in MMP enzyme incubations) over a distance of 10 mm (Figure 9.7). Hydrogels with uniform elastic modulus and immobilized biofunctionality formed using bulk photopolymerization served

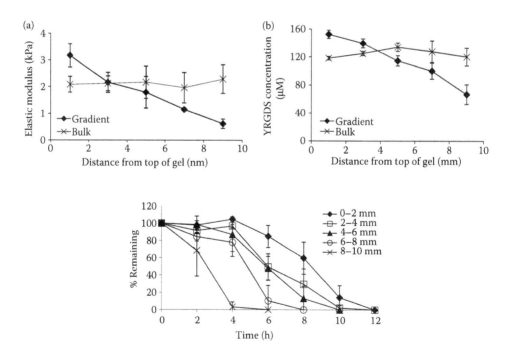

FIGURE 9.7
PBFP results in spatial variations of (a) hydrogel elastic modulus, (b) immobilized concentrations of YRGDS, and (c) gradients of MMP sensitivity and gel degradation rate. Error bars represent ± standard deviation ($n = 3$). (From Turturro MV et al. 2013a. MMP-sensitive PEG diacrylate hydrogels with spatial variations in matrix properties stimulate directional vascular sprout formation. *PLOS ONE* 8: e58897.)

as controls with values that lied within the midrange of gradient properties (an elastic modulus of ~2.13 kPa, an immobilized RGD concentration of ~125 μM, and a degradation time of ~6 h).

To investigate *in vitro* vascular sprout formation in response to the embedded gradients, a previously established coculture sprouting model of neovascularization composed of spheroidal aggregates of human umbilical vein endothelial cells (HUVECs) and human arterial smooth muscle cells (HUASMCs) was utilized (Korff et al. 2001). This *in vitro* model was chosen since it accounts for the coordinated interactions that occur between endothelial and mural cells during neovascularization that are critical to vessel assembly (Korff et al. 2001, Carmeliet and Jain 2011). The aggregates were embedded in scaffolds at different distances along the gradient. Specifically, aggregates were placed at approximate distances of 0–4 mm (top), 4–6 mm (middle), and 6–8 mm (bottom) along the gradients and consequently exposed to elastic moduli ranging from 3.17 to 2.16 kPa, 2.16 to 1.15 kPa, and 1.15 to 0.62 kPa and immobilized YRGDS concentrations between 152.2 and 139.4 mM, 139.4 and 99.9 mM, and 99.9 and 66.7 mM in the top, middle, and bottom scaffold regions, respectively. Furthermore, aggregates seeded in the top hydrogel regions were also exposed to a higher concentration of MMP sensitivity with progressive decreases in degradation time occurring in the less crosslinked regions. As shown in Figure 9.8a, aggregates seeded in control gels with isotropic properties resulted in uniform and random sprout invasion in all directions. In contrast, aggregates seeded in regions closest to the top of the gradients (top region) exhibited sprouts that curved and invaded bidirectionally along the gradients (Figure 9.8b) while aggregates seeded at intermediate (middle region) as well as furthest (bottom region) distances along the gradient resulted in progressive decreases in sprout alignment in the direction parallel to the gradient (Figure 9.8c,d). The directional response of vascular sprouts to the embedded gradients on neovascularization were quantified by (1) measuring the anisotropy index, or the ratio of the length of vessels formed in the direction parallel to the gradient as compared to the perpendicular direction, and (2) the sprout length by angle over time. The anisotropy index is defined to be one for uniform invasion and greater than one for preferential, or directed, invasion. Sprout orientation occurring along the 90° and 180° angles was indicative alignment in the direction parallel to the gradients and along 0° and 270° was depictive of alignment in the perpendicular gradient direction. Aggregates placed closest to the source of gradients (top region) demonstrated statistically significant increases in anisotropy index as compared to isotropic control scaffolds that yielded an index of one across all time points indicative of uniform sprout invasion (Figure 9.9). Aggregates seeded within intermediate (middle) and furthest distance (bottom) along the gradient also invaded directionally (anisotropy index = 1.59 ± 0.72 and 1.33 ± 0.20, respectively). Similar to the results observed for anisotropy index, the average sprout length by angle for aggregates seeded in bulk control gels (Figure 9.10a,e) was found to be equivalent in all directions at any given time point. However, aggregates seeded in the top or closest (Figure 9.10b,e), middle or intermediate (Figure 9.10c,e), and bottom or furthest (Figure 9.10d,e) regions of the gradient exhibited significantly increased sprout length at 90° and 270°, or both up and down the gradient, as compared to the perpendicular directions (0° and 180°). Overall, sprout length in the direction of the gradient was found to be similar regardless of the distance that the aggregates were placed along the gradient, but was found to progressively decrease in the direction perpendicular to the gradient as the distance of aggregate placement from the top region of the gradient was reduced.

The observed bidirectional vascular sprout response to the gradients is not entirely clear, and although similar responses have been reported in collagen scaffolds containing gradients of HA (Borselli et al. 2007), the mechanism of this response requires further

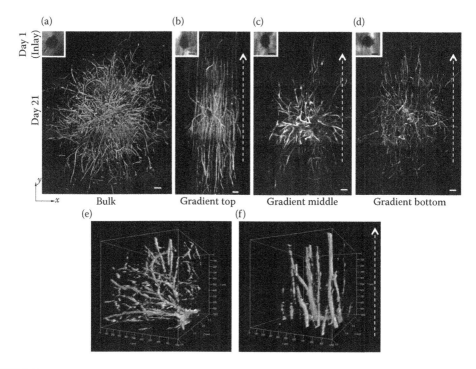

FIGURE 9.8

Three-dimensional vascular sprout invasion within PEGDA scaffolds. Flattened 3D mosaic renderings of cocul-ture aggregates seeded in (a) isotropic hydrogels as well as in scaffolds formed by PBFP in regions (b) closest (top), (c) intermediate (middle), and (d) furthest (bottom) along the gradients. Aggregates were fixed after 21 days in culture and stained for F-actin. (*Note*: The gradient runs in the y-direction in (b), (c), (d), and (e) with the increasing gradient direction depicted by the white arrow. Image; scale bar = 200 μm). (e) and (f) are 3D fluores-cent image reconstructions of vascular sprout invasion regions from (a) isotropic bulk control gels and (b) PBFP gradient hydrogels, respectively, taken at 10× magnification. (From Turturro MV et al. 2013a. MMP-sensitive PEG diacrylate hydrogels with spatial variations in matrix properties stimulate directional vascular sprout formation. *PLOS ONE* 8: e58897.)

investigation. To gain insight on the role of scaffold properties and biosignals on the observed directional vascular response, the specific role of each gradient requires further study. Independent tuning of specific gradients using PBFP may be achieved through vari-ations in the composition and flow rate of specific precursor components via a secondary feed stream (in addition to the existing photoinitiator feed stream) using programmable syringe pumps to schedule their delivery. Using this approach, individual and combined gradients of embedded biofunctionality and biomaterial properties can be systematically achieved in scaffolds in addition to alterations in gradient characteristics such as magni-tude, slope, and orientation in order to quantify these effects on 3D vessel assembly.

9.4 Conclusions and Future Directions

The process of neovascularization is highly dependent on cell response to multiple spatial and temporal changes of chemotactic, haptotactic, and mechanotactic cues that occur in

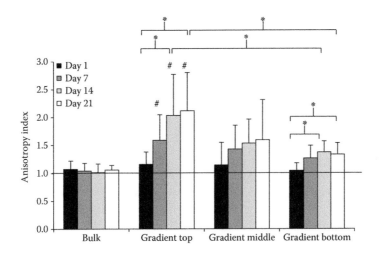

FIGURE 9.9
Anisotropy index of vascular sprouts within the top, middle, and bottom regions of PBFP gradient scaffolds as well as isotropic bulk control scaffolds over time ($n = 8$; * = $p < 0.05$; # = significant difference ($p < 0.05$) from isotropic control scaffolds at the same time point). Error bars represent ± standard deviation. (From Turturro MV et al. 2013a. MMP-sensitive PEG diacrylate hydrogels with spatial variations in matrix properties stimulate directional vascular sprout formation. *PLOS ONE* 8: e58897.)

the native extracellular microenvironment. Recapitulating this highly orchestrated and complex process in scaffolds still remains to be an extreme challenge. New engineering and biomaterial approaches are emerging at a rapid pace to create complex 3D scaffolds with embedded gradients that allow for spatial and temporal regulation of matrix properties and signaling molecules for stimulating and guiding neovascularization within 3D tissue-engineered constructs. While earlier efforts focused on vascular cell response to gradients on the surface of biomaterials, current efforts are investigating cell response and neovascularization to gradients within scaffolds, which most closely recapitulates the *in vivo* situation. Approaches have primarily focused on investigating chemotactic and haptotactic 3D cellular response for scaffold neovascularization; however, understanding the role that other anisotropic material properties play in this process results in increased complexity of scaffold design. Strategies focusing on the fabrication of controllable gradients in physical and mechanical properties that may include crosslink density, porosity, stiffness, degradation, and topographical features will provide significant insight on material features that guide 3D scaffold neovascularization in the presence of biosignals. To this end, synthetic hydrogel scaffolds offer environments for which individual and combined effects of gradients of material properties and embedded biofunctionality on 3D neovascularization can be further explored. Furthermore, gradient fabrication techniques combined with biomaterial synthesis should be aimed at designing scaffolds with gradients of controlled shape and stability as well as established techniques that allow for systematic variations in gradient magnitude and slope. While it is understood that the complex process of neovascularization cannot be completely recapitulated by any *in vitro* assay, future *in vitro* studies with gradient scaffolds should employ the use of coculture systems to more closely recapitulate the multicellular process that occurs *in vivo*. Finally, the combination of *in silico*, *in vitro*, and *in vivo* studies investigating the effects of gradient magnitude, steepness, and orientation will significantly contribute to our understanding of how these gradient characteristics influence 3D vascular assembly.

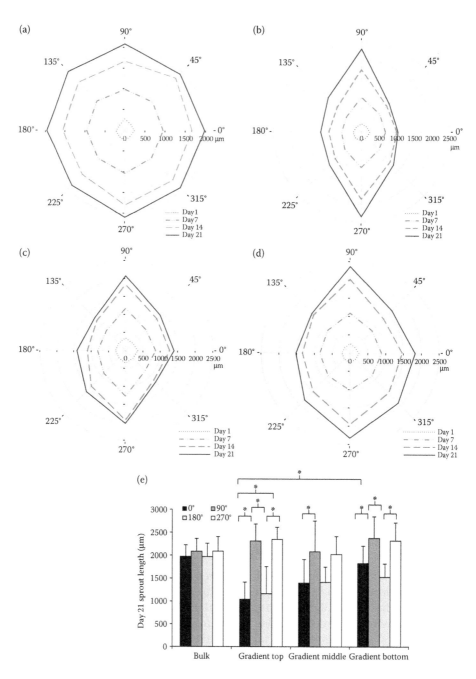

FIGURE 9.10

(a–d) Average sprout length by angle of vessels formed within isotropic bulk control scaffolds (a) as well as in scaffold regions that lie at closest (top), intermediate (middle), and furthest (bottom) distances along the gradients (b–d) over 3 weeks in culture. (*Note*: Gradients run from regions of high elastic modulus and YRGDS concentration at 90° to regions of lower elastic modulus and YRGDS concentration at 270°). (e) Day 21 sprout length as a function of location in gradient gels and scaffold type at 0°, 90°, 180°, and 270° ($n = 8$; * = $p < 0.05$; # = significant difference ($p < 0.05$) from bulk control at same time point). Error bars represent ± standard deviation. (From Turturro MV et al. 2013a. MMP-sensitive PEG diacrylate hydrogels with spatial variations in matrix properties stimulate directional vascular sprout formation. *PLOS ONE* 8: e58897.)

References

Abramsson A, Lindblom P, and Betsholtz C. 2003. Endothelial and nonendothelial sources of PDGF-B regulate pericyte recruitment and influence vascular pattern formation in tumors. *Journal of Clinical Investigation* 112: 1142–1151.

Aizawa Y, Wylie R, and Shoichet M. 2010. Endothelial cell guidance in 3D patterned scaffolds. *Advanced Materials* 22: 4831–4835.

Ausprunk DH and Folkman J. 1977. Migration and proliferation of endothelial cells in preformed and newly formed blood vessels during tumor angiogenesis. *Microvascular Research* 14: 53–65.

Benitez B and Heilshorn S. 2013. Microfluidic devices for quantifying the role of soluble gradients in early angiogenesis. In *Mechanical and Chemical Signaling in Angiogenesis,* Reinhart-King CA (ed.) pp. 47–70. Springer-Verlag Berlin Heidelberg.

Borselli C, Oliviero O, Battista S, Ambrosio L, and Netti PA. 2007. Induction of directional sprouting angiogenesis by matrix gradients. *Journal of Biomedical Materials Research Part A* 80: 297–305.

Bryant SJ and Anseth KS. 2003. Controlling the spatial distribution of ECM components in degradable PEG hydrogels for tissue engineering cartilage. *Journal of Biomedical Materials Research. Part A* 64: 70–79.

Burdick JA and Anseth KS. 2002. Photoencapsulation of osteoblasts in injectable RGD-modified PEG hydrogels for bone tissue engineering. *Biomaterials* 23: 4315–4323.

Burdick JA, Khademhosseini A, and Langer R. 2004. Fabrication of gradient hydrogels using a microfluidics/photopolymerization process. *Langmuir* 20: 5136–5156.

Cai K, Dong H, Chen C, Yang L, Jandt KD, and Deng L. 2009. Inkjet printing of laminin gradient to investigate endothelial cellular alignment. *Colloids and Surfaces. B. Interfaces* 72: 230–235.

Califano JP and Reinhart-King CA. 2010. Exogenous and endogenous force regulation of endothelial cell behavior. *Journal of Biomechanics* 43(1): 79–86: 79–86.

Carmeliet P and Jain RK. 2011. Molecular mechanisms and clinical applications of angiogenesis. *Nature* 473: 298–307.

Chan TR, Stahl PJ, and Yu SM. 2011. Matrix-bound VEGF mimetic peptides: Design and endothelial-cell activation in collagen scaffolds. *Advanced Functional Materials* 21: 4252–4262.

Chechilo NM, Khvilivitskii RJ, and Enikolopyan NS. 1972. On the phenomenon of polymerization reaction spreading. *Doklady Akademii Nauk SSSR* 204: 1180–1181.

Chen RR, Silva EA, Yuen WW, Brock AA, Fischbach C, Lin AS, Guldberg RE, and Mooney DJ. 2007a. Integrated approach to designing growth factor delivery systems. *FASEB Journal* 21: 3896–3903.

Chen RR, Silva EA, Yuen WW, and Mooney DJ. 2007b. Spatio-temporal VEGF and PDGF delivery patterns blood vessel formation and maturation. *Pharmaceutical Research* 24: 258–264.

Chung AS and Ferrara N. 2010. The extracellular matrix and angiogenesis: Role of the extracellular matrix in developing vessels and tumor angiogenesis. *Pathways* 11: 2–5.

Chung S, Sudo R, Mack PJ, Wan CR, Vickerman V, and Kamm RD. 2009. Cell migration into scaffolds under co-culture conditions in a microfluidic platform. *Lab on a Chip* 9: 269–275.

Davis GE and Senger DR. 2005. Endothelial extracellular matrix: Biosynthesis, remodeling, and functions during vascular morphogenesis and neovessel stabilization. *Circulation Research* 97: 1093–1107.

DeLong SA, Gobin AS, and West JL. 2005a. Covalent immobilization of RGDS on hydrogel surfaces to direct cell alignment and migration. *Journal of Controlled Release* 109: 139–148.

DeLong SA, Moon JJ, and West JL. 2005b. Covalently immobilized gradients of bFGF on hydrogel scaffolds for directed cell migration. *Biomaterials* 26: 3227–3234.

Diederich VE, Studer P, Kern A, Lattuada M, Storti G, Sharma RI, Snedeker JG, and Morbidelli M. 2013. Bioactive polyacrylamide hydrogels with gradients in mechanical stiffness. *Biotechnology and Bioengineering* 110: 1508–1519.

Elbert DL and Hubbell JA. 2001. Conjugate addition reactions combined with free-radical cross-linking for the design of materials for tissue engineering. *Biomacromolecules* 2: 430–441.

Francis ME, Uriel S, and Brey EM. 2008. Endothelial cell–matrix interactions in neovascularization. *Tissue Engineering Part B: Reviews*. 14: 19–32.

Gerhardt H and Betsholtz C. 2005. How do endothelial cells orientate? *EXS* 94: 3–15.

Gerhardt H, Golding M, Fruttiger M, Ruhrberg C, Lundkvist A, Abramsson A, Jeltsch M et al. 2003. VEFG guides angiogenic sprouting utilizing endothelial tip cell filopodia. *The Journal of Cell Biology* 161: 1163–1177.

Germain S, Monnot C, Muller L, and Eichmann A. 2010. Hypoxia-driven angiogenesis: Role of tip cells and extracellular matrix scaffolding. *Current Opinion in Hematology* 17: 245–251.

Geudens I and Gerhardt H. 2011. Coordinating cell behaviour during blood vessel formation. *Development* 138: 4569–4583.

Ghajar CM, Chen X, Harris JW, Suresh V, Hughes CC, Jeon NL, Putnam AJ, and George SC. 2008. The effect of matrix density on the regulation of 3-D capillary morphogenesis. *Biophysical Journal* 94: 1930–1941.

Gritli-Linde A, Lewis P, MsMahon AP, and Linde A. 2001. The whereabouts of a morphogen: Direct evidence for short-and long-range activity of hedgehog signaling peptides. *Development Biology* 236: 364–386.

Guarnieri D, Borzacchiello A, De Capua A, Ruvo M, and Netti PA. 2008. Engineering of covalently immobilized gradients of RGD peptides on hydrogel scaffolds: Effect on cell behaviour. *Macromolecular Symposia* 266: 36–40.

Guarnieri D, Capua AD, Ventre M, Borzacchiello A, Pedone C, Marasco D, Ruvo M, and Netti PA. 2010. Covalently immobilized RGD gradient on PEG hydrogel scaffold influences cell migration parameters. *Acta Biomaterialia* 6: 2532–2539.

Guo X, Elliott CG, Li Z, Xu Y, Hamilton DW, and Guan J. 2012. Creating 3D angiogenic growth factor gradients in fibrous constructs to guide fast angiogenesis. *Biomacromolecules* 13: 3262–3271.

Hadjipanayi E, Mudera V, and Brown RA. 2009. Guiding cell migration in 3D: A collagen matrix with graded directional stiffness. *Cell Motility and the Cytoskeleton* 66: 121–128.

Hahn MS, Miller JS, and West JL. 2005. Laser scanning lithography for surface micropatterning on hydrogels. *Advanced Materials* 17: 2939–2946.

Hahn MS, Miller SJ, and West JL. 2006. Three-dimensional biochemical and biomechanical patterning of hydrogels for guiding cell behavior. *Advanced Materials* 18: 2679–2684.

He J, Du Y, Villa-Uribe JL, Hwang C, Li D, and Khademhosseini A. 2010. Rapid generation of biologically relevant hydrogels containing long-range chemical gradients. *Advanced Functional Materials* 20: 131–137.

Hellström M, Kalén M, Lindahl P, Abramsson A, and Betsholtz C. 1999. Role of PDGF-B and PDGFR-beta in recruitment of vascular smooth muscle cells and pericytes during embryonic blood vessel formation in the mouse. *Development* 126: 3047–3055.

Helm CL, Fleury ME, Zisch AH, Boschetti F, and Swartz MA. 2005. Synergy between interstitial flow and VEGF directs capillary morphogenesis *in vitro* through a gradient amplification mechanism. *Proceedings of the National Academy of Sciences* 102: 15779–15784.

Hern DL and Hubbell JA. 1998. Incorporation of adhesion peptides into nonadhesive hydrogels useful for tissue resurfacing. *Journal of Biomedical Materials Research* 39: 266–276.

Isenberg BC, Dimilla PA, WM, Kim S, and Wong JY. 2009. Vascular smooth muscle cell durotaxis depends on substrate stiffness gradient strength. *Biophysical Journal* 97: 1313–1322.

Jeong GS, Han S, Shin Y, Kwon GH, Kamm RD, Lee SH, and Chung S. 2011. Sprouting angiogenesis under a chemical gradient regulated by interactions with an endothelial monolayer in a microfluidic platform. *Analytical Chemistry* 83: 8454–8459.

Keenan TM, Hsu CH, and Folch A. 2006. Microfluidic jets for generating steady-state gradients of soluble molecules on open surfaces. *Applied Physics Letters* 89: 114103–114103-3.

Kim HD and Peyton SR. 2012. Bio-inspired materials for parsing matrix physicochemical control of cell migration: A review. *Integrative Biology: Quantitative Biosciences from Nano to Macro* 4: 37–52.

Kloxin AM, Tibbitt MW, Kasko AM, Fairbairn JA, and Anseth KS. 2010. Tunable hydrogels for external manipulation of cellular microenvironments through controlled photodegradation. *Advanced Materials* 22: 61–66.

Korff T, Kimmina S, Martiny-Baron G, and Augustin HG. 2001. Blood vessel maturation in a 3-dimensional spheroidal coculture model: Direct contact with smooth muscle cells regulates endothelial cell quiescence and abrogates VEGF responsiveness. *FASEB Journal* 15: 447–457.

Lee SH, Moon JJ, and West JL. 2008. Three-dimensional micropatterning of bioactive hydrogels via two-photon laser scanning photolithography for guided 3D cell migration. *Biomaterials* 29: 2962–2968.

Lelkes PI, Hahn KA, Karmiol S, and Schmidt DH. 1998. Hypoxia/reoxygenation enhances tube formation of cultured human micro-vascular endothelial cells: The role of reactive oxygen species. In *Angiogenesis: Models, Modulators, and Clinical Applications*, Maragoudakis ME (ed.) pp. 321–336, NATO ASI Series A, Plenum Press, New York.

Lewis LL, DeBisschop CS, Pojman JA, and Volpert VA. 2005. Isothermal frontal polymerization: Confirmation of the mechanism and determination of factors affecting the front velocity, front shape, and propagation distance with comparison to mathematical modeling. *Journal of Polymer Science Part A: Polymer Chemistry* 43: 5774–5786.

Liu L, Ratner BD, Sage EH, and Jiang S. 2007. Endothelial cell migration on surface-density gradients of fibronectin, VEGF, or both proteins. *Langmuir* 23: 11168–11173.

Lo CM, Wang HB, Dembo M, and Wang YL. 2000. Cell movement is guided by the rigidity of the substrate. *Biophysical Journal* 79: 144–152.

Mac Gabhann F, Ji JW, and Popel AS. 2007. VEGF gradients, receptor activation, and sprout guidance in resting and exercising skeletal muscle. *Journal of Applied Physiology* 102: 722–734.

Mann BK, Gobin AS, Tsai AT, Schmedlen RH, and West JL. 2001. Smooth muscle cell growth in photopolymerized hydrogels with cell adhesive and proteolytically degradable domains: Synthetic ECM analogs for tissue engineering. *Biomaterials* 22: 3045–3051.

Mohan N, Dormer NH, Caldwell KL, Key VH, Berkland CJ, and Detamore MS. 2011. Continuous gradients of material composition and growth factors for effective regeneration of the osteochondral interface. *Tissue Engineering: Part A* 17: 2845–2855.

Moon JJ, Lee SH, and West JL. 2007. Synthetic biomimetic hydrogels incorporated with Ephrin-A1 for therapeutic angiogenesis. *Biomacromolecules* 8: 42–49.

Moon JJ, Saik JE, Pochéb RA, Leslie-Barbick JE, Lee SH, Smith AA, Dickinson MA, and West JL. 2010. Biomimetic hydrogels with pro-angiogenic properties *Biomaterials* 31: 3840–3847.

Moses MA. 1997. The regulation of neovascularization of matrix metalloproteinases and their inhibitors. *Stem Cells* 15: 180–189.

Nakatsu MN, Sainson RCA, Pérez-del-Pulgar S, Aoto JN, Aitkenhead M, Taylor KL, Carpenter PM, and Hughes CCW. 2003. $VEGF_{121}$ and $VEGF_{165}$ regulate blood vessel diameter through vascular endothelial growth factor receptor 2 in an *in vitro* angiogenesis model. *Laboratory Investigation* 83: 1873–1885.

Nason C, Roper T, Hoyle C, and Pojman JA. 2005. UV-induced frontal polymerization of multifunctional (Meth)acrylates. *Macromolecules* 38: 5506–5512.

Nemir S, Hayenga HN, and West JL. 2010. PEGDA hydrogels with patterned elasticity: Novel tools for the study of cell response to substrate rigidity. *Biotechnolgy and Bioengineering* 105: 636–644.

Nguyen EH, Schwartz MP, and Murphy WL. 2011. Biomimetic approaches to control soluble concentration gradients in biomaterials. *Macromolecular Bioscience* 11: 483–492.

Odedra D, Chiu LL, Shoichet M, and Radisic M. 2011. Endothelial cells guided by immobilized gradients of vascular endothelial growth factor on porous collagen scaffolds. *Acta Biomaterialia* 7: 3027–3035.

Papavasiliou G, Cheng M, and Brey EM. 2010. Strategies for vascularization of polymer scaffolds. *Journal of Investigative Medicine* 58: 838–844.

Park JE, Keller GA, and Ferrara N. 1993. The vascular endothelial growth factor (VEGF) isoforms: Differential deposition into the subepithelial extracellular matrix and bioactivity of extracellular matrix-bound VEGF. *Molecular Biology of the Cell* 4: 1317–1326.

Peret PJ and Murphy WL. 2008. Controllable soluble protein concentration gradients in hydrogel networks. *Advance Functional Materials* 18: 3410–3417.

Perumpanani AJ, Simmons DL, Gearing AJH, Miller KM, Ward G, Norbury J, Schneemann M, and Sherratt JA. 1998. Extracellular matrix-mediated chemotaxis can impede cell migration. *Proceedings of the Royal Society of London B* 265: 2347–2352.

Pojman JA. 2012. Frontal polymerization. In *Polymer Science: A Comprehensive Reference,* Matyjaszewski K and Möller M (ed.) pp. 957–980. Elsevier, Amsterdam, The Netherlands.

Pojman JA, Ilyashenko VM, and Khan AM. 1996. Free-radical frontal polymerization: Self-propagating thermal reaction waves. *Journal of the Chemical Society, Faraday Transactions* 92: 2825–2837.

Rajantie I, Ilmonen M, Alminaite A, Ozerdem U, Alitalo K, and Salven P. 2004. Adult bone marrow–derived cells recruited during angiogenesis comprise precursors for periendothelial vascular mural cells. *Blood* 104: 2084–2086.

Roam JL, Xu H, Nguyen PK, and Elbert DL. 2010. The formation of protein concentration gradients mediated by density differences of poly(ethylene glycol) microspheres. *Biomaterials* 31: 8642–8650.

Ruhrberg C. 2003. Growing and shaping the vascular tree: Multiple roles for VEGF. *Bioessays* 25: 1052–1060.

Ruhrberg C, Gerhardt H, Golding M, Watson R, Ioannidou S, Fujisawa H, Betsholtz C, and Shima DT. 2002. Spatially restricted patterning cues provided by heparin-binding VEGF-A control blood vessel branching morphogenesis. *Genes and Development* 16: 2684–2698.

Rutkowski JM and Swartz MA. 2007. A driving force for change: Interstitial flow as a morphoregulator. *Trends in cell Biology* 17: 44–50.

Saez A, Ghibaudo M, Buguin A, Silberzan P, and Ladoux B. 2007. Rigidity-driven growth and migration of epithelial cells on microstructured anisotropic substrates. *Proceedings of the National Academy of Sciences* 104: 8281–8286.

Scott EA, Nichols MD, Kuntz-Willits R, and Elbert DL. 2010. Modular scaffolds assembled around living cells using poly (ethylene glycol) microspheres with macroporation via a non-cytotoxic porogen. *Acta Biomaterialia* 6: 29–38.

Seidi A, Ramalingam M, Elloumi-Hannachi I, Ostrovidov S, and Khademhosseini A. 2011. Gradient biomaterials for soft-to-hard interface tissue engineering. *Acta Biomaterialia* 7: 1441–1451.

Seliktar D, Zisch AH, Lutolf MP, Wrana JL, and Hubbell JA. 2004. MMP-2 sensitive, VEGF-bearing bioactive hydrogels for promotion of vascular healing. *Journal of Biomedical Materials Research Part A* 68A: 704–716.

Shamloo A and Heilshorn SC. 2010. Matrix density mediates polarization and lumen formation of endothelial sprouts in VEGF gradients. *Lab on a Chip* 10: 3061–3068.

Shamloo A, Xu H, and Heilshorn S. 2012. Mechanisms of vascular endothelial growth factor-induced pathfinding by endothelial sprouts in biomaterials. *Tissue Engineering Part A* 18: 320–330.

Shevtsova VM, Melnikov DE, and Legros CJ. 2006. Onset of convection in Soret-driven instability. *Physical Review E* 73: 7302–7306.

Shin Y, Jeon JS, Han S, Jung GS, Shin S, Lee SH, Sudo R, Kamm RD, and Chung S. 2011. *In vitro* 3D collective sprouting angiogenesis under orchestrated ANG-1 and VEGF gradients. *Lab on a Chip* 11: 2175–2181.

Sieminski AL, Hebbel RP, and Gooch KJ. 2004. The relative magnitudes of endothelial force generation and matrix stiffness modulate capillary morphogenesis *in vitro. Experimental Cell Research* 297: 574–584.

Siemerink MJ, Klaassen I, Van Noorden CJF, and Schlingemann RO. 2013. Endothelial tip cells in ocular angiogenesis potential target for anti-angiogenesis therapy. *Journal of Histochemistry and Cytochemistry* 61: 101–115.

Singh M, Morris CP, Ellis RJ, Detamore MS, and Berkland C. 2008. Microsphere-based seamless scaffolds containing macroscopic gradients of encapsulated factors for tissue engineering. *Tissue Engineering: Part C* 14: 299–309.

Smith JT, Elkin, J., and Reichert WM. 2006. Directed cell migration of fibronectin gradients: Effect of gradient slope. *Experimental Cell Research* 312: 2424–2432.

Smith JT, Tomfohr JK, Wells M, Beebe T, Kelper TB, and Reichert WM. 2004. Measurement of cell migration on surface-bound fibronectin gradients. *Langmuir* 20: 8297–8286.

Smith JT, Kim DH, and Reichert WM. 2009. Haptotactic gradients for directed cell migration: Stimulation and inhibition using soluble factors. *Comb. Chem. High Throughput Screen.* 12: 598–603.

Sokic S and Papavasiliou G. 2012a. Controlled proteolytic cleavage site presentation in biomimetic PEGDA hydrogels enhances neovascularization *in vitro*. *Tissue Engineering Part A* 18: 2477–2486.

Sokic S and Papavasiliou G. 2012b. FGF-1 and proteolytically mediated cleavage site presentation influence three-dimensional fibroblast invasion in biomimetic PEGDA hydrogels. *Acta Biomaterialia* 8: 2213–2222.

Sundararaghavan HG and Burdick JA. 2012. Gradients with depth in electrospun fibrous scaffolds for directed cell behavior. *Biomacromolecules* 12: 2344–2350.

Sundararaghavan HG, Monteiro GA, Firestein BL, and Shreiber DI. 2009. Neurite growth in 3D collagen gels with gradients of mechanical properties. *Biotechnology and Bioengineering* 102: 632–643.

Swift MR and Weinstein BM. 2009. Arterial–venous specification during development. *Circulation Research* 104: 576–588.

Tayalia P and Mooney DJ. 2009. Controlled growth factor delivery for tissue engineering. *Advanced Materials* 21: 3269–3285.

Tse JR and Engler AJ. 2011. Stiffness gradients mimicking *in vivo* tissue variation regulate mesenchymal stem cell fate. *PLOS ONE* 6: e15978.

Turturro MV, Christenson MC, Larson JC, Young DA, Brey EM, and Papavasiliou G. 2013a. MMP-sensitive PEG diacrylate hydrogels with spatial variations in matrix properties stimulate directional vascular sprout formation *PLOS ONE* 8: e58897.

Turturro MV and Papavasiliou G. 2012. Generation of mechanical and biofunctional gradients in PEG diacrylate hydrogels by perfusion-based frontal photopolymerization. *Journal of Biomaterials Science Polymer Edition* 23: 917–939.

Turturro MV, Sokic S, Larson JC, and Papavasiliou G. 2013b. Effective tuning of ligand incorporation and mechanical properties in visible light photopolymerized poly(ethylene glycol) diacrylate hydrogels dictates cell adhesion and proliferation *Biomedical Materials* 8: 25001–25012.

Turturro MV, Vélez Rendón DM, Teymour F, and Papavasiliou G. 2013c. Kinetic investigation of poly(ethylene glycol) hydrogel formation via perfusion-based frontal photopolymerization: Influence of free-radical polymerization conditions on frontal velocity and swelling gradients. *Macromolecular Reaction Engineering* 7: 107–115.

Vickerman V, Blundo J, Chung S, and Kamm R. 2008. Design, fabrication and implementation of a novel multi-parameter control microfluidic platform for three-dimensional cell culture and real-time imaging. *Lab on a Chip* 8: 1468–1477.

Wong KH, Chan JM, Kamm RD, and Tien J. 2012. Microfluidic models of vascular functions. *Annual Review of Biomedical Engineering* 14: 205–230.

Zeng G, Taylor SM, McColm JR, Kappas NC, Kearney JB, Williams LH, Hartnett ME, and Bauch VL. 2007. Orientation of endothelial cell division is regulated by VEGF signaling during blood vessel formation. *Blood* 109: 1345–1352.

Zhang S and Uludağ H. 2009. Nanoparticulate systems for growth factor delivery. *Pharmaceutical Research* 26: 1561–1580.

Zisch AH, Lutolf MP, Ehrbar M, Raeber GP, Rizzi SC, Davies N, Schmokel H et al. 2003. Cell-demanded release of VEGF from synthetic, biointeractive cell-ingrowth matrices for vascularized tissue growth. *FASEB Journal* 17: 2260–2262.

Section III

Models

10

Modeling Vascularization in Tissue Engineering Scaffolds

Hamidreza Mehdizadeh, Eric Michael Brey, and Ali Cinar

CONTENTS

10.1 Introduction

Development of clinical-sized three-dimensional (3D) tissue engineering constructs requires an extensive blood vessel network to provide the cells with necessary nutrients and other factors. Hence, vascularization of engineered tissue is an essential step in generating a properly functioning tissue. While the need is clear, the underlying mechanisms of new blood vessel formation and the effects of various factors involved in the processes are far from understood, and insufficient vascularization remains one of the major challenges facing tissue engineering (Serbo and Gerecht, 2013). Over the last two decades, researchers have investigated sophisticated methods for manipulating proangiogenic factors, cells, and scaffold properties in order to overcome this challenge and promote rapid and extensive scaffold vascularization. However, optimizing the chemical, mechanical, and geometrical characteristics of scaffolds for different tissue engineering applications is a complicated and time-consuming task when relying on experimental studies alone.

Tissue engineering scaffolds are biomaterial structures engineered with characteristics to support and enhance a variety of biological responses, including vascularized tissue growth (Langer, 1999). A wide range of polymeric and ceramic materials are under investigation. Natural, synthetic, and hybrid polymeric scaffolds have shown great promise due to the flexibility in adjusting their chemical, physical, and geometrical properties (Nair and Laurencin, 2006). Typically, biodegradable scaffolds are used in order to avoid issues resulting from chronic foreign body responses or the need for surgical procedures to remove them from the body following implantation. Scaffold degradation alters its mechanical and geometrical properties dynamically, which may further influence the long-term behavior of the designed scaffolds.

Considering the complexity of the vascularization process, the development of comprehensive multidisciplinary research methods that combine experimental studies with mathematical models for investigating vascularized tissue formation could assist in the optimization process. These models could enable simulations that enhance our understanding of the role of numerous cell behaviors and scaffold properties on the process and improve the interpretation of experimental results. Theoretical models of angiogenesis can be developed to shed light on complex interactions between numerous angiogenic factors, stem cells, and the surrounding microenvironment. These models would enable the computation of variables and parameters that are difficult or impossible to determine experimentally (Wu et al., 2010), or allow testing hypotheses that are challenging to test and verify using *in vivo* or *in vitro* experiments (Vempati et al., 2011). As an example, computational models have been specifically developed to evaluate the effect of concentration gradients of angiogenic growth factors (GFs) such as vascular endothelial growth factor (VEGF) in the neighborhood of blood vessels on angiogenesis *in vivo* (Chaplain, 2000; Mac Gabhann et al., 2006b). GF gradients are difficult to measure and control *in vivo* experimentally. These models can be of great importance along with models that study the dynamic behavior of capillaries and endothelial cells (EC), as they lead to better understanding of the mechanisms of sprout formation and migration of tip cells in response to chemical signals in their microenvironment (Hashambhoy et al., 2011).

Modeling angiogenesis has been a very active area of research. Considering the high level of complexity of the angiogenesis process and the numerous and interrelated factors involved, theoretical models enable researchers to study the effects of different factors separately with precise control and to investigate the process in an organized manner. Developing theoretical models of blood vessel assembly in tissue engineering constructs is a challenging frontier of mathematical and computational modeling (Peirce et al., 2012). It includes a multitude of complex phenomena taking place at different scales: the molecular scale including gene–protein signaling pathways that lead to cellular-scale phenomena such as EC activation and differentiation of activated cells into tip cells, and eventually the macroscopic scale corresponding to sprouting and branch formation, development of new blood vessels, and coordination with parenchymal tissue development.

The ever-increasing availability of new experimental data and computational capabilities allows the development of novel theoretical models that are more comprehensive. More recent angiogenesis models are developed using novel engineering design processes that utilize a continual feedback between experiments and simulations to facilitate refinement of theoretical models as well as design of better experiments, leading to in-depth advances in angiogenesis and tissue engineering (Bentley et al., 2013). In this approach, first, the computational model is built based on available experimental knowledge, and then iterations between simulation runs and laboratory experiments are used to refine the model and gain new insight. Laboratory experiments are designed to facilitate the adjustment of model parameters and the testing of theoretical predictions. As more knowledge is gained, the complexity of the theoretical model is increased by including new knowledge in the model and then testing the results of the new model using a new set of experiments (Bentley et al., 2009; Stefanini et al., 2010; Mehdizadeh et al., 2013). This iterative approach illustrates the benefits of collaboration between theoretical and experimental research. Recent progress in computer hardware and software with fast processors and large memory, parallel computing techniques and novel user-friendly simulation toolkits has played an important role in facilitating the development of theoretical models, and this trend is expected to continue in the future.

The focus of this chapter is on the contribution of computational modeling to the study of vascularization in biomaterials and tissue engineering. Various modeling techniques that have been used for representing vascularization of tissue engineering scaffolds are introduced. An agent-based modeling platform developed in our group with the specific goal of studying angiogenesis in 3D scaffolds is then introduced, showcasing the details of an example model of vascularization of tissue engineering constructs that can be used for optimizing the design of tissue engineering scaffolds.

10.2 Introduction to Modeling Techniques

Vascularization can occur via two mechanisms: vasculogenesis and angiogenesis. Vasculogenesis occurs by differentiation of stem cells into endothelial progenitor cells and the assembly of precursor cells into vascular networks, while angiogenesis is the formation of new vessels from existing blood vessels. Angiogenesis is the mechanism that is most widely understood and models have been developed to represent both natural and pathologic angiogenesis. The majority of the models in the latter group are models investigating tumor-induced angiogenesis, where the research objective is to find ways to limit or inhibit blood vessel formation (Chaplain, 2000). The focus of this chapter is on models describing situations in which blood vessel formation is desirable and hence promoted.

Theoretical models of angiogenesis presented in the literature usually comprise either one or a combination of molecular (subcellular)-, cellular-, tissue-, and organ-level phenomena (Peirce, 2008). Models that span more than one of these levels are categorized as multiscale models. In multiscale models of angiogenesis, researchers start considering the series of events happening in either molecular or cellular levels and then observe the resulting emergent behaviors in the higher tissue or organ levels. This approach has proven very useful in explaining tissue-level phenomena based on molecular and/or cellular mechanisms.

Theoretical models can be classified as continuum or discrete, deterministic or probabilistic, or a combination (hybrid) of these approaches. Continuum models typically start from sets of nonlinear ordinary or partial differential equations such as reaction–diffusion equations that describe the spatial and temporal interactions of biological factors that regulate angiogenesis and tissue growth. In these models, cells are represented as homogeneous populations. For example, using continuum models, it has been possible to determine cell and capillary densities, number of sprouts, and sprout tip extension rate toward the angiogenic source. These models have also been used successfully for computing the concentrations of different soluble and insoluble cell substrates such as glucose, oxygen, and GFs (Edelstein and Ermentrout, 1982; Nekka et al., 1996; Chaplain, 2000; Schugart et al., 2008). Although it is not possible to represent the behavior of individual cells as independent entities in these models, they provide powerful tools for calculating the dynamic properties of cell populations and variations in intra- and extracellular factors and cell nutrients and metabolites, and provide insight into challenging questions regarding the angiogenesis process.

In discrete cell-based or agent-based models (ABM) (North et al., 2006), cells are represented as single independent entities, or agents, which interact together and with their environment to generate collective higher-level behaviors that lead to tissue patterning and morphogenesis. Discrete models can capture morphological features of the developing capillary networks more precisely than continuous models. Deterministic models do not consider variations among the cells, and outcomes are precisely determined given the model inputs; hence, a given input will always produce the same output (Sun et al., 2005). In contrast, stochastic models take into account randomness in cellular behaviors and result in a distribution of responses to similar, or even the same, input variables (Stokes and Lauffenburger, 1991). Hybrid models combine various modeling approaches (Chaplain, 2000).

ABMs are often multiscaled, span several spatial levels, and are capable of producing tissue behaviors as the result of emergent behavior of the combination of actions performed by individual cells (Zahedmanesh and Lally, 2012). Biomedical ABMs include any or a combination of several behaviors of the cells, such as growth, migration, proliferation, apoptosis, and quiescence (Walpole et al., 2013). These cell-level behaviors lead to capillary extension, branching, anastomosis, regression/remodeling, and blood flow in the vascular network. Sound theoretical models can lead to novel biological insight. As an example, it was assumed until recently that different EC sensitivity to angiogenic factors was the reason why only certain ECs along a preexisting blood vessel respond to angiogenic stimuli while neighboring ECs only microns away do not react and remain quiescent (Nekka et al., 1996). It was discovered through the combination of experimental and computational models that the process of tip cell selection is a complex series of intra- and intercellular events being controlled by the delta-like 4 (Dll4)/notch-mediated lateral inhibition (notch signaling) (Hellstrom et al., 2007; Bentley et al., 2008).

Novel modeling approaches combine continuum and discrete modeling techniques into a single framework (hybrid models) and enable scientists to develop modular modeling frameworks that include several submodels integrated into a unified platform. This provides an unmatched flexibility regarding the physiological, chemical, and mechanical cues that researchers are able to include in their models (Liu et al., 2011). Researchers can use the models to investigate a particular signal transduction pathway (in molecular level) or cell response (in cellular level) as an independent module in their model to determine its effect on the whole process using simulation runs. Furthermore, it is possible to design and validate a model and then gradually increase its complexity by adding new capabilities.

10.2.1 Equation-Based Models of Angiogenesis

Mathematical modeling of angiogenesis dates back to the 1980s, when differential equations were used to describe changes in capillary density and depths of vessel invasion. Equation-based models use fundamental equations derived from mass balances and kinetic expressions. The first models developed were one-dimensional (1D), which captured the basic features of angiogenesis at the macro-scale level (Zawicki et al., 1981; Balding and McElwain, 1985; Byrne and Chaplain, 1995). Many of these models were based on the fungal growth model of Edelstein (Edelstein and Ermentrout, 1982) taking advantage of the common features of the few well-studied processes, such as branching, anastomosis, and migration. In these models, 1D spatial modeling was made possible by averaging the dependent variables in a plane perpendicular to the direction of the propagation of the vascular front.

Stokes and Lauffenburger presented a 2D stochastic model for the analysis of random motility and chemotaxis of ECs, which could be used to predict the path of newly formed capillaries based on trajectories of the migrating tip cells (Stokes and Lauffenburger, 1991).

They studied the behavior of migrating ECs and showed that EC migration is the determining factor in capillary growth rate and vascular structure. They concluded that *in vitro* migration assays of ECs could be used in studying angiogenesis and exploring the effects of possible activators and inhibitors of angiogenesis. They defined two parameters, the magnitude of random movement accelerations and a constant decay rate for movement velocity, for quantifying the motility of ECs under persistent random walk and used their mathematical model for computing these parameters in the presence or absence of an angiogenic GF (Stokes et al., 1991). In this model, the effects of resistance to motion, random fluctuations, and chemotactic bias were combined to derive a stochastic differential equation for the rate of change of cell velocity. This model was one of the first efforts for quantifying important cellular characteristics using a combination of mathematical modeling and experiments.

Toward the end of the 1990s, more detailed 2D models of angiogenesis were introduced where partial differential equations were utilized to describe the changes in important variables such as number of sprouts, capillary density, and invasion depth (Pettet et al., 1996a,b; Orme and Chaplain, 1997; Olsen et al., 1997). These models produced a more realistic representation of the angiogenic behavior and were able to mimic the spatiotemporal distribution of newly formed capillaries within the microenvironment. The major advantage of these models was their ability to simulate reproducible results on a viewable 2D domain. Hence, their results could be used for qualitative comparison between model predictions and *in vivo* and *in vitro* experimental observations for testing, verification, and parameter estimation. Different cases and scenarios were studied using these models. These include investigating the roles of chemotaxis and haptotaxis in neovascularization (Orme and Chaplain, 1997), wound-healing management to gain an understanding as to why some wounds fail to heal (Pettet et al., 1996b), the roles of the insoluble extracellular matrix (ECM) in wound-healing angiogenesis, and ECM-mediated random motility and cell proliferation as key angiogenic processes (Olsen et al., 1997).

More complexity was added to the models through the introduction of increasing amounts of cellular-level biological knowledge. Chaplain (2009) investigated the possibility of using a mathematical model as an angiogenesis assay. His model was a combination of continuum, deterministic modeling with discrete, stochastic modeling in two and three spatial dimensions. By combining parameter estimation with independent experimental measurements, the model could be adjusted to produce predictions that were in good agreement with experimental results. A discrete model was obtained by discretizing the partial differential equation governing the rate of change of EC density and was then made stochastic by introducing coefficients to represent the probabilities of the ECs moving in different directions. These coefficients were functions of the fibronectin concentration within the ECM and angiogenic GF concentrations at neighboring sites.

Levine et al. (2001) developed a model of tumor angiogenesis, which included biochemical processes at the cell level, basing the movement of the cells on the theory of reinforced random walks and implementing standard transport equations for the diffusion of molecular species in porous media. The impact of angiostatin, a well-known antiangiogenic agent, was investigated using this model. They were able to obtain a good agreement between simulation results and previous experimental work reported elsewhere (Ausprunk and Folkman, 1977).

Current continuum models of angiogenesis typically solve a set of interrelated and nonlinear partial differential equations describing spatial and temporal interactions of several key biological factors that regulate the angiogenesis process. These factors include capillary and sprout densities, concentrations of oxygen, nutrients, and GFs, densities of various cell types, and other factors such as ECM density (Schugart et al., 2008). A very detailed continuum model was developed based on biochemical kinetics and continuum mechanics, which takes account of the behaviors of large number of cells (Levine and Nilsen-Hamilton, 2006). This population model considered the role of the kinetics of biochemical reactions, cell cycle progression, chemotaxis and haptotaxis, and contact inhibition and crowding.

A deterministic continuous framework for modeling GF-mediated angiogenesis was developed in which traditional EC density was replaced by a binary "capillary indicator function," a new concept developed to describe the capillary structure, resulting in the enhancement of the capability of the model in capturing the capillary network with higher resolution and at a cell-by-cell scale (Sun et al., 2005). This model was developed for providing a better understanding of the effects of the ECM on angiogenesis, introducing "conductivity" of the ECM as a parameter for measuring the heterogeneity and anisotropy of the matrix. The simulations could capture vascular patterning and formation of loops within the overall dendritic capillary network structure, resembling results that could be obtained from novel discrete ABMs (Figure 10.1).

In another study, a detailed 2D biophysical and molecular model was developed to study the transport of VEGF in rat muscle tissue under *in vivo* conditions (Mac Gabhann et al., 2006a). This model is especially important because of the difficulty in experimentally measuring or controlling concentration gradients of proteins *in vivo*. They were able to estimate the distribution of VEGF concentration and VEGF receptor activation throughout the tissue. Their model predicted significant VEGF gradients (average 3% VEGF/10 μm), sufficient for cell sensing, even under resting conditions. These VEGF gradients caused notable heterogeneity in VEGF-R binding in the cells, and the authors hypothesized this as the likely reason for stochastic tip cell selection and sprout formation (Bentley et al., 2008). The same group later developed a multiscale computational model of neovascularization to study the effects of blood flow, oxygen distribution, VEGF secretion and transport, and VEGF-receptor binding on various therapeutic strategies for peripheral arterial disease (Mac Gabhann et al., 2007). Using this model, they illustrated the importance of distinguishing between absolute VEGF concentration and relative VEGF concentration gradients and confirmed the previous observation that an increase in the absolute concentration of VEGF results in increased VEGF-R2 binding and proangiogenic signaling, while the magnitude of VEGF gradients play an important role in chemotactic directionality of the angiogenic sprouts (Gerhardt, 2003).

10.2.2 ABM to Describe Angiogenesis

Agent-based modeling is particularly suitable for investigating the underlying mechanisms and governing rules of complex processes. In biological systems, ABMs enable the representation of cells or vessel segments as independent, heterogeneous entities that interact with each other and their environment to generate higher-level emergent tissue- or organ-level behaviors (Thorne et al., 2007). ABMs have applications in diverse fields ranging from social sciences and supply chains to economics and biology, providing a novel modeling paradigm for dealing with the increasing complexities involved in designing computational models (Bonabeau, 2002; Behdani et al., 2011).

Advances in genetics and cellular biology have generated a vast amount of biological information, requiring a systematic approach to integrate this knowledge in a unified framework, facilitating the development of multiscale models. In a typical ABM approach, each cell is represented by a software agent with a set of states and behaviors and an internal rule base to relate them. ABMs are naturally suitable for modeling biological systems as they can be composed of discrete micro-scale constituents (e.g., cells) that form

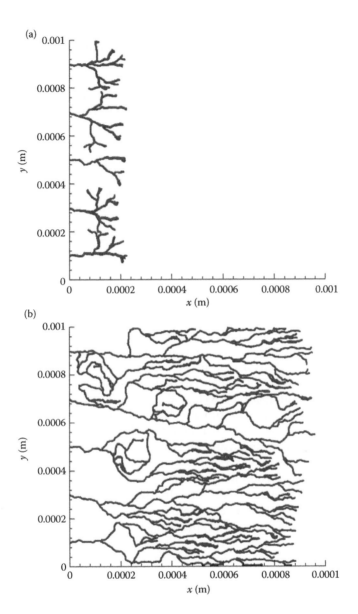

FIGURE 10.1
Sample simulation results of a deterministic continuous framework developed by Sun et al. for modeling growth factor-mediated angiogenesis after 250 time steps, for two different values of τ, the parameter introduced for limiting endothelial cell proliferation time. Proliferation speeds up by decreasing τ from (a) 1.0 (day) to (b) 0.25. Formation of different vascular patterns as a result of varying model parameters are illustrated. (From Sun, S. et al. 2005. *Bulletin of Mathematical Biology, 67,* 313–337.)

nonhomogeneous macro-scale structures (e.g., tissues and organs). Previous publications present an overview of several applications of ABM in biomedical research, highlighting published examples (Thorne et al., 2007) reviewing biomedical ABMs, outlining their value as a tool for studying highly complex biological and biomedical systems, and defining steps involved in developing biomedical ABMs (Mehdizadeh et al., 2011). ABMs have been used successfully for understanding and explaining the underlying mechanisms of growth and expansion of cancerous tumors (Athale et al., 2006) and simulating the cellular decision process in the context of a virtual brain tumor (Zhang et al., 2007). In bone tissue engineering, ABMs have been used to simulate real-time signaling induced by mechanical stimuli in osteocytic networks (Ausk et al., 2006).

ABMs offer an efficient methodology for identifying cell behavior patterns that lead to specific EC phenotypes and capillary or blood vessel network characteristics. Such a characterization is difficult to perform using conventional equation-based approaches as it is not straightforward to represent (or quantify) behaviors using mathematical equations, while in ABMs, rules are used to describe the agent behaviors and actions. An early example of angiogenesis modeling by ABMs is the 2D cellular automata (CA) developed by Peirce et al. (2004) that could quantitatively predict the behavior of a collection of cells in vessel network remodeling. CAs are simpler predecessors of ABMs. CAs are always homogeneous and uniformly densely populated on the grid (all cells are identical), whereas in ABMs the agents are heterogeneous and do not necessarily occupy all spaces within the grid. Their simulation framework incorporated more than 50 rules based on published literature data and could integrate epigenetic stimuli, GF and protein diffusion (molecular signals), and cellular behaviors to predict microvascular network patterning events such as new blood vessel formation, capillary extension, and recruitment of contractile perivascular cells. Bentley et al. (2008) used a multicell ABM to simulate the process of tip cell selection based on a proposed feedback loop that relates tip cell induction by VEGF to Dll4/notch-mediated lateral inhibition. They illustrated how such an assumption led to the appearance of "salt and pepper" patterns that were also observed experimentally (Hellstrom et al., 2007).

Qutub et al. developed a 3D multiscale ABM of the angiogenesis process and VEGF system to investigate the relationship between GF gradients, cell sprouting, proliferation and migration, and sprout extension (Qutub and Popel, 2009). This model included extensive rules and parameters that were based on literature data, and was closely related to intra- and intercellular biological mechanisms. Hence, it could be used to test and investigate the effect of these mechanisms on emergent behavior at the tissue level. Global signals and local communications were combined into a stochastic cell-level decision-making ABM framework of angiogenesis (Das et al., 2010). In this model, a cell could transit stochastically between quiescent, proliferative, apoptotic, and migrating states, leading to experimentally observed features such as cell sprouting and branch formation. This model was used to categorize different sprout characteristics as emergent behavior resulting from collective cells actions, and hence identifying factors that lead to these behaviors. They combined their simulation studies with closely regulated *in vitro* microfluidic experiments to validate their model and to investigate the impact of individual key factors on angiogenesis (Vickerman et al., 2008).

An ABM with complex logical rules and equation-based models integrated into a uniform framework for studying angiogenesis in skeletal muscle was developed by Liu et al. (2011). Their "module-based" model combined a variety of previously developed computational models, each focusing on a different aspect of the angiogenesis process and occurring at a different biological scale, ranging from the molecular to the tissue levels. The models included in their work represented microvascular blood flow, oxygen transport, vascular endothelial GF transport and EC behavior (sensing, migration, and proliferation), and combined diverse methodologies such as algebraic equations, partial differential equations, and ABM with complex logical rules. A series of computational experiments were performed to simulate angiogenesis in response to activity during a single bout of exercise in rat extensor digitorum longus (EDL) muscle. Simulations included capillary activation, cellular sprouting, and new vascular network formation. Figure 10.2 shows the

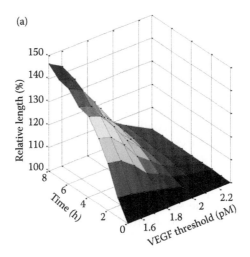

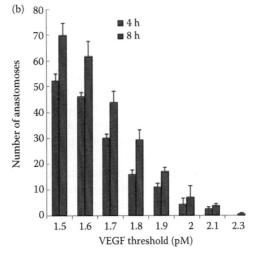

FIGURE 10.2
Sensitivity analysis studies are used to study the effect of important parameters on angiogenesis. This analysis shows the effect of VEGF threshold for EC activation on (a) the capillary length over time and (b) the number of anastomoses formed at 4 and 8 h. Simulation sample size is five for each VEGF threshold at a given time. (Reprinted with permission from Liu, G. et al. 2011. *Theoretical Biology and Medical Modelling*, 8, 6.)

results of sensitivity analysis, illustrating the significant impact of VEGF threshold level for EC activation on the capillary growth rate and the number of anastomoses, both important angiogenic attributes. The importance of this work is that it shows how the result of separate studies on different aspects of angiogenesis by different groups of researchers can be integrated in one single platform, and highlighting ABM as a methodology for making this integration possible.

Recently, Long et al. developed an integrated experimental-computational method to characterize different EC activity when stimulated by VEGF and brain-derived neurotrophic factor (BDNF). The rules-as-agents (RAA) framework used allows automated, rapid hypothesis testing (Long et al., 2013). Using this strategy, they were able to compare cell behavior hypotheses with *in vitro* angiogenesis experiments, leading to better quantitative understanding of cell behavior patterns as a function of the local microenvironment.

10.2.3 Models of Vascularization of Tissue Engineering Scaffolds and Biomaterials

In this section, models specifically developed to investigate vascularization of tissue engineering scaffolds and biomaterials are introduced. While this is a very important issue in tissue engineering and regenerative medicine, there have been few models developed to specifically address issues related to vascularization in biomaterial scaffolds. In general, the approaches that have been used consider various aspects of scaffold vascularization and address a wide spectrum of challenges encountered in promoting blood vessel growth within 3D scaffold structures.

One of the first models of angiogenesis that was specifically developed to understand and improve capillary formation within porous tissue engineering scaffolds was the equation-based model developed by Jabbarzadeh and Abrams (2007). They modeled GF release from arbitrary sources within the scaffold and its transport within porous scaffolds to investigate the effect of GF concentration profiles on EC chemotaxis and direction of capillary sprout extension (Jabbarzadeh and Abrams, 2007). Their 2D model was based on GF-mediated angiogenesis, and sprout extension was driven by the migration of tip cells toward higher GF regions within the scaffold. Dynamic variations of GF gradients resulting from release, diffusion, and degradation of the proteins within the porous structure of scaffolds was modeled, and the two models were combined to enable investigating the effect of GF concentration profiles on angiogenesis. They used their model to elaborate various degradation rates of GF by natural decay and different scenarios of GF production and were able to identify optimal GF release strategies for increased angiogenesis. The effect of these strategies on the morphology of capillary networks was also evaluated. They made attempts to quantify blood vessel growth and invasion depth and correlate it with GF concentration and release regime, and hence provided a quantifiable method for comparing various vascularization strategies.

In a different methodology, a mathematical model was developed to consider the impact of mechanical characteristics of the cellular environment on vascularization of bone implants (Checa and Prendergast, 2009). This "mechanoregulation" model allowed the study of the effect of vascular supply on bone tissue formation and a better understanding of the tissue regeneration process. They hypothesized that the amount of tissue formation is dependent on vascularization and the resulting oxygen supply. The lattice-based simulation included a tissue model composed of a number of rules governing parenchymal cell migration, proliferation, and differentiation, and a separate angiogenesis model in which capillaries were modeled as a sequence of ECs. The sprout tip cell migration was based on a combination of a persistent random walk and chemotaxis mechanisms. Using the simulation, they were able to demonstrate the impact of vascularization on the tissue formation process. Later, they extended their model to consider the effect of cell seeding and mechanical loading in porous scaffolds on angiogenesis and tissue growth. They applied the lattice-based model to the study of combined angiogenesis and bone tissue formation in porous calcium phosphate (CaP) scaffolds with regular (Checa and Prendergast, 2010) and irregular (Sandino et al., 2010) pore morphologies. These models predicted effective vascularization of the pores close to the scaffold interface and the delay of invasion to deeper regions because of steric hindrances caused by walls of the pores. Also, it was shown that the formation of vascular networks and the distribution of differentiated tissues can be regulated by the morphology of the scaffold.

A combination of experimental and theoretical models was used to study vascularization of hydrogel implants under two different conditions, a control in which the hydrogel was implanted in mice without seeded cells, and a hydrogel seeded with mouse stem cells to facilitate implant vascularization (Jain et al., 2012). In the control case, vascularization was the result of the foreign body reaction, while in the cell treatment case, vascularization was induced by angiogenic factors produced by the seeded stem cells. In experimental studies, the researchers utilized sensors containing oxygen-sensitive crystals embedded within the hydrogel to noninvasively measure local partial oxygen pressures. They coupled these experimental data with a mathematical model to study vascularization in response to oxygen requirement within the tissue engineering constructs. A series of reaction–diffusion equations were used to describe the spatiotemporal dynamics of key biophysical variables such as microvascular density, oxygen tension, angiogenic GF concentration, and cell densities. Experimental results of oxygen tension along with literature data were used to estimate model parameters. Using this combination, they were able to study vascular formation in biomedical implants in treatments with and without stem cells. The immunohistochemistry results at 10 weeks illustrate higher levels of implant oxygenation and elimination of foreign body reaction in the stem cell treatment case, indicating the positive therapeutic effect of this method of treatment.

Almeida and Bartolo developed a computational framework to enable numerical analysis of fluid flow within tissue engineering scaffolds in an attempt to design optimal scaffold structures. The aim was to design scaffolds that are less resistant to blood flow (Almeida and Bartolo, 2012). In their work, it is assumed that the empty space within the scaffold unit represents the fluid volume, resulting in the entire pore volume within the scaffold being perfused, instead of flow within a network of capillaries in the pores. Hence, their work did not take into account the dynamics of blood vessel formation. Their results indicated the possibility of inadequate blood flow in the internal sections of the scaffold, which can lead to unwanted coagulation. They also developed a software tool to evaluate mechanical properties of scaffolds with varying geometrical properties, including porosity, thickness, pore structure, and properties of the selected material (Almeida and Bártolo, 2012). Their computational model was used to predict scaffold properties, including elastic and shear moduli based on the macro- and microstructure of scaffolds, while also considering the effect of scaffold degradation. They used their model to design optimized scaffold structures suitable for various tissue engineering applications.

An "individual-based" model of angiogenesis within porous biomaterials was developed by Lemon et al. (2011). They used the term "individual-based" rather than agent-based because the cells were represented as discrete grid locations in their model, similar to automatons in CA and explicit software entities, or "agents," that explicitly represent these cells were not used in the model. The biomaterial scaffold was assumed to be implanted in the human body, and a number of rules derived from the literature were designed to define EC migration and capillary branching and anastomosis inside a virtual pore. Fibroblasts were assumed to be seeded within the pore and rules were defined to control chemotaxis and proliferation. Fibroblasts secrete GFs that diffuse within the pore, are taken up by ECs, and degrade spontaneously with a constant rate, all modeled mathematically using the approach previously described (Jabbarzadeh and Abrams, 2007). Also, the model developed by McDougall (McDougall et al., 2002) was used to describe the blood flow through the developed capillary network. Using simulations, they could illustrate that a strategy of localized seeding of fibroblasts and random seeding of ECs maximizes biomaterial vascularization. In an earlier work, they had developed a mathematical model for the vascularization of porous scaffolds using a set of coupled nonlinear ordinary differential equations to describe the time evolution of the amounts of different tissue constituents inside the scaffold (Lemon et al., 2009).

Our group developed a *multilayer* ABM framework for the investigation of the effects of geometrical properties of model 2D scaffolds, such as pore size and shape, on the rate and depth of blood vessel capillary invasion (Artel et al., 2011). Agents designed to represent ECs had variables that determine the internal state of the agent and methods that define the functions that the agent can perform. A rule base contained the logic that governs the behavior of EC agents and the actions they perform during angiogenesis within porous scaffold architecture in response to environmental factors, including sprout initiation, elongation, proliferation, tip cell migration, and anastomosis. The model was able to predict the dynamics of blood vessel invasion across various scaffold pore sizes. The modeling framework was validated by comparing simulation results with independent experimental data sets. This ABM framework was extended to simulate the process of sprouting angiogenesis within 3D porous scaffolds (Mehdizadeh et al., 2013). Figure 10.3 shows the schematic of EC agents during capillary formation. Each EC agent was represented as a cylinder, identified with front and rear nodes in space and a constant radius. The speed of vessel extension was set based on experimental studies *in vivo*. 3D scaffold models with well-defined homogeneous and heterogeneous pore architectures were designed to investigate the impact of scaffold design parameters on the vascularization process. To validate the model, results were compared with experimental results of vascularization in porous hydrogel scaffold and indicated a good agreement between simulation and experimental results. The model was used to investigate the effects of various scaffold architectural properties such as average pore size, pore size distribution, interconnectivity, and porosity on scaffold vascularization, demonstrating the ability of the model to provide insight regarding optimal scaffold properties that support vascularization of engineered tissues.

The tip cell is represented as the front node of the leading EC in every branch. Tip cell migration within the pore domain of the scaffold results in an extension of the capillary branch, while it has the ability to connect to other EC agents if they are within a defined

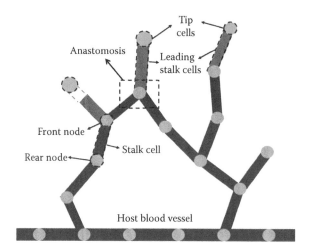

FIGURE 10.3

Schematic of the EC agent model described previously in Mehdizadeh et al. (2013). Initial host blood vessels are located at the scaffold periphery. Initial sprouts branch from host blood vessels, following the tip cell migration. Each EC is represented by one EC agent, and is assumed to be a cylinder that includes front and rear nodes with a fixed radius. The front node of the leading stalk cell represents the tip cell. The tip cell senses its local environment, and migrates in the direction of highest (GF). The leading stalk cell elongates, following the tip cell migration. A tip cell may connect to another EC agent only if it is from another branch, leading to anastomosis. (Reproduced with permission from Mehdizadeh, H. et al. 2013. *Biomaterials*, 34, 2875–2887.)

distance from the cell. The leading stalk cell follows the tip cell, elongating as the tip cell migrates. Branching probability depends on the GF concentration, with the number of branches increasing as the GF concentration is increased (Ausprunk and Folkman, 1977). The new branches extend in the direction of the highest GF gradient. Figure 10.4 shows the rule base governing the behavior of individually acting agents, leading to tip cell migration and branching, and stalk cell elongation and proliferation in this example.

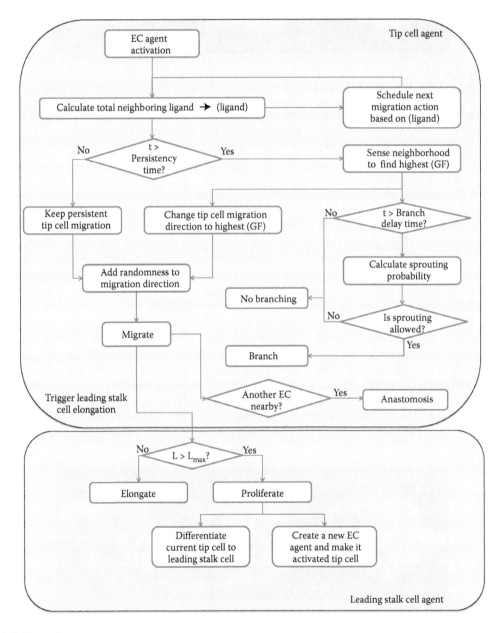

FIGURE 10.4
Flowchart depicting the main rules governing the tip cell and stalk cell behavior in ABM described in Mehdizadeh et al. (2013). Tip cell agent performs chemotactic migration, random branching, and anastomosis. Tip cell migration leads to stalk cell elongation and proliferation.

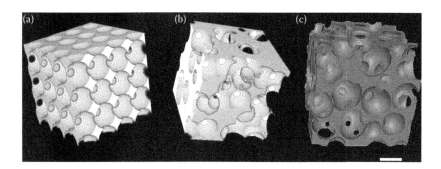

FIGURE 10.5
Renderings of (a) a homogeneous scaffold with 275 μm pore size at 60% porosity, (b) 275 μm heterogeneous pores at 60% porosity with zero pore size distribution, and (c) bioactive glass scaffold formed using a foaming technique (George et al., 2006). Bar represents 500 μm. Nonuniform distribution of pores more closely approximates actual scaffold structures. (Figure (c) is reprinted with permission from Springer Science and Business Media: *Biomaterials*, Three-dimensional modeling of angiogenesis in porous biomaterial scaffolds, 34, 2013, 2875–2887, Mehdizadeh, H. et al.)

   Random parameters were used to introduce stochastic variation into the model, leading to stochastic behavior and variance among cell populations (Mehdizadeh et al., 2013). Porous scaffolds were generated to study the effect of biomaterial architecture on vascularization. Scaffold properties, including average pore size, pore size distribution, and pore arrangement and interconnectivity, were systematically varied to study their roles in vascularization. 3D renderings of model homogeneous (Figure 10.5a) and heterogeneous (Figure 10.5b) scaffolds show that the heterogeneous pore placement agrees closely with the structure of real scaffolds (Figure 10.5c) (George et al., 2006).

   In simulation runs, the activated tip cells sprouted from the host blood vessels in response to GF gradients. The tip cells migrated and led capillary ingrowth into the pore domain of the scaffold, resulting in perfused tissue within the scaffolds. 3D images depicting scaffold vascularization after 6 weeks are shown in Figure 10.6. The majority of the pores located far from the interface were not vascularized even after 5 weeks. These results were in agreement with independent 3D cell culture model of sprouting angiogenesis that examined the invasion into the hydrogels (Chiu et al., 2011). We defined normalized pore connectivity (NPC) as the ratio of the pore throat diameter to the mean pore diameter and

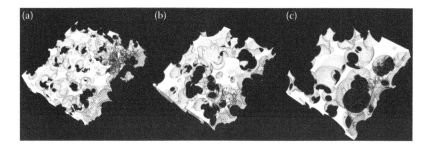

FIGURE 10.6
3D renderings of heterogeneous scaffold vascularization after 6 weeks for different pore sizes of (a) 150, (b) 275, and (c) 400 μm. Overall scaffold porosity is 80% in all pore sizes, and pore size distribution is 75 μm. The sizes of the scaffolds are 2000 × 2000 × 1000 μm. (Adapted from Mehdizadeh, H. et al. 2013. *Biomaterials*, 34, 2875–2887.)

used this parameter for comparing relative pore connectivity across different mean pore sizes (Mehdizadeh et al., 2013). NPC allows for scaling interconnectivity and resulting angiogenesis between scaffolds with significantly different mean pore sizes. This shows that computational models are useful in identifying parameters that are important and need to be considered when designing porous scaffolds.

10.3 Conclusions

Mathematical models and simulators complement experimental research by providing a computational tool to help understand, predict, and/or control *in vitro* and *in vivo* experiments. They can provide insight into complex biomedical phenomena by conducting exploratory studies rapidly at low cost and with unmatched precision and control. Modeling biomaterial scaffold vascularization aids researchers in designing better experiments and decreasing the total number of experiments required to identify optimal scaffold properties. This chapter provides an overview of various techniques for modeling vascularization of tissue engineering scaffolds. As angiogenesis is the primary mechanism for scaffold vascularization, we specifically covered models of sprouting angiogenesis. Meanwhile, it is important that future computational models also consider vasculogenesis as it is the essential mechanism enabling the fabrication of prevascularized artificial tissues through endothelial tubule formation (Reiffel et al., 2012). Specially, vascularization of larger scaffolds from peripheral blood vessels may be accelerated if the inner parts of the scaffold are prevascularized.

Computational models can be used to study the effects of scaffold mechanical and architectural properties on the angiogenesis process, making them a powerful tool for investigating optimal parameters for facilitating rapid ingrowth. While porosity and pore size have received significant attention in experimental and theoretical models, very few models have considered the effect of pore connectivity. Computational models should be improved to consider an increasing number of factors that affect vascularization. While this will increase model complexity, ABMs can be used as a novel method for developing models with a high level of modularity and facilitate the development of more complex models.

Thanks to the large amount of research in this area, our knowledge of the angiogenesis process and tissue engineering methods keeps increasing at an ever-increasing pace. While experimental studies aim at understanding the effects of one or a few factors at a time, computational models inherently do not have any limit in the number of factors they can take into account. Hence, theoretical models ideally complement experimental research, serving as a framework for incorporating our collective knowledge about the underlying processes involved in scaffold vascularization.

Acknowledgments

Funding from the National Science Foundation (CBET-0731201, IIS- 1125412) and the Veterans Administration that partially supported the work of HA, EB, and AC reported in this chapter are gratefully acknowledged.

References

Almeida, H. A. and Bártolo, P. J. 2012. Structural and vascular analysis of tissue engineering scaffolds, Part 2: Topology optimisation. *In:* Liebschner, M. A. K. (ed.) *Computer-Aided Tissue Engineering.* Humana Press.

Almeida, H. A. and Bartolo, P. J. 2012. Structural and vascular analysis of tissue engineering scaffolds, Part 1: Numerical fluid analysis. *Methods in Molecular Biology*, 868, 183–207.

Artel, A., Mehdizadeh, H., Chiu, Y. C., Brey, E. M., and Cinar, A. 2011. An agent-based model for the investigation of neovascularization within porous scaffolds. *Tissue Engineering. Part A*, 17, 2133–2141.

Athale, C., Mansury, Y., and Deisboeck, T. S. 2006. Corrigendum to "Simulating the impact of a molecular 'decision-process' on cellular phenotype and multicellular patterns in brain tumors". *Journal of Theoretical Biology*, 239, 516–517.

Ausk, B. J., Gross, T. S., and Srinivasan, S. 2006. An agent based model for real-time signaling induced in osteocytic networks by mechanical stimuli. *Journal of Biomechanics*, 39, 2638–2646.

Ausprunk, D. H. and Folkman, J. 1977. Migration and proliferation of endothelial cells in preformed and newly formed blood vessels during tumor angiogenesis. *Microvascular Research*, 14, 53–65.

Balding, D. and Mcelwain, D. L. S. 1985. A mathematical model of tumour-induced capillary growth. *Journal of Theoretical Biology*, 114, 53–73.

Behdani, B., Adhitya, A., Lukszo, Z., and Srinivasan, R. 2011. Negotiation-based approach for order acceptance in a multiplant specialty chemical manufacturing enterprise. *Industrial and Engineering Chemistry Research*, 50, 5086–5098.

Bentley, K., Gerhardt, H., and Bates, P. A. 2008. Agent-based simulation of notch-mediated tip cell selection in angiogenic sprout initialisation. *Journal of Theoretical Biology*, 250, 25–36.

Bentley, K., Jones, M., and Cruys, B. 2013. Predicting the future: Towards symbiotic computational and experimental angiogenesis research. *Experimental Cell Research*, 319, 1240–1246.

Bentley, K., Mariggi, G., Gerhardt, H., and Bates, P. A. 2009. Tipping the balance: Robustness of tip cell selection, migration and fusion in angiogenesis. *PLoS Computational Biology*, 5, e1000549.

Bonabeau, E. 2002. Agent-based modeling: Methods and techniques for simulating human systems. *Proceedings of the National Academy of Sciences of the United States of America*, 99, 7280–7287.

Byrne, H. M. and Chaplain, M. A. J. 1995. Mathematical models for tumour angiogenesis: Numerical simulations and nonlinear wave solutions. *Bulletin of Mathematical Biology*, 57, 461–486.

Chaplain, M. A. J. 2000. Mathematical modelling of angiogenesis. *Journal of Neuro-Oncology*, 50, 37–51.

Checa, S. and Prendergast, P. J. 2009. A mechanobiological model for tissue differentiation that includes angiogenesis: A lattice-based modeling approach. *Annals of Biomedical Engineering*, 37, 129–145.

Checa, S. and Prendergast, P. J. 2010. Effect of cell seeding and mechanical loading on vascularization and tissue formation inside a scaffold: A mechano-biological model using a lattice approach to simulate cell activity. *Journal of Biomechanics*, 43, 961–968.

Chiu, Y.-C., Cheng, M.-H., Engel, H., Kao, S.-W., Larson, J. C., Gupta, S., and Brey, E. M. 2011. The role of pore size on vascularization and tissue remodeling in PEG hydrogels. *Biomaterials*, 32, 6045–6051.

Das, A., Lauffenburger, D., Asada, H., and Kamm, R. D. 2010. A hybrid continuum–discrete modelling approach to predict and control angiogenesis: Analysis of combinatorial growth factor and matrix effects on vessel-sprouting morphology. *Philosophical Transactions of the Royal Society A: Mathematical, Physical and Engineering Sciences*, 368, 2937–2960.

Edelstein, L. and Ermentrout, B. 1982. The propagation of fungal colonies: A model for tissue growth. *Journal of Theoretical Biology*, 98, 679–701.

George, J., Kuboki, Y., and Miyata, T. 2006. Differentiation of mesenchymal stem cells into osteoblasts on honeycomb collagen scaffolds. *Biotechnology and Bioengineering*, 95, 404–411.

Gerhardt, H. 2003. VEGF guides angiogenic sprouting utilizing endothelial tip cell filopodia. *The Journal of Cell Biology*, 161, 1163–1177.

Hashambhoy, Y. L., Chappell, J. C., Peirce, S. M., Bautch, V. L., and Mac Gabhann, F. 2011. Computational modeling of interacting VEGF and soluble VEGF receptor concentration gradients. *Frontiers in Physiology*, 2, 62.

Hellstrom, M., Phng, L. K., Hofmann, J. J., Wallgard, E., Coultas, L., Lindblom, P., Alva, J. et al. 2007. Dll4 signalling through Notch1 regulates formation of tip cells during angiogenesis. *Nature*, 445, 776–780.

Jabbarzadeh, E. and Abrams, C. F. 2007. Strategies to enhance capillary formation inside biomaterials: A computational study. *Tissue Engineering*, 13, 2073–2086.

Jain, H. V., Moldovan, N. I., and Byrne, H. M. 2012. Modeling stem/progenitor cell-induced neovascularization and oxygenation around solid implants. *Tissue Engineering Part C: Methods*, 18, 487–495.

Langer, R. 1999. Biomaterials in drug delivery and tissue engineering: One laboratory's experience. *Accounts of Chemical Research*, 33, 94–101.

Lemon, G., Howard, D., Rose, F. R. A. J., and King, J. R. 2011. Individual-based modelling of angiogenesis inside three-dimensional porous biomaterials. *Biosystems*, 103, 372–383.

Lemon, G., Howard, D., Tomlinson, M. J., Buttery, L. D., Rose, F. R. A. J., Waters, S. L., and King, J. R. 2009. Mathematical modelling of tissue-engineered angiogenesis. *Mathematical Biosciences*, 221, 101–120.

Levine, H. 2001. Mathematical modeling of capillary formation and development in tumor angiogenesis: Penetration into the stroma. *Bulletin of Mathematical Biology*, 63, 801–863.

Levine, H. A. and Nilsen-Hamilton, M. 2006. Angiogenesis—A biochemical/mathematical perspective. *In:* Friedman, A. (ed.) *Tutorials in Mathematical Biosciences III: Cell Cycle, Proliferation, and Cancer.* Berlin: Springer-Verlag.

Liu, G., Qutub, A. A., Vempati, P., Mac Gabhann, F., and Popel, A. S. 2011. Module-based multiscale simulation of angiogenesis in skeletal muscle. *Theoretical Biology and Medical Modelling*, 8, 6.

Long, B. L., Rekhi, R., Abrego, A., Jung, J., and Qutub, A. A. 2013. Cells as state machines: Cell behavior patterns arise during capillary formation as a function of BDNF and VEGF. *Journal of Theoretical Biology*, 326, 43–57.

Mac Gabhann, F., Ji, J. W., and Popel, A. S. 2006a. Computational model of vascular endothelial growth factor spatial distribution in muscle and pro-angiogenic cell therapy. *PLoS Computational Biology*, 2, e127.

Mac Gabhann, F., Ji, J. W., and Popel, A. S. 2006b. Computational model of vascular endothelial growth factor spatial distribution in muscle and pro-angiogenic cell therapy. *PLoS Computational Biology*, 2, e127.

Mac Gabhann, F., Ji, J. W., and Popel, A. S. 2007. Multi-scale computational models of pro-angiogenic treatments in Peripheral arterial disease. *Annals of Biomedical Engineering*, 35, 982–994.

Mcdougall, S. R., Anderson, A. R., Chaplain, M. A., and Sherratt, J. A. 2002. Mathematical modelling of flow through vascular networks: Implications for tumour-induced angiogenesis and chemotherapy strategies. *Bulletin of Mathematical Biology*, 64, 673–702.

Mehdizadeh, H., Artel, A., Brey, E. M., and Cinar, A. 2011. Multi-Agent systems for biomedical simulation: Modeling vascularization of porous scaffolds. *In:* Kinny, D., Hsu, J., Governatori, G. and Ghose, A. (eds.) *Agents in Principle, Agents in Practice.* Springer: Berlin/Heidelberg.

Mehdizadeh, H., Sumo, S., Bayrak, E. S., Brey, E. M., and Cinar, A. 2013. Three-dimensional modeling of angiogenesis in porous biomaterial scaffolds. *Biomaterials*, 34, 2875–2887.

Nair, L. and Laurencin, C. 2006. Polymers as biomaterials for tissue engineering and controlled drug delivery. *In:* Lee, K. and Kaplan, D. (eds.) *Tissue Engineering I.* Springer: Berlin, Heidelberg.

Nekka, F., Kyriacos, S., Kerrigan, C., and Cartilier, L. 1996. A model of growing vascular structures. *Bulletin of Mathematical Biology*, 58, 409–424.

North, M. J., Collier, N. T., and Vos, J. R. 2006. Experiences creating three implementations of the repast agent modeling toolkit. *ACM Transactions on Modeling and Computer Simulation*, 16, 1–25.

Olsen, L., Sherratt, J. A., Maini, P. K., and Arnold, F. 1997. A mathematical model for the capillary endothelial cell-extracellular matrix interactions in wound-healing angiogenesis. *Mathematical Medicine and Biology*, 14, 261–281.

Orme, M. E. and Chaplain, M. A. 1997. Two-dimensional models of tumour angiogenesis and anti-angiogenesis strategies. *IMA Journal of Mathematics Applied in Medicine and Biology*, 14, 189–205.

Peirce, S. M. 2008. Computational and mathematical modeling of angiogenesis. *Microcirculation*, 15, 739–751.

Peirce, S. M., Gabhann, F. M., and Bautch, V. L. 2012. Integration of experimental and computational approaches to sprouting angiogenesis. *Current Opinion in Hematology*, 19, 184–191.

Peirce, S. M., Van Gieson, E. J., and Skalak, T. C. 2004. Multicellular simulation predicts microvascular patterning and in silico tissue assembly. *FASEB Journal: Official Publication of the Federation of American Societies for Experimental Biology*, 18, 731–733.

Pettet, G., Chaplain, M. A., Mcelwain, D. L., and Byrne, H. M. 1996a. On the role of angiogenesis in wound healing. *Proceedings. Biological Sciences/The Royal Society*, 263, 1487–1493.

Pettet, G. J., Byrne, H. M., Mcelwain, D. L., and Norbury, J. 1996b. A model of wound-healing angiogenesis in soft tissue. *Mathematical Biosciences*, 136, 35–63.

Qutub, A. A. and Popel, A. S. 2009. Elongation, proliferation & migration differentiate endothelial cell phenotypes and determine capillary sprouting. *BMC Systems Biology*, 3, 13.

Reiffel, A. J., Perez, J. L., Hernandez, K. A., Fullerton, N., and Spector, J. A. 2012. Optimization of vasculogenesis within naturally-derived, biodegradable hybrid hydrogel scaffolds. *Plastic and Reconstructive Surgery*, 130, 25–26 10.1097/01.prs.0000421726.25449.da.

Sandino, C., Checa, S., Prendergast, P. J., and Lacroix, D. 2010. Simulation of angiogenesis and cell differentiation in a CaP scaffold subjected to compressive strains using a lattice modeling approach. *Biomaterials*, 31, 2446–2452.

Schugart, R. C., Friedman, A., Zhao, R., and Sen, C. K. 2008. Wound angiogenesis as a function of tissue oxygen tension: A mathematical model. *Proceedings of the National Academy of Sciences*, 105, 2628–2633.

Serbo, J. and Gerecht, S. 2013. Vascular tissue engineering: Biodegradable scaffold platforms to promote angiogenesis. *Stem Cell Research and Therapy*, 4, 8.

Stefanini, M. O., WU, F. T. H., Mac Gabhann, F., and Popel, A. S. 2010. Increase of plasma VEGF after intravenous administration of bevacizumab is predicted by a pharmacokinetic model. *Cancer Research*, 70, 9886–9894.

Stokes, C. L. and Lauffenburger, D. A. 1991. Analysis of the roles of microvessel endothelial cell random motility and chemotaxis in angiogenesis. *Journal of Theoretical Biology*, 152, 377–403.

Stokes, C. L., Lauffenburger, D. A., and Williams, S. K. 1991. Migration of individual microvessel endothelial cells: Stochastic model and parameter measurement. *Journal of Cell Science*, 99, 419–430.

Sun, S., Wheeler, M. F., Obeyesekere, M., and Patrick, C. W., Jr. 2005. A deterministic model of growth factor-induced angiogenesis. *Bulletin of Mathematical Biology*, 67, 313–337.

Thorne, B. C., Bailey, A. M., and Peirce, S. M. 2007. Combining experiments with multi-cell agent-based modeling to study biological tissue patterning. *Briefings in Bioinformatics*, 8, 245–257.

Vempati, P., Popel, A., and Mac Gabhann, F. 2011. Formation of VEGF isoform-specific spatial distributions governing angiogenesis: Computational analysis. *BMC Systems Biology*, 5, 59.

Vickerman, V., Blundo, J., Chung, S., and Kamm, R. 2008. Design, fabrication and implementation of a novel multi-parameter control microfluidic platform for three-dimensional cell culture and real-time imaging. *Lab on a Chip*, 8, 1468–1477.

Walpole, J., Papin, J. A., and Peirce, S. M. 2013. Multiscale computational models of complex biological systems. *Annual Review of Biomedical Engineering*, 15, 137–154.

Wu, F. T. H., Stefanini, M. O., Gabhann, F. M., Kontos, C. D., Annex, B. H., and Popel, A. S. 2010. VEGF and soluble VEGF receptor-1 (sFlt-1) distributions in peripheral arterial disease: An in silico model. *American Journal of Physiology—Heart and Circulatory Physiology*, 298, H2174–H2191.

Zahedmanesh, H. and Lally, C. 2012. A multiscale mechanobiological modelling framework using agent-based models and finite element analysis: Application to vascular tissue engineering. *Biomechanics and Modeling in Mechanobiology*, 11, 363–377.

Zawicki, D. F., Jain, R. K., Schmid-Schoenbein, G. W., and Chien, S. 1981. Dynamics of neovascularization in normal tissue. *Microvascular Research*, 21, 27–47.

Zhang, L., Athale, C. A., and Deisboeck, T. S. 2007. Development of a three-dimensional multiscale agent-based tumor model: Simulating gene-protein interaction profiles, cell phenotypes and multicellular patterns in brain cancer. *Journal of Theoretical Biology*, 244, 96–107.

11

Multiscale Modeling of Angiogenesis

David Noren, Rahul Rekhi, Byron Long, and Amina Ann Qutub

CONTENTS

11.1 Introduction

Great strides have been made within the field of tissue engineering over the last decade. Researchers and clinicians are beginning to develop functional tissues *in vitro* that are intended to replace or repair damaged organs. However, regardless of the physiological system being developed, a critical and challenging aspect of tissue regeneration is the successful vascularization of the implanted tissue (Novosel et al. 2011, Auger et al. 2013). Without a working vascular network, tissue constructs of appreciable size, that is those exceeding 200 μm in thickness, would lack sufficient oxygen and nutrients for sustained growth. In addition, the absence of a vascular network would lead to accumulation of metabolic waste. Overcoming this challenge requires unraveling the complex system of subcellular-, cellular-, and tissue-scale interactions that take place between newly forming blood vessels and their physiological environment.

Two basic strategies exist for obtaining functional vascular networks in newly implanted tissues. First, avascular constructs can be employed in conjunction with additional treatments to promote invasion and neovascularization by the patient's own blood vessels (Isner and Asahara 1999). Second, tissue constructs can be developed with primitive vascular networks prior to implantation in hopes of decreasing the time it takes to achieve proper circulation (Laschke et al. 2006, 2009). Both approaches require integration with the host's vasculature. To accomplish this, new capillary sprouts must develop from the patient's preexisting microvasculature and invade the implanted tissue, a process known as angiogenesis.

Angiogenesis occurs in response to low oxygen tension, or hypoxia. When a tissue experiences hypoxia, affected cells respond by upregulating the transcription factor hypoxia-inducible factor 1 (HIF1) (Wang and Semenza 1993). In turn, HIF1 activates genes to facilitate the secretion of proangiogenic factors (Semenza 2004), such as vascular endothelial growth factor (VEGF), allowing hypoxic cells to signal to nearby blood vessels. These proangiogenic factors activate normally quiescent endothelial cells lining nearby blood vessels and cause them to take on distinct cellular behaviors and phenotypes. Leading cells of an angiogenic sprout, often referred to as tip cells, begin to migrate in the direction of the hypoxic region by using select proangiogenic factors, such as VEGF, as directional cues (Gerhardt et al. 2003). Trailing cells, also called stalk cells, proceed behind the tip cell and begin to proliferate to form the wall of the new vessel.

Vascular sprouting is a very complex process that is regulated by molecular, cellular, and macroscopic events. The angiogenic cell behaviors described above are regulated at the molecular level by not only VEGF, but also by other soluble signaling molecules such as transforming growth factor β (TGF-β), platelet-derived growth factor (PDGF), and angiopoietins. Structural proteins of the extracellular matrix (ECM), which developing sprouts must traverse, regulate angiogenesis by providing insoluble signaling cues. The availability of these ECM signaling proteins is in turn regulated by matrix metalloproteinases (MMPs), which are secreted by endothelial cells, making this signaling system highly interactive. Neighboring stromal cells also influence newly forming sprouts, either through secretion of soluble cues or through cell–cell contact-mediated signaling. In addition, physical forces are also known to modulate angiogenesis, for example, those originating from blood flow or mechanical loading on the surrounding tissue (Milkiewicz et al. 2001, Brown and Hudlicka 2003).

How do these factors, many occurring at different spatial scales, interact in order to regulate angiogenesis? This question can be addressed by developing a quantitative description of this complex system, that is, a systems biology approach, where each factor is described in an integrated multiscale fashion. Here, we describe the mathematical components used to build multiscale models, along with new experimental approaches for measuring model parameters and for model validation, and review how these tools have been applied to specific physiological settings.

11.2 Overview of Multiscale Modeling of Angiogenesis

Multiscale modeling encompasses a diverse class of computational methods, each one describing living systems across spatial scales and at different levels of biological organization. We define the spatial scales at the level of molecules, cells, tissues, and organs. In modeling angiogenesis, these different scales might be used to describe the distribution of oxygen or HIF1 signaling molecules, endothelial cell migration or proliferation, microvascular patterning in hypoxic tissue, and the heart vasculature, respectively (Figure 11.1a).

In this section, we review the main factors to consider when developing multiscale models of microvascular growth. There are multiple mathematical frameworks that can be selected to describe any particular biological scale. Deciding which framework(s) to use should be informed by its suitability in addressing the question of interest. Once a modeling approach is selected for each biological scale, a method to integrate components must be determined such that the simulation maintains its accuracy across spatial scales.

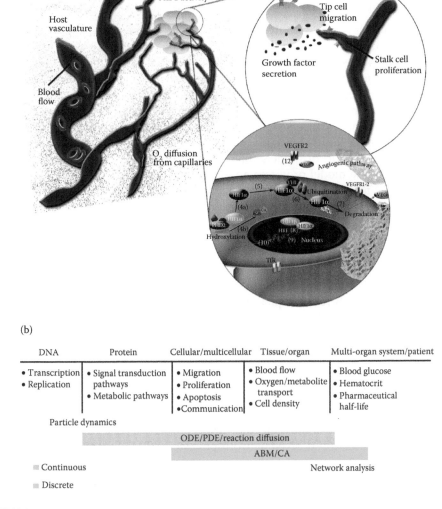

FIGURE 11.1
Modeling angiogenesis at multiple scales. (a) Illustration of the angiogenic response to hypoxic tissue at different length scales, including blood flow through vessels at the tissue level, cell migration and proliferation at the cellular level, and HIF signal transduction at the molecular level. (b) Mapping of modeling techniques to different scales. Examples of possible characteristics are given for each level of a biological system. The modeling techniques most appropriate for each scale are shown below for both continuous (yellow) and discrete (green) formulations.

Below, we define the main categories of mathematical modeling that have been applied to angiogenesis and discuss the strengths of each. In addition, we discuss one of the main challenges in computational biology: models that span multiple biological scales usually embody an enormous number of parameters. It is not uncommon for many of them to be unknown. These parameters must be defined either by optimizing the model output to fit experimental data, for example, using tools developed from fields such as machine learning, or through direct measurement. The last section in this chapter provides a case study of both approaches, where we discuss the optimization of a state machine model of angiogenesis using *in vitro* assays, and the development of high-throughput cell phenotyping to obtain high-dimensional quantitative cellular-level data.

11.2.1 Modeling Approaches

Computational models can be broadly classified as discrete or continuous (Di Ventura et al. 2006). Discrete models represent components of a biological system as individual, distinct entities, often represented by pixels (in 2D) or voxels (in 3D). In contrast, continuum models represent system components as distribution functions over continuous space, typically through the use of differential equations. Whereas a discrete model of vascular sprout formation might model individual cells as independent units, a continuous approach would represent them as a mass diffusing across the area of interest. A categorical distinction can also be made between deterministic and stochastic components of multiscale models. The former entails a group of models that behave in a fully reproducible fashion; each given simulation of the model produces an identical output every time it is generated. In contrast, stochastic models include random variables and are inherently noisy, such that two simulations with identical starting conditions can produce differing results. Numerous discrete and continuous computational frameworks are available for representing the different physiological processes occurring in angiogenesis. Here, we describe a number of modeling approaches that have been applied toward studying angiogenesis, each having different advantages and disadvantages, which define its suitability in simulating any specific biological scale. We start with those best suited for the molecular scale, namely, particle dynamics and reaction–diffusion equations; followed by approaches that best describe cellular-level events, such as cellular automaton (CA), agent-based simulation, and the cellular Potts model (CPM); and conclude with a method to model blood flow through functional vasculature on the tissue scale, namely, network flow analysis (Figure 11.1b).

11.2.1.1 Particle Dynamics

Particle-based models track molecules separately and in discrete quantities. Typically, these particles are used to represent the molecular-level components of an angiogenic process— for example, VEGF and brain-derived neurotrophic factor (BDNF) molecular motion—and are modeled using a specific set of physical laws or constraints (e.g., Newton's laws of motion). However, the particles themselves can also represent components at higher biological levels, such as cell or vessel network fragments, making this technique well suited for multiscale modeling approaches. For instance, Milde et al. (2008) developed a hybrid continuum–particle model where both growth factors and the ECM were described by continuum equations, while migrating vascular sprout tips were represented as discrete particles. Tip cells could move in an unconstrained manner and were guided by chemotactic and haptotactic gradients. This allowed developing sprouts to take on a tortuous shape, providing a close approximation to the shape of actual blood vessels.

One disadvantage of particle dynamics lies in the computational cost of simulating large numbers of discrete entities. This is particularly true of simulating molecular-scale components, where use of this method should be restricted to addressing conditions that necessitate it, for example, when a biochemical species is present at very low molecular concentrations and cannot be represented by an ensemble average. In contrast, continuum methods are less computationally demanding, and as we describe next, can provide a general framework for describing biomedical species in multiscale models.

11.2.1.2 Reaction–Diffusion Modeling

Many of the earliest models of angiogenesis described the distribution of both vascular cells and soluble angiogenic factors using variants of a continuum advection–reaction–diffusion formulation. Variables describing the density of cells and the concentration of chemical species were related using either ordinary differential equations (ODEs) or partial differential equations (PDEs). For instance, as early as the 1970s, several studies used this approach to describe the directed growth of blood vessels under the chemotactic guidance of growth factors (Deakin 1976, Balding and McElwain 1985, Chaplain et al. 1995). Although these models consider the role of soluble factors, molecular-level interactions were not described explicitly.

Models were later developed that extended this approach to include molecular-level detail. For instance, the binding of angiogenic growth factors to biological receptors could be described using ODEs, as was demonstrated by Levine et al. (2001a,b) for general growth factors and Mac Gabhann and Popel (2004) for the case of VEGF and its multiple receptors. The studies by Levine et al. also include molecular details such as degradation of the ECM by proteolytic enzymes and the secretion of proangiogenic factors from tumors and stromal cells. Cellular-level events, such as the development of vascular sprouts, were modeled as cellular densities, that is, using a continuous formulation rather than considering discrete cells. The model was used to examine the impact of antiangiogenic compounds under different assumed mechanisms, illustrating a possible application of this multiscale approach.

Overall, the use of continuum mechanics to model different aspects of angiogenesis bestows several advantages to engineers and mathematicians, including historical familiarity, well-defined parameters, deterministic model predictions, and computational efficiency. This methodology does not, however, allow for the discrete modeling of biological cells and their behaviors, which in some applications, represents a crucial facet to understanding complex physiological systems.

11.2.1.3 Cellular Automaton, Agent-Based Models, and Cellular Potts

Agent-based modeling (ABM) provides a broad computational framework in which subjects of interest, such as biological cells, can be modeled as discrete entities or agents (Maes 1993, An et al. 2009). The behavior of individual agents is governed by a set of rules that dictate how they will react to distinct events during the course of the simulation. Moreover, this methodology often embodies stochasticity, such that the system as a whole displays emergent behaviors that could not be predicted *a priori*. Thus, in the case of angiogenesis, as the agent-based simulation runs, the formation of the resultant capillary network is driven by a complex system of individually acting cells. This approach has been used to simulate multifaceted physiological systems, for instance, inflammatory cell circulation (Bailey et al. 2007), and as will be discussed in more detail later, has been applied to build computational models of angiogenesis (Qutub and Popel 2009).

CA modeling is another rule-based approach to describing discrete entities, or automatons. The behavior of these automatons is guided by a set of rules, as with ABMs, but the interaction between CA individuals is usually spatially restricted, for example, to nearest neighbors. The rules for CAs often describe simple behaviors, whereas in ABM, rules typically encompass a series of complex interactions, for example, those described by differential equations, logic-based relationships, or algebraic equations. In ABM, subsets of agents can adapt or respond to global controls, whereas the definition of a CA implies purely autonomous activity. Given these distinctions between the two methodologies, CA is often considered a subset of ABM.

CA models themselves can vary in complexity. Generally, one or more sets of uniformly programmed entities interact on a well-defined spatial grid. Simple conceptual models can be implemented under this framework, some employing very few rules, to yield complex spatial vascular patterning (Stokes and Lauffenburger 1991, Chaplain and Anderson 1996). For instance, early work by Gazit et al. (1995) used an abstract CA growth model to examine the underlying differences in the fractal-like patterning typically observed between healthy and tumor vasculature. In contrast, CA models of angiogenesis can also be quite complex, describing the interplay of many important angiogenic factors over multiple scales. Peirce et al. (2004), for example, developed a comprehensive CA model of vascular remodeling based on over 50 rules from published data. The model described several molecular signaling molecules, such as VEGF, platelet-derived growth factor (PDGF), and transforming growth factor (TGF); multiple cell types, including endothelial cells, smooth muscle cells, and perivascular cells; different cell behaviors such as differentiation, migration, and proliferation; and hemodynamic stress due to blood flow.

The CPM is another discrete approach to modeling cell behavior, which explicitly accounts for a cell's location and shape. In contrast to CA, biological cells each occupy many spaces (pixels) on a defined lattice, with the overall shape and movement of each cell being determined stochastically according to an energy minimization scheme (Glazier and Graner 1993). Cell morphology emerges through its interaction with other cells and the surrounding environment, making this methodology amenable to examining vascular patterning. This approach has been used to examine changes in endothelial cell morphology and cell–cell interactions during *in vitro* tube formation (Merks et al. 2006, 2008), as well as sprout formation and vascular branching (Bauer et al. 2007, 2009). In Bauer et al. 2007, a CPM was used to define the developing sprout, while a reaction–diffusion approach was used to describe molecular-scale effectors, such as ECM and growth factors, illustrating one way this methodology can be integrated into a multiscale simulation.

11.2.1.4 Network Flow Models

Blood flow is an integral part of angiogenic systems. Indeed, one could think of establishing blood flow as the final step in angiogenesis, reestablishing proper oxygen tension in ischemic tissue, and ending the hypoxic response that drives angiogenesis forward. Network flow models offer a simple method to simulate blood flow through both preexisting and nascent capillary networks. Under this framework, junctions between connecting vessels can be thought of as nodes in a network-like structure, while individual vessel segments are represented as edges. Flow can be estimated using simple pressure–flow–resistance relations, or can be calculated employing more physiological descriptions of blood flow (Pries et al. 1994). This approach was used by Alarcon et al. (2003), in conjunction with CA, to investigate tumor growth in an inhomogeneous oxygen field. These simulations described biological characteristics across several spatial scales, incorporating molecular oxygen diffusion through continuum equations, cancer cell growth through CA, and capillary blood flow through network flow modeling.

11.2.2 Integrating Model Components

Different model frameworks can be dynamically "linked" to form the basis of a multiscale simulation, harnessing the strengths of each modeling strategy to describe a particular spatial scale. For instance, each model framework can be linked in series, such that a reaction–diffusion model may first be employed to predict the concentration profile of specific chemical signaling agents (e.g., growth factors). These profiles would then serve as inputs for a cellular-level discrete model, which describes how endothelial cell behavior is influenced by chemical gradients. Lastly, these cell behaviors could serve as inputs into tissue- or organ-level behavior, ultimately providing predictions of how cellular-level events manifest at medically relevant scales. Note, in addition to applying different frameworks in series, there are other methods to link information from different biological scales. Simulation components can also be connected in parallel with varying degrees of integration, as discussed in more detail by Walpole et al. (2013).

11.2.3 Machine Learning: Using Statistical Learning to Refine Multiscale Models

Multiscale models can become quite complex. Within each biological scale described by the model lies the potential requirement to define hundreds of parameters. For instance, a biochemical model describing a metabolic network or signal transduction pathway may incorporate a myriad of biochemical reactions. Each reaction in turn could require one or more kinetic parameters, making the overall number of parameters score in the hundreds to thousands. A classical approach to defining these parameters is to seek estimates from literature sources. However, even in well-developed fields where the literature content is quite rich, very seldom can quantitative estimates of these parameters be found. This is particularly true for parameters needed to describe events on the cellular level, where quantitative measures of individual cell interactions and behaviors may not exist. Currently, new high-throughput methods are being developed to obtain detailed quantitative data concerning cell behaviors, an approach called cell phenotyping, which we will discuss in more detail later. However, regardless of how this data is acquired, it is often impossible to directly measure every parameter, leaving model developers the daunting task of determining these quantities by calibrating the model to match experimental results.

Fortunately, drawing from the field of machine learning, there exists a wealth of optimization schemes that can be readily applied to parameter estimation. For instance, a genetic algorithm (GA) is a widely employed technique for model construction, design, and optimization. Simply put, GAs pit competing computational models against each other, scores them for their accuracy in predicting real behavior, and then recombines the characteristics of the "winning" models to potentially produce even better "offspring." This process is repeated until the desired degree of resolution is obtained. In this way, GA "evolve" a functional solution, requiring little *ex ante* knowledge of the angiogenic process in question. Additional optimization algorithms, such as simulated annealing, have also proven very useful in estimating model parameters and we refer readers to Sun et al. (2012) for a more detailed review.

In general, machine learning techniques offer a powerful suite of tools for analysis and pattern recognition of high-throughput angiogenic data. Though diverse in scope, these methods typically involve algorithms that are "trained" on reference data to recognize specific properties in the set, as grouped into supervised, unsupervised, and semisupervised classifications based on the "learning" method employed. Thus far, machine learning methods have largely been applied toward gene profiling: either to pioneer the discovery of new angiogenic genes or to predict how variations in known gene expression affect vascular phenotypes. Such techniques could also be used at the proteomic, cellular, or tissue levels, which we will discuss later.

11.2.4 Cellular Phenotyping: Acquiring Experimental Data for Cellular-Level Predictions

Describing the behavior of single cells is a crucial part of understanding complex processes such as angiogenesis. Whether examining the response to biochemical treatments or varying hypoxic conditions, individual cells may take on different characteristics or pursue different behaviors according to their interpretation of the surrounding physiological environment. The importance of incorporating this level of detail into multiscale simulations was made evident by recent studies showing that the response of individual isogenic cells is often heterogeneous, even when observed within the same population (Niepel et al. 2009, Spencer et al. 2009).

Here, we use the term "cellular phenotyping" to denote the quantitative assessment of individual cell characteristics measured within a population of cells. One such experimental approach that can be used to accomplish this is flow cytometry, where proteomic or transcriptional information about individual cells can be correlated with biochemical treatments. Similarly, experimental techniques such as fluorescence microscopy can be used to acquire data on the level of individual cells. The latter, however, conveys the added advantage of being able to capture spatial properties and cell behaviors, such as cell–cell interactions or cell migration, which could not otherwise be measured. In addition, this approach can be used in conjunction with live cell probes, or immunostaining, allowing the simultaneous examination of signal transduction pathways and cell characteristics such as morphology. Both the amplitude and the location of signaling molecules can be correlated with cell behavior, making data obtained though this method rich in content. Many images can be taken over the course of a single experiment to capture the behavior of thousands of cells, making this approach also high in throughput.

Regardless of the method used to acquire cell phenotyping data, making a quantitative assessment of cell characteristics and behaviors requires further analysis. In the case of fluorescence microscopy, cell phenotyping requires a computational approach to extract data from images, followed by a statistical approach to understanding cell behaviors (Jones et al. 2009, Rekhi et al. 2013) (Figure 11.2). Data pertaining to individual cells can be isolated from other parts of the image using segmentation algorithms. Different cell characteristics can then be calculated in an automated fashion, including attributes such as morphology or fluorescence intensity. Cluster analysis can be applied to group these features and identify distinct phenotypes within the population of cells.

As a methodology, cellular phenotyping has the potential to uncover the insight needed to formulate descriptive rules and incorporate single cell behavior into multiscale simulations of angiogenesis. Measurements of this type, even on the level of cell pairs, can be used to obtain simple cell interaction rules, which form the basis of complex vascular patterns (Yin et al. 2008). For instance, Parsa et al. (2011) used multispectral fluorescence microscopy to observe individual and collective endothelial cell behaviors in matrigel. Here, several distinct temporal phases of cell behavior were identified during the formation of an *in vitro* vascular plexus. Moreover, it was found that heterogeneity

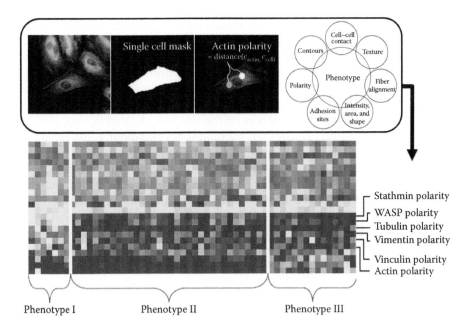

FIGURE 11.2
Schematic of the cell phenotyping process. Cells are fluorescently labeled, both to mark cell boundaries and to measure signaling activity. Images are then acquired in a high-throughput fashion, measuring thousands of cells in a single experiment. Segmentation algorithms are then applied to each image to identify individual cells. Different metrics can be used to characterize phenotype characteristics, such as cell shape and the polarity of cytoskeletal proteins. Machine learning algorithms can then be applied to cluster distinct patterns within the data and identify different cell phenotypes.

in endothelial cell behavior ultimately dictates their role in forming the final vascular structure, illustrating the importance of cell phenotyping in understanding vascular development.

11.2.5 Cells as State Machines: Integrating Theory and Experiment

State machines offer a useful framework for deriving and simulating cellular behavior. Originally conceived as a mathematical abstraction for designing computer algorithms, state machines are entities that occupy a series of fixed states—albeit only one at a time. Upon receiving a particular input, state machines can be transferred, or *transitioned*, from one state to another. The transitions can be represented mathematically as transfer functions or probability values, which describe the chance of completing the state change or staying within a particular state. In cells, for instance, these states can represent distinct cellular behaviors, while chemotactic, haptotactic, and/or mechanotactic signals can provide the requisite transition stimulus. In considering vascular sprouting, endothelial cell states can, therefore, be associated with specific angiogenic behaviors, such as migration, proliferation, apoptosis, or quiescence. This formalism allows cell states to be disentangled and analyzed for independent contributions to the angiogenic process; a task well suited for microscopy-based cellular phenotyping. This was illustrated by Rimchala et al. (2013), where this approach has been applied to measure phenotype transitions among individual endothelial cells *in vitro*. Here, distinct phenotype transition patterns were identified using statistical clustering. The prevalence of each pattern was further characterized under both VEGF and PF4 cytokine treatments, acquiring data that could eventually be used to develop a state machine model.

State machine models, along with ABM models in general, can also be developed in a manner that is highly integrated with experimental findings. One such approach was taken by Long et al. (2013), where a multiscale model of angiogenesis was developed by deriving transition state probabilities from an *in vitro* spheroid assay (Figure 11.3a). Here, the data used to build the model was on the tissue scale, that is, vascular sprouting was observed and quantitatively assessed for model development. Cells were allowed to transition between migrating, proliferating, and branching states. Rules governing the transition state probabilities of individual cells were specified in the model, which was then tested against the experimental data. The model framework was implemented such that several rule sets could be easily tested for each different cell behavior, in essence finding the model that best explained the data (Figure 11.3b). This process of model selection was driven by a GA, a machine learning process that was described earlier. Overall, this approach illustrates how model selection can be achieved through direct coupling with experimental data. Additionally, cellular-level properties, that is, transition state probabilities, are derived from tissue-scale data rather than relying on direct cell measurements, the latter being more difficult to obtain when considering *in vivo* and clinical settings.

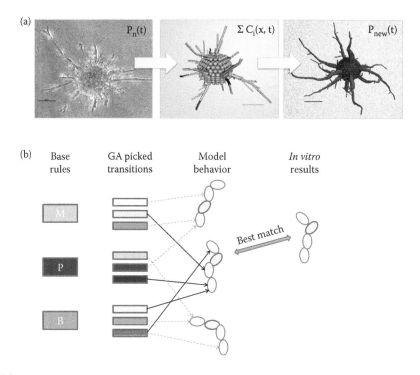

FIGURE 11.3
Building cell state machines from experimental data. (a) An *in vitro* spheroid assay is used in conjunction with a genetic algorithm to select the optimal state machine model. (From *J Theor Biol*, 326, Long et al., Cells as state machines: Cell behavior patterns arise during capillary formation as a function of BDNF and VEGF, 43–57, Copyright 2013, with permission from Elsevier.) (b) The selection process is carried out by assigning a set of behavioral rules and transition probabilities for each phenotype, simulating vascular growth, and comparing predictions with the experimental data (M = migration, P = proliferation, B = branching).

11.2.6 Multiscale Modeling of Angiogenesis in Health and Disease

Multiscale models of angiogenesis have the potential to yield insight into complex physiological conditions that might otherwise be unapproachable, for instance, predicting the concentration of proangiogenic growth factors in tissue, a quantity that remains beyond the reach of modern medical measurements and assays (Ji et al. 2007). That said, this methodology has only been applied to a limited number of diseases thus far, with the majority of models being developed to study cancer, ischemia in disease, and ischemia during exercise. A summary of the models that have been developed, along with the general components used to construct each model, is given in Table 11.1. Here, we further review a few of these application areas with a specific discussion on existing approaches from the literature and how they might be extended for further research or tissue engineering application.

11.2.6.1 Cancer

The most common pathology investigated by multiscale models of angiogenesis is that of cancer. As a tumor grows, it must "feed" its expanding metabolic demands by diverting nutrients from the bloodstream. Cancer cells accomplish this task by secreting proangiogenic signaling factors, such as VEGF and placenta growth factor (PlGF), to promote capillary sprouting from nearby vascular beds. This vascular reengineering marks a critical inflection point in the growth of the tumor, as the "leakiness" of tumor-induced vessels is thought to play an important role in the progression of cancer. Additionally, tumor invasion relies heavily on the angiogenic process; as a cancer mass becomes metastatic, it releases MMPs to dissolve away the basal membrane and ECM, such that cancer cells can migrate toward the newly formed vasculature. Once it has access to the blood vessel network, the tumor mass uses it as a conduit for transporting cancerous cells to other tissues to eventually form secondary tumors (Zetter 1998, Folkman 2002).

As described above, tumor growth is essentially interwoven with the angiogenic process, making the latter a prime target for arresting and reversing cancer progression. In particular, computational models have focused on simulations of tumor growth, invasion, and evolution, as angiogenesis plays a key mechanistic role in each of these cancer-specific behaviors (see Table 11.1). Recently, however, multiscale models of cancer angiogenesis have been utilized to elucidate how these treatments might be designed. For example, simulations that consider preexisting vessel networks surrounding tumors have been used to predict how a solid cancer mass might progress over time and respond to differing treatments (Perfahl et al. 2011). A recent review suggests that CPM and agent-based approaches might be especially suitable to examine tumor development and treatment, citing research that utilizes this technique to uncover the role of cancer stem cells in promoting tumor aggregation and antitumor treatment resistance through angiogenic growth (Szabo and Merks 2013).

Often, CPM approaches are extended by combining the lattice models with sets of ODEs or PDEs to represent diffusing tumor-relevant chemicals, growth factors, nutrients, or metabolites. Using such extensions, researchers have reproduced the relationship between VEGF gradients and tumor invasion, as well as the role of hypoxia in driving tumor progression or regression. This becomes particularly relevant in the trade-off between arresting tumor growth and increasing tumor invasiveness, or vice versa, through therapeutic modulation of angiogenesis (Frieboes et al. 2006). Here, hybrid computational models have already served as powerful tools by which to optimize potential treatment schemes,

TABLE 11.1

Angiogenesis Models with Therapeutic Application, Across Biological Scales

Application	Biological Level	Angiogenesis Modeling Methods	Main Processes Studied	Reference
Tissue engineering	General	Agent-based modeling	Vascularization through scaffold pores	Mehdizadeh et al. (2013)
	Osteogenesis	Reaction–diffusion, agent-based modeling, finite element	Fracture healing, tissue regeneration, VEGF signaling, cell growth, Dll4-Notch1 signaling, scaffold vascularization, seeding density, mechanical loading, cell differentiation, growth factor release, intracellular signaling, scaffold pore size	Geris et al. (2008), Carlier et al. (2009), Checa and Prendergast (2009), Sandino et al. (2010), Sun et al. (2013)
Cancer	General	Reaction–diffusion, pharmacokinetics, compartmental modeling, hybrid cellular automaton model, finite element, level set method	VEGF tissue distribution, delivery of anti-VEGF therapy, effects of blood flow on cancer cell colony formation, tumor invasion	Zheng et al. (2005), Billy et al. (2009), Macklin et al. (2009), Shirinifard et al. (2009), Alarcon et al. (2011), Cai et al. (2011), Perfahl et al. (2011), Stefanini et al. (2011)
	Network	Diffuse interface modeling, fractals, convection–diffusion modeling, invasion percolation	Tumor growth, hypoxia, nutrient consumption, vascular pattern formation, incorporation of progenitor cells, vascular normalization	Baish et al. (1996), Stoll et al. (2003), Amyot et al. (2005), Grizzi et al. (2005), Jain et al. (2007), Frieboes et al. (2010), Lowengrub et al. (2010), Bearer et al. (2011)
	Cell	Agent-based modeling, cellular Potts approach	Tumor growth, hypoxia, nutrient consumption	Bauer et al. (2007)
	Molecular	Reaction–diffusion models, bioinformatics	Genetics, metabolism, VEGF signaling, protein–protein interactions	Mac Gabhann and Popel (2006), Jain et al. (2008), Rivera et al. (2011)
	Brain Multiscale	Algebraic relationships, diffuse interface	Tumor growth, hypoxia, tumor invasion, heterogeneity	Zhang et al. (2007), Frieboes et al. (2010), Sun et al. (2012)

Category	Scale	Model type	Application	References
	Network	Fractals	Vascular pattern formation	Risser et al. (2007), Di Ieva et al. (2010)
	Cell	Agent-based modeling	Tumor growth, radiation effects, hypoxia	Mansury et al. (2002, 2006), Deisboeck et al. (2009), Zhang et al. (2009)
	Molecular	Reaction–diffusion models	Metabolism, delivery of antiangiogenic therapy, hypoxic response signaling	Qutub and Popel (2008)
Breast	Multiscale	Fractals, cellular automaton model, compartmental	Vascular pattern formation with Gompetzian tumor growth, VEGF whole-body distribution	Sedivy et al. (2002), Stefanini et al. (2010, 2011)
	Molecular	Reaction–diffusion models, bioinformatics	Delivery of anti-VEGF therapy, differentiating breast cancer patients by race, and angiogenic genetic pathways	Mac Gabhann and Popel (2006), Martin et al. (2009)
Lung	Multiscale	Agent-based modeling, reaction model	EGFR-TGF signaling	Wang et al. (2010)
Cardiovascular / Cardiac	Molecular	Reaction–diffusion models	FGF2 signaling and therapeutic delivery	Filion and Popel (2005), Waters et al. (2010)
Peripheral	Multiscale	Coupled convection, reaction–diffusion and agent-based models, compartmental models	Exercise effects on sprouting angiogenesis, hypoxic response, VEGF effects, whole-body VEGF distribution	Ji et al. (2007), Stefanini et al. (2010), Liu et al. (2011)
	Network	Convection–diffusion models	O_2, blood flow, membrane permeability	Ji et al. (2006)
	Cell	Agent-based modeling	Stromal cell trafficking	Bailey et al. (2009)
	Molecular	Reaction–diffusion models	VEGF signaling	Wu et al. (2010)
Macular degeneration	Molecular	Reaction–diffusion models	Ocular drug delivery	Mac Gabhann et al. (2007)
Inflammation	Multiscale	Agent-based modeling, network models	Monocyte trafficking	Bailey et al. (2007), Tang and Hunt (2010)

accounting for genetic and epigenetic heterogeneity in the solid tumor mass (Frieboes et al. 2010). Notably, there are certain preconditions that must be met for multiscale tumor models to demonstrate sufficient predictive power, like a minimum size and characteristic vessel structure to ensure an adequate simulation "sample size," such that it might be representative of larger-scale dynamics (Perfahl et al. 2011).

11.2.6.2 Cardiovascular

One of the most vibrant areas of research in angiogenesis lies in determining its role in cardiovascular disease. Despite the potentially principal role of the angiogenic process in preventing, causing, or treating such pathologies as atherosclerosis, peripheral arterial disease (PAD), and coronary heart disease, a complete understanding of its involvement is lacking. This is due, in large part, to the multilayered nature of heart disease progression—the complexity of which proves difficult to capture through *in vitro* and *in vivo* experimentation alone. Ultimately, it is also this complexity that makes intractable the design of regenerative medicine and tissue engineering therapies for cardiovascular disease. Although angiogenic therapies are being evaluated for their therapeutic efficacy against PAD, heart ischemia, and other diseases, their clinical utility remains unclear, with early results mixed.

Fortunately, there exist several promising approaches by which multiscale models of cardiac angiogenesis can be designed, constructed, and analyzed for therapeutic design (Liu et al. 2011). In considering some of the key factors involved in PAD, for instance, models have been developed for capillary blood flow (network flow); VEGF secretion, transport, and uptake (ODEs); VEGF-receptor binding (enzyme kinetics); ECM behavior (finite element modeling); muscle tissue oxygen distribution (reaction–diffusion); glycosaminoglycan sequestration (ODEs); and anatomically representative descriptions of the microvasculature (geometric modeling) (see Table 11.1). Though, individually, each computational model provides valuable insight into the mechanisms underlying angiogenesis by scale, the real strength of the multiscale modeling approach lies in coupling them together, as illustrated by Ji et al. (2007). Here, a multiscale approach was used to gain insight into angiogenic therapy following PAD. Specifically, simulations were designed around a very common animal model of PAD, namely, postfemoral ligation of rat extensor digitorum longus (EDL) muscle. The model incorporated the molecular details of key angiogenic growth factors and receptors, such as VEGF, VEGF-specific receptors, and neuropilins. Both capillaries and muscle fibers were explicitly represented, with blood flow, oxygen levels, and growth factor concentrations being predicted at the tissue scale. Close examination of the molecular-level events and their interaction with tissue scale revealed that angiogenic therapy for PAD might not only be limited to VEGF upregulation, but might also be enhanced by simultaneously upregulating neuropilins.

11.2.6.3 Regenerative Medicine and Tissue Engineering

In addition to multiscale simulations constructed for specific heart disease and cancer, numerous angiogenic models have been developed that detail mechanisms of osteogenesis (fracture healing), wound healing, exercise, and general angiogenesis, to describe a few. Thus far, however, few multiscale models have been developed to examine angiogenesis specifically for tissue engineering applications. Fortunately, angiogenesis has broad foundations across numerous physiological systems. Accordingly, many multiscale models use generic abstractions of specific growth factors or tissue so that they might represent a broad class of neovascularization processes. Though not tied to any specific disease state, these approaches provide insight into angiogenic physiology and how it might be co-opted for a wide variety of therapeutics. Moreover, many of the multiscale models in this class are designed to be generalizable, making their approaches readily tailored toward more specific pathologies. Here, we focus on the few models aimed specifically at tissue engineering along with those results that are easily extendable for tissue-engineered therapeutic advances.

The ability to incorporate functional blood vessels into new tissue grafts requires understanding the complex interactions that take place between the developing sprouts and their new biochemical environment. While soluble signaling molecules, such as growth factors, certainly play a crucial role, insoluble factors such as the scaffold material are also key in defining the angiogenic potential of a graft. Early computational studies touched on this topic by including ECM components in the multiscale model formulation (Bauer et al. 2007, Qutub and Popel 2009) or further, incorporating very detailed quantitative characterizations of the ECM and examining their role during angiogenesis (Milde et al. 2008). Only recently have studies explicitly modeled the physical characteristics of scaffold materials to determine their impact on angiogenesis. For instance, Mehdizadeh et al. (2013) created a 3D multiscale ABM of angiogenesis to examine the influence of pore architecture on sprout formation. Various pore characteristics were examined, including pore size and interconnectivity, which were ultimately predicted to have a profound influence on vessel growth.

One particularly challenging area of tissue engineering and regenerative medicine lies in the grafting and repair of bone tissues. Like any physiological process, osteogenesis is very complex and proper healing of bone tissues is intimately dependent on angiogenesis. However, given the function of bones, the design of tissue constructs for bone repair necessarily involves consideration for mechanical loading. Early approaches to this problem did not include these mechanical factors (Geris et al. 2008), but instead focused on the role of growth factors and cell interactions. More recent studies used FEM methods to simulate mechanical loading on scaffold material while simultaneously testing the effect of cell seeding density (Checa and Prendergast 2009) and mesenchymal stem cell differentiation (Sandino et al. 2010) on the vascularization of the scaffold.

The MOSAIC model is an example of a detailed multiscale angiogenesis model, formulated to examine angiogenesis and osteogenesis, which has the potential to be adapted toward other applications in tissue engineering and regenerative medicine (Carlier et al. 2012). Intracellular signaling dynamics, including a detailed description of Dll4 and Notch1 signaling, are coupled to a discrete cell-behavior-based simulation (migration, differentiation, proliferation, etc.) and a continuous tissue-level model of key nutrients and metabolites (e.g., oxygen and VEGF distribution) on a 2D grid. By tailoring its formalism to each scale, the model exhibits strong quantitative and qualitative predictive power, reproducing experimental data for key metrics of fracture healing, including bone tissue fraction, fibrous tissue fraction, and the cartilage template. Strikingly, MOSAIC simulations corroborate additional counterintuitive experimental results, such as the acceleration of cartilage resorption and the reduction in bone tissue fraction upon application of VEGF, despite earlier overall fracture healing. When combined with separate simulations of additional growth factors or VEGF isoforms, such simulations might enable a novel bench and bedside practice of "bone phenotyping": both in validating knockout animal models and proposed treatment outcomes.

Piecing together this enabled understanding, insight, and control of osteogenesis, angiogenesis, and fracture healing, researchers have engineered multiscale computational models specifically for tissue engineering design. A recent study described the simulation of bone regeneration within a 3D porous biodegradable scaffold in response to multicomponent growth factor release, molecular signaling dynamics, and nutrient transportation, as well as the physical properties of the scaffold itself (e.g., pore size, porosity) (Sun et al. 2013). Here, both osteogenesis and angiogenesis were considered. As before, the simulation featured reaction–diffusion equations to govern molecular-level dynamics (VEGF, Wnt, Osx, Runx2 signaling) coupled to an agent-based model to abstract cell-level events (osteoblast differentiation, apoptosis). This level of integration enabled the simulation to accurately predict the relationship between scaffold pore size and the growth factor release phenotype; *in silico* results were fully validated by *in vitro* data. Similar success was shown in determining the optimal pore size for angiogenesis and fracture healing (540 μm), with results fully consistent with previous and current experimental reports, for example, the irrelevance of pore size to osteogenesis. These initial studies, while few, lend credence to the power of multiscale models as tools to enable the scaling-up of tissue-engineered approaches, particularly in designing combinatorial growth factor treatments that are difficult to characterize through purely experimental means.

11.3 Discussion: Regenerative Medicine and Tissue Engineering Application

As a physiological mechanism, angiogenesis offers a useful target for diagnostic and therapeutic approaches for a wide variety of pathologies, and computational models of the process can thereby serve to catalyze the development of novel treatments at the bedside. To a large extent, this has already enabled such single-molecule entities (SME) as anti-VEGF inhibitors to shape the therapeutic landscape for cancer and macular degeneration. Ultimately, however, perhaps the greatest utility of multiscale models of angiogenesis lies in their ability to guide the development of novel approaches to regenerative medicine and tissue engineering. This catalysis of tissue engineering approaches can manifest as advances in basic science, guidance of *in vitro* and *in vivo* experimentation, or computer-aided design (CAD) tools for biosynthetic scaffolds, among other forms.

As described above, multiscale models have aided in uncovering novel insights about angiogenesis, including the tip–stalk cell dichotomy, endothelial cell chemotaxis, Delta-Notch activation, and the effects of MMPs. Findings from these models have the potential to catalyze future experimental studies, and moreover, can inspire novel biomedical applications for promoting tissue regeneration for both wound and fracture healing (Geris et al. 2010). With regard to experimental design, predictive simulations of angiogenesis from cell to organ scales can leapfrog purely trial-and-error research approaches, as computational models can be utilized to identify promising avenues of research *in silico* prior to *in vitro* or *in vivo* study. Finally, multiscale models can facilitate the design and assembly of tissue engineering scaffolds, permitting the optimization of synthetic properties such as porosity, pore structure, chemical makeup, stresses and strains, growth factor distributions, and spatial arrangements to promote controllably patterned neovascularization in bioengineered tissue and organ systems.

One additional long-term goal of this modeling work is synergistic, real-time coupling of MRI imaging and modeling of a patient's vasculature at the bedside. This technology would provide clinicians a visual snapshot of essential protein processes, predictions of angiogenesis, and local oxygen concentrations, as well as a forecast of how the vasculature will remodel in response to potential regenerative treatments or pathological perturbation (e.g., stroke, cancer, or coronary disease). Moreover, coupling mutiscale models with imaging would allow for the direct personalization of tissue-engineered therapies. Rather than designing a generic, one-size-fits-all heart scaffold, for instance, high-resolution clinical imaging could be utilized to calibrate a biosynthetic network for a patient's specific size, anatomy, and physiological needs. This approach has already shown promise in better defining the pathophysiological state of tumors (Mescam et al. 2010), with the potential for amplification by further usage of multiscale modeling.

11.4 Conclusion

Mutiscale models of angiogenesis equip researchers to devise therapeutics with targets across levels of biological organization, providing a powerful analysis tool for otherwise unattainable spatiotemporal levels. In regenerative medicine and tissue engineering, these techniques offer a pathway for designing bioartificial organ systems that are tailored to patients' specific physiological, anatomical, and medical needs. Moreover, they permit the rapid design, testing, and assembly of biosynthetic tissue—much in the way that an airplane might be simulated *in silico* prior to lengthy, expensive wind-tunnel testing for candidate designs. As such, multiscale models serve as powerful tools for biomedical investigators and clinicians alike, with the potential to make significant headway in treating diseases of the cardiovasculature, cancer, and regeneration processes.

References

Alarcon, T., Byrne, H. M., and Maini, P. K. 2003. A cellular automaton model for tumour growth in inhomogeneous environment. *J Theor Biol*, 225(2), 257–274.

An, G., Mi, Q., Dutta-Moscato, J., and Vodovotz, Y. 2009. Agent-based models in translational systems biology. *Wiley Interdiscip Rev Syst Biol Med*, 1, 159–171.

Auger, F. A., Gibot, L., and Lacroix, D. 2013. The pivotal role of vascularization in tissue engineering. *Annu Rev Biomed Eng*, 15, 177–200.

Bailey, A. M., Thorne, B. C., and Peirce, S. M. 2007. Multi-cell agent-based simulation of the microvasculature to study the dynamics of circulating inflammatory cell trafficking. *Ann Biomed Eng*, 35(6), 916–936.

Balding, D. and McElwain, D. L. 1985. A mathematical model of tumour-induced capillary growth. *J Theor Biol*, 114, 53–73.

Bauer, A. L., Jackson, T. L., and Jiang, Y. 2007. A cell-based model exhibiting branching and anastomosis during tumor-induced angiogenesis. *Biophys J*, 92(9), 3105–3121.

Bauer, A. L., Jackson, T. L., and Jiang, Y. 2009. Topography of extracellular matrix mediates vascular morphogenesis and migration speeds in angiogenesis. *PLoS Comput Biol*, 5, e1000445.

Brown, M. D. and Hudlicka, O. 2003. Modulation of physiological angiogenesis in skeletal muscle by mechanical forces: Involvement of VEGF and metalloproteinases. *Angiogenesis*, 6, 1–14.

Carlier, A., Geris, L., Bentley, K., Carmeliet, G., Carmeliet, P., and Van Oosterwyck, H. 2012. MOSAIC: A multiscale model of osteogenesis and sprouting angiogenesis with lateral inhibition of endothelial cells. *PLoS Comput Biol*, 8(10), e1002724.

Chaplain, M. A. and Anderson, A. R. 1996. Mathematical modelling, simulation and prediction of tumour-induced angiogenesis. *Invasion Metastasis*, 16, 222–234.

Chaplain, M. A., Giles, S. M., Sleeman, B. D., and Jarvis, R. J. 1995. A mathematical analysis of a model for tumour angiogenesis. *J Math Biol*, 33, 744–770.

Checa, S. and Prendergast, P. J. 2009. Effect of cell seeding and mechanical loading on vascularization and tissue formation inside a scaffold: A mechano-biological model using a lattice approach to simulate cell activity. *J Biomech*, 43(5), 961–968.

Deakin, A. S. 1976. Model for initial vascular patterns in melanoma transplants. *Growth*, 40, 191–201.

Di Ventura, B., Lemerle, C., Michalodimitrakis, K., and Serrano, L. 2006. From *in vivo* to *in silico* biology and back. *Nature*, 443, 527–533.

Folkman, J. 2002. Role of angiogenesis in tumor growth and metastasis. *Semin Oncol*, 29, 15–18.

Frieboes, H. B., Jin, F., Chuang, Y. L., Wise, S. M., Lowengrub, J. S., and Cristini, V. 2010. Three-dimensional multispecies nonlinear tumor growth-II: Tumor invasion and angiogenesis. *J Theor Biol*, 264(4), 1254–1278.

Frieboes, H. B., Zheng, X., Sun, C. H., Tromberg, B., Gatenby, R., and Cristini, V. 2006. An integrated computational/experimental model of tumor invasion. *Cancer Res*, 66, 1597–1604.

Gazit, Y., Berk, D. A., Leunig, M., Baxter, L. T., and Jain, R. K. 1995. Scale-invariant behavior and vascular network formation in normal and tumor tissue. *Phys Rev Lett*, 75, 2428–2431.

Gerhardt, H., Golding, M., Fruttiger, M., et al. 2003. VEGF guides angiogenic sprouting utilizing endothelial tip cell filopodia. *J Cell Biol*, 161, 1163–1177.

Geris, L., Gerisch, A., Sloten, J. V., Weiner, R., and Oosterwyck, H. V. 2008. Angiogenesis in bone fracture healing: A bioregulatory model. *J Theor Biol*, 251(1), 137–158.

Geris, L., Schugart, R., and Van Oosterwyck, H. 2010. *In silico* design of treatment strategies in wound healing and bone fracture healing. *Philos Trans A Math Phys Eng Sci*, 368, 2683–2706.

Glazier, J. A. and Graner, F. 1993. Simulation of the differential adhesion driven rearrangement of biological cells. *Phys Rev E Stat Phys Plasmas Fluids Relat Interdiscip Topics*, 47, 2128–2154.

Isner, J. M. and Asahara, T. 1999. Angiogenesis and vasculogenesis as therapeutic strategies for postnatal neovascularization. *J Clin Invest*, 103, 1231–1236.

Ji, J. W., Mac Gabhann, F., and Popel, A. S. 2007. Skeletal muscle VEGF gradients in peripheral arterial disease: Simulations of rest and exercise. *Am J Physiol Heart Circ Physiol*, 293(6), H3740–H3749.

Jones, T. R., Carpenter, A. E., Lamprecht, M. R. et al. 2009. Scoring diverse cellular morphologies in image-based screens with iterative feedback and machine learning. *Proc Natl Acad Sci U S A*, 106, 1826–1831.

Laschke, M. W., Harder, Y., Amon, M. et al. 2006. Angiogenesis in tissue engineering: Breathing life into constructed tissue substitutes. *Tissue Eng*, 12, 2093–2104.

Laschke, M. W., Vollmar, B., and Menger, M. D. 2009. Inosculation: Connecting the life-sustaining pipelines. *Tissue Eng Part B Rev*, 15, 455–465.

Levine, H. A., Pamuk, S., Sleeman, B. D., and Nilsen-Hamilton, M. 2001a. Mathematical modeling of capillary formation and development in tumor angiogenesis: Penetration into the stroma. *Bull Math Biol*, 63, 801–863.

Levine, H. A., Sleeman, B. D., and Nilsen-Hamilton, M. 2001b. Mathematical modeling of the onset of capillary formation initiating angiogenesis. *J Math Biol*, 42, 195–238.

Liu, G., Qutub, A. A., Vempati, P., Mac Gabhann, F., and Popel, A. S. 2011. Module-based multiscale simulation of angiogenesis in skeletal muscle. *Theor Biol Med Model*, 8, 6.

Long, B. L., Rekhi, R., Abrego, A., Jung, J., and Qutub, A. A. 2013. Cells as state machines: Cell behavior patterns arise during capillary formation as a function of BDNF and VEGF. *J Theor Biol*, 326, 43–57.

Mac Gabhann, F. and Popel, A. S. 2004. Model of competitive binding of vascular endothelial growth factor and placental growth factor to VEGF receptors on endothelial cells. *Am J Physiol Heart Circ Physiol*, 286, H153–H164.

Maes, P. 1993. Modeling adaptive autonomous agents. *Artif. Life*, 1, 135–162.

Mehdizadeh, H., Sumo, S., Bayrak, E. S., Brey, E. M., and Cinar, A. 2013. Three-dimensional modeling of angiogenesis in porous biomaterial scaffolds. *Biomaterials*, 34(12), 2875–2887.

Merks, R. M., Brodsky, S. V., Goligorksy, M. S., Newman, S. A., and Glazier, J. A. 2006. Cell elongation is key to *in silico* replication of *in vitro* vasculogenesis and subsequent remodeling. *Dev Biol*, 289, 44–54.

Merks, R. M., Perryn, E. D., Shirinifard, A., and Glazier, J. A. 2008. Contact-inhibited chemotaxis in *de novo* and sprouting blood-vessel growth. *PLoS Comput Biol*, 4, e1000163.

Mescam, M., Kretowski, M., and Bezy-Wendling, J. 2010. Multiscale model of liver DCE-MRI towards a better understanding of tumor complexity. *IEEE Trans Med Imaging*, 29, 699–707.

Milde, F., Bergdorf, M., and Koumoutsakos, P. 2008. A hybrid model for three-dimensional simulations of sprouting angiogenesis. *Biophys J*, 95, 3146–3160.

Milkiewicz, M., Brown, M. D., Egginton, S., and Hudlicka, O. 2001. Association between shear stress, angiogenesis, and VEGF in skeletal muscles *in vivo*. *Microcirculation*, 8, 229–241.

Niepel, M., Spencer, S. L., and Sorger, P. K. 2009. Non-genetic cell-to-cell variability and the consequences for pharmacology. *Curr Opin Chem Biol*, 13, 556–561.

Novosel, E. C., Kleinhans, C., and Kluger, P. J. 2011. Vascularization is the key challenge in tissue engineering. *Adv Drug Deliv Rev*, 63, 300–311.

Parsa, H., Upadhyay, R., and Sia, S. K. 2011. Uncovering the behaviors of individual cells within a multicellular microvascular community. *Proc Natl Acad Sci U S A*, 108, 5133–5138.

Peirce, S. M., Van Gieson, E. J., and Skalak, T. C. 2004. Multicellular simulation predicts microvascular patterning and *in silico* tissue assembly. *FASEB J*, 18, 731–733.

Perfahl, H., Byrne, H. M., Chen, T. et al. 2011. Multiscale modelling of vascular tumour growth in 3D: The roles of domain size and boundary conditions. *PLoS One*, 6(4), e14790.

Pries, A. R., Secomb, T. W., Gessner, T., Sperandio, M. B., Gross, J. F., and Gaehtgens, P. 1994. Resistance to blood flow in microvessels *in vivo*. *Circ Res*, 75, 904–915.

Qutub, A. A. and Popel, A. S. 2009. Elongation, proliferation and migration differentiate endothelial cell phenotypes and determine capillary sprouting. *BMC Syst Biol*, 3, 13.

Rekhi, R., Ryan, D., Zaunbrecher, B., Hu, C. W., and Qutub, A. A. 2013. Computational cell phenotyping in the lab, plant and clinic. In Zhang, G. (Ed.) *Computational Bioengineering.* CRC Press: Boca Raton.

Rimchala, T., Kamm, R. D., and Lauffenburger, D. A. 2013. Endothelial cell phenotypic behaviors cluster into dynamic state transition programs modulated by angiogenic and angiostatic cytokines. *Integr Biol (Camb)*, 5, 510–522.

Sandino, C., Checa, S., Prendergast, P. J., and Lacroix, D. 2010. Simulation of angiogenesis and cell differentiation in a CaP scaffold subjected to compressive strains using a lattice modeling approach. *Biomaterials*, 31(8), 2446–2452.

Semenza, G. L. 2004. Hydroxylation of HIF-1: Oxygen sensing at the molecular level. *Physiology (Bethesda)*, 19, 176–182.

Spencer, S. L., Gaudet, S., Albeck, J. G., Burke, J. M., and Sorger, P. K. 2009. Non-genetic origins of cell-to-cell variability in TRAIL-induced apoptosis. *Nature*, 459, 428–432.

Stokes, C. L. and Lauffenburger, D. A. 1991. Analysis of the roles of microvessel endothelial cell random motility and chemotaxis in angiogenesis. *J Theor Biol*, 152, 377–403.

Sun, J., Garibaldi, J. M., and Hodgman, C. 2012. Parameter estimation using meta-heuristics in systems biology: A comprehensive review. *IEEE/ACM Trans Comput Biol Bioinform*, 9(1), 185–202.

Sun, X., Kang, Y., Bao, J., Zhang, Y., Yang, Y., and Zhou, X. 2013. Modeling vascularized bone regeneration within a porous biodegradable CaP scaffold loaded with growth factors. *Biomaterials*, 34(21), 4971–4981.

Szabo, A. and Merks, R. M. 2013. Cellular Potts modeling of tumor growth, tumor invasion, and tumor evolution. *Front Oncol*, 3, 87.

Walpole, J., Papin, J. A., and Peirce, S. M. 2013. Multiscale computational models of complex biological systems. *Annu Rev Biomed Eng*, 15, 137–154.

Wang, G. L. and Semenza, G. L. 1993. General involvement of hypoxia-inducible factor 1 in transcriptional response to hypoxia. *Proc Natl Acad Sci U S A*, 90, 4304–4308.

Yin, Z., Noren, D., Wang, C. J., Hang, R., and Levchenko, A. 2008. Analysis of pairwise cell interactions using an integrated dielectrophoretic-microfluidic system. *Mol Syst Biol*, 4, 232.

Zetter, B. R. 1998. Angiogenesis and tumor metastasis. *Annu Rev Med*, 49, 407–424.

Section IV

Imaging

12

In Vivo *Imaging Methods for the Assessment of Angiogenesis: Clinical and Experimental Applications*

Laura Nebuloni, Gisela A. Kuhn, and Ralph Müller

CONTENTS

12.1 Introduction

Angiogenesis is the growth of new blood vessels from existing vascular networks. Under physiological conditions, it underlies almost all biological processes of morphogenesis, allowing an adequate perfusion in tissues during growth and development [1]. Further, it is a fundamental step in wound healing, where the regrowing blood vessels bring the necessary nutrients to the injured tissue [2]. Angiogenesis is also involved in pathological conditions such as psoriasis, rheumatoid arthritis, congestive heart failure, atherosclerosis, peripheral artery disease, and tumor growth. In the last case, tumor progression is usually coupled with the growth of aberrant vessel structures [3]. Moreover, the angiogenic process plays an important role in tissue engineering applications. To ensure a correct and fast integration of an artificial tissue as well as the survival of the tissue itself, if seeded with cells, formation and growth of blood vessels have to take place inside the biomaterial [4].

Angiogenesis is regulated by a complex interplay between free signaling molecules, the extracellular matrix (ECM), and the endothelial cells [1]. The actual blood vessel formation results from a delicate balance between pro- and antiangiogenic factors. The angiogenic switch is often triggered by an insufficient oxygen supply resulting in hypoxic cells [5]. The binding of the hypoxia-inducible factor (HIFα) with hypoxia response elements activates the expression of the vascular endothelial growth factor (VEGF), a key player in the angiogenic process. Other regulating factors of neovascularization are angiopoietins (Ang-1 and Ang-2), fibroblast growth factors (bFGF, aFGF), and platelet-derived endothelial growth factor (PDGF). Matrix metalloproteinases (MMPs) are proteolytic enzymes known to degrade the basement membrane and ECM, providing sufficient space for the sprouting vessels. VEGF and MMPs have been identified as favorable targets for imaging angiogenesis [6]. Furthermore, the cell adhesion molecule integrin plays a key role among angiogenic factors. Integrins expressed on endothelial cells modulate cell migration and survival during angiogenesis [7]. Among the integrins discovered to date, integrin $\alpha_v\beta_3$ is the most extensively studied. It is expressed in activated endothelium [7,8] and has been identified as one of the major targets for labeling angiogenic areas for different imaging modalities.

In the angiogenic process, two main types of vessel formation can be distinguished: sprouting and splitting angiogenesis [9]. The sprouting mechanism is the most common one (see Figure 12.1). In this case, the endothelial cells activated by biological signals known as angiogenic growth factors proliferate into the surrounding matrix and form *angiogenic sprouts*, which connect neighboring vessels. As sprouts extend toward the source of the angiogenic stimulus, they form loops to become a full-fledged vessel lumen. Sprouting occurs at a rate of several millimeters per day and enables new vessels to grow across gaps in the vasculature [10]. This mechanism is markedly different from splitting angiogenesis (also known as intussusceptions), where the capillary bed expands in size and complexity through the formation of trans-vascular tissue pillars [11].

The assessment of vascular changes is of fundamental importance in the understanding of angiogenesis and of the phenomena in which it is involved. Angiogenesis is an intrinsically time-dependent process; therefore, it is essential to monitor the changes of the vascular network in a time-lapsed fashion. While *ex vivo* representation of blood vessels is possible in the finest detail, *in vivo* visualization presents very critical aspects.

To successfully image angiogenesis, the ideal imaging method should possess the following requirements: high resolution (angiogenic sprouts have a diameter of about 5–6 µm), high penetration depth, noninvasiveness (for *in vivo* monitoring and longitudinal studies), high vascular contrast, high sensitivity (to allow the detection of the smallest changes in the vascular network), three-dimensionality (to fully understand the vascular spreading in space), and safety. Several imaging techniques based on an angiographic examination can now be applied to study the vasculature *in vivo*. X-ray fluoroscopy is one of the most traditional ones used in clinical practice. Ultrasound and optical modalities such as fluorescence and bioluminescence imaging (BLI) can also be used to visualize blood flow and perfusion in animals and humans. These techniques enable a two-dimensional visualization of the vascular network. However, the analysis of the development of blood vessels in space is essential to fully understand

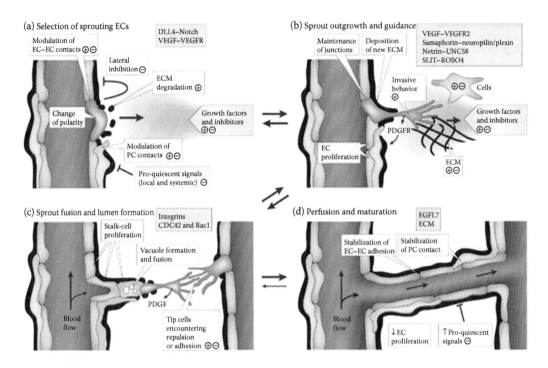

FIGURE 12.1

Schematic representation of angiogenic sprouting. (Reproduced from Adams, R.H. and K. Alitalo, Nature reviews. *Mol Cell Biol*, 2007. 8(6): 464–478.)

the vascular distribution and conformation in a tissue, which is often not clear with a simple planar representation. A three-dimensional insight can only be provided by techniques such as computed tomography (CT) and its microscaled version (micro-CT), magnetic resonance imaging (MRI), and fluorescence molecular tomography (FMT), which are able to produce a digital 3D reconstruction of the vascular tree. While both 2D and 3D imaging modalities enable the visualization of the major blood vessels, the *in vivo* representation of the thinnest ones, mainly capillaries and angiogenic sprouts, still remains unattainable. The reasons for the difficulty of this scientific challenge are: first, none of the current imaging methods used during *in vivo* experiments has the necessary resolution for the visualization of capillaries, which range around 5–6 µm in diameter. Second, the nonvoluntary movements produced by a living organism during an angiographic scan, such as breathing, heartbeat, and artery pulsation, cannot be suppressed during the measurement; this results in lower image quality. Imaging methods with higher resolution, such as synchrotron radiation micro-CT and confocal laser scanning microscopy, are available and can resolve even the smallest vessels but they cannot be applied *in vivo* [12,13].

In this chapter, the available *in vivo* imaging techniques for vascular quantification are presented and discussed on whether they can be used to monitor angiogenesis in both clinical applications and experimental research. A discussion on their advantages and disadvantages is presented, focusing on the technique that provides the highest resolution (CT).

12.2 X-Ray-Based Imaging Techniques

These modalities base their imaging principle on the use of x-rays. X-radiation is a form of electromagnetic radiation in the wavelength range of 0.01–10 nm. X-rays are ionizing radiation; therefore, the advantages of x-ray-based imaging techniques have to be carefully balanced against the risks of radiation exposure. X-rays are suitable to image hard tissues (such as bone). Soft tissues (including blood) present very low x-ray absorption; therefore, contrast agents need to be used to enhance the contrast with the surrounding tissues. Contrast enhancement is commonly employed in all x-ray-based techniques for the visualization of blood vessels.

12.2.1 X-Ray Angiography

Angiography is a medical imaging technique used to visualize the lumen of blood vessels and organs of the body. The word itself comes from the Greek words *angeion*, "vessel" and *graphein*, "to write" or "to record." The resulting image is called *angiogram* (see Figure 12.2). Contrast-enhanced x-ray fluoroscopy is the standard clinical imaging method used to perform angiographic examinations [14]. In its simplest form, a fluoroscope consists of an x-ray source and fluorescent screen between which the patient is placed. Modern fluoroscopes couple the screen to an x-ray image intensifier and CCD video camera, allowing the images to be displayed on a monitor. Using a system of guide wires and catheters, a radio-opaque contrast agent is injected in the vascular compartment. This enhances the contrast between blood vessels and the surrounding soft tissues [15]. A brief review of x-ray contrast agents used in the clinics is presented in Chapter 2.2.1. Subsequent x-ray imaging allows the visualization of the vessels filled with contrast agent. A major advantage of fluoroscopic angiography is live imaging; for this reason it is a very common technique in clinical vascular surgery. On the other hand, the contrast agent is quickly filtered from the vascular compartment; therefore, injection and imaging need to be well synchronized. Furthermore, while physicians always try to use low-radiation dose rates during

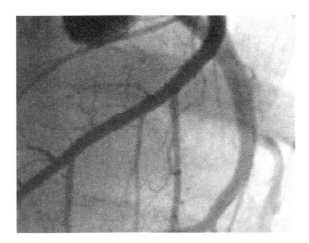

FIGURE 12.2
Coronary angiogram acquired with x-ray fluoroscopy. (Reproduced from Van Lysel, M.S., *Radiographics*: A review publication of the Radiological Society of North America, 2000. 20(6): 1769–1786.)

fluoroscopic examinations, the length of a typical procedure often results in a relatively high dose absorbed by the patient. Recent advances include flat-panel detector systems, which reduce the radiation dose. X-ray fluoroscopy is currently cheap and widespread in the clinics; it can also be applied in animal imaging but it suffers from lack of resolution (a few millimeters).

12.2.2 Computed-Tomography Angiography

Computed tomography (CT) is a noninvasive imaging technique based on x-rays that is able to produce a digital three-dimensional reconstruction of the scanned subject. CT examinations are typically used in clinical practice to retrieve anatomical information about x-ray absorbing tissues such as bone. Beams of x-rays are passed from a rotating device through the area of interest in the patient's body to obtain projection images. The multiple projections are combined using a reconstruction method that is generally based on the filtered back projection algorithm. The resultant CT image is a matrix of pixels with values proportional to the mean linear attenuation coefficient of the material.

In CTA, the vascular x-ray attenuation is enhanced by the injection of a bolus of water-based radio-opaque contrast medium via a peripheral intravenous cannula. Water-soluble contrast agents used in clinical applications are reviewed in Section 2.2.1.

Coupled with CTA, digital subtraction angiography (DSA) is now applied in clinical angiographic examinations to improve the contrast resolution of blood vessels. This technique was first shown by Mistretta et al. [16]. It is based on the acquisition of images before and after the contrast agent injection; the images acquired before are used to create a mask that is subtracted from the post-contrast images. The output image will therefore display only the distribution of the contrast agent.

Compared to catheter angiography, CTA is a much less invasive procedure. This type of examination is currently being used to screen patients for arterial disease. Its microscale version is applied for vascular imaging in small animals (see Section 2.3).

Since its introduction, CT has been regarded as an imaging technique only able to provide anatomical information. The advent of fast CT scanners has opened this imaging modality to functional imaging, which allows the monitoring of a physiological function (not only just a structure) of the body. Functional CT imaging is also referred to as dynamic or perfusion CT and is based on the analysis of the kinetics of contrast in the circulatory system (see Section 2.2.1). Functional CT not only looks at the dynamics of contrast in the major vessels, but also at tissue perfusion determined by smaller vessels (capillaries), which are below the resolution of the CT scanners. Studying tissue perfusion is very important in investigating angiogenesis in different applications. For instance, in clinical practice functional CT is applied to study perfusion in tumors. In this case, blood flow and blood volume perfusion maps can give very useful information (see Figure 12.3) [17]. An entire review on CT perfusion in oncological applications is presented by Petralia et al. [18].

12.2.2.1 X-Ray Vascular Contrast Agents for Clinical Applications

The most clinically used contrast agents for the visualization of blood vessels with CT are extracellular water-based formulations containing iodine. Iodine provides the highest x-ray absorption among nonmetal elements and is used in clinical practice because it is well tolerated by the organism. These agents are based on the tri-iodinated benzene ring, either in an ionic or nonionic form. They are rapidly cleared from the blood stream via passive filtration in the kidneys.

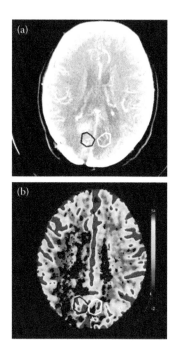

FIGURE 12.3
CT image (a) and cerebral blood volume map (b) of a patient affected by a recurrence of glioblastoma. (Reproduced from Di Nallo, A.M. et al., *J Exp Clin Cancer Res*, CR, 2009. 28: 38.)

12.2.2.1.1 Pharmacokinetic Models of Water-Soluble Contrast Agents for Functional CT Imaging

Water-soluble contrast agents are defined as extracellular fluid space markers; therefore they distribute in the extracellular space of blood (the plasma) and of tissues (the interstitial volume). The extravasation of such agents due to capillary permeability is known to happen in every tissue of the body except the brain, where it is prevented by the blood–brain barrier [19].

Several models are available that describe the distribution of contrast agents in the intravascular and extravascular space. They can be categorized into three main groups: (1) model-independent approaches based on the Fick principle or deconvolution analysis; (2) compartmental models; (3) models that account for convective transport (blood flow) and diffusional exchange (capillary permeability) with distributed parameters [20].

1. *Model-independent approaches*:

a. Fick principle

These models are based on the Fick principle. The contrast agent is considered an inert material that is not metabolized by the tissue. The mathematical expression of this model is the differential form of the principle of mass conservation:

$$\frac{dQ(t)}{dt} = FC_a(t) - FC_v(t)$$

Under the assumption of no venous outflow, blood flow can be estimated as

$$F = \frac{dQ(t)/dt}{C_a(t)} = \frac{\text{Maximum initial slope of } Q(t)}{\text{Peak height of } C_a(t)}$$

Instead, if a "local" draining vein can be identified, blood flow can be calculated as follows:

$$F = \frac{Q(t)}{\displaystyle\int_0^t C_a(t)dt - \int_0^t C_v(t)dt}$$

This estimation is correct under the assumption that there is no time shift of the contrast profile in the arterial and venous system (which in general is not true).

b. Deconvolution-based methods

These methods are based on the concept of convolution. The time attenuation curve in a tissue that follows a bolus injection of a unit mass of contrast agent is called impulse-residue function $R(t)$. If this function is known, the tissue attenuation response to a general injection profile can be obtained as a summation of scaled and time-shifted $R(t)$s. This operation is called convolution:

$$Q(t) = FC_a(t) \otimes R(t) = C_a(t) \otimes FR(t)$$

The calculation of $FR(t)$ is called deconvolution. The solution of this problem is not unique: there is more than one $FR(t)$ that would, after convolution with $C_a(t)$, give a good approximation of the tissue attenuation profile. The unstable characteristic of deconvolution is caused by noise effects. Therefore, this method is used only when noise suppression can be effectively applied.

2. *Compartmental models*: When a water-soluble contrast agent leaks in the extravascular space, a description of the tracer kinetics can be given by a multicompartment model. The most common model bases its analysis on two compartments where the contrast agent distributes homogeneously: the intravascular extracellular space (the plasma volume V_b) and the extravascular extracellular space (the interstitial volume V_e). After intravenous injection the contrast agent will leak from the blood stream to the surrounding tissues at a rate depending on a constant K. After a certain time, the tracer will move not only from the vascular to the interstitial compartment, but also the other way around at a rate depending on a constant K/V_e. However, assuming that in the initial phase the return of contrast agent to the blood stream is negligible, the distribution of the contrast agent in the two compartments is as follows:

$$Q(t) = V_b C_a(t) + K \int_0^t C_a(u)du$$

Dividing both sides by $C_a(t)$, the expression is a straight line with slope K and intercept V_b. Thus, functional maps of K and V_b can be calculated.

Considering the back-flux of contrast from the tissue to the blood compartment, the following general expression is used to derive K and V_b with nonlinear regression techniques:

$$Q(t) = V_b C_a(t) + K \int_0^t C_a(u) e^{-(K/V_e)(t-u)}$$

The fraction E of mass of contrast agent arriving at the tissue that leaks into the extravascular space in a single passage through the vasculature depends on the permeability surface area product PS:

$$E = 1 - e^{-(PS/F)}$$

3. *Distributed parameter models*: In 1966, Johnson and Wilson [21] were the first ones to propose a distributed parameter model. It includes a concentration gradient at the capillary level from the arterial inlet to the venous outlet. The solution is very complex and several computational algorithms have implemented it with different efficiency. This model allows simultaneous determination of blood flow, blood volume, mean transit time, and capillary permeability.

12.2.3 Micro-Computed Tomography Angiography (Micro-CTA)

Micro-CT scanners are designed to image small objects/animals. They are much smaller in design than clinical CTs, and they can reach imaging resolutions in the micrometer scale. Images with voxel sizes less that 5 µm are achievable, making micro-CT superior to other techniques such as ultrasound (50 µm) and MRI (25 µm) [22]. The micro-focus x-ray source produces a cone-shaped beam, which is projected through the specimen with the resultant radiographic density of the specimen projected onto a 2D detector. The cone–beam reconstruction algorithm from the 2D projections to the 3D picture was developed by Feldkamp et al. in 1984 [23]. Two micro-CT scanner setups are currently available: in the first one, the x-ray source and detector are stationary during the scan, while the sample rotates. Instead, for *in vivo* imaging, the setup where x-ray tube and detector rotate around the animal is more suitable.

Contrast resolution refers to the ability to detect differences between structures based on their different x-ray attenuation. The excellent x-ray absorption provided by hard tissues makes micro-CT a very useful and accurate imaging modality to visualize bone and its microstructure [24,25]. Nonetheless, some soft tissues such as blood vessels can be visualized *in vivo* with the intravascular injection of an x-ray contrast agent. Contrast-enhanced angiography based on micro-CT (micro-CTA) is an imaging method used in small animals to resolve the major blood vessels of the network (see Figure 12.4) [26]. The visualization of capillaries in living animals remains critical for different reasons: the available *in vivo* micro-CT scanners reach a maximum resolution of about 10 µm, thus not allowing the detection of the capillary bed. Furthermore, the physiological motion of a living animal during an *in vivo* measurement, which cannot be suppressed during

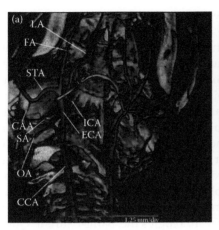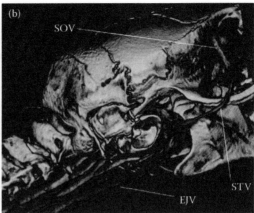

FIGURE 12.4
In vivo micro-CTA of murine cranial vessels. (Reproduced from Schambach, S.J. et al., *Stroke: J Cereb Circ*, 2009. 40(4): 1444–1450.)

anesthesia (such as breathing, heartbeat, and artery pulsation), leads to image blurring and lower image quality. Also, for *in vivo* imaging, spatial resolution is sometimes sacrificed to allow faster scanning and lower radiation dose. The radiation dose is a big concern for *in vivo* micro-CT, especially in longitudinal studies. An increase in resolution, although desirable, is difficult to achieve, due to the associated increase in radiation dose. Dose limitations for small animals are typically expressed in terms of lethal dose (LD) for a population; for example, LD 50/30 refers to the whole-body radiation dose that would kill the 50% of the exposed animals within 30 days. The LD 50/30 in mice is reported to range between 5 and 7.6 Gy [27].

As in clinical practice, micro-CT angiography can be applied together with DSA. It has been shown to be a successful imaging method to assess the vascular networks in small animals in a number of studies [28,29]. Badea et al. combined micro-CT and DSA into a single system to provide both morphologic and functional data in tumors in a single imaging session [30].

As clinical CT scanners, fast micro-CTs have been developed for functional micro-CT imaging. Several perfusion studies have been performed with micro-CT in small animals. Here, angiogenesis is investigated in terms of measurement of tumor perfusion. Imaging tumor vasculature depends on differences in permeability of vasculature of tumor and normal tissue, which cause changes in penetration of contrast agents [31].

12.2.3.1 X-Ray Vascular Contrast Agents for Experimental Research

In micro-CT imaging, a contrast agent should provide good x-ray attenuation, long half-life (time at which the agent reaches half of its initial concentration in blood) and good safety characteristics. Water-soluble contrast agents (described in Section 2.2.1) can be used in experimental research. However, due to the high heart rate of rodents (at least 10 times faster in mice than in humans), most of the micro-CT systems are not able to acquire a sufficient number of projection images during the first pass of the agent. Therefore, such contrast agents may be used if they are given at a constant infusion rate during the scanning time [32]. A custom-built microinjector has been recently described by de Lin et al.

It consists of a computer-controlled solenoid valve attached to the contrast injection catheter and a heated contrast agent reservoir [33].

Alternatively to water-soluble formulations, a new class of blood-pool contrast agents is now available for micro-CT in small animals. Such agents allow long vascular residence after a single bolus injection. To achieve the blood-pool effects, the size of the molecules in these contrast media is larger than that of capillary fenestrations. Therefore, such tracers are not supposed to leak from healthy capillaries. However, in pathological situations (namely tumor capillaries) or in angiogenic sprouts, newly formed vessels are known to have a higher permeability than the healthy ones; in this case, even such agents might leak from the blood compartment. Due to their large sizes, these molecules cannot be filtered through the renal system as water-based formulations. Instead, they are slowly excreted via the hepatobiliary system. The first blood-pool agent for micro-CT was introduced in 1998 [34]; it consisted of an iodinated contrast medium embedded in a lipidic formulation. *Fenestra VC* (ART Advanced Research Technologies Inc. Saint-Laurent, Quebec, Canada) is a commercial blood-pool agent consisting of iodinated triglycerides formulated in a stable, submicron oil-in-water lipid emulsion [35]. *Fenestra VC* has been shown to reside in the vascular compartment for more than 3 h. Therefore, optimal time points for imaging different organ systems can be selected [36]. *Fenestra VC* has been used for both cardiac micro-CT [37] and tumor imaging [28,38]. Another commercially available blood-pool contrast agent, *eXIA160XL* (Binitio Biomedical Inc., Canada), has been recently developed [39]. It consists of a long-acting aqueous colloidal poly-disperse iodinated solution. A clear advantage of blood-pool formulations compared to water-based solutions is their longer vascular residence time. Nevertheless, the drawbacks of these agents are their lower iodine content (and, therefore, their lower x-ray attenuation) and varying safety characteristics.

A new category of contrast agents that has recently been developed employs metals instead of iodine, such as gold or bismuth sulfide nanoparticles. Because of the high toxicity level of metals and their intrinsically higher x-ray contrast compared to iodine, they are used in lower concentrations. As in iodinated blood-pool agents, the dimension of the molecules of such tracers prevents their extravasation from the blood stream. A contrast agent based on gold nanoparticles coated with PEG has shown potential in imaging the vascular compartment [40]. Rabin et al. developed an agent based on bismuth-sulfide nanoparticles with a polymer coating. It was shown to have attenuation characteristics five times better than iodine-based agents, with long residence time in the vasculature and good safety characteristics [41].

Contrast agents used in micro-CT angiography usually provide x-ray absorption similar to calcified tissue. For this reason, it is difficult to distinguish bone and enhanced blood vessels. In recent years, a new technique has been developed that enables to distinguish bone from the vascular compartment enhanced with an iodinated contrast agent. This technique goes under the name of "dual or multiple energy CT" [42,43]. Briefly, after injection of the contrast medium, two or more energies of the x-ray beam (typically 140 and 80 kVp) are used during the tomographic measurement. With the advent of dual source CTs, the datasets at the two different energy levels do not have to be acquired separately, which largely excludes changes in contrast enhancement or animal movement between the acquisitions [44]. Attenuation is caused by absorption and scattering of radiation by the material under investigation. Materials can be differentiated further by applying different x-ray spectra and analyzing the differences in attenuation. This works especially well in materials with large atomic numbers due to the photo effect [45,46]. One of these materials is iodine, which is generally known to have stronger enhancement at low tube voltage settings [47]. The spectral information can therefore be used to differentiate iodine from other materials that do not show this behavior.

12.2.3.2 Gating in Micro-CTA

In cardiopulmonary imaging, physiological motion (cardiac and respiratory) affects the resolution in micro-CT. The displacement of the diaphragm or the heart through the course of one ventilatory and cardiac cycle can be more than 1 mm. This results in image blurring. Gating in micro-CT (also called 4D micro-CT, where the 4th dimension is time) represents a way to overcome this limitation: cardiac and/or respiratory states are measured and used to synchronize the acquisition of projection data at the same point in the cardiac and/or respiratory cycle. The physiological signals—the electrical activity of the heart and the ventilatory cycle—can be acquired simultaneously to perform cardiopulmonary gating. For the acquisition of the electric signal of the heart, electrodes are placed in contact with the skin and are used to record the electrocardiogram (ECG). To control the respiratory motion, different possibilities are available. Mechanical ventilation can be used [48–50]. Special mechanical systems were designed to ventilate the animals at a controlled rate and to provide the oxygen and anesthetic gas mixture, while also measuring the airway pressure via a pressure transducer in the breathing valve [51]. Endotracheal intubation is required for mechanical ventilation. This procedure requires skills to avoid trachea damage, especially in mice, due to their small size. Ford et al. have introduced an approach for gating with free-breathing rodents, which does not require intubation. An external pressure sensor is positioned on the animal's abdomen to acquire the respiratory signal. This method removes the drawbacks of intubation and mechanical ventilation, but is sensitive to changes in the respiration cycle of the animal [52]. Instead of the pressure sensor, a video camera can also be used to monitor the breathing pattern of the animal.

Gating can be performed prospectively or retrospectively. In prospective gating, the image acquisition is triggered at predefined time points of the physiological signals (ECG and/or respiratory signal). Prospective gating ensures uniform and sufficient angular sampling but can involve rather long acquisition times due to the fact that the images are triggered by the coincidence of one or two events—end-expiration in the breath cycle and/or the imaging time point (e.g., systole, diastole) in the cardiac cycle [32,53]. In retrospective gating, the physiological signals are recorded simultaneously with imaging data; after the acquisition, the projection data is sorted into sets corresponding to specific phases of respiratory and/or cardiac activity [54]. Retrospective gating for cardiac micro-CT of free-breathing mice was described by Drangova et al. [37]. It provided very fast imaging (less than 1 min) with 150 μm isotropic voxels and a temporal resolution of 12 ms. In retrospective gating, the projections are sampled at equal angular intervals independently from the cardiac and the breathing cycle. Both ECG and a respiration signal are recorded and used in post-processing to group the projections in sets corresponding to the cardiac and respiratory phases. For each of the cardiac phases, the corresponding projections have an irregular angular distribution, since many views may be missing. The number of projections for each phase may be further limited when many cardiac phases have to be reconstructed from a limited number of projections. Because of this irregular and undersampled pattern, the reconstructed images are affected by noise and artifacts. Song et al. proposed a new reconstruction algorithm able to deal with undersampling problems [55]. Another solution was provided via registration and a combination of prospective and retrospective gating [56].

12.3 Magnetic Resonance Imaging

MRI is a common imaging modality that offers high spatial resolution and penetration depth and provides both anatomical and functional information. First developed around 1980, it has currently gained a role of primary importance in medical imaging, due to its noninvasiveness and its ability to image soft tissues. It uses strong magnetic fields and nonionizing radiation in the radiofrequency (RF) range, unlike CT that uses ionizing radiation. Therefore, an advantage of MRI is that it is harmless to the patient/animal. MRI relies on the alignment of the spins of the atomic nuclei of a tissue along a preferential direction if subjected to a magnetic field (phenomenon also known as *magnetization*). If another external magnetic field (called *excitation field*) is applied in a different direction, the same nuclei undergo temporary variations in their alignments. When the excitation field is removed, the spins go back to their original alignment with temporal modalities that differ depending on the tissue. This process is called *relaxation*. The detection of different relaxation times allows the identification of different tissues [57–59].

12.3.1 Magnetic Resonance Angiography

For decades, MRA has been applied for the noninvasive visualization of blood flow [60]. Given its high spatial resolution, MR imaging is well suited for use in the assessment of angiogenesis and monitoring of its temporal changes. A variety of techniques can be used to image angiogenesis, based on the intrinsic properties of blood (NCE-MRA) or on contrast enhancement (CE-MRA) [61]. Unlike x-ray-based imaging techniques, the vascular system can be visualized by MRA in the absence of any contrast agent. Vessel density can be estimated without contrast enhancement due to the paramagnetic properties of deoxyhemoglobin within red blood cells. Such MRI techniques go under the name of noncontrast-enhanced magnetic resonance angiography (NCE-MRA). Another imaging possibility is the assessment of vascular function, which is defined as the ability of the vessels to transfer oxygen. It can be imaged by taking advantage of the MRI signal intensity changes in response to a change in blood oxygenation. Vascular function can also be monitored with NCE-MRA by following water as a tracer for perfusion. This method can used in clinical applications to monitor tumor angiogenesis.

The visualization of the vascular network without contrast enhancement is sometimes challenging because the contrast is flow-dependent and sensitive to artifacts at locations of turbulent flow. Injection of MRI contrast agents is therefore the most common method for the visualization of blood vessels with MRI (see Section 3.1.1). Contrast agents of different molecular weights as well as specific molecular markers for neovasculature can be used in contrast-enhanced magnetic resonance angiography (CE-MRA), as pictured in Figure 12.5. Similar to DSA, MRI images can be acquired before enhancement and serve as a subtraction mask on postcontrast images. Dynamic contrast-enhanced MRA (DCE-MRA) is the most widely used clinical approach to monitor angiogenesis. Different studies have employed highly paramagnetic nanoparticles to track angiogenesis by targeting $\alpha_v\beta_3$ integrins, which are activated during the initiation of the angiogenic process [62,63].

Clinical MRI scanners can be applied in animal imaging; nevertheless, to improve the signal-to-noise ratio (SNR), the spatial resolution, and to shorten the scan times, dedicated small animal scanners have been developed. Such systems are called micro-MRI or small animal MRI. Improvements in the MRI hardware include a high-field magnet (up to 16 T), a high-performance gradient system and appropriately sized RF coils (with a diameter

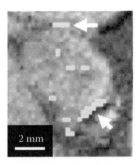

FIGURE 12.5
MR image showing angiogenic site. In yellow, MR signal enhancement due to $\alpha_v\beta_3$ integrins targeting of implanted tumor. (Reproduced from Winter, P.M. et al., *Cancer Res*, 2003. 63(18): 5838–5843.)

of 25–35 mm) [64]. This way, in-plane spatial resolutions of less than 30 μm and temporal resolutions of 5–10 ms can be reached *in vivo*.

The clinical MRI modalities previously described (NCE-MRA and CE-MRA) can also be applied to image angiogenesis in small animals. Tumor angiogenesis was quantified with noncontrast-enhanced MRA: changes in the vascularization of tumors were investigated after implanting agarose beads, multicellular spheroids, and after full thickness dermal incisions in animal models. In these cases, the correlation between the vessel density calculated *in vivo* with NCE-MRA and the one determined postmortem was highly significant [65]. Vascular function in small animals can also be assessed with NCE-MRA. The level of oxygen in blood can be modulated by changing the inhalation gas mix and by unbalancing it from normoxia toward hyperoxia or hypercapnia [65,66]. Using water as a tracer for perfusion, physiological angiogenesis was monitored in the rat ovary [67,68]. Contrast-enhanced MRA was applied in small animal imaging in a number of studies; the most common applications are the monitoring of tumor and plaque angiogenesis as well as tissue revascularization after stroke [69–71].

12.3.1.1 MR Vascular Contrast Agents

MRI contrast agents are used in CE-MRA to alter the relaxation times of tissues and body cavities where they are present. They can provide a higher (positive contrast agents) or lower (negative agents) signal [72]. For a proper evaluation of the vascular system by CE-MRA, angiographic contrast agents should have an enhanced relaxation time, a prolonged vascular residence time, and a limited extravasation to allow repeated image acquisitions after a single administration. Other prerequisite attributes are low toxicity, biological inertness, and excretability from the blood stream [73]. The most commonly used MRI positive contrast media are extracellular agents. They are typically small molecular weight compounds that distribute in the blood plasma and extracellular space of the body after administration. Due to their hydrophilicity, most of them have a rapid excretion through the kidneys with an elimination half-life of about 15–90 min. The first contrast agent approved for clinical MRI applications in humans was the anionic gadolinium-based complex (Gd-DTPA), which is now routinely used for contrast enhancement under the name of *Magnevist* (Schering, Berlin, Germany) [74]. Negative contrast agents are based on superparamagnetic nanoparticles. They are composed of an iron oxide nucleus of several nanometers in diameter. To increase their stability in aqueous media, these particles are coated with small molecules (citrate, oleate, silane, etc.) or polymers (dextran, synthetic polymers, starch, etc.) and form colloidal suspensions. Superparamagnetic nanoparticles have currently become very popular because of their strong magnetic efficacy. Furthermore, they are usually composed of biodegradable iron, which is biocompatible and can thus be recycled by cells using normal biochemical pathways for iron metabolism [75]. Small sized (7–30 nm) iron oxide nanoparticles (such as *Sinerem/Combidex*) have been shown to have optimal imaging properties. These compounds provide a larger time window for image acquisition of the vascular system, both for the first-pass and for the equilibrium-phase MRA. Major drawbacks of these applications were observed due to the superparamagnetic nature of iron oxide nanoparticles, which was responsible for the loss of the blood signal mainly when higher doses of contrast agent were injected [76].

Extracellular contrast agents used in clinical practice can be applied in MR experimental studies. Nevertheless, as with micro-CT contrast agents, several strategies were developed for the prolongation of the vascular residence time (the previously called "blood-pool effects"). Some of the MRA contrast agents mimic the circulating blood cells (liposomes or micelles) [77,78], while others mimic plasma proteins (macromolecules and colloids) [79,80] or reversibly bind to plasma proteins [78,81]. Another possibility to image blood vessels in angiogenesis relies on targeted molecular MRI imaging. The most common strategies aim at developing contrast agents that target $\alpha_v\beta_3$ integrins. Furthermore, solutions containing both iodine and gadolinium have been used as multimodality contrast agents. They can be applied for both CT and MRI [82].

12.4 Nuclear Imaging Techniques

Nuclear techniques include imaging modalities that take advantage of processes that happen in the nucleus of an atom. While MRI exploits the magnetization phenomenon, positron emission tomography (PET) and single photon emission tomography (SPECT) rely on the process of radioactive decay of isotopes (radionuclides). Radionuclides are combined with other chemical compounds or pharmaceuticals to form radiopharmaceuticals. These radiopharmaceuticals, once administered to the patient/animal, can localize specific organs or cellular receptors. This unique ability of radiopharmaceuticals allows nuclear medicine to diagnose or treat a disease based on the cellular function and physiology rather than relying on the anatomy. Because of the exceptional target specificity of radio-tracers, imaging at the molecular level is possible, such as receptor activity; this is why PET and SPECT technologies are often referred to as molecular imaging modalities.

Several potential targets for angiogenesis have been developed. At the moment, most of the work is focused on developing tracers for $\alpha_v\beta_3$ integrins, the VEGF family and MMPs [83]. One problem of tracers for VEGF is their high renal uptake due to the abundant presence of receptors for VEGF (VEGFRs). Other approaches consist in the development of PET tracers targeting the ED-B domain of fibronectin. The isoforms of fibronectin are known to be involved in vascular proliferation [84].

Until quite recently, all commercially available PET or SPECT scanners were designed primarily for human use. With increasing interest in targeted molecular imaging, dedicated small animal imaging systems have been developed. Micro-PET and micro-SPECT systems have been designed with spatial resolution and sensitivity sufficient to delineate tracer uptake in small animals [85–89].

Although both PET and SPECT imaging enable a precise localization of the radiotracer, these approaches do not provide true anatomical information. Therefore, PET and SPECT images are sometimes difficult to interpret because of the lack of anatomical structures or biological landmarks. To overcome this limitation, hybrid systems have recently been introduced including SPECT-CT and PET-CT (and their corresponding imaging systems for small animals), which offer coregistered images of the different modalities. The CT component of these systems is used to relate tracer signal with anatomical landmarks [90].

12.4.1 Positron Emission Tomography

Positron emission tomography (PET) is a medical imaging technique used in nuclear medicine that produces a three-dimensional image of functional processes in the body. Tracers based on radioisotopes are used. The radioisotope emits a positron that annihilates with an electron, producing a pair of annihilation (gamma) photons moving in almost opposite directions. These photons are detected in the scanning device by a ring of γ-cameras and reconstructed to provide the output image. Compared to other molecular imaging technologies, PET enables highly sensitive and quantitative measurements of biological or biochemical processes *in vivo* through the specific labeling of organic compounds with positron emitters, such as ^{11}C or ^{18}F. PET has the advantage that it does not have depth limitation for detecting the signal [91,92]. Therefore, it is a valuable technique in oncology, neurology, and cardiology to track angiogenesis. For instance, tumor angiogenesis was assessed in patients using ^{18}F-galacto-RGD PET imaging [93]. The main disadvantages of PET are its low resolution (about 5–7 mm in clinical PET systems), the high cost of tracers, their short half-lives (within minutes), and the risks associated with radioactivity.

Recent advances have been made in PET technology to improve resolution. Spatial resolution is determined by detector size, ring diameter, positron range, detector interactions, and reconstruction smoothing. High-resolution PET systems (micro-PET) approach a resolution of 0.2–0.3 mm, which is achieved by pinhole collimation or on a multiple pinhole approach. However, higher resolutions can be achieved but with a decrease in sensitivity. In experimental studies, VEGF receptors can be used as favorable targets to image angiogenesis. For instance, micro-PET imaging was applied in imaging VEGF receptor expression in angiogenic vessels of mouse tumors *in vivo* through ^{64}Cu-labeled VEGF, as shown in Figure 12.6 [94].

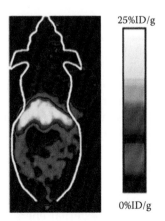

FIGURE 12.6
Micro-PET of tumor-bearing mouse after injection of ^{64}Cu-DOTA-VEGF$_{121}$. The VEGF receptor is expressed only in angiogenic vessels. (Adapted from Cai, W. et al., *J Nucl Med*, 2006. 47(12): 2048–2056.)

12.4.2 Single Photon Emission Computed Tomography

Single photon emission computed tomography (SPECT) is a nuclear medicine tomographic imaging technique that provides three-dimensional information based on gamma rays [95]. Internal radiation is administered through a low mass amount of pharmaceutical labeled with a radioactive isotope, which is inhaled, ingested, or injected. The emitted γ-rays are detected by a γ-camera. SPECT imaging is more widely available than PET imaging because the radionuclides used for SPECT are easier to prepare and usually have a longer half-life than those used for PET. However, SPECT imaging has very low detection efficiency compared to PET. The main advantage of SPECT imaging is that it allows simultaneous imaging of multiple radionuclides because the γ-rays emitted from different radioisotopes can be differentiated depending on their different energy. SPECT has been widely used in tumor imaging, infection (leukocyte) imaging, thyroid imaging, and bone imaging. As previously explained, integrins can be a favorable target in imaging angiogenesis. Blood vessels could be visualized by targeting integrins with antibodies [6,94,96].

In preclinical imaging, the resolution of micro-SPECT cameras approaches 500 μm. Significant advances have been made in detector technologies to achieve this resolution. By the use of pinhole imaging geometry with magnification, both high-resolution and high-detection efficiency can be achieved. Most of micro-SPECT systems consist of multiple detectors that surround the small animal to be imaged. The detectors are mounted on a stationary or rotating gantry and are arranged in a discrete configuration or in a close-ring configuration. To achieve high-detector intrinsic resolution, pixilated scintillation detectors with NaI(Tl) crystals and increasingly smaller pixel sizes (1 mm or less) have been developed. A ^{99m}Tc-labeled RGD peptide for targeted imaging was applied in a murine model of peripheral angiogenesis using high-resolution pinhole planar and micro-SPECT imaging [97].

To complement the strength of PET and SPECT imaging, cyclic RGD-peptide constructs with bifunctional chelators have been developed; they incorporate ^{99m}Tc and ^{64}Cu to be used in both imaging modalities [96].

12.5 Optical Imaging Modalities

Optical imaging is one of the most rapidly emerging technologies to noninvasively follow all kinds of molecular and cellular processes. This group of imaging modalities that provide nonanatomical information has recently seen an increased interest for tracking gene expression *in vivo*. However, this approach involves planar imaging of subcutaneous structures owing to the very low penetration depth and attenuation of low energy photons. While Doppler optical imaging is a useful technique in clinical imaging, fluorescence and bioluminescence are only applied in animal research.

12.5.1 Fluorescence Imaging

Fluorescence imaging (FLI) is based on the use of fluorescent proteins that emit visible photons in response to excitation by an external energy source. In near-infrared (NIR) fluorescence microscopy, high-molecular-weight fluorescent tracers (e.g., FITC-conjugated 2000 kDa dextran) are injected to demarcate blood vessels. Vessel diameter, length, surface area and volume, as well as branching patterns and intercapillary distance in a growing or regressing tumor have been described [98,99]. Amoh et al. developed animal models to study the efficacy of antiangiogenic compounds against tumor development. In this study, multicolor fluorescent proteins are used to analyze blood vessel density in tumor-induced angiogenesis. These fluorescence models are generally named AngioMouse models [100].

FMT is based on the same principle as fluorescent imaging but the resulting images are provided in a three-dimensional fashion [101]. After exposing the subject with continuous wave or pulsed light from different sources, the emitted light is captured by multiple detectors arranged in a spatially defined order in an imaging chamber. These raw data are then processed mathematically and the reconstructed tomographic image can be obtained. This technique has proved useful in the *in vivo* quantification of $\alpha_v\beta_3$ integrins in murine-activated endothelial cells [102]. In this study, a novel NIR-labeled optical agent that binds specifically to integrin-overexpressing cells showed strong selectivity toward $\alpha_v\beta_3$ integrins and therefore tumor distribution *in vivo*, allowing noninvasive and real-time quantification of angiogenesis.

12.5.2 Bioluminescence Imaging

Bioluminescence is a two-dimensional imaging method that is based on the conversion of luciferin to photon-emitting oxyluciferin. Visible photons are then detected and quantified with charge-coupled device (CCD) cameras. Bioluminescence has been employed as a reporter of many biological functions. Luminescent reporters are much dimmer than fluorescent reporters; therefore, they provide relatively modest spatial and temporal resolution. Nonetheless, they are generally more sensitive and less toxic, making them particularly useful for long-term longitudinal studies of living cells, tissues, and whole animals. A variety of "luciferase" genes (bacterial *lux*, firefly *luc*, and *Renilla Rluc*) are used for BLI in animal models. These genes need to be implanted and expressed. Their protein products catalyze the emission of light from a substrate without requiring exogenous illumination [103]. The fact that there is no need for exogenous illumination is an advantage of luminescent reporters with respect to FLI; as a matter of fact, light can perturb physiology in light-sensitive tissues (e.g., the retina) and cause phototoxic damage to cells. The traditional application of bioluminescent imaging has been to detect ATP [104], as the luciferase reaction depends on the presence of oxygen and ATP. Luciferase activity has currently been applied to detect protein–protein interactions, to track cells *in vivo*, and to monitor transcriptional and post-transcriptional regulation of functionally important genes [105].

BLI has recently been applied in imaging blood vessels. Cell adhesion molecules are highly expressed on activated endothelial cells. Specific imaging methods for visualization of endothelial activation with probes targeted to adhesion molecules such as vascular cell adhesion molecule-1 (VCAM-1) can be applied [106]. The main disadvantages of this technique are its very low penetration depth (about 1–2 mm) and its low spatial resolution (not under the millimeter scale).

12.5.3 Doppler Optical Imaging

These imaging modalities take advantage of the Doppler effect and the associated phase-resolved methods. They include Doppler optical coherence tomography (DOCT) and Doppler optical microangiography (DOMAG).

DOCT is a technique providing three-dimensional high-resolution images of biological tissues [107,108]. In DOCT, light is reflected by a moving particle and is Doppler shifted. The direction and velocity of blood flow can be assessed by measuring the intensity of

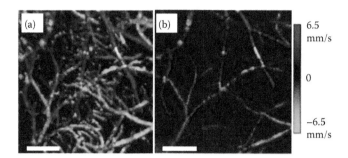

FIGURE 12.7
DOMAG (a) and DOCT (b) flow velocity image of murine cerebral blood flow. (Adapted from Wang, R.K. and L. An, *Opt Express*, 2009. 17(11): 8926–8940.)

back-reflected infrared light. Due to the different refractivity of blood plasma and cells, a detailed assessment of the vessel wall in the presence of blood is not possible. Therefore, a special imaging catheter is usually introduced into the vessels and is used to flush them with saline, thus shortening the imaging time. The first clinical DOCT systems were associated with a low-imaging acquisition speed, which precluded fast complete scans of the entire vessel. Newer systems that are currently evaluated in clinical trials are based on Fourier domain DOCT, which allows high-speed imaging of long vessel segments within a few seconds during saline flushing. Agents offering high optical absorption or scattering can be used for positive or negative contrast enhancement in DOCT. Contrast agents currently in use include single molecules, nanoparticles, and microspheres [109]. DOCT has been shown to provide a clear and detailed visualization of the vascular layers of the abdominal aorta (including the internal and external elastic membrane) of New Zealand white rabbits [110,111]. Its main advantage is the superior resolution of about 30 µm compared to other imaging techniques [108,112]. A drawback of DOCT, common also in other optical imaging modalities, is its limited penetration depth of about 1–2 mm [113].

DOMAG is a new imaging technique developed by combining optical microangiography (OMAG) with a phase-resolved method. DOMAG is reported to evaluate the velocities of blood flow within microcirculatory tissue beds with high precision [114]. The performance of traditional DOCT is severely limited by a background texture noise, due to the optical heterogeneous properties of the tissue sample. DOMAG is capable of separating the optical signals backscattered by moving tissue from the optical signals backscattered by the static tissue background. Therefore, it can provide a capability of imaging 3D flow almost free of background noise, as Figure 12.7 shows. DOMAG extracts flow velocities from OMAG flow signals, allowing the improved assessment of local fluid flow and its derived parameters, such as shear stress. Jia et al. demonstrated a dramatic improvement of DOMAG in quantifying flow-related properties within scaffolds *in situ* for functional tissue engineering compared with DOCT [115].

12.6 Ultrasound-Based Imaging Techniques

In diagnostic imaging, an intravascular ultrasound (IVUS) examination is a standard procedure to measure blood flow. Ultrasound waves (20–50 MHz) are applied in the region of interest by means of a probe. The reflected ultrasound signal consists of waves of different amplitude and frequency depending on the tissue where it has been reflected. Gray-scale IVUS only uses the information of signal amplitude. RF data analysis also uses the information of the frequency spectrum for tissue characterization. By introducing a miniaturized ultrasound transducer mounted on a specially designed catheter into the vessel compartment, cross-sectional images of the vessel can be obtained. IVUS can generate images with a spatial resolution ranging between 50 an 150 μm.

In contrast-enhanced ultrasound (CEU), specific acoustic properties of gas-filled microbubbles can be used as ultrasound contrast agents. Modulating shell properties or binding-specific probes to the microbubbles allows targeting them to structures of interest. Microbubbles can be used to target specific markers on vascular endothelium. Therefore, they allow vascular imaging of conditions like inflammation and angiogenesis. Xiao et al. used contrast-enhanced ultrasonography to evaluate angiogenesis in patients with hepatocellular carcinoma. The microvascular density was calculated and the relationship between enhancement patterns of tumor lesions and morphological characteristics of tumor microvessels was analyzed [116]. Recently, Myrset et al. developed a new targeted ultrasound contrast agent with improved binding and acoustic properties compatible with diagnostic use. Specific targeting of the inflammatory target E-selectin and the angiogenic target VEGFR2 in the presence of 100% serum was achieved [117].

Ultrasound imaging is an easy to use and cheap technique. It is highly sensitive and offers a better spatial resolution compared to others, in particular nuclear molecular imaging techniques. One of its major drawbacks is that a profound knowledge of anatomy is needed to interpret ultrasound images. Therefore, an experienced operator is necessary. To estimate the position of the actual image, certain anatomic landmarks like side-branches or bifurcations can be used. Alternatively, angiography has to be performed during acquisition of the IVUS to identify the position of the IVUS probe. IVUS and angiographic data of a vessel segment are acquired and then matched using registration techniques. The position and information of the IVUS catheter is linked to the angiographic information either in 2D or in 3D. More and more often, angiographic imaging is integrated with IVUS into a combined clinical system. Several approaches have been investigated to reach this goal. One of them is the combination of the IVUS probe with a magnetic positioning system. A magnetic array that generates a magnetic field is mounted on the angiographic system. The positioning sensor can be miniaturized and coupled with the IVUS transducer [118].

Doppler ultrasound imaging is employed not only in clinical but also in preclinical applications to monitor angiogenesis. For instance, CEU imaging has been used to detect intraplaque neovascularization in an experimental model of atherosclerosis in New Zealand white rabbits. Intraplaque neovessel density was quantified and highly correlated with contrast enhancement [119]. Again, ultrasound imaging was applied to assess the metabolic effects of angiogenic gene therapy in rabbits [120].

12.7 Summary and Conclusion

Angiogenesis is a crucial mechanism in many physiological and pathological processes. Different imaging modalities that allow imaging the neovascularization process are currently available in both clinical and preclinical applications. Their advantages and disadvantages are summarized in Table 12.1. Most of the current studies focus on imaging of angiogenesis in tumors. Little scientific research is available for monitoring vascular growth in tissue

TABLE 12.1

Summary of the Major Advantages and Disadvantages of the Current Techniques Used to Image Angiogenesis

Imaging Modality	Application	Maximum Resolution	Penetration Depth	Advantages	Disadvantages
X-ray angiography	Clinical, experimental	~1 mm	No limitation	Cheap, widespread	Low resolution, 2D, use of radiation
(Micro-)CTA	Clinical, experimental	~10 μm	No limitation	3D, very high resolution	Use of radiation, lack of targeted probes
(Micro-)MRA	Clinical, experimental	~25 μm	No limitation	3D, no radiation, high resolution	Expensive
(Micro-)PET	Clinical, experimental	~200 μm	No limitation	Very high sensitivity, use of targeted probes	Low resolution, radioactivity, expensive
(Micro-)SPECT	Clinical, experimental	~500 μm	No limitation	High sensitivity, use of targeted probes	Low resolution, radioactivity, expensive
Fluorescence, FMT	Experimental	~2 mm	~1 mm	Use of targeted probes	Low penetration depth, low resolution
Bioluminescence	Experimental	~3 mm	~1 mm	Use of targeted probes	Low penetration depth, low resolution
Doppler optical	Clinical, experimental	~30 μm	~1 mm	High resolution, precision	Low penetration depth, semi-quantitative data
Ultrasound	Clinical, experimental	~50 μm	~1 mm	Cheap, use of targeted probes, high sensitivity	Low penetration depth, lack of anatomical information

Source: Adapted from Dobrucki, L.W. and A.J. Sinusas, *Curr Opin Biotechnol*, 2007. 18(1): 90–96.)

engineering applications. The noninvasive evaluation of angiogenesis plays an important role in defining new diagnostic approaches and in evaluating new therapeutic interventions.

Acknowledgments

The authors acknowledge the European Seventh Framework Program (Grant Agreement number: 214402-2).

References

1. Karamysheva, A.F., Mechanisms of angiogenesis. *Biochemistry (Mosc)*, 2008. 73(7): 751–762.
2. Eming, S.A. et al., Regulation of angiogenesis: Wound healing as a model. *Prog Histochem Cytochem*, 2007. 42(3): 115–170.
3. Kerbel, R.S., Tumor angiogenesis. *N Engl J Med*, 2008. 358(19): 2039–2049.
4. Laschke, M.W. et al., Angiogenesis in tissue engineering: Breathing life into constructed tissue substitutes. *Tissue Eng*, 2006. 12(8): 2093–2104.

5. Carmeliet, P. and R.K. Jain, Angiogenesis in cancer and other diseases. *Nature*, 2000. 407(6801): 249–257.

6. Haubner, R., Alphavbeta3-integrin imaging: A new approach to characterise angiogenesis? *Eur J Nucl Med Mol Imaging*, 2006. 33 Suppl 1: 54–63.

7. Hood, J.D. and D.A. Cheresh, Role of integrins in cell invasion and migration. *Nat Rev Cancer*, 2002. 2(2): 91–100.

8. Xiong, J.P. et al., Crystal structure of the extracellular segment of integrin alpha Vbeta3. *Science*, 2001. 294(5541): 339–345.

9. Adams, R.H. and K. Alitalo, Molecular regulation of angiogenesis and lymphangiogenesis. Nature reviews. *Mol Cell Biol*, 2007. 8(6): 464–478.

10. Gerhardt, H., VEGF and endothelial guidance in angiogenic sprouting. *Organogenesis*, 2008. 4(4): 241–246.

11. Burri, P.H., R. Hlushchuk, and V. Djonov, Intussusceptive angiogenesis: Its emergence, its characteristics, and its significance. *Dev Dyn*, 2004. 231(3): 474–488.

12. Arribas, S.M. et al., Imaging the vascular wall using confocal microscopy. *J Physiol*, 2007. 584(Pt 1): 5–9.

13. Heinzer, S. et al., Hierarchical microimaging for multiscale analysis of large vascular networks. *Neuroimage*, 2006. 32(2): 626–636.

14. Van Lysel, M.S., The AAPM/RSNA physics tutorial for residents: Fluoroscopy: Optical coupling and the video system. *Radiographics*: A review publication of the Radiological Society of North America, 2000. 20(6): 1769–1786.

15. Crummy, A.B. et al., Computerized fluoroscopy: Digital subtraction for intravenous angiocardiography and arteriography. *AJR Am J Roentgenol*, 1980. 135(6): 1131–1140.

16. Mistretta, C.A. et al., A multiple image subtraction technique for enhancing low contrast, periodic objects. *Invest Radiol*, 1973. 8(1): 43–49.

17. Di Nallo, A.M. et al., Quantitative analysis of CT-perfusion parameters in the evaluation of brain gliomas and metastases. *J Exp Clin Cancer Res*: CR, 2009. 28: 38.

18. Petralia, G. et al., CT perfusion in oncology: How to do it. *Cancer Imaging*, 2010. 10(1): 8–19.

19. Henwood, S.M., *Clinical CT—Techniques and Practice*, ed. O.U. Press, 1999.

20. Lee, T.Y., Functional CT: Physiological models. *Trends Biotechnol*, 2002. 20(8): S3–S10.

21. Johnson, J.A. and T.A. Wilson, A model for capillary exchange. *Am J Phys*, 1966. 210(6): 1299–1303.

22. Bolland, B.J. et al., Development of *in vivo* muCT evaluation of neovascularisation in tissue engineered bone constructs. *Bone*, 2008. 43(1): 195–202.

23. Feldkamp, L.A., L.C. Davis, and J.W. Kress, Practical cone-beam algorithm. *J Opt Soc Am*, 1984. A(1): 612–619.

24. Muller, R., Hierarchical microimaging of bone structure and function. *Nat Rev Rheumatol*, 2009. 5(7): 373–381.

25. Ulrich, D. et al., The quality of trabecular bone evaluated with micro-computed tomography, FEA and mechanical testing. *Stud Health Technol Inform*, 1997. 40: 97–112.

26. Schambach, S.J. et al., Ultrafast high-resolution in vivo volume-CTA of mice cerebral vessels. *Stroke: J Cereb Circulation*, 2009. 40(4): 1444–1450.

27. Ford, N.L., M.M. Thornton, and D.W. Holdsworth, Fundamental image quality limits for microcomputed tomography in small animals. *Med Phys*, 2003. 30(11): 2869–2877.

28. Badea, C.T. et al., Tumor imaging in small animals with a combined micro-CT/micro-DSA system using iodinated conventional and blood pool contrast agents. *Contrast Media Mol Imaging*, 2006. 1(4): 153–164.

29. Lin, M.D. et al., Optimized radiographic spectra for small animal digital subtraction angiography. *Med Phys*, 2006. 33(11): 4249–4257.

30. Badea, C.T. et al., Tomographic digital subtraction angiography for lung perfusion estimation in rodents. *Med Phys*, 2007. 34(5): 1546–1555.

31. Charnley, N., S. Donaldson, and P. Price, Imaging angiogenesis. *Methods Mol Biol*, 2009. 467: 25–51.

32. Badea, C.T. et al., 4-D micro-CT of the mouse heart. *Mol Imaging*, 2005. 4(2): 110–116.
33. de Lin, M. et al., A high-precision contrast injector for small animal x-ray digital subtraction angiography. *IEEE Trans Biomed Eng*, 2008. 55(3): 1082–1091.
34. Weichert, J.P. et al., Lipid-based blood-pool CT imaging of the liver. *Acad Radiol*, 1998. 5 Suppl 1: S16–S19; discussion S28–30.
35. Vera, D.R. and R.F. Mattrey, A molecular CT blood pool contrast agent. *Acad Radiol*, 2002. 9(7): 784–792.
36. Ford, N.L. et al., Time-course characterization of the computed tomography contrast enhancement of an iodinated blood-pool contrast agent in mice using a volumetric flat-panel equipped computed tomography scanner. *Invest Radiol*, 2006. 41(4): 384–390.
37. Drangova, M. et al., Fast retrospectively gated quantitative four-dimensional (4D) cardiac micro computed tomography imaging of free-breathing mice. *Invest Radiol*, 2007. 42(2): 85–94.
38. Graham, K.C. et al., Noninvasive quantification of tumor volume in preclinical liver metastasis models using contrast-enhanced x-ray computed tomography. *Invest Radiol*, 2008. 43(2): 92–99.
39. Willekens, I. et al., Time-course of contrast enhancement in spleen and liver with Exia 160, Fenestra LC, and VC. *Mol Imaging Biol*, 2009. 11(2): 128–135.
40. Kim, D. et al., Antibiofouling polymer-coated gold nanoparticles as a contrast agent for in vivo X-ray computed tomography imaging. *J Am Chem Soc*, 2007. 129(24): 7661–7665.
41. Rabin, O. et al., An X-ray computed tomography imaging agent based on long-circulating bismuth sulphide nanoparticles. *Nat Mater*, 2006. 5(2): 118–122.
42. Johnson, T.R. et al., Material differentiation by dual energy CT: Initial experience. *Eur Radiol*, 2007. 17(6): 1510–1517.
43. Tran, D.N. et al., Dual-energy CT discrimination of iodine and calcium: Experimental results and implications for lower extremity CT angiography. *Acad Radiol*, 2009. 16(2): 160–171.
44. Flohr, T.G. et al., First performance evaluation of a dual-source CT (DSCT) system. *Eur Radiol*, 2006. 16(2): 256–268.
45. Kruger, R.A. et al., Digital K-edge subtraction radiography. *Radiology*, 1977. 125(1): 243–245.
46. Riederer, S.J. and C.A. Mistretta, Selective iodine imaging using K-edge energies in computerized x-ray tomography. *Med Phys*, 1977. 4(6): 474–481.
47. Nakayama, Y. et al., Abdominal CT with low tube voltage: Preliminary observations about radiation dose, contrast enhancement, image quality, and noise. *Radiology*, 2005. 237(3): 945–951.
48. Cavanaugh, D. et al., *in vivo* respiratory-gated micro-CT imaging in small-animal oncology models. *Mol Imaging*, 2004. 3(1): 55–62.
49. Cody, D.D. et al., Murine lung tumor measurement using respiratory-gated micro-computed tomography. *Invest Radiol*, 2005. 40(5): 263–269.
50. Namati, E. et al., *In vivo* micro-CT lung imaging via a computer-controlled intermittent isopressure breath hold (IIBH) technique. *Phys Med Biol*, 2006. 51(23): 6061–6075.
51. Hedlund, L.W. and G.A. Johnson, Mechanical ventilation for imaging the small animal lung. *ILAR J*, 2002. 43(3): 159–174.
52. Ford, N.L. et al., Prospective respiratory-gated micro-CT of free breathing rodents. *Med Phys*, 2005. 32(9): 2888–2898.
53. Badea, C.T. et al., Cardiac micro-computed tomography for morphological and functional phenotyping of muscle LIM protein null mice. *Mol Imaging*, 2007. 6(4): 261–268.
54. Johnson, K., Introduction to rodent cardiac imaging. *ILAR J*, 2008. 49(1): 27–34.
55. Song, J. et al., Sparseness prior based iterative image reconstruction for retrospectively gated cardiac micro-CT. *Med Phys*, 2007. 34(11): 4476–4483.
56. Badea, C.T. et al., *In vivo* small-animal imaging using micro-CT and digital subtraction angiography. *Phys Med Biol*, 2008. 53(19): R319–R350.
57. Evanochko, W.T., R.C. Reeves, and G.M. Pohost, Biomedical nuclear magnetic resonance: Principles and progress. *Cardiovasc Clin*, 1986. 17(1): 129–143.
58. Paushter, D.M. et al., Magnetic resonance. Principles and applications. *Med Clin North Am*, 1984. 68(6): 1393–1421.

59. Hurrell, M.A., Nuclear magnetic resonance: Principles and prospects. *N Z Med J*, 1982. 95(715): 622–623.
60. Mills, C.M. et al., Nuclear magnetic resonance: principles of blood flow imaging. *AJR Am J Roentgenol*, 1984. 142(1): 165–170.
61. Graves, M.J., Magnetic resonance angiography. *Br J Radiol*, 1997. 70: 6–28.
62. Winter, P.M. et al., Molecular imaging of angiogenesis in nascent Vx-2 rabbit tumors using a novel alpha(nu)beta3-targeted nanoparticle and 1.5 tesla magnetic resonance imaging. *Cancer Res*, 2003. 63(18): 5838–5843.
63. Winter, P.M. et al., Molecular imaging of angiogenesis in early-stage atherosclerosis with alpha(v)beta3-integrin-targeted nanoparticles. *Circulation*, 2003. 108(18): 2270–2274.
64. de Kemp, R.A. et al., Small-animal molecular imaging methods. *J Nucl Med*, 2010. 51 Suppl 1: 18S–32S.
65. Abramovitch, R., D. Frenkiel, and M. Neeman, Analysis of subcutaneous angiogenesis by gradient echo magnetic resonance imaging. *Magn Reson Med*, 1998. 39(5): 813–824.
66. Griffiths, T.L. et al., The effect of dipyridamole and theophylline on hypercapnic ventilatory responses: The role of adenosine. *Eur Respir J*, 1997. 10(1): 156–160.
67. Tempel, C. and M. Neeman, Spatial and temporal modulation of perfusion in the rat ovary measured by arterial spin labeling MRI. *J Magn Reson Imaging*, 1999. 9(6): 794–803.
68. Tempel, C. and M. Neeman, Perfusion of the rat ovary: Application of pulsed arterial spin labeling MRI. *Magn Reson Med*, 1999. 41(1): 113–123.
69. Calcagno, C. et al., Dynamic contrast enhanced (DCE) magnetic resonance imaging (MRI) of atherosclerotic plaque angiogenesis. *Angiogenesis*, 2010. 13(2): 87–99.
70. Pathak, A.P., M.F. Penet, and Z.M. Bhujwalla, MR molecular imaging of tumor vasculature and vascular targets. *Adv Genet*, 2010. 69: 1–30.
71. Seevinck, P.R., L.H. Deddens, and R.M. Dijkhuizen, Magnetic resonance imaging of brain angiogenesis after stroke. *Angiogenesis*, 2010. 13(2): 101–111.
72. Burtea, C. et al., Contrast agents: Magnetic resonance. *Handb Exp Pharmacol*, 2008(185 Pt 1): 135–165.
73. Bogdanov, A.A., M. Lewin, and R. Weissleder, Approaches and agents for imaging the vascular system. *Adv Drug Deliv Rev*, 1999. 37(1–3): 279–293.
74. Weinmann, H.J. et al., A new lipophilic gadolinium chelate as a tissue-specific contrast medium for MRI. *Magn Reson Med*, 1991. 22(2): 233–237; discussion 242.
75. Bulte, J.W. and D.L. Kraitchman, Iron oxide MR contrast agents for molecular and cellular imaging. *NMR Biomed*, 2004. 17(7): 484–499.
76. Wang, Y.X., S.M. Hussain, and G.P. Krestin, Superparamagnetic iron oxide contrast agents: physicochemical characteristics and applications in MR imaging. *Eur Radiol*, 2001. 11(11): 2319–2331.
77. Anelli, P.L. et al., Mixed micelles containing lipophilic gadolinium complexes as MRA contrast agents. *MAGMA*, 2001. 12(2–3): 114–120.
78. Parac-Vogt, T.N. et al., Synthesis, characterization, and pharmacokinetic evaluation of a potential MRI contrast agent containing two paramagnetic centers with albumin binding affinity. *Chemistry*, 2005. 11(10): 3077–3086.
79. Corot, C. et al., Physical, chemical and biological evaluations of CMD-A2-Gd-DOTA. A new paramagnetic dextran polymer. *Acta Radiol Suppl*, 1997. 412: 91–99.
80. Gaillard, S. et al., Safety and pharmacokinetics of p792, a new blood-pool agent: Results of clinical testing in nonpatient volunteers. *Invest Radiol*, 2002. 37(4): 161–166.
81. Parmelee, D.J. et al., Preclinical evaluation of the pharmacokinetics, biodistribution, and elimination of MS-325, a blood pool agent for magnetic resonance imaging. *Invest Radiol*, 1997. 32(12): 741–747.
82. Zheng, J. et al., Multimodal contrast agent for combined computed tomography and magnetic resonance imaging applications. *Invest Radiol*, 2006. 41(3): 339–348.
83. Dobrucki, L.W. and A.J. Sinusas, Imaging angiogenesis. *Curr Opin Biotechnol*, 2007. 18(1): 90–96.

84. Haubner, R., Noninvasive tracer techniques to characterize angiogenesis. *Handb Exp Pharmacol*, 2008(185 Pt 2): 323–339.

85. Chatziioannou, A. et al., Detector development for microPET II: a 1 microl resolution PET scanner for small animal imaging. *Phys Med Biol*, 2001. 46(11): 2899–2910.

86. Chatziioannou, A.F. et al., Performance evaluation of microPET: A high-resolution lutetium oxyorthosilicate PET scanner for animal imaging. *J Nucl Med*, 1999. 40(7): 1164–1175.

87. Knoess, C. et al., Performance evaluation of the microPET R4 PET scanner for rodents. *Eur J Nucl Med Mol Imaging*, 2003. 30(5): 737–747.

88. Tai, C. et al., Performance evaluation of the microPET P4: A PET system dedicated to animal imaging. *Phys Med Biol*, 2001. 46(7): 1845–1862.

89. Yang, Y. et al., Optimization and performance evaluation of the microPET II scanner for *in vivo* small-animal imaging. *Phys Med Biol*, 2004. 49(12): 2527–2545.

90. Bockisch, A. et al., Hybrid imaging by SPECT/CT and PET/CT: Proven outcomes in cancer imaging. *Semin Nucl Med*, 2009. 39(4): 276–289.

91. Gambhir, S.S., Molecular imaging of cancer with positron emission tomography. *Nat Rev Cancer*, 2002. 2(9): 683–693.

92. Sharma, V., G.D. Luker, and D. Piwnica-Worms, Molecular imaging of gene expression and protein function in vivo with PET and SPECT. *J Magn Reson Imaging*, 2002. 16(4): 336–351.

93. Wagner, B. et al., Noninvasive characterization of myocardial molecular interventions by integrated positron emission tomography and computed tomography. *J Am Coll Cardiol*, 2006. 48(10): 2107–2115.

94. Cai, W. et al., PET of vascular endothelial growth factor receptor expression. *J Nucl Med*, 2006. 47(12): 2048–2056.

95. Peremans, K. et al., A review of small animal imaging planar and pinhole spect Gamma camera imaging. *Vet Radiol Ultrasound*, 2005. 46(2): 162–170.

96. Liu, S., Radiolabeled multimeric cyclic RGD peptides as integrin alphavbeta3 targeted radiotracers for tumor imaging. *Mol Pharm*, 2006. 3(5): 472–487.

97. Hua, J. et al., Noninvasive imaging of angiogenesis with a 99mTc-labeled peptide targeted at alphavbeta3 integrin after murine hindlimb ischemia. *Circulation*, 2005. 111(24): 3255–3260.

98. Guccione, S., K.C. Li, and M.D. Bednarski, Vascular-targeted nanoparticles for molecular imaging and therapy. *Methods Enzymol*, 2004. 386: 219–236.

99. Mulder, W.J. et al., MR molecular imaging and fluorescence microscopy for identification of activated tumor endothelium using a bimodal lipidic nanoparticle. *FASEB J*, 2005. 19(14): 2008–2010.

100. Amoh, Y., K. Katsuoka, and R.M. Hoffman, Color-coded fluorescent protein imaging of angiogenesis: The AngioMouse models. *Curr Pharm Des*, 2008. 14(36): 3810–3819.

101. Tan, Y. and H. Jiang, Diffuse optical tomography guided quantitative fluorescence molecular tomography. *Appl Opt*, 2008. 47(12): 2011–2016.

102. Kossodo, S. et al., Dual *in vivo* quantification of integrin-targeted and protease-activated agents in cancer using fluorescence molecular tomography (FMT). *Mol Imaging Biol*, 2010. 12(5): 488–499.

103. Greer, L.F., 3rd and A.A. Szalay, Imaging of light emission from the expression of luciferases in living cells and organisms: A review. *Luminescence*, 2002. 17(1): 43–74.

104. Dumollard, R. et al., Sperm-triggered [Ca2+] oscillations and Ca2+ homeostasis in the mouse egg have an absolute requirement for mitochondrial ATP production. *Development*, 2004. 131(13): 3057–3067.

105. Welsh, D.K. and S.A. Kay, Bioluminescence imaging in living organisms. *Curr Opin Biotechnol*, 2005. 16(1): 73–78.

106. Dunehoo, A.L. et al., Cell adhesion molecules for targeted drug delivery. *J Pharm Sci*, 2006. 95(9): 1856–1872.

107. Hee, M.R. et al., Optical coherence tomography of the human retina. *Arch Ophthalmol*, 1995. 113(3): 325–332.

108. Huang, D. et al., Optical coherence tomography. *Science*, 1991. 254(5035): 1178–1181.

109. Boppart, S.A. et al., Optical probes and techniques for molecular contrast enhancement in coherence imaging. *J Biomed Opt*, 2005. 10(4): 41208.
110. Brezinski, M.E. et al., Optical coherence tomography for optical biopsy. Properties and demonstration of vascular pathology. *Circulation*, 1996. 93(6): 1206–1213.
111. Fujimoto, J.G. et al., High resolution *in vivo* intra-arterial imaging with optical coherence tomography. *Heart*, 1999. 82(2): 128–133.
112. Tearney, G.J. et al., *in vivo* endoscopic optical biopsy with optical coherence tomography. *Science*, 1997. 276(5321): 2037–2039.
113. MacNeill, B.D. et al., Intravascular modalities for detection of vulnerable plaque: current status. *Arterioscler Thromb Vasc Biol*, 2003. 23(8): 1333–1342.
114. Wang, R.K. and L. An, Doppler optical micro-angiography for volumetric imaging of vascular perfusion in vivo. *Opt Express*, 2009. 17(11): 8926–8940.
115. Jia, Y., L. An, and R.K. Wang, Doppler optical microangiography improves the quantification of local fluid flow and shear stress within 3-D porous constructs. *J Biomed Opt*, 2009. 14(5): 050504.
116. Xiao, J.D., W.H. Zhu, and S.R. Shen, Evaluation of hepatocellular carcinoma using contrast-enhanced ultrasonography: Correlation with microvessel morphology. *Hepatobiliary Pancreat Dis Int*, 2010. 9(6): 605–610.
117. Myrset, A.H. et al., Design and characterization of targeted ultrasound microbubbles for diagnostic use. *Ultrasound Med Biol*, 2011. 37(1): 136–150.
118. Hetterich, H. et al., New X-ray imaging modalities and their integration with intravascular imaging and interventions. *Int J Cardiovasc Imaging*, 2010. 26(7): 797–808.
119. Giannarelli, C. et al., Contrast-enhanced ultrasound imaging detects intraplaque neovascularization in an experimental model of atherosclerosis. *JACC Cardiovasc Imaging*, 2010. 3(12): 1256–1264.
120. Korpisalo, P. et al., Capillary enlargement, not sprouting angiogenesis, determines beneficial therapeutic effects and side effects of angiogenic gene therapy. *Eur Heart J*, 2010. 32(13): 1664–1672.

Section V

Vascularized Tissues

13

In Vivo *Techniques and Strategies for Enhanced Vascularization of Engineered Bone*

Bao-Ngoc B. Nguyen and John P. Fisher

CONTENTS

13.1 Introduction

Tissue engineering expanded rapidly throughout the 1990s based on the promise to create new organs and tissue constructs to replace diseased or damaged organs (Nerem 2006). However, two decades later, few products have successfully passed the Food and Drug Administration (FDA) clinical trials (Jaklenec et al. 2012); a feasible transplantable organ is still unattainable. More importantly, an organ shortage continues to persist because a consistent technique for providing vasculature and integration of tissue-engineered constructs into the host has not yet been realized (Phelps and García 2010).

Most tissues in the body rely on oxygen and nutrients supplied from blood vessels. It has been shown that new blood vessel formation is required once the tissue has grown beyond 100–200 µm from a nearby vasculature due to oxygen diffusion limitations (Jain et al. 2005). If implanted engineered tissues cannot obtain the appropriate amount of nutrients, the tissue experiences decreased function, nutrient deficiencies, or hypoxia, especially at the core of the construct (Lovett et al. 2009). Therefore, three-dimensional constructs depend on rapid development of new blood vessels and vascular networks to provide nutrients.

The development of mature vasculature is one of the major hurdles in the field of tissue-engineering research, preventing a successful transition from lab bench to relevant clinical applications. Without a sufficient supply of oxygen, engineered tissue scalability, survival, and integration with the host tissue is extremely limited (Phelps and García 2010). It is the hope that with advances in vascularization technology, many of the tissue-engineered constructs currently only viable *in vitro* will be able to become a reality. Tissue engineering aims to not only build artificial tissues in the lab, but also ultimately enhance or restore the function of diseased and damaged tissues. Therefore, vascularization of engineered tissues is a vital next step in moving the field toward successful regenerative medicine.

While a tremendous amount of research is conducted using *in vitro* methods of vascularization of engineered tissues, fewer advances have been made within the field of *in vivo* techniques due to the complex microenvironment, cost, and poorly understood synergy between the host and implanted cells. This chapter assesses the current state of the field by outlining fundamental approaches taken toward developing prevascularized bone tissues *in vivo*, as well as highlighting their advantages and disadvantages.

13.1.1 Bone Tissue Engineering

Tissue loss as a result of injury or disease leads to reduced quality of life, especially with an increasing aging population. However, strategies that encourage bone formation by significantly increasing bone density have yet to become available. It is a major clinical requirement that has stimulated increasing interest in the tissue-engineering field to develop new therapies that involve bone regeneration. While significant advances have been made combining biomaterials and cells for *in vitro* culture, the field has seen relatively fewer developments toward clinical trials.

The gold standard to prevent or treat a fractured nonunion is autologous bone grafting or delivering bone chips from a secondary donor site into the defect site, where they can then promote and attract other cells responsible for the bone-forming process (Nishi et al. 2013). However, significant donor morbidity such as chronic pain, hypersensitivity, infection, and paraesthesia occur in up to one-third of the patients (Weinand et al. 2007). In addition, the limited supply and relatively unpredictability of the autologous bone grafts have led to the use of alternatives, including the allograft. These can be used for larger defects but are limited by the possibility of immune rejection, disease transmission, and relatively lower incorporation rate compared to autografts (Lanza et al. 2007). In addition, allografts are associated with reduced cellularity and vascularization compared to autologous grafts, leading to poorer bone healing. Xenografts, the tissue derived from another species, are used less frequently because they have many of the same drawbacks in addition to their dissimilarity to human tissue structure and function (Goldstein 2002). This in turn led to the development and fabrication of synthetic scaffolds, which can be molded into different shapes and sizes to fit the defect site (Liu et al. 2012). The goal is to engineer constructs with similar properties as natural bone, including mechanical strength and structure. However, like many of the other tissue-engineered organs, bone constructs have been hampered by their inability to remain viable *in vivo*. Unlike other tissues, such as muscle bundles, bone tissue lacks an abundant preexisting vascular network, which is able to rapidly penetrate into the scaffold and avoid tissue necrosis at the center of the graft. Current methods of graft implantation *in vivo* show slow integration of the host's vasculature and sufficient nutrient and oxygen concentrations only at the host tissue-construct interface.

13.2 Scaffold Material

The three-dimensional architecture and design of scaffolds has been shown to have a profound effect on the rate of vascularization once implanted *in vivo*. For example, bone scaffolds are typically made out of porous, degradable materials that are able to provide the proper mechanical strength during repair and regeneration of the damaged or diseased bone. Some of the design and material property requirements for an ideal graft include biocompatibility, adequate pore size, and bioresorbability (Bose et al. 2012). Biocompatibility describes the scaffold's ability to support normal cellular activity, including molecular signaling, without exhibiting any toxic effect to the host tissue (Kim and Mooney 1998; Abshagen et al. 2009). For a bone scaffold, this means it must be conducive to cell recruitment and subsequent bone formation. It has been shown that pores need to be at least 100 μm in diameter to allow for diffusion of essential nutrients and oxygen (Rouwkema et al. 2008). Porosity also plays an important role in cell migration and physical communication between cells. Unfortunately, increased porosity reduces the mechanical strength of the scaffold, affecting the compressive as well as the degradation properties. Therefore, finding a delicate balance between architectural and mechanical properties is critical during bone scaffold fabrication. Finally, bioresorbability is another important factor because the scaffold has to exhibit similar properties as the surrounding host tissue, yet degrade at a desired rate to accommodate neovasculature and bone formation.

In the event that the construct is prevascularized in a well-supplied region of the body prior to implantation in the defect site, the scaffolding material is expected to remain intact for a significantly longer period of time than if directly implanted into the defect site. A variety of biomaterials have been utilized for this purpose, ranging from fibrin (Steffens et al. 2009; Correia et al. 2011), to poly(lactic-*co*-glycolic) acid (PLGA) (Weinand et al. 2007; Tsigkou et al. 2010; Hegen et al. 2011), to processed bovine cancellous bone (PBCB) (Steffens et al. 2009). Many of these scaffolds have been evaluated for long-term cytotoxic effects and have shown to have slow degradation rates.

Several different kinds of hydrogels have been found to facilitate osteogenesis. Of these, a few have shown to also promote formation of vascular sprouts. Matrigel™ is one of the most popular scaffolds used as a vascularization platform (Meinel et al. 2004; Li and Guan 2011; Domev et al. 2012). It is a matrix extract derived from mouse sarcoma cells and contains many important proteins such as laminin, entactin, and collagen. Cells encounter these structural components in their natural environment, which promote cell adhesion and spreading. In addition, Matrigel contains certain growth factors that can promote differentiation and proliferation for a variety of cell lines (Laschke et al. 2010; Marolt et al. 2012). However, due to source variance, the actual composition of Matrigel can fluctuate from lot to lot and results are inconsistent, making Matrigel nonideal for tissue-engineering techniques (Li and Guan 2011). In addition, its xenogenic origin results in unfavorable immune responses and could hinder blood vessel formation *in vivo*. For the purpose of bone tissue engineering, growth factor-enriched Matrigel has been used as an additive to standard scaffolds, such as PLGA, to enhance vascularization (Laschke et al. 2009). Laschke et al. demonstrated that Matrigel did not have any effect on the biocompatibility of the PLGA scaffold with the seeded cells, but instead improved the *in vivo* ingrowth of new blood vessels from the surrounding tissue.

More complicated, lab-generated scaffolds include modified polyethylene glycol diacrylate (PEGDA) hydrogels containing adhesive ligands and encapsulated growth factors (Phelps et al. 2010), or protease-sensitive PEG gels with functionalized integrin-binding sites (Moon et al. 2010). Porous scaffolds can be advantageous for vascularization of the bone tissue, including electrospun scaffolds made out of degradable poly(e-caprolactone) (PCL) (Shin et al. 2004) or silk fibroin (Ghanaati et al. 2011). Scaffolds fabricated with a more random porous architecture can lead to only partially connected pathways, impeding the formation of a dense vascular system. Therefore, scaffolds with well-defined, interconnected pores may result in better vessel formation. Such grafts can be best produced with the use of rapid-prototyping techniques, which utilize computer-aided design (CAD) templates to print the scaffold layer by layer out of a desired biomaterial (Hutmacher et al. 2004).

13.3 Stem and Progenitor Cells

A variety of cell types are involved in creating vascularized bone tissue, the most obvious being osteoprogenitor and endothelial cells. Recently, mural cells have also started to gain more attention as an important component of vascular network formation within bone tissue. This section describes different aspects of each cell type and their importance to bone vascularization.

13.3.1 Stem Cells

Stem cells have become the forerunner in the field of tissue engineering due to their capability to differentiate into a variety of cell types. Mesenchymal stem cells especially have been used more frequently in bone tissue engineering applications. Human mesenchymal stem cells (hMSCs) are often used as a source for osteoprogenitor cells because these multipotent cells are isolated from the bone marrow and have the ability to differentiate into a variety of lineages to become adipocytes, chondrocytes, or osteocytes. In addition, because they reside in the bone marrow, they can be obtained fairly easily from an adult patient by way of a bone marrow biopsy (Pittenger 1999). For the purpose of bone tissue engineering, the cells can be guided toward osteogenesis with the use of mechanical stimulation or growth factors, such as bone-morphogenic protein-2 or dexamethasone (Sittichokechaiwut et al. 2010). In terms of immune response, hMSCs are nonimmunogenic *in vitro*, as well as allogenic *in vivo*, making them an ideal candidate for bone regeneration (Zhang et al. 2010).

13.3.2 Endothelial Cells and Source

Endothelial cells are a thin layer of cells that make up the interior of blood vessels. They have shown to accelerate neovascular formation in many tissue-engineering constructs by creating blood vessels networks (Rouwkema et al. 2006; Abshagen et al. 2009; Grellier et al. 2009). In addition to their ability to improve vascularization and bone graft survival, endothelial cells are able to support osteogenesis, by mediating cell–cell communication via soluble factors (Koyama et al. 2002; Ribatti 2005; De la Riva et al. 2009) and gap junction proteins (Correia et al. 2011). Unfortunately, some of the drawbacks of endothelial cells include their limited proliferation ability and the necessity for inhibition of the apoptotic response (Schechner et al. 2000).

The source of endothelial cells can have an important effect on the success of a tissue-engineering construct. There is considerable phenotypic variation among cells depending on their source. For example, human umbilical vein endothelial cells (HUVECs) are easily obtained from discarded umbilical cords, can be easily expanded, and have been shown to be ideal for *in vitro* applications. However, they have resulted in immature and leaky vessels once implanted *in vivo* (Sefcik et al. 2008). On the other hand, embryonic stem cells have resulted in some promising outcomes when implanted *in vivo*, including formation of stable microvessels and integration with host endothelial cells (Levenberg et al. 2002; Marolt et al. 2012). However, due to the ethical concerns associated with them as well as their ability to differentiate into almost every cell type if not tightly controlled, their clinical use has been limited. Promising results have recently been seen with the use of endothelial progenitor cells (EPCs) (Lin et al. 2000). EPCs are a small population of circulating mononuclear cells that have shown to differentiate into and are able to keep their endothelial cell characteristics *in vitro* (Fedorovich et al. 2010).

13.3.3 Mural Cells

Mural cells are generally used to refer to vascular smooth cells and pericytes, both of which are involved in the process of blood vessel formation (Evensen et al. 2009). They have been shown to migrate toward sprouting endothelial cells in response to vascular endothelial growth factor (VEGF) and are able to help provide a stabilizing environment for the cells (Evensen et al. 2009). They allow for cell-to-cell communication, secrete angiogenic factors and extracellular matrix (ECM) components, and promote endothelial blood vessel maturation. The latter involves a series of steps that involve important spatial and temporal coordination of the endothelial cells and their signaling pathways, such as altering endothelial cell proliferation rate and changing its morphology (Dejana 2004). Several groups have shown the involvement of mural cells in angiogenesis (Sefcik et al. 2008; Tsigkou et al. 2010; Hegen et al. 2011), concluding that the recruitment of mural cells greatly affects endothelial behavior when cocultured with endothelial cells. Fluorescent microscopy shows integration of mural cells into the blood vessel network, which is especially enhanced with the addition of VEGF as a growth factor *in vitro* (Evensen et al. 2009). The resulting architectural layout of the neovasculature has also been shown to be more mature and dense when endothelial cells have been cultured with mural cells.

On the basis of these results, it shows that the formation of neovasculature is an orchestrated effort of several cell types. Therefore, future experiments may have more success if tissue-engineered constructs are implanted *in vivo* surrounded by a source of mural cells to enable mature blood vessel formation.

13.4 Cocultures

While singular endothelial-derived cell types have been able to demonstrate formation of vascular networks, the ultimate goal of generating prevascularized tissues may require

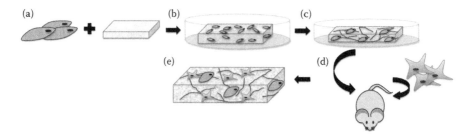

FIGURE 13.1
In vitro prevascularization using endothelial cells followed by *in vivo* coculture. Scaffolds can be prevascularized *in vitro* when seeded with endothelial cells prior to *in vivo* coculture with osteoprogenitor cells. (a) Endothelial cells seeded on the scaffold and cultured *in vitro*. (b) Endothelial sprouting and tube formation indicate a prevascularized network. (c) Prevascularized scaffold is coseeded with osteoprogenitor cells and then implanted *in vivo*. (d) The resulting scaffold shows mature blood vessel formation and bone tissue development.

the coculture of endothelial cells and the target tissue cell population. Coculture methods allow for concurrent creation of a vascular network as well as the target tissue. The use of the technique has also been shown that osteoprogenitor cells show a higher expression of alkaline phosphatase (ALP), an early osteogenic differentiation marker, when cocultured with endothelial cells. A vast amount of research has been published on coculture techniques for bone tissue engineering, ranging from preculture of endothelial cells *in vitro* followed by the addition of the osteoprogenitor cells to the scaffold but prior to *in vivo* implantation (Correia et al. 2011) (Figure 13.1), preculture of both cell types *in vitro*, and eventual implantation into an animal model (Sikavitsas et al. 2003; Shin et al. 2004; Chen et al. 2007; Pirraco et al. 2011) (Figure 13.2), to a direct coculture *in vivo* (Steffens et al. 2009; Wernike et al. 2010; Zhang et al. 2010). However, despite these efforts, there has been limited success with complete integration of the implanted scaffold with the functional perfusion of the host tissue's vascular system.

There are some general considerations to bear in mind when it comes to cocultures. In addition to the cell type, the choice of culture media, seeding density, culturing environment, and scaffold architecture are less apparent. For example, as described previously, the source and type of endothelial cells and stem cells can have a lasting effect on the engineered bone, with promising results emerging with the use of EPCs and hMSCs. Likewise, the ratio of endothelial to osteogenic culturing media as well as proportion of endothelial

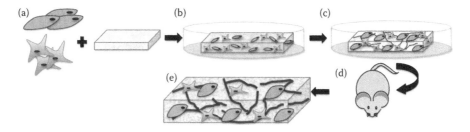

FIGURE 13.2
In vitro prevascularization via coculture of endothelial and osteoprogenitor cells. An *in vitro* coculture of endothelial cells and osteoprogenitor cells can prevascularize a scaffold prior to implantation *in vivo*. (a) Seeding and coculture of endothelial cells (red) and osteoprogenitor cells (blue) onto the scaffold. (b) *In vitro* culturing of the construct will result in (c) neovascular tube and bone tissue formation. (d) Once implanted *in vivo*, the scaffold will become further vascularized while extracellular bone matrix is formed.

to osteoprogenitor cells will affect vascular formation and osteogenic differentiation (Ma et al. 2011; Correia et al. 2011). Ma et al. determined that a 1:1 ratio of the two cell populations, cocultured only in osteogenic media, resulted in the most mineralization and angiogenic development *in vitro*.

The combination of endothelial and osteoprogenitor cells has been especially common in bone tissue engineering cocultures. Both cells types are known to secrete specific growth factors that are beneficial for growth and differentiation for each other. For example, osteoprogenitor cells secrete VEGF that can be used for proliferation and angiogenic processes by endothelial cells (Kirkpatrick et al. 2011). Likewise, endothelial cells are known to secrete growth factors such as insulin growth factor-1 (Yu et al. 2012), endothelin-1, and bone-morphogenic protein-2, promoting osteogenic growth and differentiation (Huang et al. 2010). In addition, it has been shown that the cell-to-cell communication between endothelial cells and osteoprogenitor cells can increase the production of the early osteogenic marker ALP (Rouwkema et al. 2006; Grellier et al. 2009).

Lastly, evidence from literature shows that mechanotransduction plays a synergistic role in the coculture of cells on scaffolds for bone tissue engineering. The phenomenon converts a mechanical stimulus into chemical activity, such as a signaling pathway. It is hypothesized that the fluid flow experienced by the cells imposed by loading regimes can influence cell proliferation and differentiation downstream (Geris et al. 2010). However, it is still far from being completely understood. Therefore, further research should investigate the effect of mechanical stimulation on signaling pathways within and between endothelial cells and osteoprogenitor cells that lead to enhanced angiogenic and osteogenic effects. Preliminary studies have shown that the mechanical stimulation of coupled gap junction proteins between the two cell lines can drive osteoblastic differentiation, emphasizing the importance of cell-to-cell communication (Villars et al. 2000).

13.5 Growth Factors

While scaffold properties such as porosity and degradation rates have shown to affect angiogenesis both *in vitro* and *in vivo*, a variety of different growth factors have been used in the modification of scaffolds to further promote the formation of vascular networks. Promising results have been published about combination therapies of cell and angiogenic growth factor deliveries to tissue defect sites (Huang et al. 2005). These additions to the cell culture mimic *in vivo* conditions, stabilizing the cells and protecting them from proteolytic digestion. It has been shown that successful vascular network formation requires coordination of not only the right cell type, but also the appropriate signaling factors, such as VEGF, delivered at the proper concentration and exposure times (Phelps and García 2010). So far, approaches have focused on preseeding scaffolds with growth factors prior to implantation or the incorporation of a slow release of soluble pro-angiogenic factors within the scaffolds.

VEGF is one of the most used pro-angiogenic factors and plays multifunctional roles in vascular permeability, repair, and remodeling processes. In addition, it maintains vascular structure and function. It has been extensively used in a variety of different scaffolds, at different concentrations, and at different time points during culture of endothelial cells for the purpose of bone tissue engineering (Lazarous et al. 1996; Huang et al. 2005; Peng et al. 2005; Moon et al. 2010). Studies have confirmed that VEGF plays an essential role in the neovascularization of tissue and modulates endothelial growth, proliferation, migration, and tube formation, therefore, making it an important factor in inducing angiogenesis. Bone morphogenetic protein (BMP) is another family of growth factors identified to support the use of cells for bone tissue engineering. They promote bone formation by inducing MSCs toward osteoblastic differentiation. The combination of BMP-2 and BMP-7 as osteogenic promoters has shown to be the most effective inducer of bone morphogenesis (Yilgor 2009). Yilgor et al. also showed that the incorporation of both growth factors leads to more effective differentiation than when added individually.

Although the relationship and interaction between VEGF and BMP has been thoroughly examined, angiogenesis' direct effect on osteogenesis is not yet fully understood. It is hypothesized that VEGF is able to elicit two stages (considered as early and late phases) of angiogenesis while BMP promotes osteogenesis (Hoeben et al. 2004). In addition, while undergoing osteogenic differentiation, MSCs secrete more BMP and VEGF than when cultured alone, enhancing both processes (Zhang et al. 2010).

However, even though the delivery of these growth factors is known to enhance vascularization and osteogenesis in constructs after implantation, their dosage and timing must be tightly controlled. Ozawa et al. determined that a sufficiently high local VEGF concentration of around $70 \, ng/10^6$ endothelial cells/day had a more positive impact on angiogenesis than the same concentration sustained over 28 days (Ozawa et al. 2004). In addition, the group discovered that above $100 \, ng/10^6$ endothelial cells/day resulted in unstable blood vessel formation. Excessive amounts of VEGF have also shown to lead to severe vascular leakage, tumor growth, and retinopathies in neighboring tissues (De la Riva et al. 2009). Geuze et al. explored the effect of controlled release BMP-2 and VEGF on bone formation in a large animal model when encapsulated in PLGA microparticles and implanted in critical-sized ulnar defects in dogs (Geuze et al. 2012). The group determined that ectopic bone formation was highly dependent on the dosage and speed of BMP-2 release, but independent of VEGF release.

13.6 Experimental Setups/Techniques

13.6.1 *In Vitro* Prevascularization

When cultured with angiogenic growth factors such as VEGF *in vitro*, endothelial cells are able to form prevascular structures before the construct is implanted *in vivo*. For this method, endothelial cells are usually added to the target tissue cell population, such that they are cocultured to create a prevascularized network within the tissue. After implantation, the network can then spontaneously anastomose with the host tissue's vascular system and allow for complete perfusion of the graft (Gibot et al. 2010). This method is advantageous because it does not rely on the slow integration of the surrounding blood vessel network, which often leads to tissue necrosis at the center of the graft. Instead, the preestablished endothelial system can directly connect to the surrounding network, allowing for faster blood perfusion. On the other hand, complete *in vivo* vascularization of the construct can still take days or even weeks because the construct's vascular system is not microsurgically connected to the surrounding network following implantation. The effectiveness of *in vitro* vascularization could be improved if microsurgical methods were used to help with the graft anastomosis.

Although prevascularization of endothelial cells has shown great promise for the formation of blood vessels, Ghanaati et al. demonstrated that the *in vitro* preculture period of osteoblast cells seeded on scaffolds may not play as prominent a role in vascularization as previously thought. The group cultured primary human osteoblasts (hOBs) on fibrin scaffolds for 24 h as well as 14 days before implanting the constructs subcutaneously in an immunodeficient mouse model. After 14 days *in vivo*, both groups showed significant scaffold vascularization. The 14-day preculture group resulted in significantly better penetration of the hOBs throughout the scaffold (Ghanaati et al. 2011), possibly leading to better neo-bone formation and dramatic enhancement of the host-derived vasculature. This phenomenon may be explained by the hOBs ability to create sufficient ECM and signaling factors during the preculture period, providing a strong pro-angiogenic stimulus once implanted *in vivo*.

13.6.2 *In Vivo* Prevascularization

Even though a variety of the body's complex physiological conditions may be mimicked *in vitro*, it does not provide a complete picture of a scaffold's potential and effect within a defect site. A successful *in vivo* implantation of a construct will be more revealing of the bone tissue-engineered capability to be translated into a relevant clinical application. The location and length of implantation can have a significant effect on the resulting growth and maturation of seeded cells. *In vivo* prevascularization can be completed using a variety of different techniques and the concept should be applicable to many engineered tissue types. In almost all cases, the purpose is to implant an endothelial cell- and target cell-seeded graft into a highly vascularized tissue, such as the muscle. Here, the graft will naturally become vascularized over a period of time, after which it can be harvested and implanted into the defect site.

13.6.2.1 *Cell Sheet Layering*

Another recently developed *in vivo* vascularization technique is cell sheet engineering. It utilizes the thermoresponsive properties of poly(*N*-isopropylacrylamide) (PIPAAm), which is a well-known temperature-activated polymer, to induce detachment of intact cell sheets (Okano et al. 1993). Cell sheet engineering was originally used for corneal surface reconstruction, blood vessel grafts, and myocardial tissue engineering (Lovett et al. 2009). Since then, the technique has also been utilized in some bone tissue engineering applications (Chen et al. 2012). A variety of different cells that secrete their own ECM can be cultured in the monolayer and then recovered within their own ECM without the use of a proteolytic enzyme, lifting as an entire sheet. Cell sheet-based tissue engineering has been applied for regenerative medicine for several different tissues, including myocardial, corneal epithelial, lung, and liver tissue (Shimizu et al. 2003; Nishida 2004; Kanzaki et al. 2007; Ohashi et al. 2007). Seeding the sheets on scaffolds allows for three-dimensional formation. To determine cellular behavior and study biomaterial immune response *in vivo*, the construct is subcutaneously implanted into animal models. The cell-sheet tissue-engineering technique is depicted in Figure 13.3. Pirraco et al. cocultured rat bone marrow stromal cells and HUVECs into a cell sheet using thermal-responsive culture dishes (Pirraco et al. 2011) and then implanted them subcutaneously into the dorsal flap of nude mice, demonstrating new bone formation as well as neovascularization by day 7. Others have used cell sheet layers to cover scaffolds before *in vivo* implantation. A range of scaffold materials have been used, including coral scaffolds (Dong et al. 2012), hydroxyapatite ceramic scaffolds

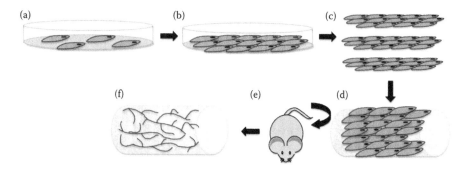

FIGURE 13.3
Cell sheet layering used for *in vivo* vascularization of tissue-engineered scaffold. Endothelial cells can be cultured into confluent cell sheets and used to cover tissue-engineered scaffolds to create a prevascularized network. (a) Endothelial cell can be cultured on special thermoresponsive tissue-culture plates. (b) When confluent, the cells will attach to each other via their ECM forming a cell sheet. (c) Several of such sheets can be combined and (d) seeded onto a scaffold. (e) The endothelial cell-sheet-covered scaffold can be implanted subcutaneously *in vivo*, resulting in (f) a prevascularized graft.

(Song et al. 2007), and polycaprolactone–calcium phosphate scaffolds (Zhang et al. 2010). Layering of cell sheets has also shown improved overall vascularization because one cell sheet is added at a time, with 1–3 days of subcutaneous vascularization culturing in between. Even though time consuming, this resulted in completely vascularized tissues after several weeks of *in vivo* culturing (Chen et al. 2012).

Even though this technique has been extensively utilized in myocardial tissue reconstruction (Shimizu et al. 2003), more research should be done to apply the technique to bone tissue vascularization where it is evident that it can provide efficient blood vessel formation within the bone tissue construct. However, further optimization of the coculture methods for repair of a bone defect would be necessary. For example, while both endothelial cells and osteoprogenitor cells have been shown to form cell sheets independently from each other, their behavior and function when cocultured together in such form has not yet been investigated. Similarly, the dense and intertwined cells within the sheet may inhibit vascularization and bone formation by the respective cell types. However, alternating cell type layers could be a possible solution for creating larger vascularized cell constructs. Lastly, a dynamic *in vitro* culture of layered cell sheets could further enhance the endothelial network formation as well as promote bone formation prior to implantation into an animal model.

13.6.2.2 Dorsal Skinfold Chamber

The implant observation chambers have been used for decades to monitor living tissues. They are frame structures surgically attached to an animal, which allow for continuous monitoring through a glass slide window. Since its invention, the dorsal skinfold chamber's use has been adapted to different animals and applications. Most advantageously, observation chambers allow for intravital microscopy of the microcirculation of the tissue (Lehr et al. 1993).

To create the dorsal skinfold chamber, a small incision is made in the dorsal region of the animal to remove the cutis skin layer, around 15 mm in mice. The remaining layers, consisting of the epidermis, subcutaneous tissue, and a thin layer of striated skin muscle are then covered with a coverslip and incorporated into a metal frame, usually composed of titanium or aluminum. The frame sandwiches the thin layers and keeps them in place using steel nuts as spacers, thus preventing compression of the nutritional blood vessels. Within these layers, pieces of biomaterial scaffold can be implanted to study their vascularization while cultured in the dorsal skinfold chamber. The cover-slipped window allows for intravital microscopy of the area and observation of the developing vascular network. Several groups have shown that the animal shows no sign of discomfort, including changes in feeding or sleeping habits after chamber implantation.

This experimental setup has been used to investigate a wide range of scaffold and cell combinations to optimize the vascularization of a tissue. For example, sphingosine 1-phosphate (S1P) is a bioactive phospholipid, which affects migration, proliferation, and survival of endothelial cells and osteoprogenitor cells (Sefcik et al. 2008) and has been tested using the dorsal skinfold chamber. The study evaluated its direct impact on the structural remodeling response of the vascularization and subsequent healing of a bone defect site and found that sustained release of S1P resulted in enhanced luminal diameter of formed vasculatures as well as increased bone formation after 2 and 6 weeks (Sefcik et al. 2008).

To image the blood vessels more effectively, a contrast agent is often injected, such as fluorescein isothiocyanate-labeled dextran (Lehr et al. 1993), which labels plasma. In addition, leukocytes can be traced after intravenous injection of fluorescent markers such as acridine orange (Lehr et al. 1993), to understand better the immune response to the implanted scaffold.

One of the major advantages of the dorsal skinfold chamber method is that it allows for repeated microscopic analysis over a long period of time, often lasting for several weeks, without causing any noticeable harm or discomfort to the animal. The subcutaneous location and finely striated tissue provides the ideal environment for the purpose of studying the development of a vascular network within a foreign biomaterial *in vivo*.

However, some limitations of the dorsal skinfold chamber method do exist. For example, due to the size limitations of the animal and the corresponding size of the surgical incision, the implanted engineered construct cannot exceed 5 mm in diameter to properly fit in the 15-mm-sized chamber (Laschke et al. 2006). In addition, the height, or thickness, of the construct cannot surpass 1 mm, so that it does not interfere with the closure of the chamber tissue by the coverslip. Such limitations may prevent the use of this *in vivo* technique for large-scale bone tissue applications. A thicker biomaterial will require a thicker layer of tissue within the chamber, inhibiting image quality. Laschke et al. (2010) utilized the dorsal skin chamber to evaluate the angiogenic response of its nano-sized hydroxyapatite particles mixed with and without poly(ester–urethane) and found that capillary sprouts were forming at the border of the scaffold, coming from the surrounding tissue, in both groups with no significant difference. Druecke et al. also utilized the dorsal skinfold method to test the neovascularization within a poly(ether ester) block–copolymer scaffold. Intravital fluorescent microscopy and quantitative data analysis of data showed that after 20 days of implantation, the microcirculation at the border of the construct resembled that of the surrounding tissue. However, the center of the graft showed significant leakage of plasma, indicating that the newly formed endothelial network was not yet mature (Druecke et al. 2004).

Several issues would have to be addressed if used for bone tissue-engineering applications. For example, beyond the size limitations, the dorsal skinfold chamber does not provide the same bone growth and signaling factors, nor does it provide any of the mechanical loading normally applied in a bone defect site. Therefore, this technique could be used as a prevascularization scaffold, but osteoprogenitor cells would require additional stimulation for bone formation.

13.6.2.3 Arteriovenous Loops

An arteriovenous (AV) loop is another common *in vivo* method used to prevascularize the scaffold prior to the injection or seeding of the target cells. This method utilizes the native blood vessel system by forming a shunt loop between an artery and a vein using a synthetic graft. This is then enclosed within a chamber that is either empty or houses a tissue-engineered scaffold, which will be vascularized over time (Dong et al. 2012). Within bone tissue engineering, several people have taken advantage of this efficient *in vivo* method to create a mature endothelial network within a scaffold (Kneser et al. 2006; Dong et al. 2012). When a PBCB matrix was implanted into an animal model using the AV loop as a vascular carrier, adequate vascular density and kinetics demonstrated that capillary sprouting was occurring in all parts of the graft, even at the center. After 8 weeks of *in vivo* culturing, the group was able to demonstrate the first successful vascularization of a solid porous matrix using the AV loop (AVL) technique, followed by bone formation throughout once implanted into a bone defect. Dong et al. evaluated vascularization of a natural coral block using the AV loop method in a rabbit model. Results showed that natural coral blocks, a biocompatible, and osteoconductive scaffold, can be vascularized using the AVL method and can be used for future bone substitutes following osteoblast seeding.

One of the disadvantages of the AVL technique, however, is that controlling the ingrowth pattern of fibrovascular cells is difficult. Therefore, the vascular network may dominate the scaffold structure, especially on the border, minimizing the space for target cell seeding. In addition, similar to the dorsal skinfold chamber method, following implantation and vascularization in the AV loop, the bone tissue-engineered scaffold will have to be removed for further culturing and transplantation into the actual bone defect, potentially resulting in donor site morbidity.

13.6.2.4 Chick Embryo Chorioallantoic Membrane

One of the innovative techniques used to accomplish prevascularization of a scaffold involves the use of the chorioallantoic membrane (CAM) of a chick egg (Figure 13.4). To obtain this thin membrane, which surrounds the inside of the shell, a fertilized and developing egg is incubated at physiological conditions for 3 days. The egg contains a special extraembryonic membrane that supports respiratory capillaries, ion exchange, and embryonic vasculature while the chick is developing. After a brief incubation, a circular window is made into the shell and is resealed with tape and cultured for an additional

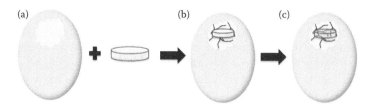

FIGURE 13.4
Scaffold prevascularization using the chick embryo CAM technique. Scaffolds can be prevascularized without the use of endothelial cells by implanting them into fertilized chick embryo eggs. (a) A small portion of the eggshell is removed, exposing the embryo's CAM. (b) A scaffold is then placed on top of the membrane and the shell is taped up. (c) After 8 days of culture, the scaffold will become vascularized as the chick embryo develops its own vascular network.

7 days. Then, the tissue construct is carefully implanted by placing it onto the surface of the CAM and then resealed it with cellotape and incubated for an additional 8 days, at which point, the scaffold is removed for analysis. This relative brief culturing period, compared to week-to-month-long periods required by other methods, is possible because the CAM undergoes significant capillary formation until day 20 of chick embryo development (Baiguera et al. 2012).

Steffens et al. performed a chick embryo CAM assay to evaluate the angiogenic potential of cell-seeded PBCB and found that vessel formation was visible after 8 days of implantation (Steffens et al. 2009). Similarly, Buschmann et al. used the CAM method to show rapid and homogeneous vascularization of human adipose-derived stem cells on PLGA/calcium phosphate nanoparticles scaffold (Buschmann et al. 2012).

However, one of the major disadvantages of the CAM method is that there may be a nonspecific inflammatory response to the implanted biomaterial, altering the seeded cell response and behavior. This hurdle may be overcome if the implantation is made very early in the development of the CAM, when its immune system is still relatively immature. In addition, the CAM method is not ideal for long-term monitoring of angiogenesis because the egg will continue to mature and prevent removal of the scaffold (Baiguera et al. 2012). The method of using chick CAMs from embryos could also be ethically controversial, although less so than the use of human embryonic stem cells. In addition, the combination of human cells with animal cells could also lead to disease transmission and rejection after the scaffold is explanted from the CAM system and implanted into a bone defect.

On the other hand, the CAM method could be advantageous because it provides the growth factors and microenvironment necessary to promote neovascularization, a process still poorly understood. The rapid vascularization of the system provides a simple, inexpensive, and effective method for evaluating the response of seeded scaffolds. With a few improvements to the current validation strategies used to evaluate the constructs while still implanted, CAM can be a great tool to study the biological response to biomaterials and cultured cells before moving onto large animal *in vivo* studies. In addition, it is a relatively easy assay and inexpensive to perform, requiring only minimally invasive techniques to obtain a prevascularized tissue-engineered construct.

13.7 Methods of Validation

After *in vivo* transplantation of cell and scaffold constructs, the development and functionality of microvascularization needs to be evaluated. The most common methods of evaluation include histological analysis, imaging techniques, or a combination of assays, imaging, and quantitative analysis.

13.7.1 Histological Analysis

Histology has become one of the most common methods of validation for all types of tissue-engineering applications. For the purpose of tracking and evaluating vascularization of bone tissue constructs, this method has been especially invaluable. It allows for clear visual validation of the blood vessel formation at the borders and the center of the graft. With protein-specific labeling, the implanted endothelial cells can be differentiated from those of the host tissue, further indicating the extent of neovascularization and integration into the surrounding tissue. Histologically processed tissues are most commonly fixed, embedded, and then sliced into micron-thick sections, after which they can be stained for proteins of interest using primary antibodies. Immunofluorescence has also become a popular technique as it allows for simultaneous staining and visualization of proteins within the tissue.

The bone formation progress can be tracked and visualized with the help of histological protein markers. Early osteogenic differentiation proteins include ALP and osteoprotegerin (OPN) (Song et al. 2007; Ilmer et al. 2009), while late markers, such as osteocalcin, osteopontin, and collagen type I are expressed during the latter phase of osteogenic differentiation (Zhang et al. 2010; Tsigkou et al. 2010). In addition, mineralization of the scaffold can be visualized using von Kossa or alizarin red stains, which are indicative of calcium deposition and early bone formation (Tsigkou et al. 2010; Zhang et al. 2010; Correia et al. 2011). The development of vascularization can also be easily followed by staining for specific angiogenic markers, such as CD31, CD34, and von Willebrand factor (vWF) (Pusztaszeri et al. 2006). Certain adhesion molecules, including vascular cellular adhesion molecule-1 (VCAM-1) and intercellular adhesion molecule-1 (ICAM-1) are indicative of the tight cell–cell junctions formed between endothelial cells (Nör et al. 2001) and increase in density as angiogenesis progresses.

The advantage of a histological analysis is that tissues can be easily visualized under bright-field or fluorescence microscopy and it can provide a three-dimensional representation based on different sections within the construct. One of the disadvantages of performing histology on *in vivo* cultured constructs, however, is that it is a terminal procedure. Therefore, the animal is usually sacrificed to remove the implanted scaffold after a certain period of time.

Histomorphometry is a subsection of histology where the specific change in morphology of a structure is especially important, allowing for a more macroscale analysis of the entire construct. Computer-aided software is able to quantify the formation of a new bone on the surface, as well as within the porous structure of the graft. More specifically, histomorphometric sections have also been used to quantify vessel diameter and vessel density in the scaffold (Roche et al. 2012). Images are taken at up to 200× magnification to see structures such as formed vessel cross sections, bone interface contact, and bone volume densities. This type of information can be used to evaluate bone growth, scaffold integration, and vascular formation. However, similar to the basic histology technique, this method is terminal and does not provide live and functional information.

13.7.2 Imaging Techniques

The combination of *in vivo* fluorescence microscopy and the dorsal skinfold chamber technique has proven to be invaluable to monitoring the inflammatory and angiogenic response of the body to a biomaterial. It also provides for a quantitative method to assess the formation of vasculature.

After capture, images can be quantitatively analyzed off-line using a variety of software packages. For example, leukocyte–endothelial interaction can be observed within the scaffold, indicating the acute inflammatory response to the implanted construct, by classifying it into adherent, rolling, or free-flowing cells (Laschke et al. 2010). In the case of nano-sized hydroxyapatite particles/poly(ester–urethane) scaffolds, the number of adherent leukocytes was comparable to the control poly(ester–urethane) scaffold, demonstrating that the body had an insignificant immune response to the implant.

Traditionally, immunostaining has been used to quantify and visualize induced vascularization *in vivo*. However, one drawback to this method is that it is a terminal procedure for the animal. Therefore, more advanced techniques have been developed that allow for functional measurement of the architecture and perfusion of the newly created vasculature.

Three-dimensional scaffolds have been scanned with resolutions of up to 10 µm using micro-CT (micro-computer tomography), giving enough detail to determine density, branching, and connectivity of the networks (Fei et al. 2010). Blood flow has also been studied using laser Doppler diffusion imaging (Phelps et al. 2010) as well as the use of transfected cells labeled with green fluorescent protein (GFP) that can be tracked using a fluorescence microscope (Tang et al. 2008).

Off-line analysis software has been extensively used to quantify microscopy images. Wernike et al. determined the functional capillary density of the scaffold postimplantation, defined as the length of perfused capillaries per area (Wernike et al. 2010). The group also looked at bone formation properties such as volume of bone pores, bone density, and the areas of interaction between the bone and vessels. In addition, vascular leakage has become an established parameter used to evaluate the success of vascularization because it indicates the maturity level of the neovasculature.

13.8 Concluding Remarks and Perspective

Overall, *in vivo* techniques have shown that integration of a bone tissue-engineered construct with the host vasculature can lead to successful anastomosis. Full or partial perfusion of the graft has been demonstrated, with endothelial sprouting and vessel formation shown after implantation.

However, the field of tissue engineering still faces some major challenges that need to be addressed before these approaches can be used for clinical applications. The importance of the cell source cannot be overlooked, especially due to the risk of infection and disease transmission. The method of prevascularization must also be carefully considered; *in vitro* preseeding of endothelial and osteoprogenitor cells onto scaffolds may lead to a basic layout for a vascular network of endothelial cells prior to *in vivo* implantation. On the other hand, an *in vivo* prevascularization technique may result in a denser vascular network, yet it requires an additional invasive surgical step. It is also important to note that if implanted for prevascularization for too long, scaffolds often degrade and become too highly integrated into the host tissue to remove.

Even though most of these bone-engineering techniques are performed *in vivo* to determine the effect of the unique microenvironment provided by the body, few actually study the direct effect of implantations into bone defects. While subcutaneous implantations into muscle pockets are valuable in terms of determining the immune- and cytotoxic effect of the constructs, they do not properly mimic the structural and dynamic environment of the bone. Experiments that transplant the tissue-engineered constructs into highly vascularized tissues such as the muscle, falsely provide high blood flow and the presence of a variety of different constructs that fail to mimic the environment of a bone defect. While successful neo-bone and vascular network formations have been demonstrated in these experimental setups, an additional transplantation into the actual bone defect site might lead to site morbidity and increased risk of infection. These *in vivo* techniques also result in a longer culturing time. Before cells can be seeded onto the scaffolds, they are cultured and conditioned *in vitro* first. Once implanted, it can take weeks to months for the entire scaffold to be completely vascularized.

Some of the advantages of *in vivo* vascularization techniques include a more accurate microenvironment to evaluate the scaffold and cell construct interactions with the surrounding tissue environment compared to the current *in vitro* methods. Even though they have significantly improved the culturing environment to closely resemble that found *in vivo*, there are still many factors that are poorly understood. Small-scale *in vivo* studies, such as the CAM assay, can provide preliminary, yet critical data on bone tissue-engineering constructs prior to larger *in vivo* experiments.

While the field of bone tissue engineering has made tremendous strides toward bringing benchtop applications toward clinical use via the use of *in vivo* experiments, there is much left to do until a fully vascularized bone construct can successfully heal a bone defect in a patient.

Acknowledgments

Research reported in this publication was supported by the National Institutes of Health under Award Number R01 AR061460. The content is solely the responsibility of the authors and does not necessarily represent the official views of the National Institutes of Health. Additionally, this work was supported by the National Science Foundation Graduate Research Fellowships (to BNBN).

References

Abshagen, K. et al., 2009. *In vivo* analysis of biocompatibility and vascularization of the synthetic bone grafting substitute nanobone. *Journal of Biomedical Materials Research. Part A*, 91(2), 557–566.

Baiguera, S., Macchiarini, P., and Ribatti, D., 2012. Chorioallantoic membrane for *in vivo* investigation of tissue-engineered construct biocompatibility. *Journal of Biomedical Materials Research. Part B, Applied Biomaterials*, 100(5), 1425–1434.

Bose, S., Roy, M., and Bandyopadhyay, A., 2012. Recent advances in bone tissue engineering scaffolds. *Trends in Biotechnology*, 30(10), 546–554.

Buschmann, J. et al., 2012. Tissue engineered bone grafts based on biomimetic nanocomposite PLGA/amorphous calcium phosphate scaffold and human adipose-derived stem cells. *Injury*, 43(10), 1689–1697.

Chen, F. et al., 2007. Engineering tubular bone constructs. *Journal of Biomechanics*, 40 Suppl 1(null), S73–S79.

Chen, Y., Zhou, N., and Huang, X., 2012. Cell sheet technology and its application in bone tissue engineering. *Zhongguo xiu fu chong jian wai ke za zhi=Zhongguo xiufu chongjian waike zazhi=Chinese Journal of Reparative and Reconstructive Surgery*, 26(9), 1122–1125.

Correia, C. et al., 2011. *In vitro* model of vascularized bone: Synergizing vascular development and osteogenesis. R. Goncalves, ed. *PloS One*, 6(12), e28352.

Dejana, E., 2004. Endothelial cell–cell junctions: Happy together. *Nature Reviews. Molecular Cell Biology*, 5(4), 261–270.

Domev, H. et al., 2012. Efficient engineering of vascularized ectopic bone from human embryonic stem cell-derived mesenchymal stem cells. *Tissue Engineering. Part A*, 18(21–22), 2290–2302.

Dong, Q. et al., 2012. Prefabrication of axial vascularized tissue engineering coral bone by an arteriovenous loop: A better model. *Material Science and Engineering: C*, 32(6), 1536–1541.

Druecke, D. et al., 2004. Neovascularization of poly(ether ester) block–copolymer scaffolds *in vivo*: Long-term investigations using intravital fluorescent microscopy. *Journal of Biomedical Materials Research. Part A*, 68(1), 10–18.

Evensen, L. et al., 2009. Mural cell associated VEGF is required for organotypic vessel formation. Y. Cao, ed. *PloS One*, 4(6), e5798.

Fedorovich, N.E. et al., 2010. The role of endothelial progenitor cells in prevascularized bone tissue engineering: Development of heterogeneous constructs. *Tissue Engineering. Part A*, 16(7), 2355–2367.

Fei, J. et al., 2010. Imaging and quantitative assessment of long bone vascularization in the adult rat using microcomputed tomography. *Anatomical Record (Hoboken, NJ: 2007)*, 293(2), 215–224.

Geris, L. et al., 2010. Mechanical loading affects angiogenesis and osteogenesis in an *in vivo* bone chamber: A modeling study. *Tissue Engineering. Part A*, 16(11), 3353–3361.

Geuze, R.E. et al., 2012. A differential effect of bone morphogenetic protein-2 and vascular endothelial growth factor release timing on osteogenesis at ectopic and orthotopic sites in a large-animal model. *Tissue Engineering. Part A*, 18(19–20), 2052–2062.

Ghanaati, S. et al., 2011. Scaffold vascularization *in vivo* driven by primary human osteoblasts in concert with host inflammatory cells. *Biomaterials*, 32(32), 8150–8160.

Gibot, L. et al., 2010. A preexisting microvascular network benefits *in vivo* revascularization of a microvascularized tissue-engineered skin substitute. *Tissue Engineering. Part A*, 16(10), 3199–3206.

Goldstein, S.A., 2002. Tissue engineering: Functional assessment and clinical outcome. *Annals of the New York Academy of Sciences*, 961, 183–192.

Grellier, M. et al., 2009. The effect of the co-immobilization of human osteoprogenitors and endothelial cells within alginate microspheres on mineralization in a bone defect. *Biomaterials*, 30(19), 3271–3278.

Hegen, A. et al., 2011. Efficient *in vivo* vascularization of tissue-engineering scaffolds. *Journal of Tissue Engineering and Regenerative Medicine*, 5(4), e52–e62.

Hoeben, A. et al., 2004. Vascular endothelial growth factor and angiogenesis. *Pharmacological Reviews*, 56(4), 549–580.

Huang, Y.-C. et al., 2005. Combined angiogenic and osteogenic factor delivery enhances bone marrow stromal cell-driven bone regeneration. *Journal of Bone and Mineral Research: The Official Journal of the American Society for Bone and Mineral Research*, 20(5), 848–857.

Huang, Z. et al., 2010. Modulating osteogenesis of mesenchymal stem cells by modifying growth factor availability. *Cytokine*, 51(3), 305–310.

Hutmacher, D.W., Sittinger, M., and Risbud, M.V., 2004. Scaffold-based tissue engineering: Rationale for computer-aided design and solid free-form fabrication systems. *Trends in Biotechnology*, 22(7), 354–362.

Ilmer, M. et al., 2009. Human osteoblast-derived factors induce early osteogenic markers in human mesenchymal stem cells. *Tissue Engineering. Part A*, 15(9), 2397–2409.

Jain, R.K. et al., 2005. Engineering vascularized tissue. *Nature Biotechnology*, 23(7), 821–823.

Jaklenec, A. et al., 2012. Progress in the tissue engineering and stem cell industry "are we there yet?". *Tissue Engineering. Part B, Reviews*, 18(3), 155–166.

Kanzaki, M. et al., 2007. Dynamic sealing of lung air leaks by the transplantation of tissue engineered cell sheets. *Biomaterials*, 28(29), 4294–4302.

Kim, B.-S. and Mooney, D.J., 1998. Development of biocompatible synthetic extracellular matrices for tissue engineering. *Trends in Biotechnology*, 16(5), 224–230.

Kirkpatrick, C.J., Fuchs, S., and Unger, R.E., 2011. Co-culture systems for vascularization—Learning from nature. *Advanced Drug Delivery Reviews*, 63(4–5), 291–299.

Kneser, U. et al., 2006. Engineering of vascularized transplantable bone tissues: Induction of axial vascularization in an osteoconductive matrix using an arteriovenous loop. *Tissue Engineering*, 12(7), 1721–1731.

Koyama, S. et al., 2002. Vascular endothelial growth factor mRNA and protein expression in airway epithelial cell lines *in vitro*. *The European Respiratory Journal: Official Journal of the European Society for Clinical Respiratory Physiology*, 20(6), 1449–1456.

De la Riva, B. et al., 2009. VEGF-controlled release within a bone defect from alginate/chitosan/PLA-H scaffolds. *European Journal of Pharmaceutics and Biopharmaceutics: Official Journal of Arbeitsgemeinschaft für Pharmazeutische Verfahrenstechnik e.V*, 73(1), 50–58.

Lanza, R., Langer, R., and Vacanti, J.P., 2007. *Principles of Tissue Engineering*, 3rd ed., Burlington, MA: Elsevier Academic Press.

Laschke, M.W. et al., 2006. Angiogenesis in tissue engineering: Breathing life into constructed tissue substitutes. *Tissue Engineering*, 12(8), 2093–2104.

Laschke, M.W. et al., 2010. *In vitro* and *in vivo* evaluation of a novel nanosize hydroxyapatite particles/poly(ester–urethane) composite scaffold for bone tissue engineering. *Acta Biomaterialia*, 6(6), 2020–2027.

Laschke, M.W. et al., 2009. *In vivo* biocompatibility and vascularization of biodegradable porous polyurethane scaffolds for tissue engineering. *Acta Biomaterialia*, 5(6), 1991–2001.

Lazarous, D.F. et al., 1996. Comparative effects of basic fibroblast growth factor and vascular endothelial growth factor on coronary collateral development and the arterial response to injury. *Circulation*, 94(5), 1074–1082.

Lehr, H.A. et al., 1993. Dorsal skinfold chamber technique for intravital microscopy in nude mice. *The American Journal of Pathology*, 143(4), 1055–1062.

Levenberg, S. et al., 2002. Endothelial cells derived from human embryonic stem cells. *Proceedings of the National Academy of Sciences of the United States of America*, 99(7), 4391–4396.

Li, Z. and Guan, J., 2011. Hydrogels for cardiac tissue engineering. *Polymers*, 3, 740–761.

Lin, Y. et al. 2000. Origins of circulating endothelial cells and endothelial outgrowth from blood. *The Journal of Clinical Investigation*, 105(1), 71–77.

Liu, X., Holzwarth, J.M., and Ma, P.X., 2012. Functionalized synthetic biodegradable polymer scaffolds for tissue engineering. *Macromolecular Bioscience*, 12(7), 911–919.

Lovett, M. et al., 2009. Vascularization strategies for tissue engineering. *Tissue Engineering. Part B, Reviews*, 15(3), 353–370.

Ma, J. et al., 2011. Coculture of osteoblasts and endothelial cells: Optimization of culture medium and cell ratio. *Tissue Engineering. Part C, Methods*, 17(3), 349–357.

Marolt, D. et al., 2012. Engineering bone tissue from human embryonic stem cells. *Proceedings of the National Academy of Sciences of the United States of America*, 109(22), 8705–8709.

Meinel, L. et al., 2004. Bone tissue engineering using human mesenchymal stem cells: Effects of scaffold material and medium flow. *Annals of Biomedical Engineering*, 32(1), 112–122.

Moon, J.J. et al., 2010. Biomimetic hydrogels with pro-angiogenic properties. *Biomaterials*, 31(14), 3840–3847.

Nerem, R.M., 2006. Tissue engineering: The hope, the hype, and the future. *Tissue Engineering*, 12(5), 1143–1150.

Nishi, M. et al., 2013. Engineered bone tissue associated with vascularization utilizing a rotating wall vessel bioreactor. *Journal of Biomedical Materials Research. Part A*, 101(2), 421–427.

Nishida, K., 2004. Corneal reconstruction with tissue-engineered cell sheets composed of autologous oral mucosal epithelium—*NEJM*. *New England Journal of Medicine*, 351(12), 1187–1196.

Nör, J.E. et al., 2001. Engineering and characterization of functional human microvessels in immunodeficient mice. *Laboratory Investigation*, 81(4), 453–463.

Ohashi, K. et al., 2007. Engineering functional two- and three-dimensional liver systems *in vivo* using hepatic tissue sheets. *Nature Medicine*, 13(7), 880–885.

Okano, T. et al., 1993. A novel recovery system for cultured cells using plasma-treated polystyrene dishes grafted with poly(*N*-isopropylacrylamide). *Journal of Biomedical Materials Research*, 27(10), 1243–1251.

Ozawa, C.R. et al., 2004. Microenvironmental VEGF concentration, not total dose, determines a threshold between normal and aberrant angiogenesis. *The Journal of Clinical Investigation*, 113(4), 516–527.

Peng, H. et al., 2005. VEGF improves, whereas sFlt1 inhibits, BMP2-induced bone formation and bone healing through modulation of angiogenesis. *Journal of Bone and Mineral Research: The Official Journal of the American Society for Bone and Mineral Research*, 20(11), 2017–2027.

Phelps, E.A. et al., 2010. Bioartificial matrices for therapeutic vascularization. *Proceedings of the National Academy of Sciences of the United States of America*, 107(8), 3323–3328.

Phelps, E.A. and García, A.J., 2010. Engineering more than a cell: Vascularization strategies in tissue engineering. *Current Opinion in Biotechnology*, 21(5), 704–709.

Pirraco, R.P. et al., 2011. Development of osteogenic cell sheets for bone tissue engineering applications. *Tissue Engineering. Part A*, 17(11–12), 1507–1515.

Pittenger, M.F., 1999. Multilineage potential of adult human mesenchymal stem cells. *Science*, 284(5411), 143–147.

Pusztaszeri, M.P., Seelentag, W., and Bosman, F.T., 2006. Immunohistochemical expression of endothelial markers CD31, CD34, von Willebrand factor, and Fli-1 in normal human tissues. *The Journal of Histochemistry and Cytochemistry: Official Journal of the Histochemistry Society*, 54(4), 385–395.

Ribatti, D., 2005. The crucial role of vascular permeability factor/vascular endothelial growth factor in angiogenesis: A historical review. *British Journal of Haematology*, 128(3), 303–309.

Roche, B. et al., 2012. Structure and quantification of microvascularisation within mouse long bones: What and how should we measure? *Bone*, 50(1), 390–399.

Rouwkema, J., De Boer, J., and Van Blitterswijk, C.A., 2006. Endothelial cells assemble into a 3-dimensional prevascular network in a bone tissue engineering construct. *Tissue Engineering*, 12(9), 2685–2693.

Rouwkema, J., Rivron, N.C., and Van Blitterswijk, C.A., 2008. Vascularization in tissue engineering. *Trends in Biotechnology*, 26(8), 434–441.

Schechner, J.S. et al., 2000. *In vivo* formation of complex microvessels lined by human endothelial cells in an immunodeficient mouse. *Proceedings of the National Academy of Sciences of the United States of America*, 97(16), 9191–9196.

Sefcik, L.S. et al., 2008. Sustained release of sphingosine 1-phosphate for therapeutic arteriogenesis and bone tissue engineering. *Biomaterials*, 29(19), 2869–2877.

Shimizu, T. et al., 2003. Cell sheet engineering for myocardial tissue reconstruction. *Biomaterials*, 24(13), 2309–2316.

Shin, M., Yoshimoto, H., and Vacanti, J.P., 2004. *In vivo* bone tissue engineering using mesenchymal stem cells on a novel electrospun nanofibrous scaffold. *Tissue Engineering*, 10(1–2), 33–41.

Sikavitsas, V.I. et al., 2003. Influence of the *in vitro* culture period on the *in vivo* performance of cell/titanium bone tissue-engineered constructs using a rat cranial critical size defect model. *Journal of Biomedical Materials Research. Part A*, 67(3), 944–951.

Sittichokechaiwut, A. et al., 2010. Short bouts of mechanical loading are as effective as dexamethasone at inducing matrix production by human bone marrow mesenchymal stem cells. *European Cells and Materials*, 20, 45–57.

Song, S.J. et al., 2007. Effects of culture conditions on osteogenic differentiation in human mesenchymal stem cells. *Journal of Microbiology and Biotechnology*, 17(7), 1113–1119.

Steffens, L. et al., 2009. *In vivo* engineering of a human vasculature for bone tissue engineering applications. *Journal of Cellular and Molecular Medicine*, 13(9B), 3380–3386.

Tang, Y. et al., 2008. Combination of bone tissue engineering and BMP-2 gene transfection promotes bone healing in osteoporotic rats. *Cell Biology International*, 32(9), 1150–1157.

Tsigkou, O. et al., 2010. Engineered vascularized bone grafts. *Proceedings of the National Academy of Sciences of the United States of America*, 107(8), 3311–3316.

Villars, F. et al., 2000. Effect of human endothelial cells on human bone marrow stromal cell phenotype: Role of VEGF? *Journal of Cellular Biochemistry*, 79(4), 672–685.

Weinand, C. et al., 2007. Comparison of hydrogels in the *in vivo* formation of tissue-engineered bone using mesenchymal stem cells and beta-tricalcium phosphate. *Tissue Engineering*, 13(4), 757–765.

Wernike, E., Montjovent, M.O., and Liu, Y., 2010. VEGF incorporated into calcium phosphate ceramics promotes vascularisation and bone formation *in vivo*. *European Cells and Materials*, 19, 30–40.

Yilgor, P., 2009. Incorporation of sequential BMP-2/BMP-7 delivery system into chitosan-based scaffolds for bone tissue engineering. *Biomaterials*, 30(21), 3551–3559.

Yu, Y. et al., 2012. Insulin-like growth factor 1 enhances the proliferation and osteogenic differentiation of human periodontal ligament stem cells via ERK and JNK MAPK pathways. *Histochemistry and Cell Biology*, 137(4), 513–525.

Zhang, Z.-Y. et al., 2010. Neo-vascularization and bone formation mediated by fetal mesenchymal stem cell tissue-engineered bone grafts in critical-size femoral defects. *Biomaterials*, 31(4), 608–620.

14

Vascularization of Encapsulated Cells

Shruti Balaji, John Patrick McQuilling, Omaditya Khanna, Eric Michael Brey, and Emmanuel C. Opara

CONTENTS

14.1 Introduction

About one-eleventh of a healthy human adult body is composed of the liquid tissue—blood (Elert, 2012), with the blood vessels as conduits that carry blood to the farthest reaches of the body. This underscores the important role blood vessels play in maintaining normal tissue homeostasis. In fact, their importance is best elucidated in diseased tissue that lack normal blood supply because of either damage to the blood vessels due to injury or the blockade of the blood vessel by atherosclerotic plaques. The loss of blood supply to such tissue leads to the tissue becoming necrotic due to the death of cells under hypoxic and nutrient stresses. It is thus apparent that adequate vascularization of implanted tissue is one of the major factors that will predict graft survival.

Vascularization is the invasion of tissue with blood vessels. When a new blood vessel or a network thereof is formed, it is known as neovascularization. One of the approaches for promoting new blood vessel formation upon tissue implantation is via the delivery of angiogenic growth factors such as vascular endothelial growth factor (VEGF), fibroblast growth factor (FGF), and platelet-derived growth factor (PDGF). Current clinical methods of bolus growth factor administration are not ideal because they require a high loading dose owing to the short half-life of many of these proteins *in vivo*, resulting in severe side effects (Gu et al., 2004). In fact, high dosages of VEGF were shown to lead to disorganized vascular structure and hyperpermeable vessels in one study (Ozawa et al., 2004). The efficacy of growth factors thus depends not only on administering the optimal dose but also on the manner in which they are administered. For example, studies have shown that sustained delivery of optimal doses over time results in adequate maintenance of stable vessel structure. Thus, continuous delivery of low levels of FGF-1 has been shown to promote new vessel formation and maintain the vascular network over time, whereas higher levels resulted in abnormal microvasculature structure formation that is accompanied by regression (Uriel et al., 2006).

The process of vascularization is complex as various tissues have different vascularization needs at different stages of development. For instance, in the developing central nervous system (CNS), embryonic blood vessels must first surround and grow outside of the tissue before penetrating the anatomical barriers of the glia and basal lamina. Once the barriers have been crossed, the blood vessels must grow within and among the developing neural tissue, all the while adapting to changes in its structural and functional needs as the mature tissue structure and function is established (Yasargil 1987). This model of peri- (outside), inter- (in between), and intra- (within) vascularization is also true of any other tissue that is implanted within the host body. This, however, also provides a route for immune recognition of foreign tissue, thus leading to either graft rejection or the need for lifelong immunosuppression therapies.

Cell encapsulation has thus become an increasingly prominent pillar of research in the field of tissue engineering due, in large part, to the need to overcome the preexisting limitation of graft rejection. Cell encapsulation therapies involve immobilizing cells within a polymeric biomaterial such that it allows for the inward diffusion of oxygen, growth factors, and other essential nutrients, while concomitantly permitting the efflux of therapeutic proteins and metabolites. As such, cell encapsulation technologies have the potential to deliver the same therapeutic benefits as organ transplants, while obviating the need for long-term immunosuppression therapy.

Vascularization of encapsulated cells becomes a complex process because certain "signals" (such as cytokines) need to be released into the surrounding circulation in order to initiate the process of neovascularization, and the addition of factors such as VEGF and FGF-1 within the capsule has only been described in the last decade (Sigrist et al., 2003; Moya et al., 2009). With immunoisolation being one of the overarching goals for tissue encapsulation, the second challenge in this field is that the pore size of the capsules should be large enough to allow for the diffusion of metabolites and therapeutic molecules produced by the encapsulated cells but small enough to prevent immune cells and molecules from crossing the capsule barrier. This would mean that any new blood vessels that form in the vicinity of the implanted capsules should be close enough to provide sufficient diffusion of oxygen, nutrients, and waste products to support and maintain viable tissue without penetrating the capsule.

This chapter will focus on the different types of cells and tissues that have been encapsulated to date and the challenges associated with their vascularization. The techniques that are being tested to surmount these challenges will also be discussed.

14.2 Encapsulation of Cells for Therapeutic Value

Immunoisolation via microencapsulation is a promising strategy for the implantation of allo- or xenogeneic cells for the long-term release of therapeutic agents. This strategy has been applied to a wide number of diseases and disorders, including heart disease, neurological disorders, endocrine disorders, cancer, kidney failure, and liver failure (see Table 14.1). While the number of therapies that utilize microencapsulation continues to increase, there are several challenges that limit the clinical success of many of these therapies. Many of these therapies have encountered technical difficulties due to mass transport problems, where a lack of oxygen or nutrients prevents long-term cell survival. In this section, we have highlighted a number of diverse microencapsulation therapies of which many are currently in clinical trials.

14.2.1 Microencapsulation as a Means of Treating Endocrine Disorders

14.2.1.1 Type 1 Diabetes Mellitus

Encapsulated islet transplantation for the regulation of blood glucose levels is a pioneering and classical example of the application of the technique of immunoisolation in medical treatment. The concept was introduced by Lim and Sun in 1980 and this was subsequently followed by a number of studies, including those in which microencapsulated islets were tested in large animals (Sun et al., 1996). The large animal studies have shown that normoglycemia can be established using this technique, but only for a limited amount of time (Soon-Shiong et al., 1992; Sun et al., 1996; Tashiro et al., 1997; Kendall et al., 2001; Dufrane et al., 2006a; Wang et al., 2008). Both mathematical modeling (Tziampazis and Sambanis, 1995; Buladi et al., 1996; Avgoustiniatos and Colton, 1997; Dulong and Legallais, 2007; Gross et al., 2007; Buchwald, 2009, 2011) and experimental evidence (Davalli et al., 1996; Avgoustiniatos and Colton, 1997; Moritz et al., 2002; Giuliani et al., 2005; Morini et al., 2006; Emamaullee and Shapiro, 2007) suggest that one of the main reasons for the limited survival of islet grafts is hypoxia. A number of strategies that are being investigated to reduce hypoxic stress in implanted microencapsulated islets will be discussed in detail later. However, it is noteworthy that microencapsulated islet transplantation as a treatment for type 1 diabetes is currently in phase 2/3 clinical trials by Living Cell Technologies (Auckland, New Zealand).

14.2.1.2 Hypoparathyroidism

Surgically acquired hypoparathyroidism and idiopathic hypoparathyroidism are two conditions that require lifelong vitamin D and calcium replacement therapy. However, this

TABLE 14.1

The Various Diseases Treated So Far by Microencapsulation Therapy

Disease State	Encapsulated Cell Type	Hormone Produced	Transplant Location
Type 1 diabetes mellitus	Islets	Insulin, C-peptide	Peritoneal cavity (Calafiore et al., 2006; Elliott et al., 2000), omentum (Kin et al., 2003; Kobayashi et al., 2006; Moya et al., 2010b,c; Opara et al., 2010), kidney capsule (Dufrane et al., 2006b), liver (Toso et al., 2005), subcutaneously (Dufrane et al., 2006b)
Hypoparathyroidism	Parathyroid	Parathyroid hormone	Brachioradial muscle (Hasse et al., 1997), forearm and leg (Cabané et al., 2009)
Parkinson's disease	Adrenal pheochromocytoma (PC12)	Dopamine	Caudate putamen (Vallbacka et al., 2001)
	Transfected myoblasts (BHK-GDNF)	Glial cell-derived neurotrophic factor (GDNF)	Striatum (Yasuhara et al., 2005)
Huntington's disease	Genetically modified baby hamster kidney (BHK)	Ciliary neurotrophic factor (CNTF)	Right lateral ventricle (Bloch et al., 2004)
Stroke (postinfarction)	Transfected kidney cells	Brain-derived neurotrophic factor (BDNF)	Left cerebrum (Katsuragi et al., 2005)
	Choroid plexus (CP) cells	BDNF, nerve growth factor (NGF), neurotrophin-3 (NT-3), fibroblast growth factor (FGF)	Both hemispheres of brain in 3–4 mm columns (Skinner et al., 2009)
	Transfected stem cells	Glucagon-like peptide-1	At infarct (Wallrapp et al., 2013)
Neuropathic pain	Adrenal chromaffin cells	Catecholamines, metenkephalin	Intrathecally (Jeon et al., 2006; Livett et al., 1981)
Alzheimer's disease	NGC0295 cells expressing human NGF	NGF	Basal forebrain nuclei (Eriksdotter-Jonhagen et al., 2012)
	Bone marrow-derived mesenchymal stem cells (MSC)	Glucagon-like peptide 1 (GLP-1)	Right lateral ventricle (Klinge et al., 2011)
	C2C12 myoblast cell line	CNTF	Right ventricle or above the cornus amonis 1 (Garcia et al., 2010)
Heart disease	Chinese hamster ovary (CHO) cells	VEGF	Directly into infarcted myocardium (Zhang et al., 2008)
	MSC	GLP-1	Coronary artery branches (Wright et al., 2012)

therapy is unable to reverse lowered calcium reabsorption and excessive urinary calcium excretion, which can result in urinary stones. Replacement of lost parathyroid function via transplantation of microencapsulated allo- or xenogenic tissue is one method that has been investigated to overcome this condition. A small animal study by Hasse et al. (1994) has demonstrated successful iso-, allo-, and xenotransplantation of microencapsulated parathyroid tissue without immunosuppression. Additionally, microencapsulated parathyroid tissue has been shown to have some success in achieving normal levels of calcium and parathyroid hormone (PTH) in humans for up to 3 months (Hasse et al., 1997; Cabané et al., 2009). However, there has not yet been any success in maintaining PTH in humans for more than 3 months. In the studies mentioned above, graft failure was, in part, a result of the use of nonpurified alginate for microencapsulation, which resulted in fibroblast infiltration. Additionally, tissue necrosis postimplantation due to hypoxic or deficient nutrient supply resulted in the need for a surplus of tissue. In small animal studies, a donor-to-recipient ratio of 10:1 was required in order to achieve physiological levels of calcium (Hasse et al., 1994).

14.2.2 Microencapsulation as a Means of Treating Neurological Disorders

14.2.2.1 Parkinson's Disease

Parkinson's disease (PD) is a progressive neurologic syndrome typically resulting from the deficiency of dopamine due to the destruction of dopaminergic neurons. The current treatment for PD is replacement of dopamine through the oral administration of levodopa (L-dopa), the precursor to dopamine. However, this therapy eventually loses effectiveness over time due to tolerance and increased sensitivity of cells to dopamine leading to the "on–off" effect. In addition, transport across the blood–brain barrier may be compromised by competition from other neural amino acids. A more permanent treatment of PD would, therefore, be to directly deliver dopamine to the brain through the implantation of microencapsulated dopamine-producing cells.

In a study by Vallbacka et al. (2001), up to 120 microcapsules containing the adrenal pheochromocytoma (PC12) cell line were delivered via cannula directly into the caudate putamen of rats. Evaluation of the capsules after 3 weeks *in vivo* demonstrated that the capsules were able to maintain their shape with minimal host response and many of the encapsulated cells were viable. However, it was found that the cells at the center of the implanted clusters did not show evidence of producing tyrosine hydroxylase, the rate-limiting enzyme required for the synthesis of dopamine. This indicates that the cells were not able to receive the needed nutrients or growth factors to maintain functionality.

14.2.2.2 Huntington's Disease

Huntington's disease is a genetic, neurodegenerative disease, which is the result of the expression of the mutated form of the protein huntingtin. While no treatment for this disorder is currently available, research has shown that the nerve growth factor (NGF) and ciliary neurotropic factor (CNTF) are able to preserve certain populations of striatal neurons, the neurons that are affected in this disease (Hefti, 1994). Unfortunately, delivery of these factors is difficult since CNTF cannot cross the blood–brain barrier and, therefore, it must be administered locally. Additionally, owing to the poor stability of CNTF, traditional drug delivery options such as minipumps are undesirable. Microencapsulated cell transplantation is thus a promising procedure that can provide a sustained and localized delivery of CNTF without requiring a storage reservoir. In a series of phase 1 clinical trials, genetically modified, CNTF-producing, baby hamster kidney (BHK) cells were perm-selectively macroencapsulated and transplanted intraventricularly in six individuals (Bloch et al., 2004). The results from the study indicated that the implanted microcapsules were able to produce CNTF for up to 6 months, at which point the microcapsules were retrieved and replaced. Additionally, several subjects receiving the implants showed significant electrophysiological improvements after transplantation. However, owing to variability in cell survival and consequently CNTF secretion by the implants, further improvement to the design of the microcapsules is needed. The high cell death in certain microcapsules was attributed to lack of contact inhibition in BHK cells. When these cells are placed *in vivo* with limited amounts of nutrients and oxygen, the high level of cell death leads to necrotic debris reducing the available space for living cells.

14.2.2.3 Alzheimer's Disease

Alzheimer's disease (AD) is one of the most common forms of dementia in individuals over the age of 60, and unfortunately there is little understanding of the underlying causes and mechanisms of this condition. Currently, there are no completely effective treatments and the current therapies simply include cholinesterase inhibitors for improving cognitive function. Recent research indicates that neurotrophins, including brain-derived neurotrophic factor (BDNF) and neurotrophin receptors in the forebrain are significantly reduced during late-stage AD (Zanin et al., 2012). Based on this knowledge, a number of strategies that utilize neurotrophin release from microencapsulated cells have been investigated. In particular, encapsulated cells that secrete CNTF (Garcia et al., 2010), GLP-1 (Klinge et al., 2011), and NGF (Emerich et al., 1994, 2006) have been studied. Recently, NsGene conducted phase 1b clinical trials using encapsulated NGF-secreting cells (Eriksdotter-Jonhagen et al., 2012). In these trials, four capsules were implanted into the basal forebrain nuclei of each patient for 12 months. The results from this study indicated that the encapsulated cells could be safely implanted and removed. However, as in other encapsulated cell therapies, there was a significant drop in NGF production by the encapsulated cells after the 12 month period.

14.2.2.4 Stroke

There are several microencapsulated cell therapies that aim to preserve neurons following a stroke. These devices include cells that secrete BDNF, neurotrophin 3 (NT-3), as well as a combination of growth factors such as FGF proteins (Heile et al., 2009; Skinner et al., 2009; Zanin et al., 2012). Phase 2 clinical trials are underway for an encapsulated cell therapy known as CellBeads, which releases GLP-1, NT-3, and VEGF simultaneously (Wallrapp et al., 2013).

14.2.3 Microencapsulation as a Means of Treating Heart Disease

There are several different therapies that utilize microencapsulation for heart regeneration currently. Stem cell transplantation to the heart following an ischemic event has the potential to improve heart function through several different mechanisms, including neovascularization; however, the administration of these cells has been a problem. Owing to the mechanical activity of the heart, only a small fraction of implanted cells remain in the heart (Hernandez et al., 2010). One solution to immobilizing cells is encapsulation, and in a study by Paul et al. (2009), they were able to retain significantly more stem cells within the heart compared to unencapsulated cells. Zhang et al. (2008) have shown improved angiogenesis and heart function in postinfarcted rats transplanted with microencapsulated CHO cells, which secreted VEGF. Encapsulated mesenchymal stem cells (MSCs) engineered to secrete GLP-1 fusion protein have also been shown to increase angiogenesis and restore left ventricular function in a porcine model of early ischemic left ventricular dysfunction (Wright et al., 2012).

14.3 Role of Encapsulation in the Emerging Field of Stem Cells

In a broad sense, stem cells are naïve cells that possess the dual properties of self-renewal and the ability to differentiate into various mature cell types given the appropriate environmental cues and conditions. They may be totipotent, multipotent, or unipotent based on the number of cell types they can differentiate into. Embryonic stem cells (ESCs), MSCs, and induced pluripotent stem cells (iPS) garner the most attention because of their multipotency, that is, they can differentiate into cells of the three different germ layers. It is this very property that makes them an attractive option for tissue transplantation and repair in the field of regenerative medicine.

Proof of concept has been provided in many cases from the differentiation of ESCs into islet-like cells (D'Amour et al., 2006) and cardiomyocytes (Otsuji et al., 2010), MSCs from various sources into neuronal cells (Mitchell et al., 2003; Fu et al., 2006) and insulin-producing cells (Chao et al., 2008), and iPS into dopamine-producing neurons (Wernig et al., 2008) and endothelial cells (Xu et al., 2009) among others. While most of these studies have been carried out in a 2D environment (as suspension or monolayer cultures), the need for 3D cultures of the same in order to mimic the physiological environment in which these cells would normally grow and differentiate is currently being explored and the encapsulation of cells offers one approach for this purpose. Encapsulation also prevents the cells from migrating from the site of administration, thus ensuring that their therapeutic value is sustained for a longer period of time. Finally, encapsulation that provides perm-selectivity would also lead to the cells being used in allogeneic transplant situations.

14.3.1 Encapsulation as a Means of Providing Stem Cells with Differentiation Cues

A lot of studies are being performed to determine the optimal conditions, including the characteristics of the encapsulating material required, to use encapsulation as a mechanism to differentiate stem cells down a specific lineage. For example, it has been shown that the myogenic differentiation of umbilical cord stem cells encapsulated specifically in alginate–fibrin microbeads, which were packed in an Arg–Gly–Asp (RGD)-modified alginate matrix (AM) and cultured in myogenic induction media, were more than threefold those of controls (Liu et al., 2012). The rigidity of the encapsulating material has also been manipulated to provide hMSCs with mechanical cues for osteogenic, neural, and myogenic differentiation (Huebsch et al., 2010; Pek et al., 2010). In these cases, encapsulation also provides a means of immunoisolation because it has been shown that differentiated human ESCs express high levels of major histocompatibility (MHC) class I proteins, which may cause them to be rejected upon transplantation (Drukker et al., 2002).

Owing to their multilineage potential, however, stem cells do not always differentiate into the desired cell type even in the presence of enabling growth factors. Park et al. (2010b) overcame this problem by exposing stem cells to primary cells in a coculture system prior to encapsulating them, and found that this approach helped in directing the differentiation of the stem cells to the desired cell type only.

14.3.2 Encapsulation of Stem Cells as a Means of Delivering Therapeutic Molecules

It has been shown that encapsulated bone marrow-derived mesenchymal stem cells (BM-MSCs) partially differentiate into hepatocytes when the capsules are implanted into the peritoneum (Liu and Chang, 2006). In this study, alginate–polylysine–alginate (APA)-encapsulated BM-MSCs were injected intraperitoneally into a Wistar rat model with 90% hepatectomy. This increased the chances of survival (survival rate was 100%) and improved the blood chemistry similar to that of bioencapsulated hepatocytes or free hepatocytes. However, this was not the case for unencapsulated BM-MSCs, in part because the unencapsulated cells did not stay immobilized within the peritoneum.

There have also been numerous studies proving the therapeutic effect of encapsulated, genetically modified stem cells. A biodegradable, synthetic extracellular matrix composed of Hystem and Extralink was used to encapsulate mouse neural stem cells (mNSCs), which were genetically modified to express the diagnostic variant of S-TRAIL (secretable tumor necrosis factor apoptosis inducing ligand; induces apoptosis in about 50% of glioblastomas). Capsules, which proved to increase mNS cell viability *in vivo*, were then implanted intracranially 1 mm from the site of a glioblastoma. Intravital microscopy revealed that although sECM-encapsulated mNSCs migrated out, they specifically homed to tumors in the brain over a period of 4 days, thus reducing the tumor load and prolonging survival of diseased mice. Here, encapsulation helped to improve implanted cell survival and function while also maintaining localization of encapsulated cells within the vicinity of the tumor where its therapeutic effect would be maximal (Kauer et al., 2012).

Similarly, BMP-2-modified MSCs encapsulated in APA microcapsules by the electrospray method were able to induce undifferentiated MSCs to become osteoblasts that had high alkaline phosphatase activity *in vitro* (Ding et al., 2007). It thus has the potential to enhance bone formation or regeneration *in vivo*.

In order to prove the applicability of encapsulated stem cells in an allogeneic or possibly even xenogeneic setting, Goren et al. (2010) encapsulated human MSCs in alginate–Poly-L-lysine (PLL) microcapsules and implanted them subcutaneously in 6-week-old female C57BL mice. Encapsulated hMSCs remained as a clear cluster with no tissue capsule surrounding them with the viability of encapsulated cells maintained even at 8 weeks postimplantation. Finally, proving their applicability in a clinical setting, the encapsulated hMSCs were seen to be significantly hypoimmunogenic, leading to a three-fold decrease in cytokine expression compared to entrapped cell lines.

14.3.3 Coencapsulation of Stem Cells as a Means of Potentiating the Action of Other Encapsulated Cells

In 2D cultures, a feeder layer of cells is often provided to support the growth of other cell types by providing metabolites and growth factors. The ability of stem cells, especially MSCs, in potentiating and even perhaps enhancing the therapeutic effect of coencapsulated cells in a similar manner but in a 3D environment has been studied. Weeks after encapsulation, hepatocytes coencapsulated with bone marrow MSCs were more viable, both *in vitro* and *in vivo*, as compared to hepatocytes that were not coencapsulated (Chang et al., 2006). In addition, they were able to maintain their metabolic capabilities *in vitro* and *in vivo*. Part of this action was due to the transdifferentiation of some of the MSCs into albumin-producing hepatocyte-like cells (Shi et al., 2009).

Furthermore, the potential of encapsulation as a means to maintain feeder-free cultures of ESCs is also being investigated. For example, hESCs encapsulated in 1.1% (w/v) calcium alginate hydrogels (Siti-Ismail et al., 2008) or hyaluronic acid hydrogels (Gerecht et al., 2007) and grown in basic maintenance medium were able to survive and undergo self-renewal even upon forming aggregates of up to 500 µm in diameter, for a period of up to 260 days, without compromising their pluripotency.

14.3.4 Role of Vascularization in the Field of Stem Cells

The vascularization of encapsulated stem cells to maintain their viability is as important as the vascularization of many other engineered tissues. One advantage of encapsulated MSCs, however, is that while secreting the required therapeutic molecule, the cells can simultaneously prevent an adverse inflammatory reaction around the capsule due to their immunomodulatory properties. This is achieved by reducing the production of proinflammatory cytokines such as interferon-γ (IFN-γ) and tumor necrosis factor-α (TNF-α) by the host immune system (Chen et al., 2011; Gebler et al., 2012). An additional cause for reduced mass transport across the capsule membrane, fibrosis, is thus reduced.

14.4 Vascularization of Encapsulated Cells

From the previous section, it is clear that the ability to control the host foreign body response toward encapsulated cells is crucial to their success *in vivo*. In addition, encapsulation presents a barrier to molecular transport by increasing both the resistance to diffusion and the distance nutrients must travel from the nearest blood vessel source to the encapsulated cells. Therefore, rapid and persistent vascularization by host cells is necessary to maintain graft functionality posttransplantation. Various biomaterials, both natural and synthetic, have been studied as a means of achieving local angiogenesis in and around implanted cell encapsulation systems.

14.4.1 General Properties of Materials Used for Encapsulation and Their Effect on Vascularization

The microarchitecture of the biomaterial used for cell encapsulation appears to influence the foreign body response it elicits upon implantation. One study has evaluated the foreign body responses to more than 150 membranes of various pore sizes and chemical compositions using a subcutaneous model *in vivo* (Brauker et al., 1995). The investigators concluded that pore size was the primary determinant of the foreign body response; membranes with pore sizes of 5–15 µm resulted in vascularization where the blood vessels were in close proximity with the membrane interface, regardless of their chemical nature. Membranes that had a pore size less than 5 µm did not vascularize and instead accumulated macrophages at the surface of the implanted biomaterial, with the blood vessels present in granulation tissue farther away.

Researchers have been able to use various porous scaffolds for cell transplantation to achieve functional efficacy in both small- and large-animal models. The fabrication of biocompatible, macroporous scaffolds from poly(dimethylsiloxane) (PDMS) has been shown to support islet grafts via implantation in the omental pouch of chemically induced diabetic syngeneic rats. The scaffolds induced intradevice vascularization that promptly led to normoglycemia (Pedraza et al., 2013). In another study, fibrin–islet composite grafts achieved effective glycemic control within an average of 3 days after subcutaneous implantation in diabetic mice, owing, in part, to highly vascularized structures that were generated within and on the surface of the grafts (Kim et al., 2012). Porous polytetrafluoroethylene (PTFE) grafts implanted into the iliac arteries of baboons, with an intermodal distance of 60 μm induced microvasculature to grow directly through the wall of the graft, with capillaries 100–500 μm apart. Conversely, grafts with small intermodal distances of less than 30 μm showed diminished endothelial coverage permeating through its surface (Clowes et al., 1986, 1987).

The precise mechanism by which membrane surface properties modulate the host cell response has not been elucidated. One theory proposes that larger pore sizes allow for a loosely packed vascular capsule, which encourages the secretion of angiogenic factors from macrophages, and prevents them from spreading on the surface of the implant (Padera and Colton, 1996). Another theory proposes that the surface irregularities of the biomaterial could cause shearing of cells at the implant–tissue interface, thus causing an inflammatory response (Rosengren et al., 1999).

14.4.2 Biomaterials: A Means of Delivering Angiogenic Growth Factors

As mentioned earlier, current clinical methods of bolus growth factor injection are not ideal. Instead, one should aim at not only administering the optimal dose but also utilize an appropriate release modality in order to maintain viable vessel structure. The use of biomaterials could thus confer the ability to achieve a localized delivery of angiogenic growth factors, which may accelerate the rate of invasion and organization of vascular networks posttransplantation of encapsulated cells.

Natural polymers such as fibrin have demonstrated success in delivering proteins and stimulating vessel formation. In one study, fibrin containing alpha-endothelial cell growth factor (alpha-ECGF) was implanted in the left ventricle of rats. Nine weeks later, the implants that contained alpha-ECGF induced angiogenesis, resulting in new blood vessel growth directed toward the heart (Fasol et al., 1994). Fibrin has also been shown to provide a controlled delivery of a whole host of other growth factors, including FGF-2, VEGF, BMP-2, and PDGF-BB (Smith et al., 2007). In order to allow for a longer, more sustained release of such growth factors that is not diffusion restrained, researchers have studied improvement in protein retention by covalent modification of the fibrin matrix (Hall and Hubbell, 2005).

Alginate has been extensively used as a delivery conduit for growth factors because of its mild gelling properties and its excellent biocompatibility. Alginate microbeads loaded with FGF-1 have been shown to provide a sustained release of protein that results in increased vascular density *in vitro* and *in vivo* (Moya et al., 2009, 2010a). These observations have provided the impetus for studies of simultaneous implantation of microbeads containing both growth factor and islets within them, as a means of stimulating neovascularization directly toward the encapsulated cells. In another approach, a multilayer alginate microencapsulation system has been proposed that serves the dual purpose of cell immunoisolation and angiogenic protein delivery for vascularization of encapsulated islet transplants as a treatment for type 1 diabetes (Khanna et al., 2010a,b). The inner alginate core serves as a three-dimensional environment for the islets, and the outer layer as a region for protein immobilization. The two layers are separated by a thin perm-selective layer of poly-ʟ-ornithine (PLO), which serves to control the permeability of solutes into the inner core of microbeads where islets are enclosed. Altering the properties of the alginate used to construct the outer layer can modulate the release of proteins such as FGF-1 for greater than 30 days *in vitro*. In one study, the use of alginate microbeads as a delivery conduit for growth factors has thus been shown to improve the efficacy of implanted islets *in vivo*. In the study, a bioengineering implant composed of a polyvinyl alcohol sponge infused with a type 1 collagen hydrogel containing donor islets, and spherical alginate spheres containing VEGF was developed. The implants that contained alginate loaded with VEGF reversed streptozotocin-induced diabetes in 100% of eight mice, whereas the implants that lacked VEGF cured only 62.5% of the mice (Vernon et al., 2012).

14.4.3 Prevascularized Constructs

In contrast to the previously described strategies that relied on inducing vascularization postimplantation, an alternate approach could be to produce a vascular network within the engineered construct prior to implantation. This preengineered microvasculature would anastomose with existing vessels in the host, thus accelerating vascularization and improving viability.

The nature of the host extracellular matrix, which produces various growth factors and cytokines, and its role in modulating the microenvironment that promotes graft survival has been described (Francis et al., 2008). The endothelium, in particular, has been shown to play an active role in cell–cell communications that support islet survival and engraftment. Using an engineered 3D microenvironment that mimics the pancreatic vasculature, the formation of tube-like vessels from endothelial cell recruitment and signaling stimuli correlated with increased islet survival. Moreover, the implantation of islets prevascularized in this vessel network led to improved engraftment profiles and improved maintenance of blood glycemic levels in diabetic mice (Kaufman-Francis et al., 2012).

Several studies have shown that allowing a period of prevascularization prior to islet implantation potentiates the cells' survival and function. In one study, investigators transplanted a prevascularized chamber into the epigastric pedicle in the groin of diabetic mice 21 days before implanting the islets within the chamber. The addition of islets into prevascularized chambers resulted in a significant reduction of blood glucose and improved glycemic control over the course of 3 weeks (Hussey et al., 2009). In another study, implantation of prevascularized islets and aggregates of pancreatic cells yielded function-specific vasculature; in fact, the prevascular constructs obtained a perfused microvasculature much more quickly (Beger et al., 1998). These observations suggest that the presence of microvasculature within implanted tissue accelerates the establishment of a stable microvasculature around the graft. Instead of prevascularizing the graft itself, an alternative approach is to prevascularize the site of transplant prior to implanting the cell graft. In fact, the ability to confirm adequate vascularization of a transplant site could be used as a criterion for determining whether it is optimal for use. For example, a polyethylene terephthalate (PET) mesh bag, containing gelatin microspheres loaded with FGF-2, was used to induce angiogenesis at the intermuscular space in diabetic rats. After confirmation of adequate new vasculature formation, islets mixed in 5% agarose were transplanted into the same site as the PET mesh bag. Further neovascularization was observed around the cells within 10 days posttransplantation, and normoglycemia was achieved and maintained for more than 35 days (Balamurugan et al., 2003).

14.4.4 Role of Stem Cells in Vascularization

The potential role of stem cells in the vascularization of implanted tissue is now gaining momentum. Stem cells can contribute to angiogenesis directly by participating in new vessel formation, or indirectly by secreting a broad spectrum of angiogenic and anti-apoptotic factors (Yang et al., 2010). Compelling evidence as to the angiogenic potential of encapsulated human MSCs comes from a study in which alginate-encapsulated MSCs that were attached to the heart of a rat myocardial infarction (MI) model via a polyethylene glycol (PEG) patch induced a large increase in microvascular density in the peri-infarct area (Levit et al., 2012). This resulted in a highly significant improvement in left ventricular function after acute MI. Additional studies have used encapsulated MSCs in a model of hind limb ischemia (Landazuri et al., 2012), with similar results. Additionally, studies show that adipose-derived mesenchymal stem cells (AMSC) might increase the vascularization of subcutaneously implanted tissue, in contrast to BM-MSC due to a higher release of VEGF *in vitro* (Vériter et al., 2011). *In vivo*, AMSCs improved the implant's oxygenation and vascularization as measured by the pO_2 within the grafts and the extent of neovascularization around the graft.

Another approach would be to genetically modify the implanted stem cell so that it secretes higher titers of VEGF that would, in turn, potentiate the process of angiogenesis. For example, one study genetically modified human MSCs and ESC-derived cells so that they had high expression of VEGF, albeit transiently (Yang et al., 2010). The investigators found that subcutaneous implantation of such genetically modified cells resulted in two-to four-fold higher vessel densities 2 weeks after implantation, compared with control cells. The cells also enhanced angiogenesis and limb salvage while reducing muscle degeneration and tissue fibrosis when injected intramuscularly into a hind limb ischemia model.

Also, coencapsulation of therapeutic cells with stem cells that have the potential to differentiate into endothelial cells has been examined. Park et al. (2010b) found that human MSCs encapsulated in thermoreversible hydrogel precultured with human umbilical vein endothelial cells (HUVEC) plus endothelial cell growth medium (EGM-2) formed tubular structures with large amounts of red blood cells (RBCs) in them, suggesting that the transplanted MSCs differentiated to endothelial cells and then formed neovascular structures (Park et al., 2010a). However, since the goal is to provide immunoisolation to the

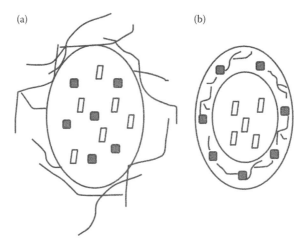

(a)   (b)

FIGURE 14.1
The suggested role of stem cells in vascularization of microcapsules can be illustrated in two ways: (a) by coencapsulating stem cells (shaded cells) with the cells of therapeutic value (blank cells), so that the stem cells can release proangiogenic factors that would increase vascularization (solid lines) around the capsule and (b) by encapsulating stem cells that have the potential to differentiate into vascular structures, in a separate chamber from the cells of therapeutic value.

therapeutic cells, it might be worth considering putting such stem cells into a separate chamber so that both cell types are isolated (Figure 14.1). This would greatly increase the chances of the therapeutic cells being surrounded by a dense vascular network that can help in overcoming issues of mass transport and oxygen insufficiency.

14.5 Conclusions

The vascularization of engrafted microcapsules is essential for the maintenance and function of the cells contained within them. Considerations for proper vascularization should include the nature of the biomaterial being used to form the capsules, pore size, site of implantation, and time and mode of angiogenic growth factor delivery. Adequate evaluation of these multiple parameters could someday lead to successful and long-term delivery of therapeutics by encapsulated cells.

References

Avgoustiniatos, E.S. and Colton, C.K. 1997. Effect of external oxygen mass transfer resistances on viability of immunoisolated tissue. *Ann N Y Acad Sci* 831, 145–167.

Balamurugan, A., Gu, Y., Tabata, Y., Miyamoto, M., Cui, W., Hori, H., Satake, A., Nagata, N., Wang, W., and Inoue, K. 2003. Bioartificial pancreas transplantation at prevascularized intermuscular space: Effect of angiogenesis induction on islet survival. *Pancreas* 26, 279–285.

Beger, C., Cirulli, V., Vajkoczy, P., Halban, P., and Menger, M. 1998. Vascularization of purified pancreatic islet-like cell aggregates (pseudoislets) after syngeneic transplantation. *Diabetes* 47, 559–565.

Bloch, J., Bachoud-Levi, A.C., Deglon, N., Lefaucheur, J.P., Winkel, L., Palfi, S., Nguyen, J.P. et al. 2004. Neuroprotective gene therapy for Huntington's disease, using polymer-encapsulated cells engineered to secrete human ciliary neurotrophic factor: Results of a phase I study. *Hum Gene Ther* 15, 968–975.

Brauker, J.H., Carr-Brendel, V.E., Martinson, L.A., Crudele, J., Johnston, W.D., and Johnson, R.C. 1995. Neovascularization of synthetic membranes directed by membrane microarchitecture. *J Biomed Mater Res* 29, 1517–1524.

Buchwald, P. 2009. FEM-based oxygen consumption and cell viability models for avascular pancreatic islets. *Theor Biol Med Model* 6, 5.

Buchwald, P. 2011. A local glucose-and oxygen concentration-based insulin secretion model for pancreatic islets. *Theor Biol Med Model* 8, 20.

Buladi, B.M., Chang, C.C., Belovich, J.M., and Gatica, J.E. 1996. Transport phenomena and kinetics in an extravascular bioartificial pancreas. *AIChE J* 42, 2668–2682.

Cabané, P., Gac, P., Amat, J., Pineda, P., Rossi, R., Caviedes, R., and Caviedes, P. 2009. Allotransplant of microencapsulated parathyroid tissue in severe postsurgical hypoparathyroidism: A case report. *Transplantation Proceedings* 41, 3879–3883.

Calafiore, R., Basta, G., Luca, G., Lemmi, A., Montanucci, M.P., Calabrese, G., Racanicchi, L., Mancuso, F., and Brunetti, P. 2006. Microencapsulated pancreatic islet allografts into nonimmunosuppressed patients with type 1 diabetes: First two cases. *Diabetes Care* 29, 137–138.

Chang Liu, Z. and Chang, T.M. 2006. Coencapsulation of hepatocytes and bone marrow cells: *In vitro* and *in vivo* studies. *Biotechnol Annu Rev* 12, 137–151.

Chao, K.C., Chao, K.F., Fu, Y.S., and Liu, S.H. 2008. Islet-like clusters derived from mesenchymal stem cells in Wharton's Jelly of the human umbilical cord for transplantation to control type 1 diabetes. *PLoS ONE* 3, e1451.

Chen, P.M., Yen, M.L., Liu, K.J., Sytwu, H.K., and Yen, B.L. 2011. Immunomodulatory properties of human adult and fetal multipotent mesenchymal stem cells. *J Biomed Sci* 18, 49.

Clowes, A.W., Kirkman, T.R., and Reidy, M.A. 1986. Mechanisms of arterial graft healing. Rapid transmural capillary ingrowth provides a source of intimal endothelium and smooth muscle in porous PTFE prostheses. *Am J Pathol* 123, 220–230.

Clowes, A.W., Zacharias, R.K., and Kirkman, T.R. 1987. Early endothelial coverage of synthetic arterial grafts: Porosity revisited. *Am J Surg* 153, 501–504.

D'Amour, K.A., Bang, A.G., Eliazer, S., Kelly, O.G., Agulnick, A.D., Smart, N.G., Moorman, M.A., Kroon, E., Carpenter, M.K., and Baetge, E.E. 2006. Production of pancreatic hormone-expressing endocrine cells from human embryonic stem cells. *Nat Biotechnol* 24, 1392–1401.

Davalli, A.M., Scaglia, L., Zangen, D.H., Hollister, J., BonnerWeir, S., and Weir, G.C. 1996. Vulnerability of islets in the immediate posttransplantation period—Dynamic changes in structure and function. *Diabetes* 45, 1161–1167.

Ding, H.F., Liu, R., Li, B.G., Lou, J.R., Dai, K.R., and Tang, T.T. 2007. Biologic effect and immunoisolating behavior of BMP-2 gene-transfected bone marrow-derived mesenchymal stem cells in APA microcapsules. *Biochem Biophys Res Commun* 362, 923–927.

Drukker, M., Katz, G., Urbach, A., Schuldiner, M., Markel, G., Itskovitz-Eldor, J., Reubinoff, B., Mandelboim, O., and Benvenisty, N. 2002. Characterization of the expression of MHC proteins in human embryonic stem cells. *Proc Natl Acad Sci U S A* 99, 9864–9869.

Dufrane, D., Goebbels, R.M., Saliez, A., Guiot, Y., and Gianello, P. 2006a. Six-month survival of microencapsulated pig islets and alginate biocompatibility in primates: Proof of concept. *Transplantation* 81, 1345–1353.

Dufrane, D., Steenberghe, M., Goebbels, R.M., Saliez, A., Guiot, Y., and Gianello, P. 2006b. The influence of implantation site on the biocompatibility and survival of alginate encapsulated pig islets in rats. *Biomaterials* 27, 3201–3208.

Dulong, J.L. and Legallais, C. 2007. A theoretical study of oxygen transfer including cell necrosis for the design of a bioartificial pancreas. *Biotechnol Bioeng* 96, 990–998.

Elert, G. 2012. The Physics Factbook™—An encyclopedia of scientific essays. In: *Volume of Blood in a Human*. http://hypertextbook.com/facts/

Elliott, R.B., Escobar, L., Garkavenko, O., Croxson, M.C., Schroeder, B.A., McGregor, M., Ferguson, G., Beckman, N., and Ferguson, S. 2000. No evidence of infection with porcine endogenous retrovirus in recipients of encapsulated porcine islet xenografts. *Cell Transplant* 9, 895–901.

Emamaullee, J.A. and Shapiro, A.M.J. 2007. Factors influencing the loss of beta-cell mass in islet transplantation. *Cell Transplantation* 16, 1–8.

Emerich, D.F., Hammang, J.P., Baetge, E.E., and Winn, S.R. 1994. Implantation of polymer-encapsulated human nerve growth factor-secreting fibroblasts attenuates the behavioral and neuropathological consequences of quinolinic acid injections into rodent striatum. *Exp Neurol* 130, 141–150.

Emerich, D.F., Thanos, C.G., Goddard, M., Skinner, S.J.M., Geany, M.S., Bell, W.J., Bintz, B. et al. 2006. Extensive neuroprotection by choroid plexus transplants in excitotoxin lesioned monkeys. *Neurobiol Dis* 23, 471–480.

Eriksdotter-Jonhagen, M., Linderoth, B., Lind, G., Aladellie, L., Almkvist, O., Andreasen, N., Blennow, K. et al. 2012. Encapsulated cell biodelivery of nerve growth factor to the Basal forebrain in patients with Alzheimer's disease. *Dement Geriatr Cogn Disord* 33, 18–28.

Fasol, R., Schumacher, B., Schlaudraff, K., Hauenstein, K.H., and Seitelberger, R. 1994. Experimental use of a modified fibrin glue to induce site-directed angiogenesis from the aorta to the heart. *J Thorac Cardiovasc Surg* 107, 1432–1439.

Francis, M., Uriel, S., and Brey, E. 2008. Endothelial cell–matrix interactions in neovascularization. *Tissue Eng Part B Rev* 14, 19–32.

Fu, Y.S., Cheng, Y.C., Lin, M.Y., Cheng, H., Chu, P.M., Chou, S.C., Shih, Y.H., Ko, M.H., and Sung, M.S. 2006. Conversion of human umbilical cord mesenchymal stem cells in Wharton's jelly to dopaminergic neurons *in vitro*: Potential therapeutic application for Parkinsonism. *Stem Cells* 24, 115–124.

Garcia, P., Youssef, I., Utvik, J.K., Florent-Bechard, S., Barthelemy, V., Malaplate-Armand, C., Kriem, B. et al. 2010. Ciliary neurotrophic factor cell-based delivery prevents synaptic impairment and improves memory in mouse models of Alzheimer's disease. *J Neurosci* 30, 7516–7527.

Gebler, A., Zabel, O., and Seliger, B. 2012. The immunomodulatory capacity of mesenchymal stem cells. *Trends Mol Med* 18, 128–134.

Gerecht, S., Burdick, J.A., Ferreira, L.S., Townsend, S.A., Langer, R., and Vunjak-Novakovic, G. 2007. Hyaluronic acid hydrogel for controlled self-renewal and differentiation of human embryonic stem cells. *Proc Natl Acad Sci U S A* 104, 11298–11303.

Giuliani, M., Moritz, W., Bodmer, E., Dindo, D., Kugelmeier, P., Lehmann, R., Gassmann, M., Groscurth, P., and Weber, M. 2005. Central necrosis in isolated hypoxic human pancreatic islets: Evidence for postisolation ischemia. *Cell Transplantation* 14, 67–76.

Goren, A., Dahan, N., Goren, E., Baruch, L., and Machluf, M. 2010. Encapsulated human mesenchymal stem cells: A unique hypoimmunogenic platform for long-term cellular therapy. *FASEB J* 24, 22–31.

Gross, J.D., Constantinidis, I., and Sambanis, A. 2007. Modeling of encapsulated cell systems. *J Theor Biol* 244, 500–510.

Gu, F., Amsden, B., and Neufeld, R. 2004. Sustained delivery of vascular endothelial growth factor with alginate beads. *J Control Release*, 96(3), 463–472.

Hall, H. and Hubbell, J.A. 2005. Modified fibrin hydrogels stimulate angiogenesis *in vivo*: Potential application to increase perfusion of ischemic tissues. *Materialwissenschaft und Werkstofftechnik* 36, 768–774.

Hasse, C., Klöck, G., Schlosser, A., Zimmermann, U., and Rothmund, M. 1997. Parathyroid allotransplantation without immunosuppression. *Lancet* 350, 1296–1297.

Hasse, C., Schrezenmeir, J., Stinner, B., Schark, C., Wagner, P.K., Neumann, K., and Rothmund, M. 1994. Successful allotransplantation of microencapsulated parathyroids in rats. *World J Surg* 18, 630–634.

Hefti, F. 1994. Neurotrophic factor therapy for nervous-system degenerative diseases. *J Neurobiol* 25, 1418–1435.

Heile, A.M., Wallrapp, C., Klinge, P.M., Samii, A., Kassem, M., Silverberg, G., and Brinker, T. 2009. Cerebral transplantation of encapsulated mesenchymal stem cells improves cellular pathology after experimental traumatic brain injury. *Neurosci Lett* 463, 176–181.

Hernandez, R.M., Orive, G., Murua, A., and Pedraz, J.L. 2010. Microcapsules and microcarriers for *in situ* cell delivery. *Adv Drug Deliv Rev* 62, 711–730.

Huebsch, N., Arany, P.R., Mao, A.S., Shvartsman, D., Ali, O.A., Bencherif, S.A., Rivera-Feliciano, J., and Mooney, D.J. 2010. Harnessing traction-mediated manipulation of the cell/matrix interface to control stem-cell fate. *Nat Mater* 9, 518–526.

Hussey, A., Winardi, M., Han, X., Thomas, G., Penington, A., Morrison, W., Knight, K., and Feeney, S. 2009. Seeding of pancreatic islets into prevascularized tissue engineering chambers. *Tissue Eng Part A* 15, 3823–3833.

Jeon, Y., Kwak, K., Kim, S., Kim, Y., Lim, J., and Baek, W. 2006. Intrathecal implants of microencapsulated xenogenic chromaffin cells provide a long-term source of analgesic substances. *Transplant Proc* 38, 3061–3065.

Katsuragi, S., Ikeda, T., Date, I., Shingo, T., Yasuhara, T., and Ikenoue, T. 2005. Grafting of glial cell line-derived neurotrophic factor secreting cells for hypoxic-ischemic encephalopathy in neonatal rats. *Am J Obstet Gynecol* 192, 1137–1145.

Kauer, T.M., Figueiredo, J.L., Hingtgen, S., and Shah, K. 2012. Encapsulated therapeutic stem cells implanted in the tumor resection cavity induce cell death in gliomas. *Nat Neurosci* 15, 197–204.

Kaufman-Francis, K., Koffler, J., Weinberg, N., Dor, Y., and Levenberg, S. 2012. Engineered vascular beds provide key signals to pancreatic hormone-producing cells. *PLoS ONE* 7, e40741.

Kendall, W.F., Jr., Collins, B.H., and Opara, E.C. 2001. Islet cell transplantation for the treatment of diabetes mellitus. *Expert Opin Biol Ther* 1, 109–119.

Khanna, O., Moya, M., Greisler, H., Opara, E C., and Brey, E. 2010a. Multilayered microcapsules for the sustained-release of angiogenic proteins from encapsulated cells. *Am J Surg* 200, 655–658.

Khanna, O., Moya, M., Opara, E.C, and Brey, E. 2010b. Synthesis of multilayered alginate microcapsules for the sustained release of fibroblast growth factor-1. *J Biomed Mater Res Part A* 95, 632–640.

Kim, J.S., Lim, J.H., Nam, H.Y., Lim, H.J., Shin, J.S., Shin, J.Y., Ryu, J.H., Kim, K., Kwon, I.C., Jin, S.M., et al. 2012. *In situ* application of hydrogel-type fibrin-islet composite optimized for rapid glycemic control by subcutaneous xenogeneic porcine islet transplantation. *J Control Release* 162, 382–390.

Kin, T., Korbutt, G.S., and Rajotte, R.V. 2003. Survival and metabolic function of syngeneic rat islet grafts transplanted in the omental pouch. *Am J Transplant* 3, 281–285.

Klinge, P.M., Harmening, K., Miller, M.C., Heile, A., Wallrapp, C., Geigle, P., and Brinker, T. 2011. Encapsulated native and glucagon-like peptide-1 transfected human mesenchymal stem cells in a transgenic mouse model of Alzheimer's disease. *Neurosci Lett* 497, 6–10.

Kobayashi, T., Aomatsu, Y., Iwata, H., Kin, T., Kanehiro, H., Hisanga, M., Ko, S., Nagao, M., Harb, G., and Nakajima, Y. 2006. Survival of microencapsulated islets at 400 days posttransplantation in the omental pouch of NOD mice. *Cell Transplantation* 15, 359–365.

Landazuri, N., Levit, R.D., Joseph, G., Ortega-Legaspi, J.M., Flores, C.A., Weiss, D., and Samabnis, A. 2012. Alginate microencapsulation of human mesenchymal stem cells as a strategy to enhance paracrine-mediated vascular recovery after hindlimb ischaemia. *J Tissue Eng Regen Med*. Doi: 10.1002/term.1680. (epub ahead of print)

Levit, R.D., Landazuri, N., Phelps, E.A., Brown, M.E., Garcia, A., Davis, M.E., Joseph, G., Long, R., and Weber, C.J. 2012. Abstract 18627: Cellular Encapsulation of Mesenchymal Stem Cells Enhances Cardiac Repair in a Rat Model of Myocardial Infarction. Paper presented at: American Heart Association (Los Angeles, California, Circulation).

Liu, J., Zhou, H., Weir, M.D., Xu, H.H., Chen, Q., and Trotman, C.A. 2012. Fast-degradable microbeads encapsulating human umbilical cord stem cells in alginate for muscle tissue engineering. *Tissue Eng Part A* 18, 2303–2314.

Liu, Z.C. and Chang, T.M. 2006. Transdifferentiation of bioencapsulated bone marrow cells into hepatocyte-like cells in the 90% hepatectomized rat model. *Liver Transpl* 12, 566–572.

Livett, B.G., Dean, D.M., Whelan, L.G., Udenfriend, S., and Rossier, J. 1981. Co-release of enkephalin and catecholamines from cultured adrenal chromaffin cells. *Nature* 289, 317–319.

Mitchell, K.E., Weiss, M.L., Mitchell, B.M., Martin, P., Davis, D., Morales, L., Helwig, B. et al. 2003. Matrix cells from Wharton's jelly form neurons and glia. *Stem Cells* 21, 50–60.

Morini, S., Braun, M., Onori, P., Cicalese, L., Elias, G., Gaudio, E., and Rastellini, C. 2006. Morphological changes of isolated rat pancreatic islets: A structural, ultrastructural and morphometric study. *J Anat* 209, 381–392.

Moritz, W., Meier, F., Stroka, D., Giuliani, M., Kugelmeier, P., Nett, P.C., Lehmann, R., Candinas, D., Gassmann, M., and Weber, M. 2002. Apoptosis in hypoxic human pancreatic islets correlates with HIF-1 alpha expression. *FASEB J* 16, 745–+.

Moya, M., Garfinkel, M., Liu, X., Lucas, S., Opara, E.C, Greisler, H., and Brey, E. 2010a. Fibroblast growth factor-1 (FGF-1) loaded microbeads enhance local capillary neovascularization. *J Surg Res* 160, 208–212.

Moya, M., Lucas, S., Francis-Sedlak, M., Liu, X., Garfinkel, M., Huang, J., Cheng, M., Opara, E.C, and Brey, E. 2009. Sustained delivery of FGF-1 increases vascular density in comparison to bolus administration. *Microvasc Res* 78, 142–147.

Moya, M.L., Cheng, M.H., Huang, J.J., Francis-Sedlak, M.E., Kao, S.W., Opara, E.C., and Brey, E.M. 2010b. The effect of FGF-1 loaded alginate microbeads on neovascularization and adipogenesis in a vascular pedicle model of adipose tissue engineering. *Biomaterials* 31, 2816–2826.

Moya, M.L., Garfinkel, M.R., Liu, X., Lucas, S., Opara, E.C., Greisler, H.P., and Brey, E.M. 2010c. Fibroblast growth factor-1 (FGF-1) loaded microbeads enhance local capillary neovascularization. *J Surg Res* 160, 208–212.

Opara, E.C., Mirmalek-Sani, S.H., Khanna, O., Moya, M.L., and Brey, E.M. 2010. Design of a bioartificial pancreas. *J Invest Med* 58, 831–837.

Otsuji, T.G., Minami, I., Kurose, Y., Yamauchi, K., Tada, M., and Nakatsuji, N. 2010. Progressive maturation in contracting cardiomyocytes derived from human embryonic stem cells: Qualitative effects on electrophysiological responses to drugs. *Stem Cell Res* 4, 201–213.

Ozawa, C., Banfi, A., and Glazer, N. 2004. Micro-environmental VEGF concentration, not total dose, determines a threshold between normal and aberrant angiogenesis. *J Clin Invest* 113, 516–527.

Padera, R.F. and Colton, C.K. 1996. Time course of membrane microarchitecture-driven neovascularization. *Biomaterials* 17, 277–284.

Park, J., Yang, H., Woo, D., Kim, H., Na, K., and Park, K. 2010a. Multi-lineage differentiation of hMSCs encapsulated in thermo-reversible hydrogel using a co-culture system with differentiated cells. *Biomaterials* 31, 7275–7287.

Park, J.S., Yang, H.N., Woo, D.G., Kim, H., Na, K., and Park, K.H. 2010b. Multi-lineage differentiation of hMSCs encapsulated in thermo-reversible hydrogel using a co-culture system with differentiated cells. *Biomaterials* 31, 7275–7287.

Paul, A., Ge, Y., Prakash, S., and Shum-Tim, D. 2009. Microencapsulated stem cells for tissue repairing: Implications in cell-based myocardial therapy. *Regen Med* 4, 733–745.

Pedraza, E., Brady, A.C., Fraker, C.A., Molano, R.D., Sukert, S., Berman, D.M., Kenyon, N.S., Pileggi, A., Ricordi, C., and Stabler, C.L. 2013. Macroporous three dimensional PDMS scaffolds for extrahepatic islet transplantation. *Cell Transplant* 22(7), 1123–1135.

Pek, Y.S., Wan, A.C., and Ying, J.Y. 2010. The effect of matrix stiffness on mesenchymal stem cell differentiation in a 3D thixotropic gel. *Biomaterials* 31, 385–391.

Rosengren, A., Danielsen, N., and Bjursten, L.M. 1999. Reactive capsule formation around soft-tissue implants is related to cell necrosis. *J Biomed Mater Res* 46, 458–464.

Shi, X.L., Zhang, Y., Gu, J.Y., and Ding, Y.T. 2009. Coencapsulation of hepatocytes with bone marrow mesenchymal stem cells improves hepatocyte-specific functions. *Transplantation* 88, 1178–1185.

Sigrist, S., Mechine-Neuville, A., Mandes, K., Calenda, V., Braun, S., Legeay, G., Bellocq, J.-P., Pinget, M., and Kessler, L. 2003. Influence of VEGF on the viability of encapsulated pancreatic rat islets after transplantation in diabetic mice. *Cell Transplantation* 12, 627–635.

Siti-Ismail, N., Bishop, A.E., Polak, J.M., and Mantalaris, A. 2008. The benefit of human embryonic stem cell encapsulation for prolonged feeder-free maintenance. *Biomaterials* 29, 3946–3952.

Skinner, S.J., Geaney, M.S., Lin, H., Muzina, M., Anal, A.K., Elliott, R.B., and Tan, P.L. 2009. Encapsulated living choroid plexus cells: Potential long-term treatments for central nervous system disease and trauma. *J Neural Eng* 6, 065001.

Smith, J.D., Melhem, M.E., Magge, K.T., Waggoner, A.S., and Campbell, P.G. 2007. Improved growth factor directed vascularization into fibrin constructs through inclusion of additional extracellular molecules. *Microvasc Res* 73, 84–94.

Soon-Shiong, P., Feldman, E., Nelson, R., Komtebedde, J., Smidsrod, O., Skjak-Braek, G., Espevik, T., Heintz, R., and Lee, M. 1992. Successful reversal of spontaneous diabetes in dogs by intraperitoneal microencapsulated islets. *Transplantation* 54, 769–774.

Sun, Y., Ma, X., Zhou, D., Vacek, I., and Sun, A.M. 1996. Normalization of diabetes in spontaneously diabetic cynomologus monkeys by xenografts of microencapsulated porcine islets without immunosuppression. *J Clin Invest* 98, 1417–1422.

Tashiro, H., Iwata, H., Warnock, G.L., Takagi, T., Machida, H., Ikada, Y., and Tsuji, T. 1997. Characterization and transplantation of agarose microencapsulated canine islets of Langerhans. *Ann Transplant* 2, 33–39.

Toso, C., Mathe, Z., Morel, P., Oberholzer, J., Bosco, D., Sainz-Vidal, D., Hunkeler, D., Buhler, L.H., Wandrey, C., and Berney, T. 2005. Effect of microcapsule composition and short-term immunosuppression on intraportal biocompatibility. *Cell Transplantation* 14, 159–167.

Tziampazis, E. and Sambanis, A. 1995. Tissue engineering of a bioartificial pancreas: Modeling the cell environment and device function. *Biotechnol Prog* 11, 115–126.

Uriel, S., Brey, E., and Greisler, H. 2006. Sustained low levels of fibroblast growth factor-1 promote persistent microvascular network formation. *Am J Surg* 192, 604–609.

Vallbacka, J.J., Nobrega, J.N., and Sefton, M.V. 2001. Tissue engineering as a platform for controlled release of therapeutic agents: Implantation of microencapsulated dopamine producing cells in the brains of rats. *J Control Release* 72, 93–100.

Vériter, S., Aouassar, N., Adnet, P., Paridaens, M., Stuckman, C., Jordan, B., Karroum, O., Gallez, B., Gianello, P., and Dufrane, D. 2011. The impact of hyperglycemia and the presence of encapsulated islets on oxygenation within a bioartificial pancreas in the presence of mesenchymal stem cells in a diabetic Wistar rat model. *Biomaterials* 32, 5945–5956.

Vernon, R., Preisinger, A., Goode, M., D'Amico, L., Yue, B., Bollyky, P., Kuhr, C., Hefty, T., Nepom, G., and Gebe, J. 2012. Reversal of diabetes in mice with a bioengineered islet implant incorporating a type I collagen hydrogel and sustained release of vascular endothelial growth factor. *Cell Transplantation* 21, 2099–2110.

Wallrapp, C., Thoenes, E., Thurmer, F., Jork, A., Kassem, M., and Geigle, P. 2013. Cell-based delivery of glucagon-like peptide-1 using encapsulated mesenchymal stem cells. *J Microencapsul* 30(4), 315–324.

Wang, T., Adcock, J., Kuhtreiber, W., Qiang, D., Salleng, K.J., Trenary, I., and Williams, P. 2008. Successful allotransplantation of encapsulated islets in pancreatectomized canines for diabetic management without the use of immunosuppression. *Transplantation* 85, 331–337.

Wernig, M., Zhao, J.P., Pruszak, J., Hedlund, E., Fu, D., Soldner, F., Broccoli, V., Constantine-Paton, M., Isacson, O., and Jaenisch, R. 2008. Neurons derived from reprogrammed fibroblasts functionally integrate into the fetal brain and improve symptoms of rats with Parkinson's disease. *Proc Natl Acad Sci U S A* 105, 5856–5861.

Wright, E.J., Farrell, K.A., Malik, N., Kassem, M., Lewis, A.L., Wallrapp, C., and Holt, C.M. 2012. Encapsulated glucagon-like peptide-1-producing mesenchymal stem cells have a beneficial effect on failing pig hearts. *Stem Cells Transl Med* 1, 759–769.

Xu, D., Alipio, Z., Fink, L.M., Adcock, D.M., Yang, J., Ward, D.C., and Ma, Y. 2009. Phenotypic correction of murine hemophilia A using an iPS cell-based therapy. *Proc Natl Acad Sci U S A* 106, 808–813.

Yang, F., Cho, S., Son, S., Bogatyrev, S., Singh, D., Green, J., Mei, Y. et al. 2010. Genetic engineering of human stem cells for enhanced angiogenesis using biodegradable polymeric nanoparticles. *Proc Natl Acad Sci U S A* 107, 3317–3322.

Yasargil, M.G. 1987. *AVM of the Brain: History, Embryology, Pathological Considerations, Hemodynamics, Diagnostic Studies, Microsurgical Anatomy.* In *Microneurosurgery*, Thieme, Stuttgart.

Yasuhara, T., Shingo, T., Muraoka, K., Kobayashi, K., Takeuchi, A., Yano, A., Wenji, Y. et al. 2005. Early transplantation of an encapsulated glial cell line-derived neurotrophic factor-producing cell demonstrating strong neuroprotective effects in a rat model of Parkinson disease. *J Neurosurg* 102, 80–89.

Zanin, M.P., Pettingill, L.N., Harvey, A.R., Emerich, D.F., Thanos, C.G., and Shepherd, R.K. 2012. The development of encapsulated cell technologies as therapies for neurological and sensory diseases. *J Control Release* 160, 3–13.

Zhang, H., Zhu, S.J., Wang, W., Wei, Y.J., and Hu, S.S. 2008. Transplantation of microencapsulated genetically modified xenogeneic cells augments angiogenesis and improves heart function. *Gene Ther* 15, 40–48.

15

Vascularization of Muscle

Tracy Criswell, Zhan Wang, Yu Zhou, and Shay Soker

CONTENTS

15.1 Introduction

Skeletal muscle has tremendous regenerative capacity, ensuring maintenance of tissue homeostasis and restoration of both structure and function after minor injury. However, adult skeletal muscle may not possess sufficient ability to recover anatomical structure and physiological functions after significant tissue loss or damage resulting from trauma, tumor ablation, prolonged denervation, or myopathies (Bach, Beier et al. 2004; Wagers and Conboy 2005; Blau 2008). In such cases, surgical intervention is often necessary. However, conventional surgical procedures, such as autologous muscle transplantation and transposition, cause significant donor site morbidity while only resulting in limited functional recovery (Bach, Beier et al. 2004).

In recent years, new therapeutic options have been developed that utilize cell therapy with myogenic cells and/or muscle tissue engineering, which afford the possibility of new approaches for addressing skeletal muscle loss or damage. Current engineered muscle constructs generally lack the organized architecture of native muscle tissue and thus only generate 1–2% of the contractile force of normal skeletal muscle (Vandenburgh 2002; Ott, Matthiesen et al. 2008). Therefore, engineering full thickness functional skeletal muscle is still a challenge that must be overcome before these therapies can be used in the clinic (Vandenburgh 2002). Alternatively, myogenic cell therapy is a relatively "simple" therapeutic approach in comparison to the *de novo* engineering of an entire segment of muscle. Multiple types of myogenic precursor cells have been tested in preclinical and early-stage clinical trials for the treatment of muscle injury, but most have failed to show significant improvement in functional recovery (Tedesco, Dellavalle et al. 2010).

There are many potential reasons for the current inability to generate fully functional engineered muscle tissue. One potential problem is attributable to the lack of consensus on the characteristics that define "myogenic progenitor stem cells" (MPCs). Current preclinical testing utilizes differentially characterized myogenic cells isolated from different tissues and whose identification varies between different research groups. A uniform definition of what constitutes an MPC is needed in order to unify this research and push this therapy into the clinic. An additional reason for the ineffectiveness of cell therapy approaches for skeletal muscle injury lies in the inability to mimic the complex muscle microenvironment *in vitro* in order to validate tissue engineering and cell therapy protocols that can be applied for use in the clinic. Skeletal muscle is a complex tissue, dependent on the simultaneous development of vascularization, innervation, and a specific extracellular matrix (ECM), which allows for the production of force and movement. The concomitant development of a functional vascular network and tissue innervation must be considered in order to successfully engineer functional skeletal muscle. Thus, the development of efficacious tissue-engineered muscle constructs that can be used in the clinic will be dependent on the identification of ideal myogenic progenitor cells and the reconstitution of the muscle microenvironment (vascularization, innervation, ECM).

In this chapter, we will summarize and discuss the current research in the field of revascularization after injury in relation to skeletal muscle tissue regeneration. We will focus on potential myogenic and angiogenic therapies that are being developed and used in the tissue engineering field to aid in the recovery of muscle volume and function after injury.

15.2 Muscle and Vascular Embryonic Development

A basic understanding of the embryonic pathways leading to the development of vascularized functional skeletal muscle is crucial for developing and guiding effective tissue engineering protocols. Many, but not all, of these pathways are recapitulated during postnatal muscle regeneration suggesting that these pathways may be viable targets for therapeutic interventions for postnatal muscle regeneration. It is important to note the key differences in these processes. Embryonic development begins with a naïve undifferentiated population of cells that must initiate *de novo* formation of all tissue. In contrast, regeneration of damaged tissue can rely on an already-existing "template" for the information required for these processes. As such, the local milieu is different for embryonic and postnatal tissue formation and these differences must be taken into consideration when trying to use this data to inform possible tissue engineering strategies. However, it is also useful to note that the cells within the injured tissue already contain the "program" necessary for all the developmental processes required for tissue formation. As scientists, we need to find ways to identify and decipher this inherent program for manipulation during the *ex vivo* engineering of tissue. Therefore, the following sections will provide a basic overview of embryonic myogenesis and vasculogenesis. More complete reviews of this subject have been previously published (Risau and Flamme 1995; Buckingham, Bajard et al. 2003; Braun and Gautel 2011; Bentzinger, Wang et al. 2012; Yusuf and Brand-Saberi 2012).

15.2.1 Embryonic Myogenesis

The muscle regenerative response to injury is thought to closely mimic the processes that occur during embryogenesis. Therefore, an understanding of embryonic skeletal muscle development can be used to inform methods used to enhance the vascularization of regenerating tissue *in vivo* and for engineering vascularized skeletal muscle tissue *in vitro*. Much of the work delineating vascular and skeletal muscle embryogenesis has been done in chick embryos due to the relative ease of access to the embryos and their relative transparency, which facilitates imaging. More recently, quail–chick and mouse–chick chimeras have allowed for cell lineage and fate tracing through all the stages of embryogenesis.

Embryonic myogenesis occurs in several distinct steps, resulting in the progressive generation of progenitor cells that ultimately results in cells committed to the myogenic pathway. Classical experiments using quail–chick chimeras have demonstrated that the embryonic development of trunk and limb skeletal muscles originate in segmented structures of paraxial mesoderm called somites (Armand, Boutineau et al. 1983; Christ and Ordahl 1995; Scaal and Christ 2004). Signals emanating from surrounding tissues, such as the neural tube and notochord, direct the migration of cells to the dorsal epithelia region of the somite, forming the dermomyotome (Cossu, Tajbakhsh et al. 1996; Scaal and Christ 2004). Cells comprising the dermomyotome express the transcription factors Pax3 and Pax7 (only Pax3 in mice), and can contribute to dermal, myogenic, and endothelial cell lineages (Christ, Jacob et al. 1983; Huang, Zhi et al. 2003; Kassar-Duchossoy, Giacone et al. 2005). The importance of Pax3 in this process is highlighted by the fact that Pax3 mutant embryos do not undergo limb muscle formation (Relaix, Polimeni et al. 2003, Relaix, Rocancourt et al. 2004). In contrast, Pax7 mutant embryos develop normally, but postnatal animals are unable to restore muscle volume after injury, suggesting a role for Pax7 in the maintenance of adult muscle stem cells.

The first wave of muscle development occurs as a population of cells, expressing the myogenic regulatory factor Myf5, migrate out of the dermomyotome to form the first primitive muscle structure, the myotome (Borycki and Emerson 2000), which is the source of the head and trunk muscles. The second step of embryonic muscle development occurs when a population of Pax3[+] epithelial cells, located within the central dermomyotome, undergo an epithelial-to-mesenchymal transition (EMT) followed by migration into the limb bud mesenchyme (Kassar-Duchossoy, Giacone et al. 2005). The mesenchyme directs the migration and correct positional patterning of the myocytes in the limb buds (Chevallier, Kieny et al. 1977; Christ, Jacob et al. 1977). The myogenic differentiation process begins after migration into the limb buds through the activation of the additional myogenic regulatory factors, MyoD and myogenin.

Lineage specification is spatially and temporally dependent on a complex coordination of signals emanating from neighboring tissues (Bonnin, Laclef et al. 2005; Gros, Manceau et al. 2005; Relaix, Rocancourt et al. 2005), which is complicated by gradient dependence, where cells most proximal to the diffusing signal develop a different fate from more distal cells. The Sonic hedgehog (Shh), Wnt, and transforming growth factor-β (TGFβ) signaling pathways, as well as their specific antagonists, have been shown to have an early influence on the development of the dermomyotome and primary myotome structures (Bonnin, Laclef et al. 2005). Further, the HGB/c-Met and fibroblast growth factor (FGF) signaling pathways have been implicated in the regulation of the migration of myocytes from the myotome into the limb bud. It should be noted that the influences of each of these signaling pathways on cell fate do not occur independently of one another. Indeed, many of these signals show temporal overlap, with spatial orientation ultimately affecting cell fate. The complex interplay and cross-talk of these cellular and microenvironmental signals are just beginning to be elucidated.

Secretion of Shh by the notochord and neural tube, in conjunction with intrinsic Wnt-1 and Wnt-3a signaling, are required for the early expression of the myogenic factors Myf5 and MyoD in cells in the dermomyotome that are destined to form the primary myotome structure (Munsterberg, Kitajewski et al. 1995; Munsterberg and Lassar 1995; Gustafsson, Pan et al. 2002). Pax3 has also been implicated in this process through direct activation of Myf5 (Bajard, Relaix et al. 2006; Daubas and Buckingham 2013). However, Pax3 is not required for initial myogenesis, as evidenced by the fact that Pax3 null embryos are still able to form the myotome structure, but lack limb muscles that are dependent on the migration of myocytes from the dermomyotome (Daston, Lamar et al. 1996; Lagha, Kormish et al. 2008). Shh signaling also maintains the expression of myogenic factors within the myotome, promotes the growth and expansion of skeletal muscle progenitor cells, and is involved in the determination of slow and fast myofibers (Duprez, Fournier-Thibault et al. 1998; Marcelle, Ahlgren et al. 1999; Kruger, Mennerich et al. 2001; Anderson, Williams et al. 2012). Conversely, the Notch signaling pathway has been shown to be anti-myogenic, thereby regulating the maintenance of the epithelial cells in the dermomyotome in a spatially controlled manner. Notch expression in the somites and dermomyotome is regulated by its antagonist, Numb. The asymmetrical localization of Numb to the dorso-medial lip of the dermomyotome assures the myogenic differentiation of cells needed to form the myotome while uninhibited Notch expression ensures the maintenance of the epithelium. The bone morphogenic proteins (BMPs) are members of the TGFβ family that have been shown to be required for the proliferation of myogenic cells, while inhibiting final myogenic differentiation. The BMP antagonist Noggin-deficient mice were shown to have defects in the development of the neural tube and somite, despite normal Shh and Pax3 expression (McMahon, Takada et al. 1998). Furthermore, blocking the action of BMPs resulted in a reduced number of muscle fibers and myogenic progenitors in chick limbs (Wang, Noulet et al. 2010). A role for BMPs has been established in the regeneration of postnatal skeletal muscle and will be discussed further later. The EMT and migration of myocytes into the distal limb buds is further regulated by HGF/c-met and FGF signaling pathways (Alexandrides, Moses et al. 1989; Florini, Ewton et al. 1993; Dietrich, Abou-Rebyeh et al. 1999). Finally, myocytes expressing the Mrf4 and myogenin bud fuse to form multinucleated contractile skeletal muscle. This "late" phase of muscle development is essential as evidenced by the fact that mice deficient for myogenin die prior to birth due to a lack of mature myofibers (Hasty, Bradley et al. 1993; Nabeshima, Hanaoka et al. 1993; Rawls, Morris et al. 1995). A subset of cells retain the expression of Pax3 and/or Pax7 and do not under myogenic differentiation. These cells are retained in the tissue as a source of muscle stem cells (satellite cells) involved in the repair of postnatal tissue.

15.2.2 Embryonic Vasculogenesis and Angiogenesis

Although the sequence of events is fairly well outlined for embryonic skeletal muscle formation, much less is known about the origins of the vascular network. Blood vessels consist of two cell types and associated ECM. Endothelial cells (ECs) line the lumen of the vessels to form a semipermeable membrane controlling the release and uptake of molecules from the surrounding tissue. This endothelial barrier also ensures the maintenance of blood flow through the secretion of antithrombotic molecules, has vasomodulatory control of the blood flow, and can regulate the cellular processes of immune and smooth muscle cells. Smooth muscle cells (SMCs), which may be differentiated from pericytes (PCs), are the second cell type associated with blood vessels and comprise the abluminal perimeter of the vessels. SMCs provide support and elasticity to the vessels. Contraction (vasocontraction) and relaxation (vasodilation) of the SMCs regulate changes in local blood flow and pressure. These cells are also motile and growth responsive to signals emanating from the vasculature after injury.

The embryonic vasculature is formed through a dynamic process including both vasculogenesis and angiogenesis, to produce vessels that undergo continuous growth, maturation, and vessel rarefication as the tissue matures. Vasculogenesis involves the *de novo* formation of blood vessels through the aggregation and arrangement of vascular precursor cells called angioblasts. During angiogenesis, new vessels are derived from the reorganization of the endothelium of preexisting vessels. Further maturation of the vessels requires branching and pruning of the growing vascular tree in order to develop an optimal tissue-specific pattern (Carmeliet 2000a; Nguyen and D'Amore 2001).

Angioblasts and hematopoietic precursor cells can be detected in the mesoderm soon after gastrulation occurs. In type I vasculogenesis, the early angioblasts assemble into cord-like structures called the vascular plexus, which ultimately forms the dorsal aorta (Pardanaud, Altmann et al. 1987; Coffin and Poole 1988). Migration of angioblasts into distant tissues with subsequent vascular formation is known as type II vasculogenesis. Vascularization of the limb is thought to occur through both vasculogenesis and angiogenesis as shown by lineage tracing experiments using quail–chick and mouse–avian chimeras (Ambler, Nowicki et al. 2001). The determination and fate of the angioblasts, as well as the development of the primitive vasculature, are regulated through the spatial and temporal expression of vascular endothelial growth factor (VEGF) and FGF by the surrounding tissues (Carmeliet 2000b; Yancopoulos, Davis et al. 2000; Poole, Finkelstein et al. 2001; Jain 2003).

Stabilization of the embryonic vasculature occurs with the recruitment of PCs and/or vascular smooth muscle cells (VSMCs). PCs are smooth-muscle-like cells restricted to the abluminal periphery of small capillaries and venules, whereas VSMCs form concentric rings around the larger vessels. The embryonic origin of mural cells (PCs and VSMCs) is still being debated, although it is beginning to appear that they may originate from several different sites, including the embryonic mesenchyme and the dorsal aorta (Hungerford, Owens et al. 1996; Sartore, Chiavegato et al. 2001). A common precursor for both endothelial cells and VSMCs has also been suggested with the establishment of cell fate determined by the presence of VEGF or PDGFβ, respectively (Yamashita, Itoh et al. 2000). However, a more recent report using lineage tracing of fluorescently labeled quail dermomyotome domains suggested separate progenitors for both endothelial cells and VSMCs differentially regulated by BMP-induced VEGFR2 (endothelial cells) or Notch (VSMCs) (Ben-Yair and Kalcheim 2008). Mural cell recruitment to the vasculature is mediated through platelet-derived growth factor (PDGF) and TGFβ signaling and angiopoietins and their receptors (Ang1/2:Tie1/2) (Pepper 1997; Thommen, Humar et al. 1997; Loughna and Sato 2001; Cho, Kozasa et al. 2003).

The angiopoietic system performs a major role in the maintenance of vascular homeostasis (Fielding, Manfredi et al. 1993; Shim, Ho et al. 2007). The Ang1 ligand is present on mural cells and binds to the tyrosine kinase receptor, Tie-2, on endothelial cells. This functions to maintain vascular integrity through tightened interactions of the mural cells and the endothelial cells (Saharinen, Eklund et al. 2008). In contrast, Ang2 is released by endothelial tip cells in response to angiogenic stimulators such as VEGF and thus acts as an antagonist to Ang1:Tie2 interactions (Shim, Ho et al. 2007).

The parallel embryonic origin of skeletal muscle and the vascular network suggests a reciprocal alliance of both tissues for proper limb development and function. Early studies suggested that endothelial progenitor cells (EPCs) in the dorsal aorta and myogenic precursor cells in the dermomyotome are derived from a common precursor population (Esner, Meilhac et al. 2006). In support of this notion, differentiation of endothelial cells to a myogenic lineage has been suggested. It was shown that CD34[+] and Flk1[+] progenitor cells, isolated from the limb muscle of embryonic mice, were able to differentiate into both endothelial cells and skeletal myofibers after culture *in vitro* (Le Grand, Auda-Boucher et al. 2004). These cells were initially negative for all myogenic markers, which became expressed over time *in vitro*, suggesting a myogenic potential for these cells if located in the correct environment. It is not yet clear whether these cells are capable of contributing to myogenesis *in vivo*. Further evidence for the close developmental association of skeletal muscle and vasculature was provided by the demonstration that Pax3[+] cells of the dermo-myotome contributed to the smooth muscle wall of the dorsal aorta (Esner, Meilhac et al. 2006). Cells fated for the dorsal aorta lost Pax3 expression during delamination from the dermomyotome epithelium, while myogenic determined cells maintained Pax3 expression (Buckingham 2007; Buckingham and Relaix 2007).

Vascular development within the limb bud occurs prior to the migration of the myogenic cells and is required for proper division and patterning of the limb muscles (Tozer, Bonnin et al. 2007). In this way, the developing vasculature directs the migration of infiltrating myogenic precursor cells. The initial migration of the angiogenic cells occurs independently of myogenic cells, as seen by normal vascular development in the limbs of Pax3 null embryos (Tozer, Bonnin et al. 2007). Furthermore, the use of VEGFA to induce hypervascularization resulted in decreased myogenesis while the inhibition of angiogenesis led to muscle fusion (Tozer, Bonnin et al. 2007). Therefore, concurrent vascularization and skeletal muscle tissue engineering will most likely be required for effective tissue formation.

15.3 Muscle/Vascular Damage and Recovery

As previously stated, many of the developmental processes in myogenesis and vasculogenesis are recapitulated during tissue regeneration after injury. However, one must take into consideration the differences in the embryonic and postnatal microenvironments when using this knowledge to inform tissue engineering procedures. In this section, we will discuss the mechanisms of endogenous skeletal muscle and vascular regeneration after injury with consideration of the similarities and differences with embryonic development.

15.3.1 Skeletal Muscle Regeneration

Postnatal skeletal muscle tissue is very stable. For example, less than 1–2% of the nuclei present in the individual myofibers turn over every week for the repair of minor injury due to normal muscle use in healthy rat tissue (Decary, Mouly et al. 1997; Schmalbruch and Lewis 2000). Mature skeletal muscle is marked by the presence of peripherally located myonuclei and a striated appearance due to the alignment of the sarcomeric proteins involved in force production and muscle contraction. Mature muscle can be identified histologically by the presence of various myogenic-specific proteins properly located with the sarcomere. These include adult myosin heavy chain (MyHC), junctophillin-1 (JP-1), and ryanodine receptor-1 (RyR-1). Embryonic forms of MyHC have been identified in regenerating myofibers, further indicating that muscle regeneration may recapitulate some aspects of embryonic myogenesis (Whalen, Harris et al. 1990).

Restoration of muscle volume and function after injury is characterized by two distinct phases: the degenerative/inflammatory phase followed by the regeneration phase. After injury, disruption of the myofiber sarcolemma results in the release of muscle-specific cytosolic proteins such as creatine kinase with a concomitant influx of calcium resulting in calcium-dependent proteolysis and myofiber (Percy, Chang et al. 1979; Armstrong 1990; Armstrong, Warren et al. 1991; Belcastro, Shewchuk et al. 1998; Alderton and Steinhardt 2000). The release of these proteins activates the inflammatory response, initially attracting neutrophils to the site of injury, which is then followed by infiltrating macrophage (Orimo, Hiyamuta et al. 1991; Fielding, Manfredi et al. 1993; Tidball 1995). Macrophage not only phagocytose degraded tissue but also play a role in the activation of myogenic cells (Almekinders and Gilbert 1986; Robertson, Maley et al. 1993; Lescaudron, Peltekian et al. 1999; Merly, Lescaudron et al. 1999; Ruffell, Mourkioti et al. 2009). This response is vital for the regeneration of damaged tissue, as it has been shown that the regenerative capacity of skeletal muscle is blocked in the absence of infiltrating immune cells (Lescaudron, Peltekian et al. 1999). Following degeneration, the muscle repair and regeneration processes take over, as myogenic cells proliferate, differentiate, and fuse to form new muscle fibers (Snow 1977, 1978; Darr and Schultz 1987) (Figure 15.1). Newly formed myofibers can

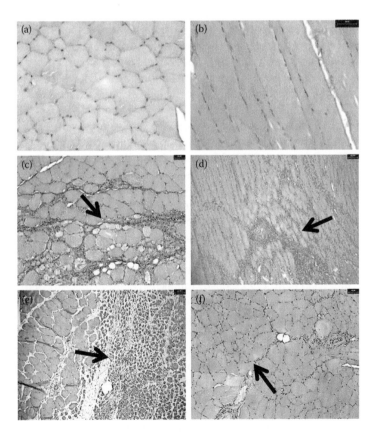

FIGURE 15.1
Degenerative and regenerative phases of skeletal muscle recovery after injury (H + E). (a,b) Uninjured skeletal muscle tissue. (c) Note the small discolored degenerating myofibers 7 days after crush injury (arrows). (d) Immune infiltrate. (e) Small, basophilic regenerating myofibers (arrows). (f) More mature regenerating myofibers showing centrally located nuclei (arrows).

be identified by the presence of centrally located myonuclei and small myofiber diameters, and are often basophilic, indicating active protein synthesis (Hall-Craggs 1974; Allbrook 1975; Charge and Rudnicki 2004).

The active regeneration process relies on a well-orchestrated sequence of events, beginning with the activation and expansion of resident tissue stem cells and ending with the fusion of lineage committed muscle progenitor cells (MPCs) into myofibers. The satellite cell is the primary stem cell involved in the myogenic process. Satellite cells were identified and described in 1961 by Alexander Mauro (Mauro 1961). These cells have been traditionally identified by their anatomical location between the muscle fibers and the plasma membrane; thus the resultant name. Satellite cells can also be identified by their expression of a number of specific cell markers, including the paired domain transcription factors Pax3 and Pax7 (Cornelison and Wold 1997). Whereas Pax3 is the major determinant of myogenic cells in the developing embryo, Pax7 is used as the canonical marker of postnatal satellite cells due to its expression in both quiescent and activated cells (Cornelison and Wold 1997). Since their identification in 1961, the concept that satellite cells contribute to postnatal muscle fiber growth and serve as a major player in muscle regeneration has been well documented (Collins, Olsen et al. 2005; Montarras, Morgan et al. 2005; Sambasivan, Yao et al. 2011; Wang and Rudnicki 2012).

After injury, quiescent satellite cells reenter the cell cycle and become activated in response to cytoplasmic proteins released from damaged myofibers. Activated Pax7+ satellite cells migrate to the site of injury, upregulate the myogenic transcription factors, Myf5 and MyoD (Fuchtbauer and Westphal 1992; Smith, Janney et al. 1994; Yablonka-Reuveni and Rivera 1994; Tajbakhsh, Bober et al. 1996; Cooper, Tajbakhsh et al. 1999), which result in the loss of Pax7 expression and the induction of Mrf4 and myogenin, genes that commit the satellite cell to myogenic differentiation and lead to cell fusion and regenerated muscle fibers (Yablonka-Reuveni and Rivera 1994; Cornelison and Wold 1997; Ten Broek, Grefte et al. 2010). The rapid proliferation of the satellite cells has been shown to provide sufficient cells for muscle repair (Snow 1977, 1978; Yin, Price et al. 2013). Myogenic committed satellite cells are referred to as myogenic precursor cells (MPCs) or myoblasts (Dhawan and Rando 2005).

Satellite cell activation is influenced by signals arising from the local and systemic environments (Yin, Price et al. 2013). Many of these signals are similar to those used during embryonic myogenesis. Extensive work has been done examining the role of the Wnt and Notch pathways in skeletal muscle regeneration. In addition to its role in myogenic cell fate determination, Wnt signaling is also involved in the establishment of cell polarity (non-canonical Wnt/PCP pathway) and cell proliferation (canonical Wnt/β-catenin pathway) (Munsterberg, Kitajewski et al. 1995; Cossu, Tajbakhsh et al. 1996; Otto, Schmidt et al. 2008). The effects of Wnt signaling are highly dependent on the level of its downstream effectors, which, interestingly, have been shown to be antagonized by Notch signaling. There is contradictory data on the exact role of Wnt in the activation of satellite cells after injury. By activating or inhibiting Wnt signaling, Brack, Conboy et al. (2008) found activation of the Wnt pathways only in late differentiating myofibers. They showed increased maturity of muscle fibers in response to Wnt signaling and a decrease in maturity when Wnt signaling was inhibited (Brack, Conboy et al. 2008). In contrast, several groups have proposed that Wnt ligands and downstream β-catenin signaling promote satellite cell proliferation (Otto, Schmidt et al. 2008; Perez-Ruiz, Ono et al. 2008). In fact, owing to the complexity of Wnt signaling and the involvement of many ligands and intracellular signaling pathways, it is conceivable that Wnt may play a role in both processes. Wnt has also recently been shown to be regulated by a number of miRNAs, further adding to its complex regulation (Snyder, Rice et al. 2013).

Notch also plays a role in cell fate determination and proliferation (Artavanis-Tsakonas, Rand et al. 1999; Yin, Price et al. 2013). Notch signaling is upregulated soon after injury and results in the activation and proliferation of satellite cells (Conboy and Rando 2002). Inhibition of Notch signaling results in the absence of satellite cell activation, while overactivation of Notch signaling results in increased satellite cell proliferation and diminished myogenic differentiation (Conboy and Rando 2002; Wen, Bi et al. 2012). Therefore, inhibition of Notch signaling during the late differentiation stage of MPC fusion is required for tissue regeneration. Similar to what occurs during embryonic myogenesis, Numb acts as a Notch inhibitor during postnatal tissue regeneration (Conboy and Rando 2002). Wnt signaling during differentiation has also shown to be able to inhibit Notch (Brack, Conboy et al. 2008).

Several key growth factors also regulate skeletal muscle regeneration and can originate from local tissue or systemic circulation after injury. The HGF (Allen, Sheehan et al. 1995; Tatsumi, Anderson et al. 1998, Tatsumi, Sheehan et al. 2001), basic FGF (bFGF) (Allen and Boxhorn 1989; Haugk, Roeder et al. 1995), insulin growth factor (IGF) (Allen and Boxhorn 1989), VEGF (Springer, Chen et al. 1998), and PDGF-BB (Doumit, Cook et al. 1993) signaling pathways all positively influence the activation and/or proliferation of satellite cells after injury. In contrast, PDFG-AA (Doumit, Cook et al. 1993), myostatin (Amthor, Huang et al. 2002) and other members of the TGFβ (Allen and Boxhorn 1989) signaling pathways maintain satellite cell quiescence, inhibit cell proliferation, or final differentiation. These pathways can be expressed in an autocrine fashion by the muscles themselves, or by infiltrating immune cells, the vasculature, or the local ECM.

15.3.2 Vascular Regeneration

An intact network of blood vessels is required for the maintenance of tissue health through the delivery of oxygen and nutrients to the tissue and the removal of waste. Conversely, impaired function of the vasculature results in many severe pathologies, including pulmonary hypertension, myocardial infarction, stroke, and cancer formation/metastasis (Carmeliet, DeSmet et al. 2009). After traumatic injury, the vascular network acts locally as an "early responder" by signaling to the immune system through the release of specific chemokines. Specifically for tissue engineering purposes, the concomitant development of a functional vascular system within the engineered tissue is of vital importance for the maintenance of cell survival, especially in thick tissue such as skeletal muscle. It has been commonly believed that all postnatal vessel formation and regeneration occurred by angiogenesis, and that vasculogenesis was solely restricted to large vessel formation in the embryo. More current data now suggest that both processes can occur in postnatal tissue in response to injury (Grenier, Scime et al. 2007; Carmeliet and Jain 2011).

Since the health and function of tissue is dependent on adequate oxygen delivery through the circulating blood, the stability of intact functional blood vessels is vitally important. Postnatal blood vessels are extraordinarily stable, and quiescent adult endothelial cells (ECs) are long-lived due to autocrine and paracrine production of prosurvival factors such as VEGF, FGF, and Ang1 (Jain 2003; Lee, Chen et al. 2007; Gaengel, Genove et al. 2009). However, a population of ECs retains the potential to become rapidly activated after injury in order to quickly establish blood flow to injured tissues. Angiogenesis proceeds through several well-defined steps that encompass "loosening" of cell–cell contacts (EC–EC and EC–PCs), degradation of the surrounding ECM, sprouting and migration of ECs into the damaged tissue, recruitment of PCs back to the vessel walls, and reestablishment of stable perfused blood vessels (Figure 15.2).

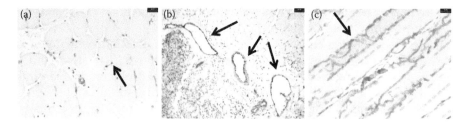

FIGURE 15.2
CD146 staining of ECs demonstrates disruption of vasculature after skeletal muscle injury. (a) Small capillaries are associated with each myofiber in uninjured tissue (arrow). (b) After injury, blood vessels are dilated and demonstrate an irregular morphology (arrow). (c) Blood vessel sprouting after injury (arrow).

Not surprisingly, the signaling pathways in regenerative angiogenesis are similar to those seen during the developmental process of vascularization, although historically this has been difficult to prove due to the embryonic lethality of mice lacking specific angiogenic genes. Deletion of VEGF, VEGF receptors, PDGF, Notch, and TGFβ pathway genes results in impaired vessel development and maturation and early embryonic lethality (Dickson, Martin et al. 1995; Fong, Rossant et al. 1995; Shalaby, Rossant et al. 1995; Carmeliet, Ferreira et al. 1996; Ferrara, Carver-Moore et al. 1996; Oshima, Oshima et al. 1996; Dumont, Jussila et al. 1998; Carmeliet, Lampugnani et al. 1999; Xue, Gao et al. 1999; Krebs, Xue et al. 2000; Aase, vonEuler et al. 2001). After injury, initial angiogenic signals, such as TGFβ and PDGF, are released from activated platelets at the site of injury (Carmeliet and Jain 2011). These signals provide a chemotactic gradient for the influx of immune cells to the wound site. Tunneling of macrophage through the ECM of damaged tissue, and the release of proangiogenic secreted factors, orchestrate initial remodeling of the vascular network (Moldovan, Goldschmidt-Clermont et al. 2000; Moldovan 2002; Kumar, Martin et al. 2013). Release of angiogenic signals from the immune cells facilitates the sprouting of an initial network of immature vessels. Pruning of the immature vessels followed by recruitment of PCs results in the reconstitution of a mature vascular network (Zawicki, Jain et al. 1981).

Minor tissue injury often results in angiogenesis dependent on the activation and rearrangement of local endothelial cells in response to autocrine and paracrine cues from the local tissue microenvironment. However, local endothelial cells may not be sufficient after a significant injury and full vascular reconstitution relies on the recruitment of cells from outside of the tissue niche. Several sources of these cells have been recently identified. Circulating endothelial progenitor cells (CEPs) have been shown to originate from EPCs in the bone marrow and can be found in the circulation after injury (Rafii and Lyden 2003). In addition to CEPs, circulating endothelial cells (CECs) may be released from the injured tissue into the systemic blood flow (Rafii and Lyden 2003). Tissue-specific EPCs that respond to their specific local microenvironment have also been recently identified (Butler, Kobayashi et al. 2010; Ding, Nolan et al. 2010; Ding, Nolan et al. 2011).

Recently, EPCs have been isolated from peripheral blood (Asahara, Murohara et al. 1997). Although these cells may hold promise for use in therapeutic angiogenesis, isolation of pure EPCs from blood is difficult since the hematopoietic and endothelial lineages share overlapping cell surface markers and the methodology involved in EPC separation is surface antigen based (Lin, Weisdorf et al. 2000; Hur, Yoon et al. 2004; Ingram, Mead et al. 2004). In addition, there is contradictory information on the true definition and characterization of EPCs, which has resulted in confusion in data interpretation (Ingram, Caplice et al. 2005; Timmermans, Plum et al. 2009; Yoder 2010).

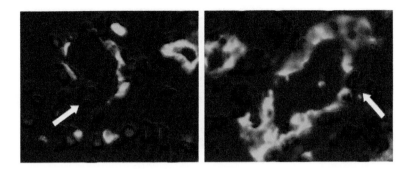

FIGURE 15.3
Transplanted human ECs integrate into host mouse vasculature. Mouse-specific CD31 (green), human-specific mitochondria (red), and nuclei (DAPI-blue).

Isolation of a homogeneous population of CD31+ EPCs has shown that this population of cells has the capacity for vessel formation *in vivo*. These cells have a typical EC/EPC phenotype, exhibit prolonged proliferation *in vitro*, and possess the ability to integrate into nascent vasculature *in vivo* (Melero-Martin, Khan et al. 2007, Melero-Martin, DeObaldia et al. 2008). However, the extremely small number of EPCs that can be isolated from peripheral blood has made it difficult to obtain sufficient cells for study using standard culture and sorting protocols (Ingram, Mead et al. 2004). Reinisch and colleagues have reported a method of whole blood culture for EPC in order to increase the success of EPC cultures. This method appears to result in four times as many EPC colonies as traditional methods (Reinisch, Hofmann et al. 2009), and may prove beneficial for future therapeutic purposes.

One would expect that the combination delivery of EPCs and mural cells/PCs could lead to longer maintenance of stable vasculature structures. This hypothesis is supported by recent *in vitro* and *in vivo* studies. Mesenchymal stem cells (MSCs) and human umbilical vein endothelial cells (HUVECs) cocultured together *in vitro*, resulted in the organization of the HUVECs into branched tubular networks and the deposition of condensed basement proteins (Sorrell, Baber et al. 2007, 2009; Criswell, Corona et al. 2013) (Figure 15.3). *In vivo*, Melero-Martin, DeObaldia et al. (2008) reported that when MSCs and EPCs isolated from cord blood were injected subcutaneously into nude mice, they formed vascular networks that lasted for 4 weeks. These networks consisted of an endothelial lumen that was derived from the implanted cells and SMA+ (smooth muscle actin) PCs. Au, Tam et al. (2008) reported that engineered blood vessels derived from HUVECs and human MSCs remained stable and functional for more than 130 days *in vivo*. These findings could guide future practices in tissue engineering and regenerative medicine, allowing physicians to form stable and long-lasting vasculature for engineered tissue.

15.3.3 Muscle–Vessel Interactions after Injury

In postnatal tissue, satellite cells and endothelial cells are in close proximity, and interact during regeneration and physiological procedures (Christov et al. 2007). Satellite cells are located close to capillaries, and in normal muscle, the number of associated satellite cells is correlated to the vascularization of the myofiber. The variation of satellite cells in physiological and pathological situations is correlated with capillary density (Christov, Chretien et al. 2007).

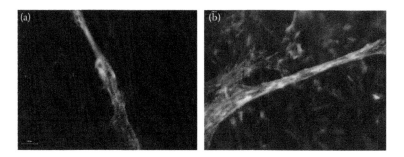

FIGURE 15.4
ECs (red) and MPCs (green) are closely associated when cultured together *in vitro* on (a) matrigel or (b) bladder acellular matrix scaffold.

It has been long known that the MPC "niche" in healthy muscle tissue is closely associated with the vasculature and thus, the processes of myogenesis and angiogenesis occur simultaneously during the regeneration process (Scholz et al. 2003; Abou-Khalil, Mounier et al. 2010; Best, Gharaibeh et al. 2012).

Signals arising from the ECs within the vasculature play a vital role in the maintenance of MPC viability and are able to stimulate MPC growth through the secretion of soluble growth factors (Delo, Eberli et al. 2008; Deasy, Feduska et al. 2009; Frey, Jansen et al. 2012; Gianni-Barrera, Trani et al. 2013). Moreover, our work and that of others demonstrated increased efficacy of MPC therapy after injury, as well as increased growth of tissue-engineered muscle in the presence of ECs or increased VEGF expression (De Coppi, Delo et al. 2005; Delo, Eberli et al. 2008; Deasy, Feduska et al. 2009). Further work demonstrated that MPCs and ECs were closely associated when cultured *in vitro* (Figure 15.4) and that subcutaneous implantation of co-seeded scaffolds into nude mice resulted in larger tissue-engineered muscle and increased numbers of patent blood vessels and nerve bundles (Criswell, Corona et al. 2013). Furthermore, the use of MPCs engineered to express VEGF was able to counteract the decline in the ability to engineer functional muscle tissue using cells isolated from older animals (Delo, Eberli et al. 2008). This data emphasize the importance of concurrent vascularization during skeletal muscle regeneration.

15.4 Tissue Engineering Approaches for Vascularized Muscle Tissue Formation

Current approaches for integrating a vasculature within tissue-engineered muscle utilize angiogenic growth factors and cells and/or tissue-engineered scaffolds that instruct and guide vessel formation. The use of EC progenitors alone or in combination with MPCs have been shown to enhance vascularization of injured tissue in preclinical animal models of Duchene muscular dystrophy, myocardial infarction, and hind limb ischemia (Van Den Bos and Taylor 2003; Aranguren, Pelacho et al. 2011; Chen, Rathbone et al. 2011; Leistner, Fischer-Rasokat et al. 2011; Ota, Uehara et al. 2011). However, the optimal angiogenic cell population for use in tissue engineering is still controversial and highly debated. The use of the angiogenic growth factor VEGF has also been shown to expedite endogenous blood vessel development and anastomoses of vessels derived from transplanted ECs with the host tissue vasculature (Brandao, Costa et al. 2011; Frey, Jansen et al. 2012; Zhou, He et al. 2013). However, caution must be exercised with the use of angiogenic growth factors for blood vessel development in injured tissue since prolonged exposure to these factors can result in excessive bleeding and hematoma formation. Bioengineered scaffolds can be designed for timed and controlled release of growth factors, which more closely mimics the process in native tissue. Moreover, scaffold can be engineered to guide EC migration and blood vessel development in predetermined configurations. Ultimately, it will most likely be a combination of cell therapy and tissue engineering approaches that deliver fully functional vascularized tissue-engineered skeletal muscle.

Despite their relative paucity in skeletal muscle tissue, satellite cells have tremendous potential to repair damaged muscle fibers and repopulate the myogenic stem cell pool in muscle tissue (Collins, Olsen et al. 2005; Rossi, Flaibani et al. 2011; Parker, Loretz et al. 2012; Yin, Price et al. 2013). While adult satellite cells are recognized as the main source of skeletal muscle stem cells, other cell types with myogenic potential have also been reported. These cells have been reported to arise from distal bone marrow or connective tissue in the vicinity of muscle fibers, and include bone marrow-derived hematopoietic and mesenchymal stem cells (Dezawa, Ishikawa et al. 2005; Au, Tam et al. 2008), skeletal muscle resident side population cells (muSP) (Asakura, Seale et al. 2002), muscle-derived stem cells (MDSCs) (Qu-Petersen, Deasy et al. 2002; Cao, Zheng et al. 2003), mesoangioblasts (Minasi, Riminucci et al. 2002; Cossu and Bianco 2003), and PCs (Dellavalle, Sampaolesi et al. 2007). Myogenic endothelial cells (MECs) are a subpopulation of cells derived from adult blood vessels that have been shown to contribute to myofiber formation in injured and dystrophic skeletal muscle (Zheng, Cao et al. 2007).

Mesoangioblasts, a subset of cells derived from the embryonic dorsal aorta, have shown the ability to differentiate into multiple mesodermal lineages, and appear to possess the ability to contribute to myofiber formation in dystrophic mice when implanted via intraarterial injection (Minasi, Riminucci et al. 2002). PCs derived from skeletal muscle can differentiate into myotubes *in vitro* when they are cocultured with myoblast cell lines or when cultured alone in muscle differentiation media (Dellavalle, Sampaolesi et al. 2007). Bone marrow cells, including single hematopoietic stem cells, have also been reported to contribute to myofibers formation (Camargo, Green et al. 2003; Corbel, Lee et al. 2003). However, the contribution of these alternate myogenic cells to myofiber formation is rare (1–10%) (LaBarge and Blau 2002; Abedi, Greer et al. 2004).

Despite decades of ongoing research into therapeutic angiogenesis, only limited clinical advantage has been demonstrated. Currently, the clinical uses of therapeutic angiogenesis has been largely limited to avascular or thin tissues, such as skin, bladder, or cartilage, relying predominately on infiltration of blood vessels from local host tissue (Bland, Dreau et al. 2012). This suggests that despite some success of vascular formation within thin tissues, these technologies are insufficient for the vascularization of thick engineered tissue such as skeletal muscle. Early research using MPCs engineered to express VEGF increased numbers of blood vessels in the tissue receiving the transplanted cells (Springer, Ozawa et al. 2003). However, disorganized and "leaky" vessels, similar to tumor vasculature, are a prominent defect in vascularized engineered tissues, especially in response to the addition of exogenous angiogenic growth factors (Dellian, Witwer et al. 1996; Jain 2003; Jain, Au et al. 2005; Lovett, Lee et al. 2009).

The ability to closely control the dosage, gradient, and time release of angiogenic factors is becoming more feasible through the use of hydrogels and growth factor-linked scaffolds (Lee, Peters et al. 2000; Ennett, Kaigler et al. 2006; Kaigler, Wang et al. 2006; Doi, Ikeda et al.

2007; Vulic and Shoichet 2012; Wang, Shansky et al. 2012). It is becoming increasingly clear that the results of angiogenic growth factors are dose dependent. Tissues have a biphasic response to VEGF doses. Low tissue concentration of VEGF results in increased vascular permeability, while high concentrations can result in hemangioma formation (Carmeliet 2000b). VEGF gradients within the tissue provide directionality for the sprouting tip cells during the angiogenesis procedure (Gerhardt 2008). As an example, *in vitro* microbead-based sprouting assays have demonstrated that ECs demonstrate maximal alignment in a 0–100 ng/mL VEGF gradient. Overly high VEGF concentrations saturated receptor binding and sprouting directionality was lost (Nehls and Drenckhahn 1995; Dietrich and Lelkes 2006; Chen, Silva et al. 2007). Furthermore, microfabrication techniques produced endothelial cell-lined tree-shaped channels that resembled the natural vasculature bed, but it was not clear if they would be able to support the tissue's vascular demand (Kaihara, Borenstein et al. 2000; Fidkowski, Kaazempur-Mofrad et al. 2005). Finally, the use of multiple growth factors may be warranted in order to mimic the endogenous angiogenic process (Borselli, Storrie et al. 2010).

Coculturing ECs along with MPCs has been shown to be an effective means of enhancing vascular formation in engineered skeletal muscle tissue (Levenberg, Rouwkema et al. 2005; Criswell, Corona et al. 2013). Several groups have convincingly shown the ability of ECs to enhance the development of engineered skeletal muscle constructs *in vitro* and *in vivo* (De Coppi, Delo et al. 2005; Levenberg, Rouwkema et al. 2005; Delo, Eberli et al. 2008; Deasy, Feduska et al. 2009; Criswell, Corona et al. 2013). The presence of ECs also enhanced the number of nerve bundles associated with the engineered tissue (Criswell, Corona et al. 2013). Likewise, it has been suggested that certain populations of MPCs may be able to enhance angiogenesis in skeletal muscle after injury (Ota, Uehara et al. 2011).

15.5 Conclusions

To date, there has been little success in the engineering of functional skeletal muscle tissue that can be used for clinical purposes. This is most likely due to the complex structure of this tissue and its dependence on the integration of additional microenvironmental components. The fabrication of functional muscle, whether *in vitro*, *ex vivo*, or *in vivo* is dependent on many factors, including the integration of a functional vascular network. An understanding of the cell–cell and cell–microenvironmental interactions as well as the signaling pathways that regulate normal tissue development can be used to inform the approaches used in tissue engineering. It is important to recognize the fact that the myogenic and vasculogenic processes occur concurrently and in close approximation during development. Therefore, it is intriguing to speculate that the adult tissue already possesses the "program" necessary for full regeneration, and may only require appropriate signals for reactivation. In fact, this is what occurs during tumorigenesis, as terminally differentiated cells assume a more primitive state. Of course, careful control of these processes would be required for successful tissue engineering in order to avoid potential tumor formation. The newly emerging field of regenerative pharmacology aims to identify mechanisms that can be targeted *in vivo* in order to activate endogenous pathways of tissue regeneration, perhaps altogether bypassing the need for cell therapy.

References

Aase, K., G. von Euler et al. 2001. Vascular endothelial growth factor-B-deficient mice display an atrial conduction defect. *Circulation* 104(3): 358–364.

Abedi, M., D. A. Greer et al. 2004. Robust conversion of marrow cells to skeletal muscle with formation of marrow-derived muscle cell colonies: A multifactorial process. *Exp Hematol* 32(5): 426–434.

Abou-Khalil, R., R. Mounier et al. 2010. Regulation of myogenic stem cell behavior by vessel cells: The menage a trois of satellite cells, periendothelial cells and endothelial cells. *Cell Cycle* 9(5): 892–896.

Alderton, J. M. and R. A. Steinhardt. 2000. How calcium influx through calcium leak channels is responsible for the elevated levels of calcium-dependent proteolysis in dystrophic myotubes. *Trends Cardiovasc Med* 10(6): 268–272.

Alexandrides, T., A. C. Moses et al. 1989. Developmental expression of receptors for insulin, insulin-like growth factor I (IGF-I), and IGF-II in rat skeletal muscle. *Endocrinology* 124(2): 1064–1076.

Allbrook, D. 1975. Transplantation and regeneration of striated muscle. *Ann R Coll Surg Engl* 56(6): 312–324.

Allen, R. E. and L. K. Boxhorn. 1989. Regulation of skeletal muscle satellite cell proliferation and differentiation by transforming growth factor-beta, insulin-like growth factor I, and fibroblast growth factor. *J Cell Physiol* 138(2): 311–315.

Allen, R. E., S. M. Sheehan et al. 1995. Hepatocyte growth factor activates quiescent skeletal muscle satellite cells *in vitro*. *J Cell Physiol* 165(2): 307–312.

Almekinders, L. C. and J. A. Gilbert. 1986. Healing of experimental muscle strains and the effects of nonsteroidal antiinflammatory medication. *Am J Sports Med* 14(4): 303–308.

Ambler, C. A., J. L. Nowicki et al. 2001. Assembly of trunk and limb blood vessels involves extensive migration and vasculogenesis of somite-derived angioblasts. *Dev Biol* 234(2): 352–364.

Amthor, H., R. Huang et al. 2002. The regulation and action of myostatin as a negative regulator of muscle development during avian embryogenesis. *Dev Biol* 251(2): 241–257.

Anderson, C., V. C. Williams et al. 2012. Sonic hedgehog acts cell-autonomously on muscle precursor cells to generate limb muscle diversity. *Genes Dev* 26(18): 2103–2117.

Aranguren, X. L., B. Pelacho et al. 2011. MAPC transplantation confers a more durable benefit than AC133+ cell transplantation in severe hind limb ischemia. *Cell Transplant* 20(2): 259–269.

Armand, O., A. M. Boutineau et al. 1983. Origin of satellite cells in avian skeletal muscles. *Arch Anat Microsc Morphol Exp* 72(2): 163–181.

Armstrong, R. B. 1990. Initial events in exercise-induced muscular injury. *Med Sci Sports Exerc* 22(4): 429–435.

Armstrong, R. B., G. L. Warren et al. 1991. Mechanisms of exercise-induced muscle fibre injury. *Sports Med* 12(3): 184–207.

Artavanis-Tsakonas, S., M. D. Rand et al. 1999. Notch signaling: Cell fate control and signal integration in development. *Science* 284(5415): 770–776.

Asahara, T., T. Murohara et al. 1997. Isolation of putative progenitor endothelial cells for angiogenesis. *Science* 275(5302): 964–967.

Asakura, A., P. Seale et al. 2002. Myogenic specification of side population cells in skeletal muscle. *J Cell Biol* 159(1): 123–134.

Au, P., J. Tam et al. 2008. Bone marrow-derived mesenchymal stem cells facilitate engineering of long-lasting functional vasculature. *Blood* 111(9): 4551–4558.

Bach, A. D., J. P. Beier et al. 2004. Skeletal muscle tissue engineering. *J Cell Mol Med* 8(4): 413–422.

Bajard, L., F. Relaix et al. 2006. A novel genetic hierarchy functions during hypaxial myogenesis: Pax3 directly activates Myf5 in muscle progenitor cells in the limb. *Genes Dev* 20(17): 2450–2464.

Belcastro, A. N., L. D. Shewchuk et al. 1998. Exercise-induced muscle injury: A calpain hypothesis. *Mol Cell Biochem* 179(1–2): 135–145.

Ben-Yair, R. and C. Kalcheim. 2008. Notch and bone morphogenetic protein differentially act on dermo-myotome cells to generate endothelium, smooth, and striated muscle. *J Cell Biol* 180(3): 607–618.

Bentzinger, C. F., Y. X. Wang et al. 2012. Building muscle: Molecular regulation of myogenesis. *Cold Spring Harb Perspect Biol* 4(2). DOI:10.1101/cshperspect.a008342.

Best, T. M., B. Gharaibeh et al. 2012. Stem cells, angiogenesis and muscle healing: A potential role in massage therapies? *Br J Sports Med* 47: 556–560.

Bland, E., D. Dreau et al. 2012. Overcoming hypoxia to improve tissue-engineering approaches to regenerative medicine. *J Tissue Eng Regen Med* 7: 505–514.

Blau, H. M. 2008. Cell therapies for muscular dystrophy. *N Engl J Med* 359(13): 1403–1405.

Bonnin, M. A., C. Laclef et al. 2005. *Six1* is not involved in limb tendon development, but is expressed in limb connective tissue under Shh regulation. *Mech Dev* 122(4): 573–585.

Borselli, C., H. Storrie et al. 2010. Functional muscle regeneration with combined delivery of angiogenesis and myogenesis factors. *Proc Natl Acad Sci USA* 107(8): 3287–3292.

Borycki, A. G. and C. P. Emerson, Jr. 2000. Multiple tissue interactions and signal transduction pathways control somite myogenesis. *Curr Top Dev Biol* 48: 165–224.

Brack, A. S., I. M. Conboy et al. 2008. A temporal switch from notch to Wnt signaling in muscle stem cells is necessary for normal adult myogenesis. *Cell Stem Cell* 2(1): 50–59.

Brandao, D., C. Costa et al. 2011. Endogenous vascular endothelial growth factor and angiopoietin-2 expression in critical limb ischemia. *Int Angiol* 30(1): 25–34.

Braun, T. and M. Gautel. 2011. Transcriptional mechanisms regulating skeletal muscle differentiation, growth and homeostasis. *Nat Rev Mol Cell Biol* 12(6): 349–361.

Buckingham, M. 2007. Skeletal muscle progenitor cells and the role of Pax genes. *C R Biol* 330(6–7): 530–533.

Buckingham, M., L. Bajard et al. 2003. The formation of skeletal muscle: From somite to limb. *J Anat* 202(1): 59–68.

Buckingham, M. and F. Relaix. 2007. The role of Pax genes in the development of tissues and organs: Pax3 and Pax7 regulate muscle progenitor cell functions. *Annu Rev Cell Dev Biol* 23: 645–673.

Butler, J. M., H. Kobayashi et al. 2010. Instructive role of the vascular niche in promoting tumour growth and tissue repair by angiocrine factors. *Nat Rev Cancer* 10(2): 138–146.

Camargo, F. D., R. Green et al. 2003. Single hematopoietic stem cells generate skeletal muscle through myeloid intermediates. *Nat Med* 9(12): 1520–1527.

Cao, B., B. Zheng et al. 2003. Muscle stem cells differentiate into haematopoietic lineages but retain myogenic potential. *Nat Cell Biol* 5(7): 640–646.

Carmeliet, P. 2000a. Mechanisms of angiogenesis and arteriogenesis. *Nat Med* 6(4): 389–395.

Carmeliet, P. 2000b. VEGF gene therapy: Stimulating angiogenesis or angioma-genesis? *Nat Med* 6(10): 1102–1103.

Carmeliet, P., F. De Smet et al. 2009. Branching morphogenesis and antiangiogenesis candidates: Tip cells lead the way. *Nat Rev Clin Oncol* 6(6): 315–326.

Carmeliet, P., V. Ferreira et al. 1996. Abnormal blood vessel development and lethality in embryos lacking a single VEGF allele. *Nature* 380(6573): 435–439.

Carmeliet, P. and R. K. Jain. 2011. Molecular mechanisms and clinical applications of angiogenesis. *Nature* 473(7347): 298–307.

Carmeliet, P., M. G. Lampugnani et al. 1999. Targeted deficiency or cytosolic truncation of the VE-cadherin gene in mice impairs VEGF-mediated endothelial survival and angiogenesis. *Cell* 98(2): 147–157.

Charge, S. B. and M. A. Rudnicki. 2004. Cellular and molecular regulation of muscle regeneration. *Physiol Rev* 84(1): 209–238.

Chen, R. R., E. A. Silva et al. 2007. Integrated approach to designing growth factor delivery systems. *FASEB J* 21(14): 3896–3903.

Chen, X. K., C. R. Rathbone et al. 2011. Treatment of tourniquet-induced ischemia reperfusion injury with muscle progenitor cells. *J Surg Res* 170(1): e65–73.

Chevallier, A., M. Kieny et al. 1977. Limb-somite relationship: Origin of the limb musculature. *J Embryol Exp Morphol* 41: 245–258.

Cho, H., T. Kozasa et al. 2003. Pericyte-specific expression of Rgs5: Implications for PDGF and EDG receptor signaling during vascular maturation. *FASEB J* 17(3): 440–442.

Christ, B., H. J. Jacob et al. 1977. Experimental analysis of the origin of the wing musculature in avian embryos. *Anat Embryol (Berl)* 150(2): 171–186.

Christ, B., M. Jacob et al. 1983. On the origin and development of the ventrolateral abdominal muscles in the avian embryo. An experimental and ultrastructural study. *Anat Embryol (Berl)* 166(1): 87–101.

Christ, B. and C. P. Ordahl. 1995. Early stages of chick somite development. *Anat Embryol (Berl)* 191(5): 381–396.

Christov, C., F. Chretien et al. 2007. Muscle satellite cells and endothelial cells: Close neighbors and privileged partners. *Mol Biol Cell* 18(4): 1397–1409.

Coffin, J. D. and T. J. Poole. 1988. Embryonic vascular development: Immunohistochemical identification of the origin and subsequent morphogenesis of the major vessel primordia in quail embryos. *Development* 102(4): 735–748.

Collins, C. A., I. Olsen et al. 2005. Stem cell function, self-renewal, and behavioral heterogeneity of cells from the adult muscle satellite cell niche. *Cell* 122(2): 289–301.

Conboy, I. M. and T. A. Rando. 2002. The regulation of Notch signaling controls satellite cell activation and cell fate determination in postnatal myogenesis. *Dev Cell* 3(3): 397–409.

Cooper, R. N., S. Tajbakhsh et al. 1999. *In vivo* satellite cell activation via Myf5 and MyoD in regenerating mouse skeletal muscle. *J Cell Sci* 112 (Pt 17): 2895–2901.

Corbel, S. Y., A. Lee et al. 2003. Contribution of hematopoietic stem cells to skeletal muscle. *Nat Med* 9(12): 1528–1532.

Cornelison, D. D. and B. J. Wold. 1997. Single-cell analysis of regulatory gene expression in quiescent and activated mouse skeletal muscle satellite cells. *Dev Biol* 191(2): 270–283.

Cossu, G. and P. Bianco. 2003. Mesoangioblasts—Vascular progenitors for extravascular mesodermal tissues. *Curr Opin Genet Dev* 13(5): 537–542.

Cossu, G., S. Tajbakhsh et al. 1996. How is myogenesis initiated in the embryo? *Trends Genet* 12(6): 218–223.

Criswell, T. L., B. T. Corona et al. 2013. The role of endothelial cells in myofiber differentiation and the vascularization and innervation of bioengineered muscle tissue *in vivo*. *Biomaterials* 34(1): 140–149.

Darr, K. C. and E. Schultz. 1987. Exercise-induced satellite cell activation in growing and mature skeletal muscle. *J Appl Physiol* 63(5): 1816–1821.

Daston, G., E. Lamar et al. 1996. Pax-3 is necessary for migration but not differentiation of limb muscle precursors in the mouse. *Development* 122(3): 1017–1027.

Daubas, P. and M. E. Buckingham. 2013. Direct molecular regulation of the myogenic determination gene Myf5 by Pax3, with modulation by Six1/4 factors, is exemplified by the -111kb-Myf5 enhancer. *Dev Biol* 376(2): 236–244.

De Coppi, P., D. Delo et al. 2005. Angiogenic gene-modified muscle cells for enhancement of tissue formation. *Tissue Eng* 11(7–8): 1034–1044.

Deasy, B. M., J. M. Feduska et al. 2009. Effect of VEGF on the regenerative capacity of muscle stem cells in dystrophic skeletal muscle. *Mol Ther J Am Soc Gene Ther* 17(10): 1788–1798.

Decary, S., V. Mouly et al. 1997. Replicative potential and telomere length in human skeletal muscle: Implications for satellite cell-mediated gene therapy. *Hum Gene Ther* 8(12): 1429–1438.

Dellavalle, A., M. Sampaolesi et al. 2007. Pericytes of human skeletal muscle are myogenic precursors distinct from satellite cells. *Nat Cell Biol* 9(3): 255–267.

Dellian, M., B. P. Witwer et al. 1996. Quantitation and physiological characterization of angiogenic vessels in mice: Effect of basic fibroblast growth factor, vascular endothelial growth factor/ vascular permeability factor, and host microenvironment. *Am J Pathol* 149(1): 59–71.

Delo, D. M., D. Eberli et al. 2008. Angiogenic gene modification of skeletal muscle cells to compensate for ageing-induced decline in bioengineered functional muscle tissue. *BJU Int* 102(7): 878–884.

Dezawa, M., H. Ishikawa et al. 2005. Bone marrow stromal cells generate muscle cells and repair muscle degeneration. *Science* 309(5732): 314–317.

Dhawan, J. and T. A. Rando. 2005. Stem cells in postnatal myogenesis: Molecular mechanisms of satellite cell quiescence, activation and replenishment. *Trends Cell Biol* 15(12): 666–673.

Dickson, M. C., J. S. Martin et al. 1995. Defective haematopoiesis and vasculogenesis in transforming growth factor-beta 1 knock out mice. *Development* 121(6): 1845–1854.

Dietrich, F. and P. I. Lelkes. 2006. Fine-tuning of a three-dimensional microcarrier-based angiogenesis assay for the analysis of endothelial-mesenchymal cell co-cultures in fibrin and collagen gels. *Angiogenesis* 9(3): 111–125.

Dietrich, S., F. Abou-Rebyeh et al. 1999. The role of SF/HGF and c-Met in the development of skeletal muscle. *Development* 126(8): 1621–1629.

Ding, B. S., D. J. Nolan et al. 2010. Inductive angiocrine signals from sinusoidal endothelium are required for liver regeneration. *Nature* 468(7321): 310–315.

Ding, B. S., D. J. Nolan et al. 2011. Endothelial-derived angiocrine signals induce and sustain regenerative lung alveolarization. *Cell* 147(3): 539–553.

Doi, K., T. Ikeda et al. 2007. Enhanced angiogenesis by gelatin hydrogels incorporating basic fibroblast growth factor in rabbit model of hind limb ischemia. *Heart Vessels* 22(2): 104–108.

Doumit, M. E., D. R. Cook et al. 1993. Fibroblast growth factor, epidermal growth factor, insulin-like growth factors, and platelet-derived growth factor-BB stimulate proliferation of clonally derived porcine myogenic satellite cells. *J Cell Physiol* 157(2): 326–332.

Dumont, D. J., L. Jussila et al. 1998. Cardiovascular failure in mouse embryos deficient in VEGF receptor-3. *Science* 282(5390): 946–949.

Duprez, D., C. Fournier-Thibault et al. 1998. Sonic Hedgehog induces proliferation of committed skeletal muscle cells in the chick limb. *Development* 125(3): 495–505.

Ennett, A. B., D. Kaigler et al. 2006. Temporally regulated delivery of VEGF *in vitro* and *in vivo*. *J Biomed Mater Res A* 79(1): 176–184.

Esner, M., S. M. Meilhac et al. 2006. Smooth muscle of the dorsal aorta shares a common clonal origin with skeletal muscle of the myotome. *Development* 133(4): 737–749.

Ferrara, N., K. Carver-Moore et al. 1996. Heterozygous embryonic lethality induced by targeted inactivation of the VEGF gene. *Nature* 380(6573): 439–442.

Fidkowski, C., M. R. Kaazempur-Mofrad et al. 2005. Endothelialized microvasculature based on a biodegradable elastomer. *Tissue Eng* 11(1–2): 302–309.

Fielding, R. A., T. J. Manfredi et al. 1993. Acute phase response in exercise. III. Neutrophil and IL-1 beta accumulation in skeletal muscle. *Am J Physiol* 265(1 Pt 2): R166–172.

Florini, J. R., D. Z. Ewton et al. 1993. IGFs and muscle differentiation. *Adv Exp Med Biol* 343: 319–326.

Fong, G. H., J. Rossant et al. 1995. Role of the Flt-1 receptor tyrosine kinase in regulating the assembly of vascular endothelium. *Nature* 376(6535): 66–70.

Frey, S. P., H. Jansen et al. 2012. VEGF improves skeletal muscle regeneration after acute trauma and reconstruction of the limb in a rabbit model. *Clin Orthop Relat Res* 470(12): 3607–3614.

Fuchtbauer, E. M. and H. Westphal. 1992. MyoD and myogenin are coexpressed in regenerating skeletal muscle of the mouse. *Dev Dyn* 193(1): 34–39.

Gaengel, K., G. Genove et al. 2009. Endothelial-mural cell signaling in vascular development and angiogenesis. *Arterioscler Thromb Vasc Biol* 29(5): 630–638.

Gerhardt, H. 2008. VEGF and endothelial guidance in angiogenic sprouting. *Organogenesis* 4(4): 241–246.

Gianni-Barrera, R., M. Trani et al. 2013. VEGF over-expression in skeletal muscle induces angiogenesis by intussusception rather than sprouting. *Angiogenesis* 16(1): 123–136.

Grenier, G., A. Scime et al. 2007. Resident endothelial precursors in muscle, adipose, and dermis contribute to postnatal vasculogenesis. *Stem Cells* 25(12): 3101–3110.

Gros, J., M. Manceau et al. 2005. A common somitic origin for embryonic muscle progenitors and satellite cells. *Nature* 435(7044): 954–958.

Gustafsson, M. K., H. Pan et al. 2002. Myf5 is a direct target of long-range Shh signaling and Gli regulation for muscle specification. *Genes Dev* 16(1): 114–126.

Hall-Craggs, E. C. 1974. The regeneration of skeletal muscle fibres per continuum. *J Anat* 117(Pt 1): 171–178.

Hasty, P., A. Bradley et al. 1993. Muscle deficiency and neonatal death in mice with a targeted mutation in the myogenin gene. *Nature* 364(6437): 501–506.

Haugk, K. L., R. A. Roeder et al. 1995. Regulation of muscle cell proliferation by extracts from crushed muscle. *J Anim Sci* 73(7): 1972–1981.

Huang, R., Q. Zhi et al. 2003. The relationship between limb muscle and endothelial cells migrating from single somite. *Anat Embryol (Berl)* 206(4): 283–289.

Hungerford, J. E., G. K. Owens et al. 1996. Development of the aortic vessel wall as defined by vascular smooth muscle and extracellular matrix markers. *Dev Biol* 178(2): 375–392.

Hur, J., C. H. Yoon et al. 2004. Characterization of two types of endothelial progenitor cells and their different contributions to neovasculogenesis. *Arterioscler Thromb Vasc Biol* 24(2): 288–293.

Ingram, D. A., N. M. Caplice et al. 2005. Unresolved questions, changing definitions, and novel paradigms for defining endothelial progenitor cells. *Blood* 106(5): 1525–1531.

Ingram, D. A., L. E. Mead et al. 2004. Identification of a novel hierarchy of endothelial progenitor cells using human peripheral and umbilical cord blood. *Blood* 104(9): 2752–2760.

Jain, R. K. 2003. Molecular regulation of vessel maturation. *Nat Med* 9(6): 685–693.

Jain, R. K., P. Au et al. 2005. Engineering vascularized tissue. *Nat Biotechnol* 23(7): 821–823.

Kaigler, D., Z. Wang et al. 2006. VEGF scaffolds enhance angiogenesis and bone regeneration in irradiated osseous defects. *J Bone Miner Res* 21(5): 735–744.

Kaihara, S., J. Borenstein et al. 2000. Silicon micromachining to tissue engineer branched vascular channels for liver fabrication. *Tissue Eng* 6(2): 105–117.

Kassar-Duchossoy, L., E. Giacone et al. 2005. Pax3/Pax7 mark a novel population of primitive myogenic cells during development. *Genes Dev* 19(12): 1426–1431.

Krebs, L. T., Y. Xue et al. 2000. Notch signaling is essential for vascular morphogenesis in mice. *Genes Dev* 14(11): 1343–1352.

Kruger, M., D. Mennerich et al. 2001. Sonic hedgehog is a survival factor for hypaxial muscles during mouse development. *Development* 128(5): 743–752.

Kumar, A. H., K. Martin et al. 2013. Role of CX3CR1 receptor in monocyte/macrophage driven neovascularization. *PLoS One* 8(2): e57230.

LaBarge, M. A. and H. M. Blau. 2002. Biological progression from adult bone marrow to mononucleate muscle stem cell to multinucleate muscle fiber in response to injury. *Cell* 111(4): 589–601.

Lagha, M., J. D. Kormish et al. 2008. Pax3 regulation of FGF signaling affects the progression of embryonic progenitor cells into the myogenic program. *Genes Dev* 22(13): 1828–1837.

Le Grand, F., G. Auda-Boucher et al. 2004. Endothelial cells within embryonic skeletal muscles: A potential source of myogenic progenitors. *Exp Cell Res* 301(2): 232–241.

Lee, K. Y., M. C. Peters et al. 2000. Controlled growth factor release from synthetic extracellular matrices. *Nature* 408(6815): 998–1000.

Lee, S., T. T. Chen et al. 2007. Autocrine VEGF signaling is required for vascular homeostasis. *Cell* 130(4): 691–703.

Leistner, D. M., U. Fischer-Rasokat et al. 2011. Transplantation of progenitor cells and regeneration enhancement in acute myocardial infarction (TOPCARE-AMI): Final 5-year results suggest long-term safety and efficacy. *Clin Res Cardiol* 100(10): 925–934.

Lescaudron, L., E. Peltekian et al. 1999. Blood borne macrophages are essential for the triggering of muscle regeneration following muscle transplant. *Neuromuscul Disord* 9(2): 72–80.

Levenberg, S., J. Rouwkema et al. 2005. Engineering vascularized skeletal muscle tissue. *Nat Biotechnol* 23(7): 879–884.

Lin, Y., D. J. Weisdorf et al. 2000. Origins of circulating endothelial cells and endothelial outgrowth from blood. *J Clin Invest* 105(1): 71–77.

Loughna, S. and T. N. Sato. 2001. Angiopoietin and Tie signaling pathways in vascular development. *Matrix Biol* 20(5–6): 319–325.

Lovett, M., K. Lee et al. 2009. Vascularization strategies for tissue engineering. *Tissue Eng Part B Rev* 15(3): 353–370.

Marcelle, C., S. Ahlgren et al. 1999. *In vivo* regulation of somite differentiation and proliferation by Sonic Hedgehog. *Dev Biol* 214(2): 277–287.

Mauro, A. 1961. Satellite cell of skeletal muscle fibers. *J Biophys Biochem Cytol* 9: 493–495.

McMahon, J. A., S. Takada et al. 1998. Noggin-mediated antagonism of BMP signaling is required for growth and patterning of the neural tube and somite. *Genes Dev* 12(10): 1438–1452.

Melero-Martin, J. M., M. E. De Obaldia et al. 2008. Engineering robust and functional vascular networks *in vivo* with human adult and cord blood-derived progenitor cells. *Circ Res* 103(2): 194–202.

Melero-Martin, J. M., Z. A. Khan et al. 2007. *In vivo* vasculogenic potential of human blood-derived endothelial progenitor cells. *Blood* 109(11): 4761–4768.

Merly, F., L. Lescaudron et al. 1999. Macrophages enhance muscle satellite cell proliferation and delay their differentiation. *Muscle Nerve* 22(6): 724–732.

Minasi, M. G., M. Riminucci et al. 2002. The meso-angioblast: A multipotent, self-renewing cell that originates from the dorsal aorta and differentiates into most mesodermal tissues. *Development* 129(11): 2773–2783.

Moldovan, N. I. 2002. Role of monocytes and macrophages in adult angiogenesis: A light at the tunnel's end. *J Hematother Stem Cell Res* 11(2): 179–194.

Moldovan, N. I., P. J. Goldschmidt-Clermont et al. 2000. Contribution of monocytes/macrophages to compensatory neovascularization: The drilling of metalloelastase-positive tunnels in ischemic myocardium. *Circ Res* 87(5): 378–384.

Montarras, D., J. Morgan et al. 2005. Direct isolation of satellite cells for skeletal muscle regeneration. *Science* 309(5743): 2064–2067.

Munsterberg, A. E., J. Kitajewski et al. 1995. Combinatorial signaling by Sonic hedgehog and Wnt family members induces myogenic bHLH gene expression in the somite. *Genes Dev* 9(23): 2911–2922.

Munsterberg, A. E. and A. B. Lassar. 1995. Combinatorial signals from the neural tube, floor plate and notochord induce myogenic bHLH gene expression in the somite. *Development* 121(3): 651–660.

Nabeshima, Y., K. Hanaoka et al. 1993. Myogenin gene disruption results in perinatal lethality because of severe muscle defect. *Nature* 364(6437): 532–535.

Nehls, V. and D. Drenckhahn. 1995. A novel, microcarrier-based *in vitro* assay for rapid and reliable quantification of three-dimensional cell migration and angiogenesis. *Microvasc Res* 50(3): 311–322.

Nguyen, L. L. and P. A. D'Amore. 2001. Cellular interactions in vascular growth and differentiation. *Int Rev Cytol* 204: 1–48.

Orimo, S., E. Hiyamuta et al. 1991. Analysis of inflammatory cells and complement C3 in bupivacaine-induced myonecrosis. *Muscle Nerve* 14(6): 515–520.

Oshima, M., H. Oshima et al. 1996. TGF-beta receptor type II deficiency results in defects of yolk sac hematopoiesis and vasculogenesis. *Dev Biol* 179(1): 297–302.

Ota, S., K. Uehara et al. 2011. Intramuscular transplantation of muscle-derived stem cells accelerates skeletal muscle healing after contusion injury via enhancement of angiogenesis. *Am J Sports Med* 39(9): 1912–1922.

Ott, H. C., T. S. Matthiesen et al. 2008. Perfusion-decellularized matrix: Using nature's platform to engineer a bioartificial heart. *Nat Med* 14(2): 213–221.

Otto, A., C. Schmidt et al. 2008. Canonical Wnt signalling induces satellite-cell proliferation during adult skeletal muscle regeneration. *J Cell Sci* 121(Pt 17): 2939–2950.

Pardanaud, L., C. Altmann et al. 1987. Vasculogenesis in the early quail blastodisc as studied with a monoclonal antibody recognizing endothelial cells. *Development* 100(2): 339–349.

Parker, M. H., C. Loretz et al. 2012. Activation of Notch signaling during *ex vivo* expansion maintains donor muscle cell engraftment. *Stem Cells* 30(10): 2212–2220.

Pepper, M. S. 1997. Transforming growth factor-beta: Vasculogenesis, angiogenesis, and vessel wall integrity. *Cytokine Growth Factor Rev* 8(1): 21–43.

Percy, M. E., L. S. Chang et al. 1979. Serum creatine kinase and pyruvate kinase in Duchenne muscular dystrophy carrier detection. *Muscle Nerve* 2(5): 329–339.

Perez-Ruiz, A., Y. Ono et al. 2008. Beta-Catenin promotes self-renewal of skeletal-muscle satellite cells. *J Cell Sci* 121(Pt 9): 1373–1382.

Poole, T. J., E. B. Finkelstein et al. 2001. The role of FGF and VEGF in angioblast induction and migration during vascular development. *Dev Dyn* 220(1): 1–17.

Qu-Petersen, Z., B. Deasy et al. 2002. Identification of a novel population of muscle stem cells in mice: Potential for muscle regeneration. *J Cell Biol* 157(5): 851–864.

Rafii, S. and D. Lyden. 2003. Therapeutic stem and progenitor cell transplantation for organ vascularization and regeneration. *Nat Med* 9(6): 702–712.

Rawls, A., J. H. Morris et al. 1995. Myogenin's functions do not overlap with those of MyoD or Myf-5 during mouse embryogenesis. *Dev Biol* 172(1): 37–50.

Reinisch, A., N. A. Hofmann et al. 2009. Humanized large-scale expanded endothelial colony-forming cells function *in vitro* and *in vivo*. *Blood* 113(26): 6716–6725.

Relaix, F., M. Polimeni et al. 2003. The transcriptional activator PAX3-FKHR rescues the defects of Pax3 mutant mice but induces a myogenic gain-of-function phenotype with ligand-independent activation of Met signaling *in vivo*. *Genes Dev* 17(23): 2950–2965.

Relaix, F., D. Rocancourt et al. 2004. Divergent functions of murine Pax3 and Pax7 in limb muscle development. *Genes Dev* 18(9): 1088–1105.

Relaix, F., D. Rocancourt et al. 2005. A Pax3/Pax7-dependent population of skeletal muscle progenitor cells. *Nature* 435(7044): 948–953.

Risau, W. and I. Flamme. 1995. Vasculogenesis. *Annu Rev Cell Dev Biol* 11: 73–91.

Robertson, T. A., M. A. Maley et al. 1993. The role of macrophages in skeletal muscle regeneration with particular reference to chemotaxis. *Exp Cell Res* 207(2): 321–331.

Rossi, C. A., M. Flaibani et al. 2011. *In vivo* tissue engineering of functional skeletal muscle by freshly isolated satellite cells embedded in a photopolymerizable hydrogel. *FASEB J* 25(7): 2296–2304.

Ruffell, D., F. Mourkioti et al. 2009. A CREB-C/EBPbeta cascade induces M2 macrophage-specific gene expression and promotes muscle injury repair. *Proc Natl Acad Sci USA* 106(41): 17475–17480.

Saharinen, P., L. Eklund et al. 2008. Angiopoietins assemble distinct Tie2 signalling complexes in endothelial cell-cell and cell-matrix contacts. *Nat Cell Biol* 10(5): 527–537.

Sambasivan, R., R. Yao et al. 2011. Pax7-expressing satellite cells are indispensable for adult skeletal muscle regeneration. *Development* 138(17): 3647–3656.

Sartore, S., A. Chiavegato et al. 2001. Contribution of adventitial fibroblasts to neointima formation and vascular remodeling: From innocent bystander to active participant. *Circ Res* 89(12): 1111–1121.

Scaal, M. and B. Christ. 2004. Formation and differentiation of the avian dermomyotome. *Anat Embryol (Berl)* 208(6): 411–424.

Schmalbruch, H. and D. M. Lewis. 2000. Dynamics of nuclei of muscle fibers and connective tissue cells in normal and denervated rat muscles. *Muscle Nerve* 23(4): 617–626.

Scholz, D., S. Thomas et al. 2003. Angiogenesis and myogenesis as two facets of inflammatory post-ischemic tissue regeneration. *Mol Cell Biochem* 246(1–2): 57–67.

Shalaby, F., J. Rossant et al. 1995. Failure of blood-island formation and vasculogenesis in Flk-1-deficient mice. *Nature* 376(6535): 62–66.

Shim, W. S., I. A. Ho et al. 2007. Angiopoietin: A TIE(d) balance in tumor angiogenesis. *Mol Cancer Res* 5(7): 655–665.

Smith, C. K., 2nd, M. J. Janney et al. 1994. Temporal expression of myogenic regulatory genes during activation, proliferation, and differentiation of rat skeletal muscle satellite cells. *J Cell Physiol* 159(2): 379–385.

Snow, M. H. 1977. Myogenic cell formation in regenerating rat skeletal muscle injured by mincing. II. An autoradiographic study. *Anat Rec* 188(2): 201–217.

Snow, M. H. 1978. An autoradiographic study of satellite cell differentiation into regenerating myotubes following transplantation of muscles in young rats. *Cell Tissue Res* 186(3): 535–540.

Snyder, C. M., A. L. Rice et al. 2013. MEF2A regulates the Gtl2-Dio3 microRNA mega-cluster to modulate WNT signaling in skeletal muscle regeneration. *Development* 140(1): 31–42.

Sorrell, J. M., M. A. Baber et al. 2007. A self-assembled fibroblast-endothelial cell co-culture system that supports *in vitro* vasculogenesis by both human umbilical vein endothelial cells and human dermal microvascular endothelial cells. *Cells Tissues Organs* 186(3): 157–168.

Sorrell, J. M., M. A. Baber et al. 2009. Influence of adult mesenchymal stem cells on *in vitro* vascular formation. *Tissue Eng Part A* 15(7): 1751–1761.

Springer, M. L., A. S. Chen et al. 1998. VEGF gene delivery to muscle: Potential role for vasculogenesis in adults. *Mol Cell* 2(5): 549–558.

Springer, M. L., C. R. Ozawa et al. 2003. Localized arteriole formation directly adjacent to the site of VEGF-induced angiogenesis in muscle. *Mol Ther* 7(4): 441–449.

Tajbakhsh, S., E. Bober et al. 1996. Gene targeting the myf-5 locus with nlacZ reveals expression of this myogenic factor in mature skeletal muscle fibres as well as early embryonic muscle. *Dev Dyn* 206(3): 291–300.

Tatsumi, R., J. E. Anderson et al. 1998. HGF/SF is present in normal adult skeletal muscle and is capable of activating satellite cells. *Dev Biol* 194(1): 114–128.

Tatsumi, R., S. M. Sheehan et al. 2001. Mechanical stretch induces activation of skeletal muscle satellite cells *in vitro*. *Exp Cell Res* 267(1): 107–114.

Tedesco, F. S., A. Dellavalle et al. 2010. Repairing skeletal muscle: Regenerative potential of skeletal muscle stem cells. *J Clin Invest* 120(1): 11–19.

Ten Broek, R. W., S. Grefte et al. 2010. Regulatory factors and cell populations involved in skeletal muscle regeneration. *J Cell Physiol* 224(1): 7–16.

Thommen, R., R. Humar et al. 1997. PDGF-BB increases endothelial migration on cord movements during angiogenesis *in vitro*. *J Cell Biochem* 64(3): 403–413.

Tidball, J. G. 1995. Inflammatory cell response to acute muscle injury. *Med Sci Sports Exerc* 27(7): 1022–1032.

Timmermans, F., J. Plum et al. 2009. Endothelial progenitor cells: Identity defined? *J Cell Mol Med* 13(1): 87–102.

Tozer, S., M. A. Bonnin et al. 2007. Involvement of vessels and PDGFB in muscle splitting during chick limb development. *Development* 134(14): 2579–2591.

Van Den Bos, E. J. and D. A. Taylor. 2003. Cardiac transplantation of skeletal myoblasts for heart failure. *Minerva Cardioangiol* 51(2): 227–243.

Vandenburgh, H. H. 2002. Functional assessment and tissue design of skeletal muscle. *Ann N Y Acad Sci* 961: 201–202.

Vulic, K. and M. S. Shoichet. 2012. Tunable growth factor delivery from injectable hydrogels for tissue engineering. *J Am Chem Soc* 134(2): 882–885.

Wagers, A. J. and I. M. Conboy. 2005. Cellular and molecular signatures of muscle regeneration: Current concepts and controversies in adult myogenesis. *Cell* 122(5): 659–667.

Wang, H., F. Noulet et al. 2010. Bmp signaling at the tips of skeletal muscles regulates the number of fetal muscle progenitors and satellite cells during development. *Dev Cell* 18(4): 643–654.

Wang, L., J. Shansky et al. 2012. Design and fabrication of a biodegradable, covalently crosslinked shape-memory alginate scaffold for cell and growth factor delivery. *Tissue Eng Part A* 18(19–20): 2000–2007.

Wang, Y. X. and M. A. Rudnicki. 2012. Satellite cells, the engines of muscle repair. *Nat Rev Mol Cell Biol* 13(2): 127–133.

Wen, Y., P. Bi et al. 2012. Constitutive Notch activation upregulates Pax7 and promotes the self-renewal of skeletal muscle satellite cells. *Mol Cell Biol* 32(12): 2300–2311.

Whalen, R. G., J. B. Harris et al. 1990. Expression of myosin isoforms during notexin-induced regeneration of rat soleus muscles. *Dev Biol* 141(1): 24–40.

Xue, Y., X. Gao et al. 1999. Embryonic lethality and vascular defects in mice lacking the Notch ligand Jagged1. *Hum Mol Genet* 8(5): 723–730.

Yablonka-Reuveni, Z. and A. J. Rivera. 1994. Temporal expression of regulatory and structural muscle proteins during myogenesis of satellite cells on isolated adult rat fibers. *Dev Biol* 164(2): 588–603.

Yamashita, J., H. Itoh et al. 2000. Flk1-positive cells derived from embryonic stem cells serve as vascular progenitors. *Nature* 408(6808): 92–96.

Yancopoulos, G. D., S. Davis et al. 2000. Vascular-specific growth factors and blood vessel formation. *Nature* 407(6801): 242–248.

Yin, H., F. Price et al. 2013. Satellite cells and the muscle stem cell niche. *Physiol Rev* 93(1): 23–67.

Yoder, M. C. 2010. Is endothelium the origin of endothelial progenitor cells? *Arterioscler Thromb Vasc Biol* 30(6): 1094–1103.

Yusuf, F. and B. Brand-Saberi. 2012. Myogenesis and muscle regeneration. *Histochem Cell Biol* 138(2): 187–199.

Zawicki, D. F., R. K. Jain et al. 1981. Dynamics of neovascularization in normal tissue. *Microvasc Res* 21(1): 27–47.

Zheng, B., B. Cao et al. 2007. Prospective identification of myogenic endothelial cells in human skeletal muscle. *Nat Biotechnol* 25(9): 1025–1034.

Zhou, W., D. Q. He et al. 2013. Angiogenic gene-modified myoblasts promote vascularization during repair of skeletal muscle defects. *J Tissue Eng Regen Med*. DOI:10.1002/term.1692.

16

Vascularization in Engineered Adipose Tissue

Marcella K. Vaicik, Ronald N. Cohen, and Eric Michael Brey

CONTENTS

16.1 Introduction

Approximately, 20% of the total body weight in healthy adult men and 25% in women is attributed to the adipose tissue (Tanzi and Fare, 2009). In the past, adipose tissue was primarily associated with energy storage. However, in more recent years, it has become clear that adipose tissue serves many other important functions, including the regulation of glucose tolerance and insulin sensitivity, feeding behavior, and inflammatory responses (Trayhurn and Wood, 2004). Adipose tissue is constantly remodeled throughout life, and the appropriate adipose development and function are essential for long-term health.

Over 5 million reconstruction procedures are annually performed in the United States alone (www.plasticsurgery.org), and the number of these procedures continues to increase both domestically and internationally. Adipose for soft-tissue reconstruction is required following trauma and tumor resection, for treatment of congenital defects, and for composite tissue reconstruction. Surgical methods generally involve autologous tissue transplantation to the site of the injured tissue. However, these methods are particularly challenging when applied to adipose tissue. In addition, techniques are limited by tissue availability, difficulty in transfer of the tissue, and donor site morbidity. Tissue-engineered biomaterials that aid in the formation or regeneration of properly functioning adipose tissue could provide an alternative approach for soft-tissue reconstruction. Biomaterials that more closely match the structure and function of healthy adipose tissue could provide surgeons with new options for overcoming the current treatment limitations of soft-tissue injuries and defects.

In addition to the replacement and repair of damaged tissue, engineered adipose tissue could also provide *in vitro* preclinical three-dimensional (3D) tissue models for developing therapeutics, testing pharmaceuticals, and studying the basic physiological and pathological processes. Traditionally, researchers have used two-dimensional (2D) substrates to investigate cell behavior. However, cells *in vivo* exist in a 3D environment surrounded by a complex extracellular matrix (ECM). Studies with a number of other cell types suggest that cell function approaches physiological conditions more closely when an appropriate 3D culture system is used. Recent studies suggest that this is also true for adipocytes (Chun et al., 2006, Stacey et al., 2009).

Regardless of the application, engineering anatomically and functionally correct adipose tissue requires the formation of microvascular networks within the tissue. The networks allow exchange of vital nutrients, cells, and signals required for adipose function. In this chapter, we first review the structure and composition of adipose tissue and discuss the functional roles it plays in the human body. We then discuss the importance of vasculature to the development, expansion, and maintenance of adipose tissue. Finally, we summarize ongoing research focused on engineering vascularized adipose tissue, identifying limitations, and suggesting areas for further investigation.

16.2 Adipose Tissue

Adipose tissue is a highly specialized connective tissue of mesodermal origin. The human body consists of two main types of adipose tissue, white adipose tissue (WAT) and brown adipose tissue (BAT). The more prevalent adipose type in the human body is WAT. Total WAT ranges from 10 to 35 kg in adults (Mott et al., 1999), while adults have on the level of 10 g (0.01 kg) of BAT (Cypess et al., 2009). All WAT depots are not connected and each depot may have specific functions. Two of the more studied, and important, WAT depots are subcutaneous and visceral (or intraperitoneal) WAT. The distribution of adipose in each depot varies by individual and is influenced by age, sex, and health. Increased visceral WAT is associated with increased risk for cardiovascular disease and type 2 diabetes whereas subcutaneous WAT volume may be associated with lower risk for these health problems (Gesta et al., 2007).

While WAT is involved in energy storage and hormone regulation, BAT is involved in energy expenditure in the regulation of thermogenesis. There is evidence that brown adipocytes develop from a progenitor distinct from white adipocytes, although this is controversial. Historically, it was felt that BAT played an important role in infants and other species, but not in adult humans. However, recent research has challenged this view, and the possibility of targeting BAT to increase energy utilization has been considered as a potentially therapeutic modality in the treatment of obesity. While BAT may serve important functional roles in the body, the large volume of WAT in adults and its important contributions to metabolic function and clinical disorders (such as obesity and lipodystrophy) has made it the more widely studied type of adipose tissue. To our knowledge, there have not been any efforts to engineer BAT. For this reason, we will focus on WAT throughout the chapter.

16.2.1 Cellular Composition

The primary cell type in adipose tissue is the adipocyte. Adipocytes (specifically, white adipocytes) are specialized cells that synthesize and store triglycerides as a lipid droplet

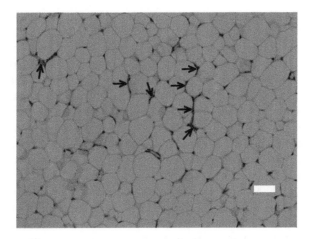

FIGURE 16.1
WAT stained with hematoxylin and eosin (H&E). Large lipid-loaded adipocytes are seen throughout with some ECM observed in between. ECM is indicated by arrows. Scale bar is 50 μm.

for later release to provide energy. In addition to their role in energy storage, adipocytes release several adipokines that contribute to the function of adipose as an endocrine organ. These adipokines will be described in more detail in a later section. In fully differentiated, or mature, adipocytes, the majority of the central cytoplasm contains a single large lipid droplet with the nucleus located at the periphery (Figure 16.1). Each adipocyte is surrounded by a thin network of proteins known as the basement membrane (BM) and is in close proximity to capillaries (Huang and Greenspan, 2012) (Figure 16.2).

While adipocytes are the predominant cell type, WAT is a heterogeneous organ composed of several distinct cell types. Adipocytes can increase or decrease in size through

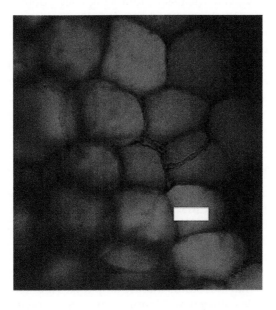

FIGURE 16.2
Confocal microscopy image of epididymal fat pad in which adipocytes are stained green with bodipy and vasculature stained red with lectin perfusion. Scale bar is 25 μm.

the process of lipid loading and energy release but do not proliferate. Preadipocytes are the precursor cells to adipocytes and they are able to proliferate prior to differentiation. In culture, preadipocytes have a morphology similar to fibroblasts. The preadipocytes are thought to reside in the connective tissue-rich areas of adipose tissue near the vasculature and are recruited to differentiate into mature adipocytes either for tissue renewal (replacement of existing adipocytes) or for adipose tissue expansion (increasing the number of adipocytes and capacity for energy storage) (Tang et al., 2008).

Adipose tissue also contains adipose-derived stem cells (ASCs). ASCs are a heterogeneous population of multipotent mesenchymal cells and possibly endothelial progenitor cells that can be stimulated to differentiate into a variety of cell types, including adipocytes, osteoblasts, and, in some cases, endothelial cells (ECs), smooth muscle cells, and cardiomyocytes (Madonna et al., 2009). The term ASC was adopted by the International Fat Applied Technology Society to identify the isolated, plastic-adherent, multipotent cell population; a specific portion of the cells isolated from the stromal vascular fraction of adipose tissue (Gimble et al., 2007). A few studies suggest depot specific differences in the amount and type of ASCs present (van Harmelen et al., 2004, Prunet-Marcassus et al., 2006). These cells may take on a perivascular location within adipose tissue (Tang et al., 2008).

Adipose tissue also contributes to immune function and contains lymphocytes, T-cells, and macrophages (Ahima and Flier, 2000, Hotamisligil, 2006). These cells secrete cytokines that can contribute to adipose tissue metabolism. In addition, pro-inflammatory cytokines increase systemic inflammation and are often elevated in obese individuals with metabolic dysfunction (Hotamisligil, 2006, Ouchi et al., 2011). Both pro- and anti-inflammatory macrophages can be found in WAT (Aron-Wisnewsky et al., 2009). The relative levels of these cells are related to the overall physiology with a balance toward an anti-inflammatory state in healthy individuals and pro-inflammatory states in obese individuals.

Adipose tissue is highly vascularized indicating the presence of vascular cells, including ECs, mural cells, and smooth muscle cells. The blood vessels are responsible for transporting nutrients and oxygen to the tissue, removing waste, delivering cells, and serve as a pipeline for circulating adipokines (Ouchi et al., 2011). These vascular cells can respond to adipokines and secrete factors that directly communicate with adipocytes and preadipocytes to promote growth and expansion (Cao, 2007, 2010).

Adipose tissue also consists of a variety of different types of fibroblasts. Fibroblasts along with preadipocytes and adipocytes secrete ECM proteins required for the stability and function of the tissue. However, unlike adipocytes, the fibroblasts are found in the connective tissue regions of adipose tissue and can increase in number and ECM expression in response to disease conditions.

16.2.2 Extracellular Matrix

In addition to cells, adipose tissue consists of organized insoluble networks of proteins and polysaccharides known as ECMs. Adipose ECM can be placed in two categories based on location and composition: (1) interstitial or stromal ECM and (2) BMs. Stromal ECMs are networks of supramolecular structures found in the connective tissue of adipose. They can consist of a number of different components, but fibronectin and collagen type VI make up a significant fraction of the molecules present. BMs are a thin (<1 μm), sheet-like network of proteins that surrounds mature adipocytes and blood vessels (Huang and Greenspan, 2012). Type IV collagen and laminins are components of the BM. In adipose tissue, the BM sheet surrounds adipocytes while stromal ECM lies between regions of multiple adipocytes and do not directly contact the adipocytes (Huang and Greenspan,

2012). The stromal ECM can influence the behavior of many other cell types or can signal to adipocytes by acting through the BM.

While the components of adipose tissue ECM may be present in other tissues, the concentration and relative composition is unique. As adipose tissue continuously undergoes growth and expansion, there is a dynamic ECM remodeling process throughout life. Adipose tissue contains ECM collagens I, III, IV, and VI, and laminins, fibronectin, heparin sulfate, and proteoglycans (Uriel et al., 2008, 2009). Fibril collagen VI is the most prevalent of collagen types found in adipose tissue, while very little collagen I is present (Khan et al., 2009a). Collagen VI strongly binds with BM collagen IV (Mariman and Wang, 2010a). The ECM composition of adipose can significantly vary between individuals and based on tissue health. In metabolic diseases, collagen levels are often increased in the stromal ECM contributing to fibrosis (Khan et al., 2009b, Divoux et al., 2010).

The ECM has been shown to play an important role in regulating cell growth, differentiation, and migration, and influencing tissue development and repair in a number of tissues (Aumailley and Krieg, 1996, Aumailley and Gayraud, 1998, Huang and Greenspan, 2012). While little is known about how the ECM influences adipocyte behavior, recent observations suggest that the function and expansion of adipose are influenced by ECM properties. The composition of the ECM changes as preadipocytes differentiate into adipocytes. The ECM composition transitions from fibronectin rich to rich in BM proteins laminin and collagen IV surrounding the mature adipocyte (Divoux and Clément, 2011). ECM changes are biphasic; the first phase is focused on cell commitment and differentiation with decreased collagen synthesis followed by the second phase of cell lipid loading and BM deposition (Mariman and Wang, 2010a).

Interactions between cell surface receptors and BM components possibly play a direct role in regulating adipogenesis through mechanical interactions. Mechanical stretching of preadipocytes has been shown to inhibit differentiation (Mariman and Wang, 2010a). Stretching preadipocytes in 2D prevents the typical morphological change observed during differentiation in which adipocytes convert from an elongated fibroblast-like cell into a round adipocyte (Mariman and Wang, 2010b). The ECM has been shown to be involved with cell morphological development necessary for differentiation from preadipocyte to mature lipid-containing adipocyte (Mariman and Wang, 2010a). Mice that are deficient in membrane-anchored metalloproteinase, MT1-MMP (MMP14) are lipodystrophic (Chun et al., 2006). Interestingly, preadipocytes from MMP14 null mice are able to differentiate into mature adipocytes in 2D but not in 3D (Chun et al., 2006). This study suggests that the ability of preadipocytes to degrade their surrounding ECM impacts their ability to morphologically change shape and differentiate into adipocytes (Chun et al., 2006). In addition, the ECM may help to protect the lipid core of the adipocyte from external mechanical forces (Divoux and Clément, 2011).

Recent data suggest that the mechanical properties of the ECM also aid adipocytes in lipid storage. In a prolonged positive energy balance as seen in obesity, adipocytes continue to load the lipid until they reach a maximum capacity or "critical size." At this point, the adipocyte no longer increases in lipid content or size (DiGirolamo et al., 1998). The milestone of an adipocyte cell reaching this "critical size" is thought to trigger preadipocytes to proliferate and then differentiate to form more adipocytes for storing additional lipid (Hausman et al., 2001). The specific "critical size" achieved by an individual adipocyte may depend on the mechanical properties of the surrounding ECM. When adipocytes enlarge, they increase the force that they exert on the ECM. More rigid ECM may resist this expansion, allowing less expansion and increasing differentiation of local preadipocytes. The restrictions on the adipocyte may also trigger a pro-inflammatory phenotype. Adipose ECM in collagen VI knockout mice has been shown to be less rigid, resulting in enlarged adipocytes without the necrosis and inflammation typically associated with larger cells (Khan et al., 2009b). In obese human adipose tissue, pericellular fibrosis is increased possibly resulting in more rigid ECM and altered mechanical forces on individual adipocytes (Clément et al., 2004). These studies suggest that the rigidity of the ECM may contribute to the pro-inflammatory state of adipose tissue in obese individuals, creating a cycle of increasing fibrosis leading to greater inflammation and ECM accumulation. In addition, the ability of adipocytes to expand appropriately in the setting of positive energy balance may be important in the regulation of systemic insulin sensitivity, since the presence of ectopic lipid in the liver and muscle leads to insulin resistance (Kim et al., 2007).

16.3 Function

Historically, studies of adipose tissue function focused on its role in energy storage, insulation, and thermal regulation. In 1995, with the discovery of leptin, the view of adipose tissue function expanded to that of an endocrine organ. Other growth factors and hormones were previously known to be secreted from adipose tissue before leptin was discovered. However, it was the identification of the leptin hormone and its numerous physiological functions, energy homeostasis and metabolism, angiogenesis, bone metabolism, coagulation and fibrinolysis, vasoconstriction/vasorelaxation, sexual maturation, kidney function, modulation of the immune system, and hematopoiesis (Adamczak and Wiecek, 2013), which led to adipose tissue being defined as an endocrine organ. Recent research has examined the secretory and metabolic functions of adipose tissue. Adipocytes secrete numerous biologically active substances (proteins, peptides, hormones, etc.) into their ECM environment and some of these substances are transported into the circulatory system where they affect the overall metabolic state of the individual. While each adipocyte produces a small amount of bioactive substances, the combined amount impacts the overall body function since adipose tissue is one of the largest organs in the human body.

16.4 Microvasculature

The vasculature in adipose tissue supplies oxygen and nutrients, provides cytokines and growth factors, and supplies cells. Additionally, the vascular portion of adipose tissue facilitates infiltration of inflammatory cells, removes metabolic waste, and distributes adipokines, fatty acids, or glycerol to fuel other cells. Blood flow is important not just for providing oxygen and nutrients to the cells in adipose tissue but also for supplying and storing energy in the form of free fatty acids. When the body has a positive energy balance, excess energy is stored in a lipid reservoir of triglycerol in the adipocyte. When the body has a negative energy balance, the adipocyte releases triglycerol in the form of fatty acids and glycerol into the blood stream to be used for energy by other organs through a process known as lipolysis.

Similar to most tissues, adipose consists of two microvascular systems: lymphatic and blood. Several adipokines secreted by adipocytes are found in both lymphatic and blood vasculature. The lymph system in adipose tissue may contribute to inflammation in obese adipose tissue and has recently gained more research attention (Bastard and Fáeve, 2013). However, here, we will focus on the blood circulatory system. For a review on the lymphatic system, please see Chakraborty et al. (2010).

Adipose expansion or regression requires coordination of adipogenesis and angiogenesis. The relationship between these two processes is not well understood. Clearly, the vasculature is needed for oxygen and nutrient supply as the tissue increases in volume, but the vasculature is also thought to help maintain adipose tissue volume by coordinating the process of replacing apoptotic adipocytes with new adipocytes. Adipogenesis along with angiogenesis may result in an increase in the total amount of adipocytes, and, in some cases, more adipocytes than are needed to renew and maintain the tissue. This hyperplasia can lead to obesity. However, adipose expansion does not just occur through adipogenesis. It may also result from hypertrophy, or an increase in adipocyte size due to increasing lipid loading within the cell.

The hormones, growth factors, matrix proteins, enzymes, and cytokines secreted by adipocytes can regulate vascular function (Poulos et al., 2010). These factors may contribute to blood flow regulation, lipid and cholesterol metabolism, and immune system function (Poulos et al., 2010). In addition, several of the adipocyte cell-secreted factors called adipocytokines or adipokines are directly involved with angiogenesis and vascularization.

Several factors involved in regulating angiogenesis both positively and negatively are produced by adipose tissue (Table 16.1). Two important angiogenic factors secreted by adipocytes are vascular endothelial growth factor (VEGF) and hepatocyte growth factor (HGF) (Christiaens and Lijnen, 2010). ECs are stimulated to proliferate and migrate by VEGF (Christiaens and Lijnen, 2010). Cells in the stroma and mature adipocytes express VEGF (Ledoux et al., 2008). HGF secreted by adipocytes has been shown to promote EC tube formation *in vitro* (Saiki et al., 2006). FGF-2 possibly produced by vascular cells in adipose (Poulos et al., 2010) can stimulate both adipocyte differentiation and the proliferation and migration of ECs (Kawaguchi et al., 1998). Secreted protein acidic and rich in cysteine (SPARC) is a matricellular protein involved in cell–matrix interactions. It binds to VEGF and inhibits FGF-2 resulting in decreased EC proliferation (Christiaens and Lijnen, 2010). Also, SPARC is associated with insulin resistance and ECM fibrosis (Kos and Wilding, 2010). The angiopoietins contribute to the maintenance and stabilization of blood vessels (Hausman and Richardson, 2004, Cao, 2007). The role of the receptors for the angiopoietins, TIE1 and TIE2, in adipose angiogenesis are not well known; however, expression is increased during fat pad development (Neels et al., 2004). Leptin is involved in food intake, insulin sensitivity, and angiogenesis (Bouloumie et al., 1998, Ronti et al., 2006). It is proangiogenic, upregulates VEGF expression, and induces the specific formation of fenestrated capillaries (Christiaens and Lijnen, 2010). Adiponectin has had separate findings of both anti-angiogenic and pro-angiogenic effects (Christiaens and Lijnen, 2010). Thrombospondins can inhibit angiogenesis by impacting EC migration, proliferation, apoptosis, and tube formation (Carlson et al., 2008). Several inhibitors of angiogenesis are produced in adipose tissue as well, including thrombospondins, SPARC, resistin, interferon-gamma, and tissue inhibitors of metalloproteasis (TIMPS) (Cao, 2010, Christiaens and Lijnen, 2010). Understanding of the process by which adipose controls the complex balance of pro- and anti-angiogenic factors to regulate vascularization will provide new insight into adipose expansion.

TABLE 16.1

List of Factors Secreted in Adipose Tissue That Regulate Angiogenesis

Adipose Tissue-Secreted Factor	Angiogenesis-Related Function	Function	References
Adiponectin	Anti-inflammatory and pro-angiogenic effects reported	Enhances insulin sensitivity, inhibits monocyte adhesion, and stimulates angiogenesis	Kershaw and Flier (2004), Christiaens and Lijnen (2010)
Adipsin		Required for production of acylation stimulating protein needed for lipid and glucose metabolism	Kershaw and Flier (2004)
Angiotensinogen		Related to angiotensin concentration	Poulos et al. (2010)
Apelin	Pro-angiogenic	Responsive to metabolic changes in insulin	Poulos et al. (2010)
Apolipoprotein E		Metabolic role in lipid transport	Poulos et al. (2010)
Fibroblast growth factor (FGF)-2	Pro-angiogenic	Stimulates adipocyte differentiation	Rophael et al. (2007), Christiaens and Lijnen (2010)
Hepatocyte growth factor (HGF)	Pro-angiogenic	Promotes HUVEC tube formation *in vitro*	Christiaens and Lijnen (2010)
Hypoxia-inducible factor (HIF) 1-alpha		Stimulates fibrosis in hypoxic WAT	Rutkowski et al. (2009)
Interleukin 6 (IL-6)	Inflammatory marker	Plays a role in metabolic syndrome diseases	Poulos et al. (2010)
Leptin	Pro-angiogenic	Impairs nitric oxide sensitivity and production, and promotes migration of ECs	Rutkowski et al. (2009), Cao (2010), Kershaw and Flier (2004), Christiaens and Lijnen (2010), Bouloumie et al. (1998)
Matrix metallo-proteases (MMPs)	Pro- and anti-angiogenesis	Involved in ECM remodeling	Christiaens and Lijnen (2010), Poulos et al. (2010)
Macrophages and monocyte chemoattractant protein (MCP-1)	Pro-inflammatory	Recruits monocytes to inflammation sites; induces insulin resistance	Kershaw and Flier (2004), Poulos et al. (2010)
Plasminogen activator inhibitor-1 (PAI-1)	Pro-angiogenic	Associated with cardiovascular disease	Kershaw and Flier (2004), Poulos et al. (2010)
Platelet-derived growth factor (PDGF)-beta	Pro-angiogenic	Involved in pericyte recruitment around capillaries	Rophael et al. (2007)
Resistin	Anti-angiogenic		Kershaw and Flier (2004)
Secreted protein acidic and rich in cysteine (SPARC)	Anti-angiogenic		Christiaens and Lijnen (2010), Kos and Wilding (2010)
Thrombospondins	Anti-angiogenic	ECM remodeling	Christiaens and Lijnen (2010)
Tissue factor (TF)	Pro-angiogenic	Initiator of coagulation cascade	Christiaens and Lijnen (2010)

TABLE 16.1 (continued)

List of Factors Secreted in Adipose Tissue That Regulate Angiogenesis

Adipose Tissue-Secreted Factor	Angiogenesis-Related Function	Function	References
Tissue inhibitors of metalloproteasis (TIMPS)	Anti-angiogenic	Involved in ECM remodeling	Bastard and Fáeve (2013), Poulos et al. (2010)
Tumor necrosis factor (TNF)-alpha	Pro-inflammatory	Increases endothelial-immune cell adhesion molecules	Rutkowski et al. (2009), Poulos et al. (2010)
Vascular endothelial growth factor (VEGF)	Pro-angiogenic	Stimulates migration and proliferation of ECs	Christiaens and Lijnen (2010), Bastard and Fáeve (2013)
Visfatin	Pro-inflammatory	Induces expression of inflammatory cytokines	Poulos et al. (2010)

16.5 Vascularization during Adipose Tissue Development and Expansion

In human development, vascular structures appear prior to the first adipocyte (Bastard and Fáeve, 2013). In rodent development, adipocytes are observed in the epididymal depot only after the initial vascular plexus establishes blood flow (Bastard and Fáeve, 2013). In adults, angiogenesis has been shown to be essential for adipogenesis (Crandall et al., 1997, Cao, 2007, Rophael et al., 2007, Christiaens and Lijnen, 2010). Blood vessel sprouting occurs near clusters of differentiating adipocytes (Nishimura et al., 2007), and VEGF levels in adipose tissue correlate with adipose weight (Miyazawa-Hoshimoto et al., 2005).

Confocal imaging of bodipy-stained adipocytes and lectin-perfused vessels has been used to show that adipogenesis *in vivo* takes place within adipogenic/angiogenic cell clusters that contain blood vessels and various stromal cells (Nishimura et al., 2007). Anti-VEGF treatment was shown to inhibit both angiogenesis and adipogenesis, resulting in smaller epididymal fat pads compared to control animals (Nishimura et al., 2007). This study is consistent with earlier studies showing that inhibition of angiogenesis reduces adipose tissue mass. In the previous studies, angiogenic inhibitors such as TNP-470, angiostatin, endostatin, Bay-129566, an adipose vasculature- targeted proapoptotic peptide (Kolonin et al., 2004), and thalidomide were administered to an obese mouse model resulting in decreased body weight and adipose tissue mass in a dose-dependent manner (Liu et al., 2003, Rupnick et al., 2002). These findings support angiogenesis as having an important and possibly even a critical role in adipose tissue expansion.

Interestingly, the extent of vascularization appears to vary with the specific adipose depot. Through transcriptomic analysis, the visceral depot has been shown to have more specific angiogenic-related pathways compared to subcutaneous adipose tissue (Hocking et al., 2010). *In vivo* 3D imaging approaches have shown that visceral depots are more vascularized than subcutaneous depots (Villaret et al., 2010). Since an expansion of visceral adipose depots is particularly detrimental to systemic insulin sensitivity, this increased angiogenic capacity may play an important pathophysiologic role in the development of obesity-related metabolic abnormalities, although this remains to be conclusively established.

16.6 Tissue Engineering Adipose

Soft-tissue defects caused by congenital malformation, tumor resection, or trauma can impact function and structure. Clinical options for treating these issues are limited. Engineering functional adipose tissue may serve as an alternative treatment for adipose regeneration or replacement. However, the volume of the tissue that can be engineered is limited without vascularization. Coordinating the processes of vascularization and adipogenesis within the engineered tissues is critical for generating adipose of clinical volume and appropriate function.

Building an extensive microvascular network within engineered tissues is a significant challenge. Most approaches to engineering adipose tissue are currently limited in size because diffusion is used for supplying nutrients. Diffusion effectively supplies nutrients to cells at a maximum distance of approximately 100–200 μm from a capillary. This diffusion restriction has limited the ability to create larger 3D constructs. Adipose tissue also relies on a vascular network to support its function. While adipose tissue engineering has been an active area of research for the past two decades, there have been little efforts to specifically address vascularization simultaneously along with adipogenesis. In general, the following approaches have been explored for engineering vascularized adipose tissue (Ahmed et al., 2008, Tanzi and Fare, 2009):

1. Biomaterial scaffolds have been combined with bioactive factors that stimulate the recruitment of adipocyte cells and blood vessels from the surrounding tissue following implantation.
2. Scaffolds have been prepopulated with preadipocytes, ECs, and/or stem cells that provide biochemical or mechanical cues from the scaffold to direct vessel formation and differentiation of adipocytes.
3. Biomaterials with angiogenic factors have been implanted around a vascular pedicle and are in contact with the existing adipose tissue to engineer a vascularized, surgically transferable adipose tissue.

To successfully engineer vascularized adipose tissue for any of these applications, the cells, scaffold materials, inductive factors and implant location must be carefully considered. The following section discusses the previous work on optimizing the engineered tissue design focused on the formation of vascularized adipose tissue.

16.6.1 Cell Seeding

Mature adipocytes do not proliferate. In addition, mature adipocytes float in media and are difficult to seed onto a scaffold. Therefore, preadipocytes or ASC is generally utilized for adipose tissue engineering. These cells proliferate and, given the appropriate biochemical and/or mechanical cues, can differentiate into mature adipocytes. In a few studies (mostly *in vitro*), ECs have been cocultured with preadipocytes/adipocytes/ASCs in biomaterials in an attempt to approximate vascularized adipose tissue. Preadipocytes and human dermal microvascular ECs (HDMVEC) were cocultured in fibrin gels *in vitro* (Borges et al., 2007). The fibrin scaffold with preadipocytes and ECs had a significant decrease in the number and length of sprouts compared to scaffolds with only ECs (Borges et al., 2007). When media preconditioned from preadipocytes was used for culturing EC spheroids in fibrin, it resulted in a 10-fold increase in EC migration indicating the value of angiogenic factors secreted by preadipocytes (Borges et al., 2007). In a different study, ASC was differentiated on a cell culture plate (2D) for 2 weeks. Then, "vascular support cells," which included fibroblasts, ASC, and mesenchymal stem cells (MSCs), were added to the culture plate along with abscorbate for 1 week (Sorrell et al., 2011). Next, vascular cells were added as a combination of human umbilical vein endothelial cell (HUVEC) and MSC and cultured for 10–14 days (Sorrell et al., 2011). Then, dermal fibroblasts were added and cultured for 1 week (Sorrell et al., 2011). Finally, the cell constructs were detached from the plate and two constructs were stacked together to form an *in vitro* adipose construct (Sorrell et al., 2011). These elaborate constructs contained cell-secreted ECM and were an attempt to recreate a complex cellular environment, but were not evaluated for vascular density or function (Sorrell et al., 2011).

Another coculture model using adipocytes and ECs on silk fibroin scaffolds found that adipocytes in combination with ECs were insulin responsive. The adipocytes were stimulated by both high insulin dosages (hyperinsulinemic conditions) and normal insulin levels and evaluated for function in lipogenesis and lipolysis. Hyperinsulinemic conditions resulted in increased glycerol secretion similar to the physiologic condition of insulin resistance. The coculture model had differential response to high insulin levels compared to normal insulin level (Choi et al., 2010b). The adipocyte-only scaffolds lacked insulin responsiveness and did not show stimulation by high or normal insulin dosage (Choi et al., 2010b), indicating that the combination of ECs and adipocytes possibly contributes to adipocyte insulin resistance. A different study encapsulated ASCs in a collagen and alginate microbead scaffold with the outside of the scaffold seeded with HUVECs. These scaffolds were subcutaneously injected into immune suppressed mice (Yao et al., 2013). Blood vessel growth connected the host vasculature to the constructs and the ASC preseeded scaffold significantly contained more adipocytes than control groups that were simply a mixture of all the components such as ASC, HUVEC, collagen, and alginate without the microbead scaffold structure (Yao et al., 2013). The study did not investigate differences between scaffolds containing ASCs with HUVECs compared to ASCs without HUVECs.

A modified chorioallantoic membrane model has been used to evaluate the presence of a vascular network on vascular ingrowth into a cell-seeded scaffold. In this model, an incubated fertilized egg that has had a piece of the shell was removed to expose the chorioallantoic membrane. Borges et al. seeded a fibrin matrix with HDMVEC spheroids and preadipocytes and implanted it into the adapted chorioallantoic membrane model (Borges et al., 2003). The scaffold was put into direct contact with the vascularized membrane, covered with a cap, and incubated for 7 days. Results showed the formation of a capillary network from the tissue-engineered composite to the host vascular system without applying exogenous angiogenic growth factors (Borges et al., 2003). In total, cell studies suggest that a combination of vascular progenitor cells and preadipocytes could promote vascularized adipose formation *in vitro* and *in vivo*.

16.6.2 Scaffold Materials/Conditions

Several different materials have been investigated for adipose tissue engineering (Hemmrich and von Heimburg, 2006, Tanzi and Fare, 2009, Choi et al., 2010a). The biomaterials used often exhibit inherent provascularization properties or are supplemented with additional factors to enhance vascularization. In a number of studies, the materials have been combined with a flow through pedicle chamber model to enhance vascularization. The pedicle provides a vascular source that could also be used for free tissue transfer following tissue development. In a typical implementation, a chamber made from a relatively inert, nondegradable material is implanted around the epigastric pedicle in the groin area or the femoral arteriovenous pedicle. The chamber volume is filled with the desired scaffold supplemented with cells or growth factors. New blood vessels sprout from the flow through the pedicle and/or surrounding tissue to invade the scaffold to form new vascular networks.

Tissue-engineering approaches investigated in the pedicle chamber include Matrigel, a commercially available protein solution extracted from a spontaneous mouse sarcoma that undergoes gelation at 37°C, supplemented with a variety of growth factors, including FGF-2 (Kelly et al., 2006, Stillaert et al., 2007, Rophael et al., 2007) $VEGF_{120}$ (Rophael et al., 2007), and PDGF-BB (Rophael et al., 2007). In addition, polycaprolactone scaffolds (Wiggenhauser et al., 2012), sponge-like polyurethane scaffolds (Wiggenhauser et al., 2012), type I collagen with controlled release of FGF-2 from gelatin microspheres (Vashi et al., 2006), dermis-derived hydrogels (Cheng et al., 2010), and adipose-derived hydrogels (Uriel et al., 2008) have been examined in this model. It is not clear what the best approach is, but when the tissues are in contact with the existing adipose, all of them appear to permit vascularized adipose formation when supplemented with an angiogenic factor (Kelly et al., 2006). However, adipose-based materials have been shown to promote extensive vascularized adipose formation, even in the absence of angiogenic stimuli (Uriel et al., 2008). This suggests that biomaterials with adipose-specific factors and composition may further enhance adipose formation.

Alginate microbeads loaded with FGF-1 and suspended in type I collagen were implanted in the same pedicle model and placed in contact with adipose tissue. While this system increased vascular density, it did not result in significant adipogenesis. This was an important finding indicating that the ability to induce neovascularization does not in itself result in adipogenesis (Moya et al., 2010). Adipogenic cues are needed in addition for adipogenesis to occur. Scaffold selection is important and can play an instructive role to help recruit host cells. Adipocytes can also be seeded on the scaffolds to stimulate adipogenesis. However, in this model, adipogenesis and angiogenesis result from the host tissue source and not from the grafted cells (Stillaert et al., 2007). This suggests that the seeded cells stimulate adipogenesis through paracrine interactions and not direct differentiation.

A seminal study with Matrigel supplemented with basic fibroblast factor (bFGF) and subcutaneously injected into nude mice (Kawaguchi et al., 1998) showed that the stimulation of vascularization can increase de novo adipogenesis. After 5 weeks, the constructs were explanted and histological analysis showed the extensive presence of both adipocytes and neovascularization (Kawaguchi et al., 1998). This study showed that a biomaterial with bioactive cues could be utilized to recruit host cells to form de novo adipose tissue. Similar to the pedicle studies, these materials were implanted in contact with adipose (subcutaneous) tissue. The importance of bioactive factors was further exhibited by the injection of photocured styrenated gelatin microspheres releasing FGF-1, an angiogenic factor, and insulin-like growth factor (IGF-1), an adipogenic factor, into the backs of rats. The injected factors resulted in de novo adipogenesis. The triglyceride content was significantly higher in the group that received both FGF-1 and IGF compared to the group that received only IGF-1, the adipogenic factor (Masuda et al., 2004).

The extent of adipogenesis is also influenced by scaffold architecture. An *in vivo* study used poly(ethylene glycol) (PEG) hydrogels with channels containing an angiogenic factor bFGF and encapsulated human mesenchymal stem cells (hMSCs-) derived adipocytes (Stosich et al., 2007). The PEG hydrogels were subcutaneously implanted into the dorsal skin of severe combined immune deficiency mice (Stosich et al., 2007). This study showed that the host tissue invaded into the channels after 12 weeks (Stosich et al., 2007). While the study suggests that the channels were important for vascularized adipose formation, they did not systematically investigate the role of the scaffold properties on adipogenesis and vasularization.

16.7 Conclusion

Tissue-engineering strategies to produce functional adipose tissue are needed. The ability to engineer clinical volumes of adipose with appropriate function requires coordinating this process with vascularization. While there has been a focus on scaffolds for adipose tissue engineering, there has been little effort to specifically coordinate the processes of vascularization and adipogenesis. Recent work with 3D EC/adipocyte coculture models provides insight into this relationship and potentially leads to more sophisticated approaches to engineering vascularized adipose tissues. In addition, controlled studies into how the presence of adipogenic and angiogenic factors within biomaterial scaffolds influences adipose tissue development are needed.

References

Adamczak, M. and Wiecek, A. 2013. The adipose tissue as an endocrine organ. *Seminars in Nephrology*, 33, 2–13.

Ahima, R. S. and Flier, J. S. 2000. Leptin. *Annual Review of Physiology*, 62, 413–437.

Ahmed, T. A. E., Dare, E. V., and Hincke, M. 2008. Fibrin: A versatile scaffold for tissue engineering applications. *Tissue Engineering Part B—Reviews*, 14, 199–215.

Aron-Wisnewsky, J., Tordjman, J., Poitou, C., Darakhshan, F., Hugol, D., Basdevant, A., Aissat, A., Guerre-Millo, M., and Clement, K. 2009. Human adipose tissue macrophages: M1 and M2 cell surface markers in subcutaneous and omental depots and after weight loss. *Journal of Clinical Endocrinology and Metabolism*, 94, 4619–4623.

Aumailley, M. and Gayraud, B. 1998. Structure and biological activity of the extracellular matrix. *Journal of Molecular Medicine—Jmm*, 76, 253–265.

Aumailley, M. and Krieg, T. 1996. Laminins: A family of diverse multifunctional molecules of basement membranes. *Journal of Investigative Dermatology*, 106, 209–214.

Bastard, J.-P. and Fáeve, B. 2013. *Physiology and Physiopathology of Adipose Tissue*. Paris; New York: Springer.

Borges, J., Mueller, M. C., Momeni, A., Stark, G. B., and Torio-Padron, N. 2007. *In vitro* analysis of the interactions between preadipocytes and endothelial cells in a 3D fibrin matrix. *Minimally Invasive Therapy and Allied Technologies*, 16, 141–148.

Borges, J., Mueller, M. C., Padron, N. T., Tegtmeier, F., Lang, E. M., and Stark, G. B. 2003. Engineered adipose tissue supplied by functional microvessels. *Tissue Engineering*, 9, 1263–1270.

Bouloumie, A., Drexler, H. C. A., Lafontan, M., and Busse, R. 1998. Leptin, the product of Ob gene, promotes angiogenesis. *Circulation Research*, 83, 1059–1066.

Cao, Y. 2007. Angiogenesis modulates adipogenesis and obesity. *Journal of Clinical Investigation*, 117, 2362–2368.

Cao, Y. H. 2010. Adipose tissue angiogenesis as a therapeutic target for obesity and metabolic diseases. *Nature Reviews Drug Discovery*, 9, 107–115.

Carlson, C. B., Gunderson, K. A., and Mosher, D. F. 2008. Mutations targeting intermodular interfaces or calcium binding destabilize the thrombospondin-2 signature domain. *Journal of Biological Chemistry*, 283, 27089–27099.

Chakraborty, S., Zawieja, S., Wang, W., Zawieja, D. C., and Muthuchamy, M. 2010. Lymphatic system: A vital link between metabolic syndrome and inflammation. *Annual NewYork Academy Science*, 1207(Suppl 1), E94–E102.

Cheng, M., Uriel, S., Moya, M., Francis-Sedlak, M., Wang, R., Huang, J., Chang, S., and Brey, E. 2010. Dermis-derived hydrogels support adipogenesis *in vivo*. *Journal of Biomedical Materials Research Part A*, 92A, 852–858.

Choi, J. H., Gimble, J. M., Lee, K., Marra, K. G., Rubin, J. P., Yoo, J. J., Vunjak-Novakovic, G., and Kaplan, D. L. 2010a. Adipose tissue engineering for soft tissue regeneration. *Tissue Engineering Part B—Reviews*, 16, 413–426.

Choi, J. H., Gimble, J. M., Vunjak-Novakovic, G., and Kaplan, D. L. 2010b. Effects of hyperinsulinemia on lipolytic function of three-dimensional adipocyte/endothelial co-cultures. *Tissue Engineering Part C—Methods*, 16, 1157–1165.

Christiaens, V. and Lijnen, H. R. 2010. Angiogenesis and development of adipose tissue. *Molecular and Cellular Endocrinology*, 318, 2–9.

Chun, T. H., Hotary, K. B., Sabeh, F., Saltiel, A. R., Allen, E. D., and Weiss, S. J. 2006. A pericellular collagenase directs the 3-dimensional development of white adipose tissue. *Cell*, 125, 577–591.

Clément, K., Viguerie, N., Poitou, C., Carette, C., Pelloux, V., Curat, C. A., Sicard, A. et al. 2004. Weight loss regulates inflammation-related genes in white adipose tissue of obese subjects. *Faseb Journal*, 18, 1657–1669.

Crandall, D. L., Hausman, G. J., and Kral, J. G. 1997. A review of the microcirculation of adipose tissue: Anatomic, metabolic and angiogenic perspectives. *Microcirculation—London*, 4, 211–232.

Cypess, A. M., Lehman, S., Williams, G., Tal, I., Rodman, D., Goldfine, A. B., Kuo, F. C. et al. 2009. Identification and importance of brown adipose tissue in adult humans. *New England Journal of Medicine*, 360, 1509–1517.

Digirolamo, M., Fine, J. B., Tagra, K., and Rossmanith, R. 1998. Qualitative regional differences in adipose tissue growth and cellularity in male Wistar rats fed *ad libitum*. *American Journal of Physiology—Regulatory Integrative and Comparative Physiology*, 274, R1460–R1467.

Divoux, A. and Clément, K. 2011. Architecture and the extracellular matrix: The still unappreciated components of the adipose tissue. *Obesity Reviews*, 12, e494–e503.

Divoux, A., Tordjman, J., Lacasa, D., Veyrie, N., Hugol, D., Aissat, A., Basdevant, A. et al. 2010. Fibrosis in human adipose tissue: Composition, distribution, and link with lipid metabolism and fat mass loss. *Diabetes*, 59, 2817–2825.

Gesta, S., Tseng, Y.-H., and Kahn, C. R. 2007. Developmental origin of fat: Tracking obesity to its source. *Cell*, 131, 242–256.

Gimble, J. M., Katz, A. J., and Bunnell, B. A. 2007. Adipose-derived stem cells for regenerative medicine. *Circulation Research*, 100, 1249–1260.

Hausman, D. B., Digirolamo, M., Bartness, T. J., Hausman, G. J., and Martin, R. J. 2001. The biology of white adipocyte proliferation. *Obesity Reviews*, 2, 239–254.

Hausman, G. J. and Richardson, R. L. 2004. Adipose tissue angiogenesis. *Journal of Animal Science*, 82, 925–934.

Hemmrich, K. and Von Heimburg, D. 2006. Biomaterials for adipose tissue engineering. *Expert Review of Medical Devices*, 3, 635–645.

Hocking, S. L., Wu, L. E., Guilhaus, M., Chisholm, D. J., and James, D. E. 2010. Intrinsic depot-specific differences in the secretome of adipose tissue, preadipocytes, and adipose tissue-derived microvascular endothelial cells. *Diabetes*, 59, 3008–3016.

Hotamisligil, G. S. 2006. Inflammation and metabolic disorders. *Nature*, 444, 860–867.

Huang, G. and Greenspan, D. S. 2012. ECM roles in the function of metabolic tissues. *Trends in Endocrinology and Metabolism*, 23, 16–22.

Kawaguchi, N., Toriyama, K., Nicodemou-Lena, E., Inou, K., Torii, S., and Kitagawa, Y. 1998. De novo adipogenesis in mice at the site of injection of basement membrane and basic fibroblast growth factor. *Proceedings of the National Academy of Sciences of the United States of America,* 95, 1062–1066.

Kelly, J. L., Findlay, M. W., Knight, K. R., Penington, A., Thompson, E. W., Messina, A., and Morrison, W. A. 2006. Contact with existing adipose tissue is inductive for adipogenesis in Matrigel. *Tissue Engineering,* 12, 2041–2047.

Kershaw, E. E. and Flier, J. S. 2004. Adipose tissue as an endocrine organ. *Journal of Clinical Endocrinology and Metabolism,* 89, 2548–2556.

Khan, T., Muise, E. S., Iyengar, P., Wang, Z. V., Chandalia, M., Abate, N., Zhang, B. B., Bonaldo, P., Chua, S., and Scherer, P. E. 2009a. Metabolic dysregulation and adipose tissue fibrosis: Role of collagen VI. *Molecular and Cellular Biology,* 29, 1575–1591.

Khan, T., Muise, E. S., Iyengar, P., Wang, Z. V., Chandalia, M., Abate, N., Zhang, B. B., Bonaldo, P., Chua, S., and Scherer, P. E. 2009b. Metabolic dysregulation and adipose tissue fibrosis: Role of collagen VI. *Molecular and Cellular Biology,* 29, 1575–1591.

Kim, J. Y., Van De Wall, E., Laplante, M., Azzara, A., Trujillo, M. E., Hofmann, S. M., Schraw, T. et al. 2007. Obesity-associated improvements in metabolic profile through expansion of adipose tissue. *Journal of Clinical Investigation,* 117, 2621–2637.

Kolonin, M. G., Saha, P. K., Chan, L., Pasqualini, R., and Arap, W. 2004. Reversal of obesity by targeted ablation of adipose tissue. *Nature Medicine,* 10, 625–632.

Kos, K. and Wilding, J. P. H. 2010. Sparc: A key player in the pathologies associated with obesity and diabetes. *Nature Reviews Endocrinology,* 6, 225–235.

Ledoux, S., Queguiner, I., Msika, S., Calderari, S., Rufat, P., Gasc, J. M., Corvol, P., and Larger, E. 2008. Angiogenesis associated with visceral and subcutaneous adipose tissue in severe human obesity. *Diabetes,* 57, 3247–3257.

Liu, L. P., Meydani, M., and Mayer, J. 2003. Angiogenesis inhibitors may regulate adiposity. *Nutrition Reviews,* 61, 384–387.

Madonna, R., Geng, Y.-J., and De Caterina, R. 2009. Adipose tissue-derived stem cells characterization and potential for cardiovascular repair. *Arteriosclerosis Thrombosis and Vascular Biology,* 29, 1723–1729.

Mariman, E. C. and Wang, P. 2010a. Adipocyte extracellular matrix composition, dynamics and role in obesity. *Cellular and Molecular Life Sciences,* 67, 1277–1292.

Mariman, E. C. M. and Wang, P. 2010b. Adipocyte extracellular matrix composition, dynamics and role in obesity. *Cellular and Molecular Life Sciences,* 67, 1277–1292.

Masuda, T., Furue, M., and Matsuda, T. 2004. Photocured, styrenated gelatin-based microspheres for de novo adipogenesis through corelease of basic fibroblast growth factor, insulin, and insulin-like growth factor I. *Tissue Engineering,* 10, 523–535.

Miyazawa-Hoshimoto, S., Takahashi, K., Bujo, H., Hashimoto, N., Yagui, K., and Saito, Y. 2005. Roles of degree of fat deposition and its localization on VEGF expression in adipocytes. *American Journal of Physiology—Endocrinology and Metabolism,* 288, E1128–E1136.

Mott, J. W., Wang, J., Thornton, J. C., Allison, D. B., Heymsfield, S. B., and Pierson, R. N. 1999. Relation between body fat and age in 4 ethnic groups. *American Journal of Clinical Nutrition,* 69, 1007–1013.

Moya, M. L., Cheng, M.-H., Huang, J.-J., Francis-Sedlak, M. E., Kao, S.-W., Opara, E. C., and Brey, E. M. 2010. The effect of FGF-1 loaded alginate microbeads on neovascularization and adipogenesis in a vascular pedicle model of adipose tissue engineering. *Biomaterials,* 31, 2816–2826.

Neels, J. G., Thinnes, T., and Loskutoff, D. J. 2004. Angiogenesis in an *in vivo* model of adipose tissue development. *Faseb Journal,* 18, 983–985.

Nishimura, S., Manabe, I., Nagasaki, M., Hosoya, Y., Yamashita, H., Fujita, H., Ohsugi, M. et al. 2007. Adipogenesis in obesity requires close interplay between differentiating adipocytes, strontal cells, and blood vessels. *Diabetes,* 56, 1517–1526.

Ouchi, N., Parker, J. L., Lugus, J. J., and Walsh, K. 2011. Adipokines in inflammation and metabolic disease. *Nature Reviews Immunology,* 11, 85–97.

Poulos, S. P., Hausman, D. B., and Hausman, G. J. 2010. The development and endocrine functions of adipose tissue. *Molecular and Cellular Endocrinology*, 323, 20–34.

Prunet-Marcassus, B., Cousin, B., Caton, D., André, M., Pénicaud, L., and Casteilla, L. 2006. From heterogeneity to plasticity in adipose tissues: Site-specific differences. *Experimental Cell Research*, 312, 727–736.

Ronti, T., Lupattelli, G., and Mannarino, E. 2006. The endocrine function of adipose tissue: An update. *Clinical Endocrinology*, 64, 355–365.

Rophael, J. A., Craft, R. O., Palmer, J. A., Hussey, A. J., Thomas, G. P. L., Morrison, W. A., Penington, A. J., and Mitchell, G. M. 2007. Angiogenic growth factor synergism in a murine tissue engineering model of angiogenesis and adipogenesis. *American Journal of Pathology*, 171, 2048–2057.

Rupnick, M. A., Panigrahy, D., Zhang, C. Y., Dallabrida, S. M., Lowell, B. B., Langer, R., and Folkman, M. J. 2002. Adipose tissue mass can be regulated through the vasculature. *Proceedings of the National Academy of Sciences of the United States of America*, 99, 10730–10735.

Rutkowski, J. M., Davis, K. E., and Scherer, P. E. 2009. Mechanisms of obesity and related pathologies: The macro- and microcirculation of adipose tissue. *Febs Journal*, 276, 5738–5746.

Saiki, A., Watanabe, F., Murano, T., Miyashita, Y., and Shirai, K. 2006. Hepatocyte growth factor secreted by cultured adipocytes promotes tube formation of vascular endothelial cells *in vitro*. *International Journal of Obesity*, 30, 1676–1684.

Sorrell, J. M., Baber, M. A., Traktuev, D. O., March, K. L., and Caplan, A. I. 2011. The creation of an *in vitro* adipose tissue that contains a vascular–adipocyte complex. *Biomaterials*, 32, 9667–9676.

Stacey, D. H., Hanson, S. E., Lahvis, G., Gutowski, K. A., and Masters, K. S. 2009. *In vitro* adipogenic differentiation of preadipocytes varies with differentiation stimulus, culture dimensionality, and scaffold composition. *Tissue Engineering Part A*, 15, 3389–3399.

Stillaert, F., Findlay, M., Palmer, J., Idrizi, R., Cheang, S., Messina, A., Abberton, K., Morrison, W., and Thompson, E. W. 2007. Host rather than graft origin of Matrigel-induced adipose tissue in the murine tissue-engineering chamber. *Tissue Engineering*, 13, 2291–2300.

Stosich, M. S., Bastian, B., Marion, N. W., Clark, P. A., Reilly, G., and Mao, J. J. 2007. Vascularized adipose tissue grafts from human mesenchymal stem cells with bioactive cues and microchannel conduits. *Tissue Engineering*, 13, 2881–2890.

Tang, W., Zeve, D., Suh, J. M., Bosnakovski, D., Kyba, M., Hammer, R. E., Tallquist, M. D., and Graff, J. M. 2008. White fat progenitor cells reside in the adipose vasculature. *Science*, 322, 583–586.

Tanzi, M. C. and Fare, S. 2009. Adipose tissue engineering: State of the art, recent advances and innovative approaches. *Expert Review of Medical Devices*, 6, 533–551.

Trayhurn, P. and Wood, I. S. 2004. Adipokines: Inflammation and the pleiotropic role of white adipose tissue. *British Journal of Nutrition*, 92, 347–355.

Uriel, S., Huang, J.-J., Moya, M. L., Francis, M. E., Wang, R., Chang, S.-Y., Cheng, M.-H., and Brey, E. M. 2008. The role of adipose protein derived hydrogels in adipogenesis. *Biomaterials*, 29, 3712–3719.

Uriel, S., Labay, E., Francis-Sedlak, M., Moya, M., Weichselbaum, R., Ervin, N., Cankova, Z., and Brey, E. 2009. Extraction and assembly of tissue-derived gels for cell culture and tissue engineering. *Tissue Engineering Part C—Methods*, 15, 309–321.

Van Harmelen, V., Röhrig, K., and Hauner, H. 2004. Comparison of proliferation and differentiation capacity of human adipocyte precursor cells from the omental and subcutaneous adipose tissue depot of obese subjects. *Metabolism*, 53, 632–637.

Vashi, A. V., Abberton, K. M., Thomas, G. P., Morrison, W. A., O'connor, A. J., Cooper-White, J. J., and Thompson, E. W. 2006. Adipose tissue engineering based on the controlled release of fibroblast growth factor-2 in a collagen matrix. *Tissue Engineering*, 12, 3035–3043.

Villaret, A., Galitzky, J., Decaunes, P., Esteve, D., Marques, M.-A., Sengenes, C., Chiotasso, P. et al. 2010. Adipose tissue endothelial cells from obese human subjects: Differences among depots in angiogenic, metabolic, and inflammatory gene expression and cellular senescence. *Diabetes*, 59, 2755–2763.

Wiggenhauser, P. S., Mueller, D. F., Melchels, F. P. W., Egana, J. T., Storck, K., Mayer, H., Leuthner, P. et al. 2012. Engineering of vascularized adipose constructs. *Cell and Tissue Research*, 347, 747–757.

Yao, R., Zhang, R. J., Lin, F., and Luan, J. 2013. Biomimetic injectable HUVEC–adipocytes/collagen/ alginate microsphere co-cultures for adipose tissue engineering. *Biotechnology and Bioengineering*, 110, 1430–1443.

17

In Vivo *Vascularization for Large-Volume Soft Tissue Engineering*

Geraldine M. Mitchell and Wayne A. Morrison

CONTENTS

17.1 Introduction

17.1.1 Clinical Relevance

Healing of extensively damaged tissues and organs occurs by a combination of wound contracture, fibrosis, and scarring, and tissue that is lost is generally not reconstituted. Soft tissue or bone defects can be replaced by grafting like-tissue from another part of the body, "robbing Peter to pay Paul," but complex three-dimensional (3D) tissue deficits and organ loss can only be restored by transplantation from another donor.

Hence, the idea of manipulating the body to grow its own tissues or organs—tissue engineering—is an attractive alternative, and precedence does exist that makes the concept tangibly feasible. Embryogenesis, the development of a complete organism from a minimum number of cellular elements is the ultimate expression of tissue engineering. Similarly, early trimester fetal tissue damage regenerates new tissue rather than repairing by scar and fibrosis as occurs in adults (Metcalfe and Ferguson 2007). Certain species such as salamanders (Roy and Gatien 2008; Yokoyama 2008; Menger et al. 2010) retain this capacity throughout life. In humans, there are select tissues and organs that do have a remarkable regenerative capacity notably the liver, but also bone and the surface layers of the skin and gut. Clues for the mechanism of this self-renewal or regeneration have been sought from such examples and stem cells and inflammatory cells (Godwin et al. 2013) are clearly essential elements.

17.1.2 Traditional Tissue Engineering

Unlike "regenerative medicine," which is essentially cell therapy, "tissue engineering" has the added complexity of scale. This is a considerably more challenging task and 3D tissue engineering requires conceptually different design concepts from single-cell or cell cluster therapy. The traditional 3D model as articulated by Vacanti (Vacanti et al. 1991, Vacanti and Vacanti 2000) and others proposed three essential elements for tissue growth: (1) cells, (2) scaffolds, and (3) growth factors. Cells were seeded onto or throughout the scaffold in culture medium and expanded to the desired density with a view to subsequent implantation into the living environment. Scaffold design prototypes were based on meticulous analysis and replication of nature's extracellular matrices (ECMs). Bone and cartilage were early targets. Much was discussed and debated regarding scaffold structure, size, elasticity, strength, interconnectivity, pore size, and so on and ingenious materials and techniques were developed to satisfy these theoretical requirements. Scaffolds were surface coated with natural proteins or functionalized in various ways to facilitate cell adhesion and/ or to bind growth factors that were scheduled to be released according to an orchestrated cascade to induce specific processes, such as angiogenesis, differentiation, and so on. Most of the basic testing of these materials and their biocompatibility profiles were with simple cells, cultured *in vitro*. *In vivo* animal model testing was more difficult to model but quickly exposed the greater complexity of cell/scaffold behavior when exposed to the circulation and the inflammatory response. *In vivo* behavior was essentially different to *in vitro* behavior and one did not predict the other. Many scaffolds induced intense inflammatory reactions *in vivo* and were hostile to cell and vascular invasion. It became apparent that biological matrices appropriate for a particular cell-type behavior vary according to the stage of differentiation of the cell, and that different tissues require different matrices. Biodynamic behavior in tissue also influences cell differentiation and stability.

Growth factors attached to scaffolds eluted rapidly or remained locked in the scaffold and were inaccessible due to difficulty in designing predictable biodegradation time lines. The complexity of the changing growth factor and cytokine types and concentrations that were required over the time frame of cell differentiation made attempts at replicating the process with loaded scaffolds simplistic.

Cell expansion could be achieved by sophisticated bioreactors *in vitro*, minimizing the *in vivo* environment, but loading scaffolds with large densities of cells also proved conceptually unsound once delivered into the living environment because of the critical need for rapid vascularization from the implanted bed, termed extrinsic vascularization. Unless the cells were within 150–200 microns of a capillary in the recipient tissue, the implanted cells died (Folkman and Hochberg 1973). Thus, the issues related to the vascularization of large 3D constructs *in vivo* although belatedly addressed became preeminent. Various techniques have subsequently been used to increase the vascularization and therefore the size of 3D tissue engineering constructs—these techniques and the various factors that influence vascularization are discussed in the following sections.

17.2 Three-Dimensional Tissue Engineering *In Vivo*

17.2.1 Clinical Success with Tissue-Engineered Constructs

A limited number of clinical translations of tissue-engineered materials have been successful, and all are either avascular tissues such as cartilage (Paige and Vacanti 1995; Takazawa et al. 2012) or very thin tissues, including skin (Auger et al. 2004; Shevchenko et al. 2010) and elements of the urinary tract (Park et al. 2000; Raya-Rivera et al. 2011), which can survive on diffusion from recipient site capillaries prior to new capillaries growing into the construct.

17.2.2 Extrinsic Vascularization

Thus, the pivotal issue that significantly impedes the generation of large 3D organs and tissues is the slow speed and incomplete ingrowth of blood vessels into a tissue engineering construct. The traditional method by which tissue engineering constructs become vascularized is a process termed extrinsic vascularization whereby insertion of a scaffold with cells *in vivo* promotes capillary invasion into the scaffold from the peripheral (extrinsic) tissue bed. Areas such as under the kidney capsule (Saito et al. 2011), subdermally (Mercuri et al. 2013), and within skeletal muscle (Balamurugan et al. 2003) have frequently been selected for construct insertion—as these sites possess a reasonably extensive capillary network available for infiltration into the scaffold.

17.2.2.1 Angiogenesis and Vasculogenesis

Blood vessel ingrowth into an avascular region is the usual response of the adult mammal and although there may be a number of stimulatory factors, hypoxic signals are the major drivers of new blood vessel growth (Rouwkema et al. 2008). Capillary growth into avascular/ischemic tissue (such as in wound healing, and the female reproductive tract) may occur by two general processes: either microvessel sprouting from preexisting blood vessels termed angiogenesis, or by *in situ* capillary plexus formation from circulating endothelial precursor cells (EPCs) termed vasculogenesis (Asahara et al. 1997). Vasculogenesis was initially described in the formation of blood vessels in embryonic development (Risau and Flamme 1995), but is recognized to occur at sites requiring new vessel formation in the adult (Asahara et al. 1999); however, the contribution of vasculogenesis is generally considerably less than angiogenesis in most situations of adult new blood vessel network formation (Lokmic and Mitchell 2008a) and in tissue engineering (Simcock et al. 2009). Angiogenic capillary network formation may occur through one of a number of mechanisms, which include capillary *elongation*; *inosculation*—the interconnection of capillary networks involving anatomical joining of capillaries from separate microcirculatory networks so that their lumens and vessel walls become continuous; *interssusception*—involvesthe protrusion of opposite microvascular walls into the capillary lumen creating a contact zone between previously opposite endothelial cells. The endothelial bilayer remodels with an interstitial core, thus creating two blood vessels (Djonov et al. 2003); and *sprouting*—the activation of endothelial cells in capillaries or veins, whereby the surrounding ECM is dissolved as the activated endothelial cells proliferate and migrate as solid capillary sprouts that subsequently develop a lumen and mature with the attachment of pericytes and/or smooth muscle cells (Darland and D'Amore 1999). Sprouting angiogenesis is the most common mechanism for adult angiogenic growth of new blood vessel networks (Carmeliet 2005).

The process of extrinsic vascularization is slow—depending on the size of the construct this process can take up to several weeks to complete (Rouwkema et al. 2008). As a consequence centrally implanted cells in the construct, which are >150–200 microns from a capillary will die (Folkman et al. 1973), and for large constructs there is a considerable chance that centrally situated cells will die before new capillary growth can supply these cells with oxygen and nutrients. Thus with this commonly used extrinsic vascularization approach the main limitation on the size of the construct and its ultimate functional viability is the speed of ingrowth of new capillaries. Consequently, construct volume is generally limited to 3 mm^3 (Lokmic and Mitchell 2008a).

17.2.3 Intrinsic Vascularization

The isolation of vascular pedicles and their transfer to adjacent sites to vascularize tissues, including new skin territories had its origins in the concept of tissue prefabrication. Prefabrication involving a vascular pedicle was first proposed by Hori et al. (1979) and Erol and Spira (1980). In the latter case, an arteriovenous loop (AVL) was created by dividing an AV pedicle and joining the artery to the vein thus creating a fistula. This was then laid over an adjacent nonvascularized, nongraftable surface (such as bare bone) where it progressively covered the surface with new angiogenic tissue, which could then support a skin graft. This concept has been expanded and used in experimental and clinical reconstructive surgery research (Morrison et al. 1990; Khouri et al. 1992, 1994) to "prefabricate" new vascularized territories of skin and fat (flaps) or composite tissues of skin/bone or cartilage by implanting AV pedicles within these tissues and allowing the angiogenic stimulus from the surgery to vascularize these prefabricated tissues (Figure 17.1). The pedicle and associated newly vascularized tissue could then be transferred by microsurgical means to the site where the new flap was required (Morrison et al. 1990, 1997; Takato et al. 1993; Pribaz et al. 1999a,b; Guo and Pribaz 2009).

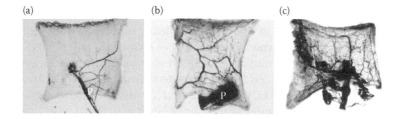

(a)   (b)   (c)

FIGURE 17.1
India ink perfusion and whole mount preparation of a ligated arteriovenous pedicle (P) that has been surgically moved from its normal position and implanted subdermally. Figures illustrate the pedicle and increasing angiogenic sprouting from it at (a) 1 week, (b) 2 weeks, and (c) 3 weeks.

The early work of Khouri et al. (1994) and subsequent work of Tanaka and Morrison in introducing a plastic chamber around the isolated vascular pedicle led them to observe that this model of an AVL in a protected space not only promoted angiogenesis but also encouraged fibroblastic tissue growth; a primitive form of tissue engineering. Rat and mouse vascularized tissue engineering chambers were developed (Mian et al. 2000; Tanaka et al. 2000; Cronin et al. 2004). In the rat model, the most common vascular configuration was the creation of an AVL based on the femoral artery and vein, which were cut in the groin and an autologous femoral vein graft harvested from the opposite leg inserted between the cut femoral vessel ends to create an AVL or shunt (Figure 17.2). The AVL was then inserted in a custom-made polycarbonate (plastic) chamber manufactured in the Department of Chemical and Biomolecular Engineering, University of Melbourne having a volume of approximately 0.5 mL.

The angiogenic process associated with the AVL inside the chamber has been extensively analyzed by Lokmic et al. (2007) over a time course of 4 months. The AVL vascular configuration immediately generated an endogenous fibrin scaffold that filled approximately 2/3 to 3/4 of the chamber space. From the viewpoint of studying the vascularization process, it is a significant advantage of this model that vascularization occurs even in the absence of an added ECM (Mian et al. 2000). The origin of the endogenous fibrin is thought to be leakage from the AVL anastomoses. At approximately 3–5 days, capillary sprouting commences initially from the recipient vein segment of the loop (Figure 17.3). New capillary network generation proceeds from this sprouting—initially, this network surrounds the AVL and then extends throughout the fibrin clot. At 7–10 days, sprouting of small vessels from the recipient femoral artery, and pericytes and vascular smooth muscle cells attach abluminally around the endothelial network to form arterioles, capillaries, and venules. Capillary sprouts were not observed from the vein graft segment of the AVL in the first 2 weeks of development (Figure 17.3). A very extensive capillary network is generated in the following few weeks being maximal at 10 days, and largely maintained for the subsequent 16 weeks, although some blood vessel regression was observed at later time points. Blood vessel growth in the chamber appears to cease when the periphery of the fibrin clot is reached, thus the distance that the network will grow is limited by the volume of the fibrin clot.

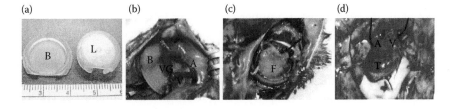

FIGURE 17.2
(a–d) Macroscopic views of significant stages of development of the AVL chamber construct: (a) Chamber base (B) and chamber lid (L). (b) The surgically created AVL on the chamber base (B) prior to closing the chamber lid. A: femoral artery; VG: vein graft, V: femoral vein of the AVL. (c) The chamber lid has been removed from a 3-day construct. A large volume of fibrin matrix (F) surrounds the AVL, obscuring it. Arrow: position where the femoral artery and vein enter and leave the construct. (d) At 112 days, a flap of healthy fibrovascular tissue (T) covers the AVL, which is obscured. The femoral artery (A) and vein (V) entering and leaving the construct are visible. (Republished with permission from Figure 1 of Lokmic, Z. et al. G.M. 2007. An arterio-venous loop in a protected space generates a permanent, highly vascular, tissue engineered construct. *FASEB J* 21:511–22.)

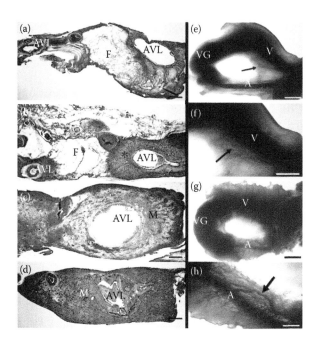

FIGURE 17.3

(a–d) Toluidine blue staining of vertical sections through the AVL construct: (a) At 3 days, the AVL is surrounded by fibrin matrix (F). (b) At 7 days, two cross sections through the AVL, now surrounded by a cuff of new connective tissue (*) and fibrin (F). (c) At 14 days, the AVL is surrounded by a wide cuff of mature vascularized connective tissue (M). The fibrin matrix (F) is barely visible at the periphery. (d) At 28 days, the AVL is encompassed by mature vascularized connective tissue (M). No fibrin is evident. Scale bars: 500 μm; (e–h) Resin casts of the AVL construct at day 10 (e, f) and day 14 (g, h). No capillaries or arterioles sprout from the vein graft (VG) (e, g) at either time point. Capillaries are seen sprouting (arrows) from the femoral vein (V) at 10 days (e, f). Occasional arterioles (thick arrows) branch from the femoral artery (A) at day 10 (e) but were far more numerous at day 14 (g, h). Scale bars: e–h: 2 mm. (Republished with permission from Figure 2 of Lokmic, Z. et al. 2007. An arterio-venous loop in a protected space generates a permanent, highly vascular, tissue engineered construct. *FASEB J* 21:511–22.)

17.2.3.1 *Alternative Vascular Configurations Used with the Intrinsic Vascularization Approach*

Tanaka et al., in 2003, tested a number of different intrinsic vascular configurations of the rat femoral pedicle (artery and vein) within a folded artificial dermis sheet. This sheet was covered externally with a silicone layer to exclude the influence of extrinsic vascularization. A chamber was not used in these experiments. The AVL shunt (1) was compared to a ligated blindly ending AV pedicle (2), and to a flow through configuration where the artery and vein remained in continuity simply passing through the chamber (3). The volume of new tissue (new blood vessels and new supporting connective tissue) generated was significantly greater in the AVL group compared to both the flow through pedicle and the ligated pedicle ($p < 0.01$). Although the AVL demonstrated significant increases in new tissue growth—the two other pedicle types (ligated and flow through) also demonstrated good tissue growth around the pedicle—indicating these configurations could be used in 3D tissue engineering when vein grafts are not available. Furthermore, both the ligated pedicle and flow through configurations are much simpler and more practical for clinical translation.

Tilkorn et al., in 2010, using a model devised by Alberto Bedogni, avoided the need for a vein graft in the rat AVL chamber model by lengthening the venous limb by employing the epigastric vein branch of the femoral as a connecting vascular segment to the cut femoral artery. This chamber model supported myoblast survival and enabled bilateral chambers to be created in the rat.

Whichever configuration of vessels is used, it is clear from these multiple experiments that vascular patency at least in the short term is essential for angiogenic sprouting and tissue growth. Where the vessels were found to be thrombosed in the first week, minimal growth was observed. Similarly, in control experiments, unless a space is maintained to allow tissue volume expansion either by a protective chamber or an ECM that does not resorb before its volume is replaced with new tissue, tissue growth is stifled due to the surrounding wound contracture and fibrosis.

Zdolsek et al., in 2011, further demonstrated the versatility of the AVL model when they inserted an "off-the-shelf" cold-stored venous allograft (stored for 4 weeks at 4°C, which removes the cellular component but maintains the connective tissue framework of the vascular graft) into the loop and compared these to autologous fresh vein grafts in the AVL construct in terms of construct growth and angiogenesis at 2 and 6 weeks, respectively. The study found that the insertion of the acellular allograft did not compromise the overall construct growth (weight or volume) nor angiogenic growth (percent and absolute vascular volume) in the chamber at 6 weeks. This further supports the concept that it is the femoral vein and later the femoral artery that sprouts new vessels into the chamber space rather than the vein graft. Although in this model there was an incidence of late AVL thrombosis, intensive angiogenic sprouting and the overall tissue growth was not adversely affected by the late ischemia.

A mouse model using a flow through pedicle based on the epigastric vessels enclosed in a silicon sleeve (volume 45–50 µL) (Figure 17.4) was developed by Cronin et al. in 2004. This model allows bilateral chambers to be implanted in opposite groins, but generates very little fibrin matrix for angiogenic growth and therefore requires an added ECM to fill the chamber and encourage angiogenic migration. For experimental purposes, the mouse-derived hydrogel Matrigel™ ideally fulfills these requirements (Figure 17.4). Other matrices, including collagen and pluronic gel, have also been used in this murine chamber model (Vashi et al. 2006, 2008) and poly (lactic-*co*-glycolic acid) (PLGA) (Cronin et al. 2004). Although all matrices used have supported sprouting angiogenesis, many resorbed or retracted, thereby limiting angiogenic migration to the outer reaches of the chamber capacity. Matrigel would appear to be the best matrix for maintaining its original injected volume in the chamber space and offer the best potential for a chamber to fill with new tissue. Cronin et al. (2004) were also able to determine in their study that the flow through pedicle was more successful than the ligated epigastric pedicle in generating new capillary networks and in overall tissue production.

17.2.4 Combined Intrinsic and Extrinsic Vascularization Approach

Using both intrinsic and extrinsic approaches, whereby the construct vascularizes simultaneously from a central vascular pedicle, and from peripheral ingrowth via perforations in the external chamber has also been described (Tanaka et al. 2006). In each case where this combination has been employed, a significant increase in construct weight and volume has been observed (Dolderer et al. 2007; Zdolsek et al. 2011). Zdolsek also observed significant increases in percent and absolute connective tissue volume in the perforated chambers, compared to intrinsic vascularization chambers alone, and although increases in percent and vascular volume were observed in their study, these differences were not significant. Arkudas et al. (2012) also combined the two approaches

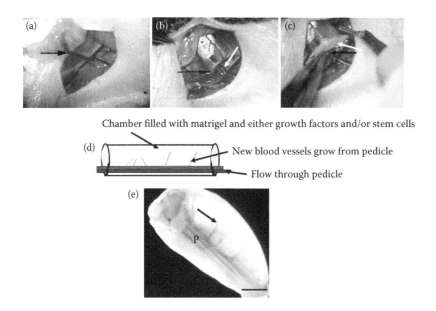

FIGURE 17.4
(a–c) Photographs illustrating the surgery involved in creating the mouse flow through pedicle chamber: (a) The superficial inferior epigastric pedicle (arrow) branching off the femoral vessels. (b) The cylindrical silicone chamber sutured into the groin (black arrow). The epigastric vascular pedicle is entering the silicon chamber (white arrow). (c) Bone wax sealing the base of the chamber (white arrow). The black arrow indicates the epigastric pedicle entering chamber. (d) Diagram illustrating all the elements of the chamber construct. (e) Macroscopic appearance of an angiogenic growth factor-treated construct, after removal from the chamber at 6 weeks. P, pedicle; arrow: new blood vessels. Scale bar in (e): 1 mm. (Republished with permission from Figure 1 of Rophael, J.A. et al. 2007. Angiogenic growth factor synergism in a murine tissue-engineering model of angiogenesis and adipogenesis. *Am J Pathol* 171:2048–57.)

and found that 83% of extrinsic vessels were connected to the AVL at 2 weeks, and 97% at 8 weeks.

17.2.5 Growing Specific Mesenchymal Tissues and Organs Using Intrinsic Vascularized Chamber Models

Both the rat and the mouse chamber models have been used to implant tissues, and stem or progenitor cells to generate specific tissues. These include adipose tissue flaps—connected directly to a vascular pedicle (Dolderer et al. 2007, 2011; Findlay et al. 2011), cardiomyoblasts (Morritt et al. 2007; Choi et al. 2010; Tee et al. 2012), pituitary stem cells (Lepore et al. 2007), thymus tissue (Seach et al. 2010), skeletal muscle tissue (Messina et al. 2005), myoblasts (Tilkorn et al. 2010, 2012), liver progenitor cells (Forster et al. 2011; Yap et al. 2013), and islet of Langerhans pancreatic tissue (Hussey et al. 2009, 2010; Forster et al. 2011). In addition, several of these studies successfully transplanted the growing chamber tissue on its vascular pedicle to another site in the body and demonstrated posttransplant survival of the engineered tissue (Tanaka et al. 2006; Dolderer et al. 2011; Tee et al. 2012). Other research groups have also used intrinsic vascularization in AVL chamber models to grow skeletal muscle (Bach et al. 2006), bone (Kneser et al. 2006; Polykandriotis et al. 2007; Boos et al. 2012), and liver tissue (Fiegel et al. 2009, 2010), and others have grown cardiomyocytes in a flow through pedicled chamber in rats (Birla et al. 2009).

17.2.6 Upscaling: Engineering Large Three-Dimensional Constructs Using the Intrinsic/Extrinsic Vascularization Approach

A number of studies have attempted to significantly expand the volume of constructs generated using the intrinsic vascularization approach by providing a much *larger chamber space*. Our group has tested the combined intrinsic/extrinsic vascularization concepts using an AVL in a chamber in rats, rabbits, pigs, and humans. Hofer et al. (2003) in the rat model increased the polycarbonate chamber volume to 1.9 mL but with the same-sized AVL as used in smaller chambers (0.5 mL). The AVL was sandwiched between two disks of porous PLGA, which essentially filled this expanded chamber space. Histological examination of the constructs up to 8 weeks revealed increased tissue growth into the PLGA over time, and concluded that larger quantities of tissue can be generated from the same-sized AVL by using larger-size growth chambers, which are filled with an ECM. Tanaka et al. (2006) significantly increased construct size by increasing perforated chamber size using the same vascular pedicle in small and large chambers in rabbits.

Beier et al. (2009) attempted upscaling of the AVL chamber system in a large animal model in sheep by enclosing an AVL based on the saphenous artery and vein pedicle in a chamber that was 2.8 × 1.8 × 1.8 cm (a volume of 9 mL approximately). The authors present a study of the angiogenesis and tissue growth in such a system to 6 weeks. The volume of tissue that formed over this period was not detailed.

We have trialed a similar chamber model with a much larger volume (78.5 mL) and external perforations in the pig model (Figure 17.5) in an attempt to generate tissue of the approximate

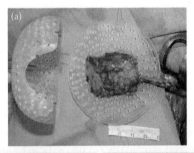

FIGURE 17.5

Large-volume chambers containing a fat flap and pedicle in the pig: (a) Fat flap (5 mL volume) based on the superficial circumflex iliac vessels placed on the base of the chamber at chamber insertion. (b) Tissue growth in a 6-week fat flap chamber with the long superficial circumflex iliac pedicle entering it. Note the multiple small knobs of tissue that result from tissue growth from outside the chamber through the chamber pores into the chamber. (c) Cut surface of 6-week non-poly(l-lactide-*co*-glycolide) fat flap chamber with central adipose tissue growth (*) surrounded by fibrovascular tissue. (Republished with permission from Figure 1 of Findlay, M.W. et al. 2011. Tissue-engineered breast reconstruction: Bridging the gap towards large-volume tissue engineering in humans. *Plast Reconstr Surg* 128:1206–15.)

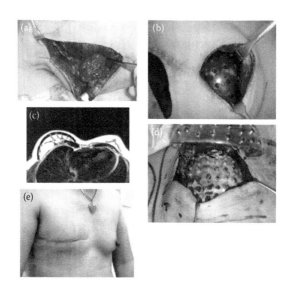

FIGURE 17.6
Creating a human breast (Neopec 2012): (a) Fat flap on the thoraco-dorsal vascular pedicle in the right axilla. (b) Fat flap transposed into the chamber shell *in situ* at the site of the breast. (c) MRI at 12 months showing tissue forming in the curved chamber (on the right-hand side). (d) Chamber being removed showing chamber space filled with new tissue. (e) Six months postremoval of the chamber showing the new breast.

volume of a human breast (Findlay et al. 2011). In initial preliminary studies using only a ligated pedicle in the axilla without any added ECM/scaffold, very little tissue (2 mL approximately) was generated around the pedicle. However, when a pedicled fat flap of 5- mL volume from the groin based on the superficial circumflex iliac vessels was transposed into this large chamber space either alone or sandwiched between two disks of poly(L-lactide-*co*-glycolide) sponge, the chambers largely filled with tissue by 6 weeks. Approximately 35% of this volume was fat tissue having expanded from an initial 5-mL volume over that time ($N = 8$). In two animals assessed at 22 weeks, the volume had remained constant. Thus large 3D vascularized adipose tissue flaps on their own vascular pedicle were generated in this model, which in one case was transferred surgically on its pedicle to another site. Thus, experimentally, there appeared to be proof of principle that 3D adipose tissue volumes appropriate for human breast reconstruction could be generated in large animals.

We proceeded to a pilot study in humans using the same concepts of a small vascularized fat flap pedicled inside a hollow chamber without matrix to test the proof of principle found in animals that such a flap could expand and attempt to fill the chamber space (reported at the Sydney International Breast Cancer Conference, October 2012). In one of four patients tested, a perforated chamber 180 cc in volume completely filled with new tissue in 12 months (Figure 17.6). This tissue comprised fibrovascular tissue approximately 20% of which was fat and this volume remained stable over the following 6 months. In the remaining three patients, the inserted fat flap did not expand and was densely contracted by scar encapsulation within the chamber (Neopec: VSA Grant by Victorian State Government).

17.2.7 Influence of the ECM in Determining Three-Dimensional Construct Size with Intrinsic Vascularization

Despite encouraging results using various vascularization models, there is clearly a limit on the vascularization process to continually expand especially into an ever-increasingly large space. The process is possibly different in different species and at different ages. Nevertheless, to enhance the capacity for scale-up of both vascularization and tissue, a matrix that is vasculogenic seems essential.

Lokmic et al. (2008b) demonstrated that inhibition of fibrin clot with enoxaparin sodium significantly reduced construct size in the AVL chamber model. (The construct is defined as the AVL loop and surrounding tissue in the chamber.) Construct weight and volume were significantly reduced at 10 and 21 days (final time point examined) compared to untreated constructs. Angiogenesis with enoxaparin treatment was delayed and absolute vascular volume generated around the AVL significantly decreased at 10 days. The volume of the fibrin matrix and the quality of the fibrin matrix are likely to determine construct size and degree of vascularization in this model.

Cassell et al. (2001) monitored tissue growth in rat AVL chambers up to 4 weeks when filled with commercially available fibrin matrices, PLGA, Matrigel, or no added ECM. The results indicate that PLGA generated the largest constructs and presented good tissue infiltration compared to Matrigel and fibrin. However, a later study (Cao et al. 2006) indicated that although good tissue ingrowth occurred in the first few weeks, in PLGA scaffolds, a foreign body response occurred—as is the case with many synthetic scaffolds used *in vivo*, and as the PLGA degraded, the release of lactic acid killed the vascularized tissue in the smallest PLGA pores. The study concluded that only tissue pores greater than 300 μm supported sustained vascularized tissue ingrowth at 8 weeks in PLGA.

Thus, the physical/chemical/biocompatibility properties of the ECM used in tissue engineering scaffolds can greatly influence tissue growth and the ultimate volume and vascularization of tissue engineering constructs.

Although biologically derived scaffolds such as collagen, fibrin (as Cassell et al. 2002), and various tissue-derived scaffolds, including pig intestinal mucosa—Surgisis (Franklin et al. 2002; Yohannes et al. 2002), skeletal muscle-derived Myogel (Abberton et al. 2008; Francis et al. 2009), adipose tissue-derived matrices (Uriel et al. 2008; Choi et al. 2009; Flynn 2010; Poon et al. 2013), as well as part or whole organ connective tissue decellularized matrices (Ott et al. 2008; Ji et al. 2012), and biological scaffolds of nonanimal origin, such as chitosan (Jiang et al. 2008; Costa-Pinto et al. 2011) and silk (Mauney et al. 2007; Bellas et al. 2013) can be used in tissue engineering, synthetic 3D biomaterial scaffolds have traditionally been preferred. This is despite the fact that many promote a foreign body response *in vivo*, although this can be modified and its significance may diminish over time. Rouwkema (2008) and others have documented the influence of synthetic scaffold design in promoting tissue ingrowth and vascularization. Particularly influential is the pore size of the scaffold and the interconnectivity of the pores. Larger pore sizes and complete interconnectivity between pores (that is few if any blind ending pores) enhance vascular ingrowth. Newer scaffold design systems, for example, the solid free-form fabrication systems, allow complex scaffold design with specific architecture. The structure of the scaffold can be varied internally for different specific tissue growth in various zones of individual scaffolds. Rouwkema describes as one of these systems the rapid prototyping or fiber deposition technology by application of different fiber layers using CAD (computer-aided design) (Vaquette et al. 2012). Various materials, including polymers, metals, ceramics, and gels with encapsulated cells, can be used in these fiber layers. Such structurally accurate designs are essential for weight-bearing tissue engineering for bone and cartilage, but soft tissues and organs require very different moduli and fiber frameworks. As previously mentioned, matrix ECMs are key to growth factor release and cell signaling and orchestrating tissue growth stimuli in a patterned sequence. Native ECMs are influenced by biochemical as well as biomechanical signals and none of these can be replicated by purely synthetic scaffolds. Self-assembly is an accepted phenomenon where a cell lays down its own matrix and as the cell's needs change, so too does the type of matrix that it produces. Progressively cell clusters form appropriate to their environment. Clinically, this concept has been long recognized and is seen in "creeping substitution" of bone grafts.

17.3 Promoting Blood Vessel Growth in Tissue Engineering

17.3.1 Angiogenic Growth Factors

There have been numerous studies that have endeavored to increase the rate of capillary growth both extrinsically or intrinsically into tissue engineering constructs, generally by applying angiogenic growth factors. Prior to their use in tissue engineering, growth factors such as FGF-2 and PDGF B were used successfully to significantly enhance tissue growth around an implanted pedicle in prefabrication experiments (Khouri et al. 1994; Hickey et al. 1998). Tanaka et al. (2006) were able to significantly increase the size of new tissue around a pedicle by utilizing two techniques: (1) a larger chamber volume around the same-sized pedicle and (2) the application of FGF-2 within the ECM inside the chamber. This group compared small and large perforated chambers into which collagen sheets were wrapped around a saphenous AV pedicle. Chambers in each small- and large-chamber group were treated with or without FGF-2 and examined at 4 weeks. Larger chambers with or without FGF-2 produced twice the construct volume as small chambers. FGF-2 stimulated large-volume constructs that completely filled the chamber while no-FGF-2 chambers were only half filled.

The principal growth factors used to promote angiogenesis within tissue engineering constructs are FGF-2 and VEGF-A. Both are potent promoters of endothelial cell proliferation, but have quite short half-lives, suggesting that they would exert their influence over only a few days to weeks (Lokmic and Mitchell 2008a). Growth factors can be simply added to the construct (Kawaguchi et al. 1998; Rophael et al. 2007) or can be incorporated into scaffolds, or on the surface of scaffolds (Tabata et al. 2000; Nair et al. 2010; Ayvazyan et al. 2011; MacDonald et al. 2011), or onto the surface of degrading beads (Elçin et al. 1996; Peters et al. 1998; Vashi et al. 2006; Gandhi et al. 2013), where the growth factor is slowly released as the beads or scaffolds degrade. Genetic manipulation of implanted cells so that they specifically produce angiogenic growth factors post-implantation to promote angiogenesis has also been used in regenerative medicine (Song et al. 2010).

Several research groups have concurrently used two angiogenic growth factors to significantly increase blood vessel ingrowth in *in vivo* tissue engineering constructs. Richardson et al. (2001) used VEGF and PDGF, while Rophael et al. (2007) used VEGF, FGF, and PDGF. The role of PDGF is to further promote capillary maturation after capillary cord proliferation. It is likely that optimal use of angiogenic growth factors in tissue engineering requires a spatiotemporal controlled cocktail of factors (stimulatory and maturation growth factors) applied at various times during construct growth to generate a permanent vascular network (Lokmic and Mitchell 2008a). Whether the use of growth factors has a long-term benefit has been questioned by Rophael et al. (2007) who observed a significant increase in angiogenesis at 2 weeks, which declined to control levels at 6 weeks. However, even this transitory increase in vascularization may be positive for the survival of the construct if it occurs at an early time point and enhances implanted cell survival in the first 7–10 days when cells are most vulnerable to ischemic death.

17.3.2 Preforming Vascular Networks *In Vitro*

Injection of endothelial cells *in vivo* has been suggested for enhancing capillary network formation. Our group injected human endothelial precursor cells (EPCs) intracardially into rats and observed their appearance in the groin chamber construct where their numbers were substantially increased by SDF-1 infused into the chamber. However, they could not be convincingly identified within the new vasculature (Simcock et al. 2009). Alternatively, there is growing evidence that capillary networks can be formed *in vitro* and transplanted *in vivo* to form part of the blood vessel network of a tissue engineering construct. The initial work in this area was by Black et al. (1998), Koike et al. (2004), and Levenberg et al. (2005). It is well known that endothelial cells are capable of forming capillary networks in 3D fibrin gels (Chen et al. 2009). It has also been described that endothelial cells in multicellular spheroids placed *in vitro* in 3D gels sprout capillary-like structures (Wenger et al. 2005; Finkenzeller et al. 2009). HUVEC spheroids implanted *in vivo* in Matrigel/fibrin with high doses of VEGF-A and FGF-2 form new capillary networks that inosculate with recipient site (host) capillaries and are functional (Alajati et al. 2008).

The addition of vascular support cells (fibroblasts or vascular smooth muscle cells, mesenchymal stem cells) with endothelial cells in scaffolds appears to accelerate preformed capillary formation and maturation (Levenberg et al. 2005; Berthod et al. 2006; Sorrell et al. 2007; Au et al. 2008; Chen et al. 2009; Newman et al. 2011). These preformed *in vitro* networks have been successfully transferred *in vivo* and inosculated (Laschke et al. 2009) with recipient site capillaries, are functional, and promote larger *in vivo* construct growth (Levenberg et al. 2005; Chen et al. 2009).

More recent developments in preformed vasculature involve endothelial cells sandwiched within or between specific tissue cell layer sheets *in vitro*, which were transferred successfully *in vivo* (Sekine et al. 2013; Sugibayashi et al. 2013), with the demonstration of growth of capillary networks from the endothelial cell-seeded sheets throughout the construct, and inosculation with recipient site capillaries. Recently, the formation of capillary networks from individual endothelial cells aligned on a ridged matrix that formed capillaries and then inosculated *in vitro* with the severed ends of artery and vein segments also implanted in the gel has been demonstrated. This arterio-capillary-venous network was then successfully transplanted *in vivo* (Chui et al. 2012).

In vitro preformed networks have the potential to be transferred into either extrinsically or intrinsically vascularized *in vivo* tissue engineering constructs. Regardless of the vascularization approach taken, the preformed network is capable of inosculation with capillary networks from the recipient site (Laschke and Menger 2012) provided the recipient capillary bed is in close proximity. Early blood circulation within the construct can be achieved, as it is known that inosculation of capillaries within skin grafts can take as little as 48–72 h to occur (O'Ceallaigh et al. 2007; Calcagni et al. 2011), although if the recipient site capillaries are not close to the preformed network, this time frame will lengthen as the recipient site capillaries will take longer time to grow and reach the preformed network.

17.4 Conclusions

It is clear that scale-up is the key to successful clinical 3D tissue engineering. Although an intrinsic (supported by extrinsic) blood supply expanding *pari passu* with the newly growing tissue construct is an attractive concept, there are limitations to its potential for continuous expansion and tissue nourishment. The ischemic drive weakens over time and fibrotic encapsulation of the newly expanding tissue prevents its migration to the outer confines of the proposed space.

Progressive space enlargement at a pace just ahead of the expanding construct, such that its advancing front is always in contact with peripheral vascularized tissue, repeated angiogenic boosts, serial ECM refills, serial cell seeding (angiogenic or specific tissue precursors), cell preconditioning, and antifibrotics are just some of the strategies that may be efficacious to increasing the potential volume of the tissue-engineered product.

The chamber is a contrivance to maintain space for expansion of the newly growing tissue within it and to prevent closure of the space by fibrosis of the surrounding tissues as occurs in a standard wound. Once closed, the angiogenic stimulus ceases and scar maturation follows. The chamber however is unlikely to be an attractive clinical option even if it can be made biodegradable and a more practical alternative would be an injectable hydrogel, which maintains its volume over the time period of the projected tissue growth. Into this could be incorporated specific cells, serially boosted as needed. In fact, this injectable prototype model is in current clinical use in the form of fat grafting and is the most common and successful form of tissue engineering to date (Coleman et al. 1997; Yoshimura et al. 2008). Lipoaspirate (fat tissue vacuum extracted via a cannula from the subcutaneous tissue layer) is a mixture of matrix ECMs, cells (differentiated adipocytes, undifferentiated precursor cells, endothelial cells, macrophages, and others), and blood. When injected into the recipient site, the cells rely on rapid nutritional sustenance, which must derive from extrinsic vascularization. It is becoming increasingly accepted that the adipocytes in transplanted lipoaspirate do not survive, but the injected tissue volume remains and acts as an ECM spacer into which stem cells within the injected lipoaspirate or from the host migrate and differentiate into new fat to replace the ECM scaffold space. Ischemia and inflammation drive the angiogenesis and cell signaling in the regenerating fat (Lilja et al. 2013).

Despite our perseveration and commitment to *in vivo* tissue engineering with its close parallels to reconstructive surgical principles with which we are familiar, the holy grail of 3D tissue engineering may ultimately involve *in vitro* techniques.

Tissues seeded on scaffolds with an incorporated preformed microvasculature (Levenberg et al. 2005), which can be inosculated to a macrovessel *ex vivo* (Chui et al. 2012), will allow immediate transfer of this product into a recipient animal by microsurgical vessel anastomosis overcoming the problems of a critical time frame for revascularization and permitting ultimate 3D tissue scale-up. Commencing *in vitro* with a cell seeded,decellularized organ ECM or tissue ECM represents one possible avenue of this conceptual pathway (Ott et al. 2008).

Acknowledgments

The authors acknowledge several National Health & Medical Research Council of Australia Project Grants whose work contributed to some of the studies described in this review, funding from the Australian Catholic University/O'Brien Institute Tissue Engineering Centre and the Victorian State Government's Department of Innovation, Industry and Regional Development's Operational Infrastructure Support Program.

References

Abberton, K.M., Bortolotto, S.K., Woods, A.A. et al. 2008. Myogel, a novel, basement membrane-rich, extracellular matrix derived from skeletal muscle, is highly adipogenic *in vivo* and *in vitro*. *Cells Tissues Organs* 188:347–58.

Alajati, A., Laib, A.M., Weber, H. et al. 2008. Spheroid engineering of a human vasculature. *Nat Meth* 5:439–45.

Arkudas, A., Pryymachuk, G., Beier, J.P. et al. 2012. Combination of extrinsic and intrinsic pathways significantly accelerates axial vascularization of bioartificial tissues. *Plast Reconstr Surg* 129:55e–65e.

Asahara, T., Murohara, T., Sullivan, A. et al. 1997. Isolation of putative progenitor endothelial cells for angiogenesis. *Science* 275(5302):964–7.

Asahara, T., Masuda, H., Takahashi, T. et al. 1999. Bone marrow origin of endothelial progenitor cells responsible for post natal vasculogenes is in physiological and pathological neovascularization. *Circ Res* 85:221–8.

Au, P., Tam, J., Fukumura, D., and Jain, R.K. 2008. Bone marrow-derived mesenchymal stem cells facilitate engineering of long-lasting functional vasculature. *Blood* 111:4551–8.

Auger, F.A., Berthod, F., Moulin, V., Pouliot, R., and Germain, L. 2004. Tissue-engineered skin substitutes: From *in vitro* constructs to *in vivo* applications. *Biotechnol Appl Biochem* 39:263–75.

Ayvazyan, A., Morimoto, N., Kanda, N. et al. 2011. Collagen-gelatin scaffold impregnated with bFGF accelerates palatal wound healing of palatal mucosa in dogs. *J Surg Res* 171: e247–57.

Bach, A.D., Arkudas, A., Tjiawi, J. et al. 2006. A new approach to tissue engineering of vascularized skeletal muscle. *J Cell Mol Med* 10:716–26.

Balamurugan, A.N., Gu, Y., Tabata, Y. et al. 2003. Bioartificial pancreas transplantation at prevascularized intermuscular space: Effect of angiogenesis induction on islet survival. *Pancreas* 26:279–85.

Beier, J.P., Horch, R.E., Arkudas, A. et al. 2009. De novo generation of axially vascularized tissue in a large animal model. *Microsurgery* 29:42–51.

Bellas, E., Panilaitis, B.J., Glettig, D.L. et al. 2013. Sustained volume retention *in vivo* with adipocyte and lipoaspirate seeded silk scaffolds. *Biomaterials* 34:2960–8.

Berthod, F., Germain, L., Tremblay, N., and Auger, F.A. 2006. Extracellular matrix deposition by fibroblasts is necessary to promote capillary-like tube formation *in vitro*. *J Cell Phys* 207:491–8.

Birla, R.K., Dhawan, V., Dow, D.E., Huang, Y.C., and Brown, D.L. 2009. Cardiac cells implanted into a cylindrical, vascularized chamber in vivo: Pressure generation and morphology. *Biotechnol Lett* 31:191–201.

Black, A.F., Berthod, F., L'Heureux, N., Germain, L., and Auger, F.A. 1998. *In vitro* reconstruction of a human capillary-like network in a tissue-engineered skin equivalent. *FASEB J* 12:1331.

Boos, A.M., Loew, J.S., Weigand, A. et al. 2012. Engineering axially vascularized bone in the sheep arteriovenous-loop model. *J Tissue Eng Regen Med* 2012 Mar 22.

Calcagni, M., Althaus, M.K., Knapik, A.D. et al. 2011. *In vivo* visualization of the origination of skin graft vasculature in a wild-type/GFP crossover model. *Microvasc Res* 82:237–45.

Cao, Y., Mitchell, G., Messina, A. et al. 2006. The influence of architecture on degradation and tissue ingrowth into three dimensional poly(lactic-*co*-glycolic acid) scaffolds *in vitro* and *in vivo*. *Biomaterials* 27:2854–64.

Carmeliet, P. 2005. Angiogenesis in life, disease and medicine. *Nature* 438(7070):932–6.

Cassell, O.C., Hofer, S.O., Morrison, W.A., and Knight, K.R. 2002. Vascularisation of tissue-engineered grafts: The regulation of angiogenesis in reconstructive surgery and in disease states. *Br J Plast Surg* 55:603–10.

Cassell, O.C., Morrison, W.A., Messina, A. et al. 2001. The influence of extracellular matrix on the generation of vascularized, engineered, transplantable tissue. *Ann NY Acad Sci* 944:429–42.

Chen, X., Aledia, A.S, Ghajar, C.M. et al. 2009. Prevascularization of a fibrin based tissue construct accelerates the formation of functional anastomoses with host vasculature. *Tiss Eng Part A* 15:1363–71.

Chiu, L.L.Y., Montgomery, M., Liang, Y., Liu, H., and Radisic M. 2012. Perfusable branching microvessel bed for vascularization of engineered tissues. *PNAS* 109:E3414–23.

Choi, J.S., Yang, H.J., Kim, B.S. et al. 2009. Human extracellular matrix powders for injectable cell delivery and adipose tissue engineering. *J Cont Release* 139:2–7.

Choi, Y.S., Matsuda, K., Dusting, G.J., Morrison, W.A., and Dilley, R.J. 2010. Engineering cardiac tissue *in vivo* from human adipose-derived stem cells. *Biomaterials* 31:2236–42.

Coleman, S.R. 1997. Facial recontouring with lipostructure. *Clin Plast Surg* 24:347e–547e.

Costa-Pinto, A.R., Reis, R.L., and Neves, N.M. 2011. Scaffolds based bone tissue engineering: The role of chitosan. *Tissue Eng Part B* 17:331–47.

Cronin, K.J., Messina, A., Knight, K.R. et al. 2004. New murine model of spontaneous autologous tissue engineering, combining an arteriovenous pedicle with matrix materials. *Plast Reconstr Surg* 113:260–9.

Darland, D.C. and D'Amore, P.A. 1999. Blood vessel maturation: Vascular development comes of age. *J Clin Invest* 103:157–8.

Djono, V., Baum, O., and Burri, P.H. 2003. Vascular remodeling by intussusceptive angiogenesis. *Cell Tissue Res* 314:107–17.

Dolderer, J.H., Abberton, K.M., Thompson, E.W. et al. 2007. Spontaneous large volume adipose tissue generation from a vascularized pedicled fat flap inside a chamber space. *Tissue Eng* 13:673–81.

Dolderer, J.H., Thompson, E.W., Slavin, J. et al. 2011. Long-term stability of adipose tissue generated from a vascularized pedicled fat flap inside a chamber. *Plast Reconstr Surg* 127:2283–92.

Elçin, Y.M., Dixit, V., and Gitnick, G. 1996. Controlled release of endothelial cell growth fact or from chitosan-album in microspheres for localized angiogenesis: *In vitro* and *in vivo* studies. *Artif Cells Blood Substit Immobil Biotechnol* 24:257–71.

Erol, O.O. and Spira, M. 1980. New capillary bed formation with a surgically constructed arteriovenous fistula. *Plast Reconstr Surg* 66:109–15.

Fiegel, H.C., Kneser, U., Kluth, D. et al. 2009. Development of hepatic tissue engineering. *Pediatr Surg Int* 25:667–73.

Fiegel, H.C., Pryymachuk, G., Rath, S. et al. 2010. Foetal hepatocyte transplantation in a vascularized AV-loop transplantation model in the rat. *J Cell Mol Med* 14:267–74.

Findlay, M.W., Dolderer, J.H., Trost, N. et al. 2011. Tissue-engineered breast reconstruction: Bridging the gap toward large-volume tissue engineering in humans. *Plast Reconstr Surg* 128:1206–15.

Finkenzeller, G., Graner, S., Kirkpatrick, C.J., Fuchs, S., and Stark, G.B. 2009. Impaired *in vivo* vasculogenic potential of endothelial progenitor cells in comparison to human umbilical vein endothelial cells in a spheroid-based implantation model. *Cell Prolif* 42:498–505.

Flynn LE. 2010. The use of decellularized adipose tissue as an inductive microenvironment for the adipogenic differentiation of human adipose-derived stem cells. *Biomaterials* 31:4715–24.

Folkman, J. and Hochberg, M. 1973. Self-regulation of growth in three dimensions. *J Exp Med* 138:745–53.

Forster, N., Palmer, J.A., Yeoh, G. et al. 2011. Expansion and hepatocytic differentiation of liver progenitor cells *in vivo* using a vascularized tissue engineering chamber in mice. *Tissue Eng Part C* 17:359–66.

Forster, N.A., Penington, A.J., Hardikar, A.A. et al. 2011. A prevascularized tissue engineering chamber supports growth and function of islets and progenitor cells in diabetic mice. *Islets* 3:271–83.

Francis, D., Abberton, K., Thompson, E., and Daniell, M. 2009. Myogel supports the *ex-vivo* amplification of corneal epithelial cells. *Exp Eye Res* 88:339–46.

Franklin, M.E., Gonzalez, J.J., Michaelson, R.P., Glass, J.L., and Chock, D.A. 2002. Preliminary experience with new bioactive prosthetic material for repair of hernias in infected fields. *Hernia* 6:171–4.

Gandhi, J.K., Opara, E.C., and Brey, E.M. 2013. Alginate-based strategies for therapeutic vascularization. *Ther Deliv* 4:327–41.

Godwin, J.W., Pinto, A.R., and Rosenthal, N.A. 2013. Macrophages are required for adult salamander limb regeneration. *Proc Natl Acad Sci USA.* 110(23):9415–20.

Guo, L. and Pribaz, J.J. 2009. Clinical flap prefabrication. *Plast Reconstr Surg* 124(6 Suppl):e340–50.

Hickey, M.J., Wilson, Y., Hurley, J.V., and Morrison, W.A. 1998. Mode of vascularization of control and basic fibroblast growth factor-stimulated prefabricated skin flaps. *Plast Reconstr Surg* 101:1296–304.

Hofer, S.O., Knight, K.M., Cooper-White, J.J. et al. 2003. Increasing the volume of vascularized tissue formation in engineered constructs: An experimental study in rats. *Plast Reconstr Surg* 111:1186–92.

Hori, Y., Tamai, S., Okuda, H. et al. 1979. Blood vessel transplantation to bone. *J Hand Surg* 4:23–33.

Hussey, A.J., Winardi, M., Han, X.L. et al. 2009. Seeding of pancreatic islets into prevascularized tissue engineering chambers. *Tiss Eng Part A* 12:3823–33.

Hussey, A.J., Winardi, M., Wilson, J. et al. 2010. Pancreatic islet transplantation using vascularised chambers ameliorates hyperglycaemia. *Cells Tiss Org* 191:382–93.

Ji, R., Zhang, N., You, N. et al. 2012. The differentiation of MSCs into functional hepatocyte-like cells in a liver biomatrix scaffold and their transplantation into liver-fibrotic mice. *Biomaterials* 33:8995–9008.

Jiang, T., Kumbar, S.G., Nair, L.S., and Laurencin CT. 2008. Biologically active chitosan systems for tissue engineering and regenerative medicine. *Curr Top Med Chem* 8:354–64.

Kawaguchi, N., Toriyama, K., Nicodemou-Lena, E. et al. 1998. De novo adipogenesis in mice at the site of injection of basement membrane and basic fibroblast growth factor. *Proc Natl Acad Sci USA* 95:1062–6.

Khouri, R.K., Hong, S.P., Deune, E.G. et al. 1994. De novo generation of permanent neovascularized soft tissue appendages by platelet-derived growth factor. *J Clin Invest* 94:1757–63.

Khouri, R.K., Upton, J., and Shaw, W.W. 1992. Principles of flap prefabrication. *Clin Plast Surg* 19:763–71.

Kneser, U., Polykandriotis, E., Ohnolz, J. et al. 2006. Engineering of vascularized transplantable bone tissues: Induction of axial vascularization in an osteoconductive matrix using an arteriovenous loop. *Tissue Eng* 12:1721–31.

Koike, N., Fukumura, D., Gralla, O. et al. 2004. Tissue engineering: Creation of long-lasting blood vessels. *Nature* 428:138–9.

Laschke, M.W., Vollmar, B., and Menger, M.D. 2009. Inosculation: Connecting the life-sustaining pipelines. *Tissue Eng Part B* 15:455–65.

Laschke, M.W. and Menger, M.D. 2012. Vascularization in tissue engineering: Angiogenesis versus inosculation. *Eur Surg Res* 48:85–92.

Lepore, D.A., Thomas, G.P., Knight, K.R. et al. 2007. Survival and differentiation of pituitary colony-forming cells *in vivo*. *Stem Cells* 25:1730–6.

Levenberg, S., Rouwkema, J., Macdonald, M. et al. 2005. Engineering vascularized skeletal muscle. *Nat Biotechnol* 23:879–84.

Lilja, H.E., Morrison, W.A., Han, X.L. et al. 2013. An adipoinductive role of inflammation in adipose tissue engineering: Key factors in the early development of engineered soft tissues. *Stem Cells Dev* 22:1602–13.

Lokmic, Z. Stillaert, F., Morrison, W.A., Thompson, E.W., and Mitchell, G.M. 2007. An arterio-venous loop in a protected space generates a permanent, highly vascular, tissue engineered construct. *FASEB J* 21:511–22.

Lokmic, Z. and Mitchell, G.M. 2008a. Engineering the microcirculation. *Tissue Eng Part B* 14B:87–103.

Lokmic, Z., Thomas, J.L., Morrison, W.A., Thompson, E.W., and Mitchell, G.M. 2008b. An endogenously deposited fibrin scaffold determines construct size in the surgically created arterio-venous loop chamber model of tissue engineering. *J Vasc Surg*, 48: 974–85.

Macdonald, M.L., Samuel, R.E., Shah, N.J. et al. 2011. Tissue integration of growth factor-eluting layer-by-layer polyelectrolyte multilayer coated implants. *Biomaterials* 32:1446–53.

Mauney, J.R., Nguyen, T., Gillen, K. et al. 2007. Engineering adipose-like tissue *in vitro* and *in vivo* utilizing human bone marrow and adipose-derived mesenchymal stem cells with silk fibroin 3D scaffolds. *Biomaterials* 28:5280–90.

Menger, B., Vogt, P.M., Kuhbier, J.W., and Reimers K. 2010. Applying amphibian limb regeneration to human wound healing: A review. *Ann Plast Surg* 65:504–10.

Mercuri, J.J., Patnaik, S., Dion, G. et al. 2013. Regenerative potential of decellularized porcine nucleus pulposus hydrogel scaffolds: Stem cell differentiation, matrix remodeling, and biocompatibility studies. *Tissue Eng Part A* 19:952–66.

Messina, A., Bortolotto, S.K., Cassell, O.C., Abberton, K.M., and Morrison, W.A. 2005. Generation of a vascularized organoid using skeletal muscle as the inductive source. *FASEB J* 19:1570–2.

Metcalfe, A.D. and Ferguson, M.W. 2007. Tissue engineering of replacement skin: The crossroads of biomaterials, wound healing, embryonic development, stem cells and regeneration. *JR Soc Interface* 4:413–37.

Mian, R., Morrison, W.A., Hurley, J.V. et al. 2000. Formation of new tissue from an arteriovenous loop in the absence of added extracellular matrix. *Tiss Eng* 6:595–603.

Morrison, W.A., Dvir, E., Doi, K. et al. 1990. Prefabrication of thin transferable axial-pattern skin flaps: An experimental study in rabbits. *Br J Plast Surg* 43:645–54.

Morrison, W.A., Penington, A.J., Kumpta, S.K., and Callan, P. 1997. Clinical applications and technical limitations of prefabricated flaps. *Plast Reconstr Surg* 99:378–85.

Morritt, A.N., Bortolotto, S.K., Dilley, R.J. et al. 2007. Cardiac tissue engineering in an *in vivo* vascularized chamber. *Circulation* 115:353–60.

Nair, A., Thevenot, P., Dey, J. et al. 2010. Novel polymeric scaffolds using protein microbubbles as porogen and growth factor carriers. *Tiss Eng Part C* 16:23–32.

Newman, A.C., Nakatsu, M.N., Chou, W., Gershon, P.D., and Hughes, C.C. 2011. Fibroblasts in angiogenesis: Fibroblast-derived matrix proteins are essential for endothelial cell lumen formation. *Mol Biol Cell* 22:3791–800.

O'Ceallaigh, S., Herrick, S.E., Bennett, W.R. et al. 2007. Perivascular cells in asking raft are rapidly repopulated by host cells. *J Plast Reconstr Aesthet Surg* 60:864–75.

Ott, H.C., Matthiesen, T.S., Goh, S.K. et al. 2008. Perfusion-decellularized matrix: Using nature's platform to engineer a bioartificial heart. *Nat Med* 14:213–21.

Paige, K.T. and Vacanti, C.A. 1995. Engineering new tissue: Formation of neo-cartilage. *Tissue Eng* 1:97–106.

Park, K.D., Kwon, I.K., and Kim, Y.H. 2000. Tissue engineering of urinary organs. *Yonsei Med J* 41:780–8.

Peters, M.C., Isenberg, B.C., Rowley, J.A., and Mooney DJ. 1998. Release from alginate enhances the biological activity of vascular endothelial growth factor. *J Biomater Sci Polym Ed* 9:1267–78.

Polykandriotis, E., Arkudas, A., Beier, J.P. et al. 2007. Intrinsic axial vascularization of an osteoconductive bone matrix by means of an arteriovenous vascular bundle. *Plast Reconstr Surg* 120:855–68.

Poon, C.J., Pereira, E., Cotta, M.V. et al. 2013. Preparation of an adipogenic hydrogel from subcutaneous adipose tissue. *Acta Biomater* 9:5609–20.

Pribaz, J.J., Fine, N., and Orgill, D.P. 1999a. Flap prefabrication in the head and neck: A 10-year experience. *Plast Reconstr Surg* 103:808–20.

Pribaz, J.J., Weiss, D.D., Mulliken, J.B., and Erikson, E. 1999b. Prelaminated free flaps reconstruction of complex central facial defects, *Plast Reconstr Surg* 104:357–365.

Raya-Rivera, A., Esquiliano, D.R., Yoo, J.J. et al. 2011. Tissue-engineered autologous urethras for patients who need reconstruction: An observational study. *Lancet* 377:1175–82.

Richardson, T.P., Peters, M.C. Ennett, A.B., and Mooney, D.J. 2001. Polymeric system for dual growth factor delivery. *Nat Biotechnol.* 19:1029–34.

Risau, W. and Flamme, I. 1995. Vasculogenesis. *Annu Rev Cell Dev Biol* 11:73–91.

Rophael, J.A., Craft, R.O., Palmer, J.A. et al. 2007 Angiogenic growth factor synergism in a murine tissue engineering model of angiogenesis and adipogenesis angiogenic. *Am J Pathology* 171:2048–57.

Rouwkema, J., Rivron, N.C., and van Blitterswijk, C.A. 2008. Vascularization in tissue engineering. *Trends Biotechnol.* 26:434–41.

Roy, S. and Gatien, S. 2008. Regeneration in axolotls: A model to aim for! *Exp Gerontol* 43:968–73.

Saito, R., Ishii, Y., Ito, R. et al. 2011. Transplantation of liver organoids in the omentum and kidney. *Artif Organs* 35:80–3.

Seach, N., Mattesich, M., Abberton, K. et al. 2010. Vascularized tissue engineering mouse chamber model supports thymopoiesis of ectopic thymus tissue grafts. *Tissue Eng Part C* 16:543–51.

Sekine, H., Shimizu, T., Sakaguchi K, Dobashi, I., Wada, M., Yamato, M., Kobayashi, E., Umezu, M., and Okano, T. 2013. *In vitro* fabrication of functional three-dimensional tissues with perfusable blood vessels. *Nat Commun* 4:1399. Doi: 10.1038/ncomms2406.

Shevchenko, R.V., James, S.L., and James, S.E. 2010. A review of tissue-engineered skin bioconstructs available for skin reconstruction. *J R Soc Interface* 7:229–58.

Simcock, J.W., Penington, A.J. Morrison, W.A., Thompson, E.W, and Mitchell, G.M. 2009. Endothelial precursor cells home to a vascularised tissue engineering chamber by application of the angiogenic chemokine CXCL12. *Tissue Eng Part A* 15:655–64.

Song, H., Song, B.W., Cha, M.J., Choi, I.G., and Hwang, K.C. 2010. Modification of mesenchymal stem cells for cardiac regeneration. *Expert Opin Biol Ther* 10:309–19.

Sorrell, J.M., Baber, M.A., and Caplan, A.I. 2007. A self-assembled fibroblast-endothelial cell co-culture system that supports *in vitro* vasculogenesis by both human umbilical vein endothelial cells and human dermal microvascular endothelial cells. *Cells Tissues Organs* 186:157–68.

Sugibayashi, K., Kumashiro, Y., Shimizu, T., Kobayashi, J., and Okano, T. 2013. A molded hyaluronic acid gel as a micro-template for blood capillaries. *J Biomater Sci Polym Ed* 24(2):135–47.

Tabata, Y., Miyao, M., Ozeki, M., and Ikada, Y. 2000. Controlled release of vascular endothelial growth factor by use of collagen hydrogels. *J Biomater Sci Polym Ed* 11:915–30.

Takato, T., Komuro, Y., Yonehara, H., and Zuker, R.M. 1993. Prefabricated venous flaps: An experimental study in rabbits. *Br J Plast Surg.* 46:122–6.

Takazawa, K., Adachi, N., Deie, M. et al. 2012. Evaluation of magnetic resonance imaging and clinical outcome after tissue-engineered cartilage implantation: Prospective 6-year follow-up study. *J Orthop Sci* 17:413–24.

Tanaka, Y., Tsutsumi, A., Crowe, D.M., Tajima, S., and Morrison, W.A. 2000. Generation of an autologous tissue (matrix) flap by combining an arteriovenous shunt loop with artificial skin in rats: Preliminary report. *Br J Plast Surg* 53:51–7.

Tanaka, Y., Sung, K.C., Tsutsumi, A. et al. 2003. Tissue engineering skin flaps: Which vascular carrier, arteriovenous shunt loop or arteriovenous bundle, has more potential for angiogenesis and tissue generation? *Plast Reconstr Surg* 112:1636–44.

Tanaka, Y., Sung, K.C., Fumimoto, M. et al. 2006. Prefabricated engineered skin flap using an arteriovenous vascular bundle as a vascular carrier in rabbits. *Plast Reconstr Surg* 117:1860–75.

Tee, R., Morrison, W.A., Dusting, G.J. et al. 2012. Transplantation of engineered cardiac muscle flaps in syngeneic rats. *Tissue Eng Part A* 18:1992–9.

Tilkorn, D., Bedogni, A., Keramidaris, E. et al. 2010. Implanted myoblast survival is dependent on the degree of vascularization in a novel delayed implantation/prevascularization tissue engineering model. *Tissue Eng Part A* 16:165–78.

Tilkorn, D.J., Davies, E.M., Keramidaris, E et al. 2012. *In vitro* myoblast preconditioning enhances subsequent survival post *in vivo* implantation into a tissue engineering chamber. *Biomaterials* 33:3868–79.

Uriel, S., Huang, J.J., Moya, M.L. et al. 2008. The role of adipose protein derived hydrogels in adipogenesis. *Biomaterials* 2008; 29(27):3712–9.

Vacanti, C.A., Langer, R., Schloo, B., and Vacanti, J.P. 1991. Synthetic polymers seeded with chondrocytes provide a template for new cartilage formation. *Plast Reconstr Surg* 88:753–9.

Vacanti, C.A. and Vacanti, J.P. 2000. The science of tissue engineering. *Orthop Clin North Am* 31:351–6.

Vaquette, C, Fan, W, Xiao, Y et al. 2012. A biphasic scaffold design combined with cell sheet technology for simultaneous regeneration of alveolar bone/periodontal ligament complex. *Biomaterials* 33:5560–73.

Vashi, A.V., Abberton, K.M., Thomas, G.P. et al. 2006. Adipose tissue engineering based on the controlled release of fibroblast growth factor-2 in a collagen matrix. *Tiss Eng* 12:3035–43.

Vashi, A.V., Keramidaris, E., Abberton, K.M. et al. 2008. Adipose differentiation of bone marrow-derived mesenchymal stem cells using PluronicF-127 hydrogel *in vitro*. *Biomaterials* 29:573–9.

Wenger, A., Kowalewski, N., Stahl, A. et al. 2005. Development and characterization of a spheroidal coculture model of endothelial cells and fibroblasts for improving angiogenesis in tissue engineering. *Cells Tissues Organs* 181:80–8.

Yap, K.K., Dingle, A.M., Palmer, J.A et al. 2013. Enhanced liver progenitor cell survival and differentiation *in vivo* by spheroid implantation in a vascularized tissue engineering chamber. *Biomaterials* 34:3992–4001.

Yohannes, P., Rotariu, P., Liatsikos, E. et al. 2002. Role of a cellular collagen matrix surgisis in the endoscopic management of ureteropelvic junction obstruction. *J Endourol* 16:549–56.

Yokoyama, H. 2008. Initiation of limb regeneration: The critical steps for regenerative capacity. *Dev Growth Differ* 50:13–22.

Yoshimura, K., Sato, K., Aoi, N. et al. 2008. Cell-assisted lipo transfer for facial lipoatrophy: Clinical use of adipose-derived stem cells. *Derm. Surg* 34:1178–85.

Zdolsek, J.M., Morrison, W.A., Dingle, A.M., Penington, A.J., and Mitchell, G.M., 2011. An "off the shelf" vascular allograft supports angiogenic growth in three dimensional tissue engineering. *J Vasc Surg* 53:435–44.

Index

Printed and bound by CPI Group (UK) Ltd, Croydon, CR0 4YY

18/10/2024

01776254-0009